NETWORK THEORY

McGRAW-HILL ELECTRICAL AND ELECTRONIC ENGINEERING SERIES

Frederick Emmons Terman, *Consulting Editor*
W. W. Harman and J. G. Truxal, *Associate Consulting Editors*

ANGELAKOS AND EVERHART · Microwave Communications
ANGELO · Electronic Circuits
ANGELO · Electronics · BJTs FETs, and Microcircuits
ASELTINE · Transform Method in Linear System Analysis
ATWATER · Introduction to Microwave Theory
BERANEK · Acoustics
BRACEWELL · The Fourier Transform and Its Application
BRENNER AND JAVID · Analysis of Electric Circuits
BROWN · Analysis of Linear Time-invariant Systems
BRUNS AND SAUNDERS · Analysis of Feedback Control Systems
CARLSON · Communication Systems: An Introduction to Signals and Noise in Electrical Communication
CHEN · The Analysis of Linear Systems
CHEN · Linear Network Design and Synthesis
CHIRLIAN · Analysis and Design of Electronic Circuits
CHIRLIAN · Basic Network Theory
CHIRLIAN AND ZEMANIAN · Electronics
CLEMENT AND JOHNSON · Electrical Engineering Science
CUNNINGHAM · Introduction to Nonlinear Analysis
D'AZZO AND HOUPIS · Feedback Control System Analysis and Synthesis
ELGERD · Control Systems Theory
EVELEIGH · Adaptive Control and Optimization Techniques
FEINSTEIN · Foundations of Information Theory
FITZGERALD, HIGGINBOTHAM, AND GRABEL · Basic Electrical Engineering
FITZGERALD AND KINGSLEY · Electric Machinery
FRANK · Electrical Measurement Analysis
FRIEDLAND, WING, AND ASH · Principles of Linear Networks
GEHMLICH AND HAMMOND · Electromechanical Systems
GHAUSI · Principles and Design of Linear Active Circuits
GHOSE · Microwave Circuit Theory and Analysis
GLASFORD · Fundamentals of Television Engineering
GREINER · Semiconductor Devices and Applications
HAMMOND · Electrical Engineering
HANCOCK · An Introduction to the Principles of Communication Theory

HARMAN · Fundamentals of Electronic Motion
HARMAN · Principles of the Statistical Theory of Communication
HARMAN AND LYTLE · Electrical and Mechanical Networks
HAYASHI · Nonlinear Oscillations in Physical Systems
HAYT · Engineering Electromagnetics
HAYT AND KEMMERLY · Engineering Circuit Analysis
HILL · Electronics in Engineering
JAVID AND BROWN · Field Analysis and Electromagnetics
JOHNSON · Transmission Lines and Networks
KOENIG AND BLACKWELL · Electromechanical System Theory
KOENIG, TOKAD, AND KESAVAN · Analysis of Discrete Physical Systems
KRAUS · Antennas
KRAUS · Electromagnetics
KUH AND PEDERSON · Principles of Circuit Synthesis
KUO · Linear Networks and Systems
LEDLEY · Digital Computer and Control Engineering
LEPAGE · Complex Variables and the Laplace Transform for Engineering
LEPAGE AND SEELY · General Network Analysis
LEVI AND PANZER · Electromechanical Power Conversion
LEY, LUTZ, AND REHBERG · Linear Circuit Analysis
LINVILL AND GIBBONS · Transistors and Active Circuits
LITTAUER · Pulse Electronics
LYNCH AND TRUXAL · Introductory System Analysis
LYNCH AND TRUXAL · Principles of Electronic Instrumentation
LYNCH AND TRUXAL · Signals and Systems in Electrical Engineering
MCCLUSKEY · Introduction to the Theory of Switching Circuits
MANNING · Electrical Circuits
MEISEL · Principles of Electromechanical-energy Conversion
MILLMAN AND HALKIAS · Electronic Devices and Circuits
MILLMAN AND TAUB · Pulse, Digital, and Switching Waveforms
MINORSKY · Theory of Nonlinear Control Systems
MISHKIN AND BRAUN · Adaptive Control Systems
MOORE · Traveling-wave Engineering
MURDOCH · Network Theory
NANAVATI · An Introduction to Semiconductor Electronics

Oberman · Disciplines in Combinational and Sequential Circuit Design
Pettit · Electronic Switching, Timing, and Pulse Circuits
Pettit and McWhorter · Electronic Amplifier Circuits
Pfeiffer · Concepts of Probability Theory
Reza · An Introduction to Information Theory
Reza and Seely · Modern Network Analysis
Ruston and Bordogna · Electric Networks: Functions, Filters, Analysis
Ryder · Engineering Electronics
Schilling and Belove · Electronic Circuits: Discrete and Integrated
Schwartz · Information Transmission, Modulation, and Noise
Schwarz and Friedland · Linear Systems
Seely · Electromechanical Energy Conversion
Seely · Electron-tube Circuits
Seely · Introduction to Electromagnetic Fields
Seifert and Steeg · Control Systems Engineering
Shooman · Probabilistic Reliability: An Engineering Approach
Siskind · Direct-current Machinery
Skilling · Electric Transmission Lines
Stevenson · Elements of Power System Analysis
Stewart · Fundamentals of Signal Theory
Strauss · Wave Generation and Shaping
Su · Active Network Synthesis
Terman · Electronic and Radio Engineering
Terman and Pettit · Electronic Measurements
Thaler · Elements of Servomechanism Theory
Thaler and Brown · Analysis and Design of Feedback Control Systems
Thaler · Analysis and Design of Nonlinear Feedback Control Systems
Thaler and Pastel · Analysis and Design of Nonlinear Feedback Control Systems
Tou · Digital and Sampled-data Control Systems
Tou · Modern Control Theory
Truxal · Automatic Feedback Control System Synthesis
Tuttle · Electric Networks: Analysis and Synthesis
Valdes · The Physical Theory of Transistors
Van Bladel · Electromagnetic Fields
Weeks · Antenna Engineering
Weinberg · Network Analysis and Synthesis

NETWORK THEORY

J. B. Murdoch

Professor and Chairman of Electrical Engineering
University of New Hampshire

McGraw-Hill Book Company

New York, St. Louis, San Francisco, London, Sydney, Toronto, Mexico, Panama

Network Theory

Library of Congress Catalog Card Number 70-81610

44049

1 2 3 4 5 6 7 8 9 0 MAMM 7 6 5 4 3 2 1 0 6 9

This book was set in Times Roman by The Universities Press, and printed on permanent paper and bound by The Maple Press Company. The designer was Richard P. Kluga; the drawings were done by J. & R. Technical Services. Inc. The editors were B. G. Dandison. Jr. and Madelaine Eichberg. Adam Jacobs supervised the production.

To Ann and
Joe, Pete, Amy, Annie, and Tommy
Who continue to make it all so very worthwhile

PREFACE

This book is intended for an advanced undergraduate or beginning graduate course in network theory. At the undergraduate level, for the student with reasonable facility in mathematics, it can be used following a basic course in network or circuit analysis. At the graduate level it should be used during the first semester of the student's graduate program.

The University of New Hampshire, as is typical of many small colleges and universities, has a modest master's degree program in electrical engineering. The students enrolled in this program, often on an extension or part-time basis, generally have a different purpose or background from students at larger institutions with sizeable resident programs offering the doctorate. Also they may have been out of college for several years while working in industry. In particular their backgrounds in networks and associated mathematics are often somewhat weak in one way or another. There are at least two ways by which this deficiency may be corrected:

1. A remedial course can be offered which covers the prerequisite undergraduate network theory and mathematics. One disadvantage of this

seems to be that the incoming graduate students are not homogeneous in their deficiencies. One student may be able to handle Laplace transforms well but know little about matrices; a second student's background may be just the reverse. Thus a remedial course, in trying to serve everyone, may actually not serve anyone properly. A second disadvantage is that the student's graduate program is set back a semester. This discourages many industrial students from pursuing a master's degree.

2. A first graduate course can be offered that reviews undergraduate network theory but does it from a different point of view, more rigorously, and in greater depth. Ideally, the deficient student (assuming his background is not too weak) erases his deficiency and at the same time develops new methods and tools of mathematics and network analysis, whereas the already reasonably prepared student deepens his understanding of network theory, polishes his math, develops the habit of reading the literature, and acquires a thesis topic.

This second approach of reviewing in depth and extending has been used at the University of New Hampshire for several years. Incoming electrical-engineering graduate students take two required courses in their first semester, one in field theory and the other in network theory. These courses seek to bring the student to a reasonable level of proficiency in fields and networks while at the same time sharpening the mathematical tools he needs for subsequent graduate courses. Thus partial differential equations, boundary-value problems, Bessel and Legendre functions, and conformal transformations are included in the fields course, whereas matrices and determinants, systems of equations, linear-graph theory, Laplace and Fourier transforms, and complex and state variables are stressed in the networks course.

This book has evolved from notes for this first graduate course in network theory and has been classroom-tested for three years. Roughly the first half of the book concentrates on the steady-state analysis of linear networks from a matrix and topological point of view, and the last half of the book considers the free and forced behavior of linear networks, using Laplace and Fourier transforms and state and complex variables. Chapter 8, on Natural Frequencies, serves as a bridge between these two areas. Sufficient depth is provided in each section of the book to ensure a student being able to read related material in the literature intelligently. At the same time, sufficient review is included to ensure that the deficient student is not overwhelmed. Chapter problems are divided into three categories in keeping with this philosophy: drill (a graduate student does need some), theory and proofs, and application.

The book is too long to be covered completely in a single semester. However, there is adequate time in a single semester to cover either one of the two major topics in detail while treating the second lightly. Alternatively, both major topics can be covered with some thoroughness in one semester by omitting several special and advanced topics. In terms of chapter and section coverage, the three possibilities are as follows:

Accent on steady-state analysis, topology, and matrices Chaps. 1 to 7, Secs. 8.2, 8.5, 8.9 to 8.11, 9.2, 9.3, 10.5, 10.6, 11.4, and 11.7

Accent on free and forced response, Laplace and Fourier transforms, and state and complex variables Chaps. 1 and 2, Secs. 3.1 to 3.5, 7.1 to 7.4, Chap. 8 (omit 8.5), Chaps. 9 and 10, Chap. 11 (omit 11.4), Chap. 12

Balanced emphasis on each topic Chap. 1 (omit 1.11 and 1.14), Chap. 2 (omit 2.9 and 2.10), Chap. 3 (omit 3.6 to 3.10), Chap. 4 (omit 4.9 to 4.13), Chap. 5 (omit 5.7 to 5.11), Chap. 7 (omit 7.5 and 7.6), Chap. 8 (omit 8.9 to 8.11), Chap. 9, Chap. 10 (omit 10.2, 10.5 and 10.7), Chap. 11 (omit 11.4 and 11.6 to 11.8), Chap. 12 (omit 12.9 to 12.16)

Of course, the book can be used for a full-year course, topology and matrices being accented during the first semester and transforms and state and complex variables during the second. At the graduate level, the instructor will probably have time each semester to supplement the book with topics of his own choosing and interest. At the undergraduate level, the book itself should suffice.

It is a pleasure to express thanks to the two people who introduced me to the fascinating business of networks during my Ph.D. days, Dr. Dov Hazony and Dr. James D. Schoeffler, of Case Institute of Technology. I am also indebted to Prof. Donald W. Melvin, of the University of New Hampshire, who classroom-tested and commented most constructively on various sections of the book. To Mrs. Barbara Keating, who typed the early chapters, and to Mrs. Mary Ann Stickney, who came along at just the right time and devoted herself to the typing of the later chapters and the preparation of the final manuscript, I acknowledge a sincere debt of gratitude. Finally, to my graduate students, who suffered through uncorrected ditto copy, my appreciation for suggestions and corrections.

J. B. MURDOCH

CONTENTS

7.

NETWORK THEOREMS, DUALITY, AND SENSITIVITY 209

8.

NATURAL FREQUENCIES AND FREE RESPONSE 245

9.

TRANSFORM METHODS 293

10. FORCED AND STEADY-STATE RESPONSES 356

11. STATE VARIABLES AND EQUATIONS 420

12. COMPLEX-VARIABLE THEORY 462

1 MATRICES AND DETERMINANTS

1.1 INTRODUCTION

A linear system is one that can be characterized by a set of linear differential or algebraic equations. For a linear electric network, these equations are the Kirchhoff law equations resulting from a loop or node analysis of the network. The reader is assumed to have a working knowledge of the formulation and solution of such equations. Specifically, it is assumed that he can write Kirchhoff's voltage law (KVL) equations around the various loops of a network and Kirchhoff's current law (KCL) equations at the various nodes and, presuming the right number of equations has been written, that he can solve for the unknown currents or voltages using determinants and Cramer's rule.

Simple networks are easily analyzed in the manner just described. Questions about the number of necessary equations seldom arise. However, as the complexity of the network increases, a more systematic approach is needed. It is the purpose of the first several chapters to develop such an approach based on linear graph theory. Matrices and determinants are particularly useful mathematical tools in this development. We therefore devote the first chapter to their study.

The reader is assumed to have some familiarity with matrices and determinants. Our purpose here is to review their properties, develop a facility in and understanding of their use, and emphasize those features and special forms which are of particular importance in the study of linear graph theory.

1.2 DEFINITION OF A MATRIX

Consider a set of m linear algebraic equations in n unknowns

$$\begin{aligned} a_{11}x_1 + a_{12}x_2 + \cdots + a_{1n}x_n &= y_1 \\ a_{21}x_1 + a_{22}x_2 + \cdots + a_{2n}x_n &= y_2 \\ \cdots\cdots\cdots\cdots\cdots\cdots\cdots\cdots \\ a_{m1}x_1 + a_{m2}x_2 + \cdots + a_{mn}x_n &= y_m \end{aligned} \tag{1.2.1}$$

We ask: How can these equations be written more concisely? To answer the question, we rewrite Eq. (1.2.1) in the *matrix* form

$$\begin{bmatrix} a_{11} & a_{12} & \cdots & a_{1n} \\ a_{21} & a_{22} & \cdots & a_{2n} \\ \cdots & \cdots & \cdots & \cdots \\ a_{m1} & a_{m2} & \cdots & a_{mn} \end{bmatrix} \begin{bmatrix} x_1 \\ x_2 \\ \cdot \\ \cdot \\ \cdot \\ x_n \end{bmatrix} = \begin{bmatrix} y_1 \\ y_2 \\ \cdot \\ \cdot \\ \cdot \\ y_m \end{bmatrix} \tag{1.2.2}$$

which can be shortened to

$$AX = Y \tag{1.2.3}$$

Equations (1.2.2) and (1.2.3) are definitions. Each is a shorthand form of Eq. (1.2.1). Matrix A is an ordered array of coefficients which transforms a set of n variables in x into a set of m variables in y, as prescribed by Eq. (1.2.1). It has m rows and n columns and is said to be of *order* $m \times n$. Similarly X and Y are *column* matrices of order $n \times 1$ and $m \times 1$, respectively.

1.3 INDEX NOTATION

Thus far we have represented a matrix either by a single capital letter or by a bracketed array of single- or double-subscripted coefficients. A third, very useful representation is provided by index notation. Let

$$
\begin{aligned}
A &\equiv [a_{ij}] \\
X &\equiv [x_i] \qquad (1.3.1)\\
Y &\equiv [y_i]
\end{aligned}
$$

In this notation, a_{ij} is the element in the ith row and jth column of A, and $[a_{ij}]$ is the matrix composed of these elements. Similarly x_i is the element in the ith row of the matrix X (there being only one column), and likewise for y_i.

The definitions of Eq. (1.3.1) can be substituted into Eq. (1.2.3) to obtain

$$[a_{ij}][x_i] = [y_i] \qquad (1.3.2)$$

Equation (1.3.2) makes little sense in terms of the instructions provided by the indices. The index i is repeated on the left of Eq. (1.3.2) but not on the right; there is nothing in the notation to guide us in multiplying the two matrices on the left together.

To make Eq. (1.3.2) meaningful, let us write Eq. (1.2.1) as a summation using index notation

$$\sum_{k=1}^{n} a_{ik}x_k = y_i \qquad i = 1, 2, \ldots, m \qquad (1.3.3)$$

For each value of i, one equation of Eq. (1.2.1) is developed as k proceeds from 1 through n. Putting Eq. (1.3.3) in matrix form yields the entire set of equations

$$\underset{i=1}{\overset{m}{}}\left[\sum_{k=1}^{n} a_{ik}x_k\right] = \underset{i=1}{\overset{m}{}}[y_i] \qquad (1.3.4)$$

The brackets in Eq. (1.3.4) tell us to let i take on all its permitted values from 1 through m.

The repeated subscript k in Eqs. (1.3.3) and (1.3.4) is known as a *dummy index*, indicating a summation over the n columns of A and rows of X. Clearly the number of columns in the first matrix and rows in the second matrix must be the same for this matrix product to be defined. Matrices satisfying this requirement are said to be *conformable* in the order given.

There is no reason to retain the summation sign in Eq. (1.3.4) if we are willing to adopt the convention that a repeated index indicates a summation. (In the few cases where a repeated index does not indicate a summation we shall use a Greek letter for the repeated index.) Also, the ranges of i and k are customarily omitted [as they are in Eq. (1.2.3)]. With these changes, Eq. (1.3.4) becomes

$$[a_{ik}x_k] = [y_i] \qquad (1.3.5)$$

Let us now return to our difficulties with Eq. (1.3.2) and change this equation to read

$$[a_{ik}][x_k] = [y_i] \tag{1.3.6}$$

The form in Eq. (1.3.6) looks much better. Formally, we have replaced the second (column) index of the first matrix and the first (row) index of the second matrix by a dummy summation index (in this case k).

Now comparing Eqs. (1.3.5) and (1.3.6), we obtain

$$[a_{ik}][x_k] = [a_{ik}x_k] \tag{1.3.7}$$

which shows, in compact index-notation form, exactly what is meant by the product of two matrices.

1.4 MULTIPLICATION

To develop further what is meant by matrix multiplication and show how several matrices can be multiplied together (*cascaded*), let B be an $n \times p$ matrix given by

$$B = [b_{ij}] \tag{1.4.1}$$

and let Z be a $p \times 1$ column matrix given by

$$Z = [z_i] \tag{1.4.2}$$

Now let the n variables in x in Eq. (1.2.1) be defined by n equations in the p variables in z by

$$[b_{il}][z_l] = [x_i] \qquad \text{or} \qquad BZ = X \tag{1.4.3}$$

where l is a dummy summation index like k in Eq. (1.3.2) but runs from 1 through p. Substituting Eq. (1.4.3), with i replaced by k, into Eq. (1.3.6) yields

$$[a_{ik}][b_{kl}][z_l] = [y_i] \tag{1.4.4}$$

which, according to Eq. (1.3.7), can be written as

$$[a_{ik}b_{kl}][z_l] = [y_i] \tag{1.4.5}$$

On the left in Eqs. (1.4.4) and (1.4.5) there are two dummy summation indices, namely, k and l, and there is one nonrepeated index i. The summation indices do not appear on the right; the nonrepeated index does. This is always the case.

Now define

$$[c_{il}] = [a_{ik}b_{kl}] \tag{1.4.6}$$

which yields

$$[c_{il}][z_l] = [y_i] \tag{1.4.7}$$

From Eq. (1.4.6), we can formulate a general procedure for taking a matrix product AB. First A and B must be conformable. Here A is an $m \times n$ matrix and B is an $n \times p$ matrix. They are conformable, and their product is an $m \times p$ matrix. Second, each entry in the C matrix is formed by letting k take on all values from 1 through n, with i and l fixed, and summing the products. For example,

$$c_{12} = a_{11}b_{12} + a_{12}b_{22} + \cdots + a_{1n}b_{n2} \tag{1.4.8}$$

Example 1.1 Let the matrices A, B, Z, and Y in Eq. (1.4.4) be given by

$$A = \begin{bmatrix} a_{11} & a_{12} \\ a_{21} & a_{22} \end{bmatrix} \qquad B = \begin{bmatrix} b_{11} & b_{12} \\ b_{21} & b_{22} \end{bmatrix} \qquad Z = \begin{bmatrix} z_1 \\ z_2 \end{bmatrix} \qquad Y = \begin{bmatrix} y_1 \\ y_2 \end{bmatrix}$$

We can write any of the y_i's by inspection using index notation and Eq. (1.4.4). For example, to write y_2, we note that $i = 2$ and k and l each take on the values 1 and 2. Thus

$$\begin{aligned} y_2 &= a_{2k}b_{kl}z_l \\ &= a_{21}b_{11}z_1 + a_{21}b_{12}z_2 + a_{22}b_{21}z_1 + a_{22}b_{22}z_2 \end{aligned}$$

Note that by using index notation, we did not have to perform the multiplication piecemeal, i.e., by first obtaining AB and then ABZ, or BZ and then ABZ. This can be an advantage when several matrices are to be multiplied together.

Example 1.2 The two networks shown in Fig. 1.1 are to be cascaded as indicated by the dashed lines. We wish to form the overall network matrix.

The voltage-current equations† for each network are

$$\begin{aligned} e_1 &= ae_2 - bi_2 \qquad & e_3 &= a'e_4 - b'i_4 \\ i_1 &= ce_2 - di_2 \qquad & i_3 &= c'e_4 - d'i_4 \end{aligned}$$

When the networks are cascaded,

$$e_3 = e_2 \qquad \text{and} \qquad i_3 = -i_2$$

Hence the equations for the second network become

$$\begin{aligned} e_2 &= a'e_4 - b'i_4 \\ -i_2 &= c'e_4 - d'i_4 \end{aligned}$$

† The equations are given in terms of the chain parameters of each network. Two-port network parameters are discussed in Chap. 3.

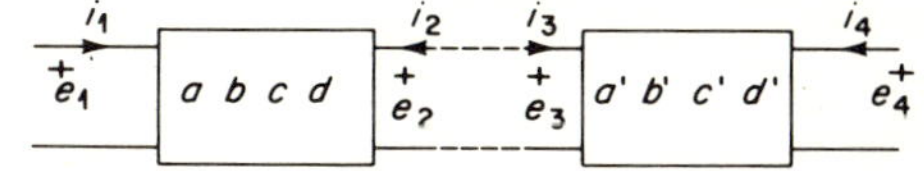

Fig. 1.1 Cascaded networks.

Now the network equations can be written in matrix form to give

$$\begin{bmatrix} e_1 \\ i_1 \end{bmatrix} = \begin{bmatrix} a & b \\ c & d \end{bmatrix} \begin{bmatrix} e_2 \\ -i_2 \end{bmatrix} \qquad \begin{bmatrix} e_2 \\ -i_2 \end{bmatrix} = \begin{bmatrix} a' & b' \\ c' & d' \end{bmatrix} \begin{bmatrix} e_4 \\ -i_4 \end{bmatrix}$$

Combining these equations yields

$$\begin{bmatrix} e_1 \\ i_1 \end{bmatrix} = \begin{bmatrix} a & b \\ c & d \end{bmatrix} \begin{bmatrix} a' & b' \\ c' & d' \end{bmatrix} \begin{bmatrix} e_4 \\ -i_4 \end{bmatrix}$$

The expanded form is

$$\begin{bmatrix} e_1 \\ i_1 \end{bmatrix} = \begin{bmatrix} aa' + bc' & ab' + bd' \\ ca' + dc' & cb' + dd' \end{bmatrix} \begin{bmatrix} e_4 \\ -i_4 \end{bmatrix}$$

1.5 ADDITION

The addition of two matrices A and B is possible only if they are of the same order. Assuming this to be the case, their sum is given by

$$[a_{ij}] + [b_{ij}] = [a_{ij} + b_{ij}] \tag{1.5.1}$$

Note that the sum is formed simply by adding corresponding elements in each matrix. Subtraction of two matrices follows the same pattern.

Example 1.3 The two networks in Fig. 1.1 are now to be connected so that they have common input and common output currents, as shown in Fig. 1.2.

The individual network voltage-current equations, this time in terms of z parameters, are

$$\begin{bmatrix} e_1 \\ e_2 \end{bmatrix} = \begin{bmatrix} z_{11} & z_{12} \\ z_{21} & z_{22} \end{bmatrix} \begin{bmatrix} i_1 \\ i_2 \end{bmatrix} \qquad \begin{bmatrix} e_3 \\ e_4 \end{bmatrix} = \begin{bmatrix} z'_{11} & z'_{12} \\ z'_{21} & z'_{22} \end{bmatrix} \begin{bmatrix} i_3 \\ i_4 \end{bmatrix}$$

Assuming that the network parameters do not change when the networks are interconnected†

$$i_3 = i_1 \qquad \text{and} \qquad i_4 = i_2$$

Now each network matrix multiplies the same current matrix, and thus the matrices can be added to give the desired input-output relations

$$\begin{bmatrix} e_1 + e_3 \\ e_2 + e_4 \end{bmatrix} = \begin{bmatrix} z_{11} + z'_{11} & z'_{12} + z'_{12} \\ z_{21} + z'_{21} & z'_{22} + z'_{22} \end{bmatrix} \begin{bmatrix} i_1 \\ i_2 \end{bmatrix}$$

† Limitations are discussed in Chap. 3.

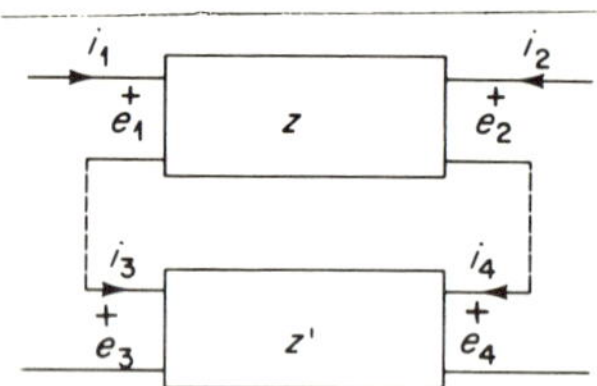

Fig. 1.2 Series-connected networks.

1.6 MATHEMATICAL PROPERTIES

Let A and B be two matrices of the same order. A and B are equal if and only if $a_{ij} = b_{ij}$ for all i and j. Thus equality means equal term by term.

If a matrix A is multiplied by a scalar quantity k, each element in A is multiplied by k. Thus

$$kA = [ka_{ij}] \tag{1.6.1}$$

To conjugate a matrix A, each element in the matrix is conjugated

$$A^* = [a^*_{ij}] \tag{1.6.2}$$

Let us next examine the associative, commutative, and distributive laws of mathematics as they apply to matrices. Let A, B, and C be three matrices. Then, assuming the operations of multiplication and addition are defined,

$$A + (B + C) = (A + B) + C \qquad \text{associative law of addition} \tag{1.6.3}$$

$$A(BC) = (AB)C \qquad \text{associative law of multiplication} \tag{1.6.4}$$

$$A + B = B + A \qquad \text{commutative law of addition} \tag{1.6.5}$$

$$AB \neq BA \qquad \text{commutative law of multiplication} \tag{1.6.6}$$

$$A(B + C) = AB + AC \qquad \text{distributive law} \tag{1.6.7}$$

Only the commutative law of multiplication fails to hold in general for matrices.

Example 1.4 Given a matrix $A = \begin{bmatrix} 1 & 2 \\ -1 & 3 \end{bmatrix}$, find a matrix B such that $AB = BA$.

First we note that B must be a square 2×2 matrix. Let it be given by

$$B = \begin{bmatrix} a & b \\ c & d \end{bmatrix}$$

Then

$$AB = \begin{bmatrix} a + 2c & b + 2d \\ -a + 3c & -b + 3d \end{bmatrix} \qquad \text{and} \qquad BA = \begin{bmatrix} a - b & 2a + 3b \\ c - d & 2c + 3d \end{bmatrix}$$

For the matrix products to be equal, they must be equal term by term. Thus

$$a + 2c = a - b \qquad b + 2d = 2a + 3b$$

$$-a + 3c = c - d \qquad -b + 3d = 2c + 3d$$

Solutions are given by

$$d = a + b \qquad b = -2c$$

Any two of the four parameters (except b and c together. may be assigned values. Choosing $c = 1$ and $a = 1$ yields one acceptable B matrix as

$$B = \begin{bmatrix} 1 & -2 \\ 1 & -1 \end{bmatrix}$$

1.7 SUBMATRICES AND PARTITIONING

In the matrix analysis of linear networks, it is often possible to reduce computational labor by partitioning a given matrix into two or more *submatrices.* Consider two matrices A and B

$$A = \left[\begin{array}{cc:c} a_{11} & a_{12} & a_{13} \\ a_{21} & a_{22} & a_{23} \\ \hdashline a_{31} & a_{32} & a_{33} \end{array}\right] \qquad B = \left[\begin{array}{cc} b_{11} & b_{12} \\ b_{21} & b_{22} \\ \hdashline b_{31} & b_{32} \end{array}\right] \tag{1.7.1}$$

The product AB can be obtained by direct expansion. However, let us partition each matrix into submatrices as shown and rewrite them in the form

$$A = \begin{bmatrix} A_{11} & A_{12} \\ A_{21} & A_{22} \end{bmatrix} \qquad B = \begin{bmatrix} B_{11} \\ B_{21} \end{bmatrix} \tag{1.7.2}$$

where

$$A_{11} = \begin{bmatrix} a_{11} & a_{12} \\ a_{21} & a_{22} \end{bmatrix} \qquad A_{12} = \begin{bmatrix} a_{13} \\ a_{23} \end{bmatrix} \qquad A_{21} = [a_{31} \quad a_{32}] \quad A_{22} = a_{33} \tag{1.7.3}$$

Now the matrix product is

$$AB = \begin{bmatrix} A_{11}B_{11} + A_{12}B_{21} \\ A_{21}B_{11} + A_{22}B_{21} \end{bmatrix} \tag{1.7.4}$$

which is a 2×1 matrix in submatrix form. The elements of each submatrix can be substituted in Eq. (1.7.4) and the multiplications carried out to obtain the expanded form.

Just as two matrices must be conformable for their product to be defined, so must two submatrices. Thus the partition lines in Eq. (1.7.1) are not arbitrary. To ensure conformability of all submatrices, for each vertical partition line separating *columns* r and $r + 1$ in the first matrix, there must be one horizontal partition line separating *rows* r and $r + 1$ in the second matrix, and no others.

Example 1.5 The impedance, current, and voltage matrices of a network are given by

$$Z = \left[\begin{array}{cc:cc:c} 2s & s & 0 & 0 & 0 \\ s & 3s & 0 & 0 & 0 \\ \hdashline 0 & 0 & 1 & 0 & 0 \\ 0 & 0 & 0 & 2 & 0 \\ \hdashline 0 & 0 & 0 & 0 & \frac{1}{s} \end{array}\right] \qquad J = \left[\begin{array}{c} j_1 \\ j_2 \\ \hdashline j_3 \\ j_4 \\ \hdashline j_5 \end{array}\right] \qquad V = \left[\begin{array}{c} v_1 \\ v_2 \\ \hdashline v_3 \\ v_4 \\ \hdashline v_5 \end{array}\right]$$

It is desired to express the voltage matrix $V = ZJ$ such that the contributions due to inductance, resistance, and capacitance are separate.

Let the Z, J, and V matrices be partitioned as shown and rewrite them as

$$Z = \begin{bmatrix} sL & 0 & 0 \\ 0 & R & 0 \\ 0 & 0 & \dfrac{1}{sC} \end{bmatrix} \qquad J = \begin{bmatrix} J_a \\ J_b \\ J_c \end{bmatrix} \qquad V = \begin{bmatrix} V_a \\ V_b \\ V_c \end{bmatrix}$$

The matrix product of Z and J yields

$$\begin{bmatrix} V_a \\ V_b \\ V_c \end{bmatrix} = \begin{bmatrix} sLJ_a \\ RJ_b \\ \dfrac{1}{sC} J_c \end{bmatrix}$$

The submatrices can be expanded to give the individual voltages in terms of the individual currents. For example

$$[V_a] = \begin{bmatrix} v_1 \\ v_2 \end{bmatrix} = [sLJ_a] = \begin{bmatrix} 2s & s \\ s & 3s \end{bmatrix} \begin{bmatrix} j_1 \\ j_2 \end{bmatrix} = \begin{bmatrix} 2sj_1 + sj_2 \\ sj_1 + 3sj_2 \end{bmatrix}$$

In submatrix form, V is a 3×1 matrix, whereas in expanded form it is a 5×1 matrix. It should be noted that partitioning is especially useful in this case because of the large number of off-diagonal zeros in the Z matrix (the network has very little coupling).

1.8 SPECIAL MATRICES: I

In this section, and a later one, certain special matrices that are used considerably in network theory are discussed.

TRANSPOSE OF A MATRIX

The *transpose* of a matrix A is obtained by interchanging the rows and columns of A. Let A be given by $[a_{ij}]$. Then

$$A^T = [\alpha_{ij}] \qquad \text{where } \alpha_{ij} = a_{ji} \tag{1.8.1}$$

The T indicates transpose, and the index notation shows that A^T is made up of elements α_{ij} taken from the jith locations of A.

Example 1.6 Show that if the product AB of two matrices A and B is defined, then $(AB)^T = B^T A^T$.

This proof affords us an excellent opportunity to use index notation. To begin, the ijth element of $(AB)^T$ is the jith element of AB. Thus

$$[(AB)^T]_{ij} = (AB)_{ji}$$

But $(AB)_{ji}$ can be written in terms of the elements of A and B as

$$(AB)_{ji} = a_{jk} b_{ki}$$

where the summation on k is once again understood. The individual element products commute (although the matrices as a whole do not). Therefore

$$a_{jk}b_{ki} = b_{ki}a_{jk}$$

The elements of b_{ki} and a_{jk} may be written as transposes of b_{ik} and a_{kj}, respectively, to give

$$b_{ki}a_{jk} = b_{ik}{}^T a_{kj}{}^T = (B^T A^T)_{ij}$$

This shows that the ijth element of $(AB)^T$ is equal to the ijth element of $B^T A^T$. Since the equality holds for all i and j,

$$(AB)^T = B^T A^T$$

There are two general methods of proving matrix equalities. The first deals with the individual elements of the matrices, and the second deals with the matrices themselves. The proof just completed illustrates the first method. We shall use the second method in later examples.

SYMMETRIC AND SKEW-SYMMETRIC MATRICES

A square matrix whose elements satisfy the condition

$$a_{ij} = \pm a_{ji} \tag{1.8.2}$$

is called a *symmetric* matrix if the plus sign holds for all i and j and a *skew-symmetric* matrix if the minus sign holds for all i and j. A symmetric matrix and its transpose are identical, whereas a skew-symmetric matrix and its transpose differ only in sign. A skew-symmetric matrix has all zeros on its main diagonal (upper left to lower right).

Any square matrix with real entries can be written as the sum of a symmetric and a skew-symmetric matrix. The proof is requested in Prob. 1.14.

Example 1.7 Let A and B be symmetric matrices of order n. Prove that the matrix product AB is symmetric if and only if A and B commute.

We shall carry out the proof in two ways, (*a*) with the matrices as a whole and (*b*) with the elements of the matrices.

(*a*)
$$\begin{aligned} AB &= A^T B^T && \text{symmetry} \\ &= (BA)^T && \text{transpose property} \\ &= (AB)^T && \text{commutative} \end{aligned}$$

Symmetry of the individual matrices permits the first equality. The second equality results from the property of transposes proved in Example 1.6. The third equality holds only if A and B commute.

(*b*)
$$\begin{aligned} (AB)_{ij} &= a_{ik}b_{jk} = a_{ki}b_{jk} && \text{symmetry} \\ &= b_{jk}a_{ki} && \text{elements commutative} \\ &= a_{jk}b_{ki} && \text{matrices commutative} \\ &= (AB)_{ji} \end{aligned}$$

Here we deal with the ijth element of AB. The symmetry of the individual matrices is used first. Next it is noted that the element products are always commutative. Finally, the matrices must commute.

These proofs verify the "only if" portion of the theorem, the "necessary" condition; i.e., for AB to be symmetric, it is necessary that A and B commute. To verify the "if" portion, the "sufficient" condition, the equalities are considered in reverse order starting with the assumed commutation of A and B. Then it can be said that for AB to be symmetric it is sufficient that A and B commute.

DIAGONAL AND UNIT MATRICES

A *diagonal* matrix is a square matrix having entries on the main diagonal and zeros everywhere else. If the entries on the main diagonal are all 1s, the matrix is called a *unit* or *identity* matrix.

Let a diagonal matrix of order n be given by

$$D = \begin{bmatrix} d_1 & 0 & \cdots & 0 \\ 0 & d_2 & \cdots & 0 \\ \cdot & \cdot & \cdot & \cdot \\ 0 & 0 & \cdots & d_n \end{bmatrix} \tag{1.8.3}$$

To write this matrix in index notation, we use the *Kronecker delta*, defined by

$$\delta_{ij} = \begin{cases} 1 & \text{for } i = j \\ 0 & \text{for } i \neq j \end{cases} \tag{1.8.4}$$

Then D can be written as

$$D = [d_{i\alpha}] \qquad \text{where } d_{i\alpha} = \delta_{i\alpha} d_\alpha \tag{1.8.5}$$

In Eq. (1.8.5) the Greek letter α is used instead of j for the second index to show that no summation is intended, as discussed in Sec. 1.3. The reader should carry out the operation of finding each $d_{i\alpha}$ to convince himself that the form given in Eq. (1.8.5) is indeed correct without a summation on α.

Premultiplication of a matrix A by D (the product DA) multiplies the ith row of A by d_i. *Postmultiplication* of A by D (the product AD) multiplies the jth column of A by d_j. A proof of these statements is requested in Prob. 1.15.

A unit or identity matrix is defined by

$$U = \begin{bmatrix} 1 & 0 & \cdots & 0 \\ 0 & 1 & \cdots & 0 \\ \cdot & \cdot & \cdot & \cdot \\ 0 & 0 & \cdots & 1 \end{bmatrix} \tag{1.8.6}$$

Premultiplication or postmultiplication of a matrix A by U does not change A. Thus

$$UA = AU = A \tag{1.8.7}$$

HERMITIAN MATRICES

A square matrix whose elements satisfy the condition

$$a_{ij} = \pm a_{ji}^* \tag{1.8.8}$$

is called a *hermitian* matrix if the plus sign holds for all i and j and a *skew-hermitian* matrix if the minus sign holds for all i and j. It follows that

$$H^T = \pm H^* \tag{1.8.9}$$

where again the plus sign holds for the hermitian case and the minus sign for the skew-hermitian case.

The main diagonal elements of a hermitian matrix must be real or zero; those of a skew-hermitian matrix must be imaginary or zero.

Any square matrix with complex entries can be expressed as the sum of a hermitian and a skew-hermitian matrix. The proof is requested in Prob. 1.14.

Hermitian matrices appear quite frequently in network theory, particularly in power and energy relationships.

1.9 DEFINITION OF A DETERMINANT

The reader has undoubtedly been introduced to determinants and cofactors while finding solutions to n equations in n unknowns. He knows how to find and use these quantities in obtaining such solutions but has perhaps given little consideration to their underlying definitions.

Our procedure here will be first to define a determinant and a cofactor and then to explore their properties and uses in terms of these definitions. This study will lead us quite naturally to the inverse of a matrix.

Let A be a square matrix of order n, given by

$$A = \begin{bmatrix} a_{11} & a_{12} & \cdots & a_{1n} \\ a_{21} & a_{22} & \cdots & a_{2n} \\ \cdot & \cdot & \cdot & \cdot \\ a_{n1} & a_{n2} & \cdots & a_{nn} \end{bmatrix} \tag{1.9.1}$$

The determinant of A, denoted by $|A|$, is a number associated with A and given by either of the following two product expressions:

$$|A| = e_{i_1 i_2 \cdots i_n} a_{1i_1} a_{2i_2} \cdots a_{ni_n} \qquad \text{row summation} \tag{1.9.2}$$

or

$$|A| = e_{i_1 i_2 \cdots i_n} a_{i_1 1} a_{i_2 2} \cdots a_{i_n n} \qquad \text{column summation} \tag{1.9.3}$$

where

$$e_{i_1 i_2 \cdots i_n} = \begin{cases} 0 & \text{if any two indices are alike} \\ +1 & \text{if an even number of interchanges of adjacent indices places the indices in numerical order} \\ -1 & \text{if an odd number of interchanges of adjacent indices places the indices in numerical order} \end{cases} \tag{1.9.4}$$

Each i index in Eqs. (1.9.2) and (1.9.3) is a repeated dummy index. Therefore Eqs. (1.9.2) and (1.9.3) are each summations of n^n products obtained by letting each i index take on all possible integer values from 1 through n. The e index determines the sign of a product and whether or not a product vanishes. Two examples will serve to clarify the notation.

Example 1.8 Find the value of the determinant of the following matrix:

$$A = \begin{bmatrix} a_{11} & a_{12} \\ a_{21} & a_{22} \end{bmatrix}$$

Using Eq. (1.9.2),

$$|A| = e_{i_1 i_2} a_{1 i_1} a_{2 i_2}$$

which, when expanded, gives

$$|A| = e_{11} a_{11} a_{21} + e_{12} a_{11} a_{22} + e_{21} a_{12} a_{21} + e_{22} a_{12} a_{22}$$

Terms 1 and 4 vanish since two indices are alike. No interchanges of adjacent indices are required in term 2 to place the indices in numerical order. Hence term 2 carries a plus sign. One interchange is necessary in term 3; thus term 3 has a minus sign. The final result is

$$|A| = a_{11} a_{22} - a_{12} a_{21}$$

We can also obtain $|A|$ from Eq. (1.9.3)

$$\begin{aligned} |A| &= e_{i_1 i_2} a_{i_1 1} a_{i_2 2} \\ &= e_{11} a_{11} a_{12} + e_{12} a_{11} a_{22} + e_{21} a_{21} a_{12} + e_{22} a_{21} a_{22} \\ &= a_{11} a_{22} - a_{21} a_{12} \end{aligned}$$

The results of Example 1.8 suggest the familiar rule: To find the determinant of a 2×2 matrix, subtract the product of the off-diagonal terms from the product of the main-diagonal terms.

Example 1.9 Find the value of the determinant of the following matrix:

$$A = \begin{bmatrix} a_{11} & a_{12} & a_{13} \\ a_{21} & a_{22} & a_{23} \\ a_{31} & a_{32} & a_{33} \end{bmatrix}$$

Using Eq. (1.9.2),

$$|A| = e_{i_1 i_2 i_3} a_{1i_1} a_{2i_2} a_{3i_3}$$

The expansion of $|A|$ gives 27 terms, 21 of which vanish because two or more indices are alike. Including only the nonvanishing terms, we obtain

$$\begin{aligned} |A| &= e_{123}a_{11}a_{22}a_{33} + e_{132}a_{11}a_{23}a_{32} + e_{213}a_{12}a_{21}a_{33} \\ &\quad + e_{231}a_{12}a_{23}a_{31} + e_{312}a_{13}a_{21}a_{32} + e_{321}a_{13}a_{22}a_{31} \end{aligned}$$

Terms 1, 4, and 5 require an even number of index interchanges and thus have a plus sign. In term 4, for example, $e_{231} = -e_{213} = e_{123}$. Two interchanges were necessary; thus the sign is plus. Terms 2, 3, and 6 require an odd number of index interchanges and carry a minus sign. The final result is

$$\begin{aligned} |A| &= a_{11}a_{22}a_{33} + a_{12}a_{23}a_{31} + a_{13}a_{21}a_{32} \\ &\quad - a_{11}a_{23}a_{32} - a_{12}a_{21}a_{33} - a_{13}a_{22}a_{31} \end{aligned}$$

Equation (1.9.3) yields the same answer.

The results of Example 1.9 suggest a second familiar rule for determining $|A|$ directly from A when A is a 3×3 matrix: Append the first two columns of A to A. This yields the array

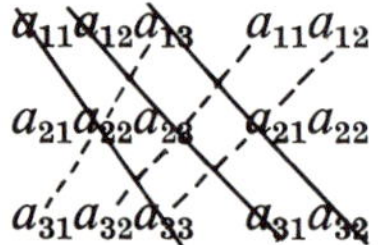

Now sum the products indicated by the diagonal solid lines and subtract the sum of the products indicated by the diagonal dashed lines to obtain $|A|$. This procedure is not applicable to determinants of higher order (a rank-4 determinant has 24 nonvanishing terms, but an extension of the procedure just outlined would yield only 8 terms).

The evaluation of a third-order determinant can be interpreted in another way which *is* applicable to higher-order determinants. Rewrite the final result of Example 1.9 in the form

$$|A| = a_{11}(a_{22}a_{33} - a_{23}a_{32}) - a_{12}(a_{21}a_{33} - a_{23}a_{31}) + a_{13}(a_{21}a_{32} - a_{22}a_{31}) \tag{1.9.5}$$

Each term in parentheses is a second-order determinant called a *minor* of A (denoted by M_{ij}). For example, the minor associated with a_{11} is M_{11}, and it is the determinant of the matrix that remains after deleting row 1 and column 1 from A. If a given minor is multiplied by the sign appearing in front of the element associated with it in the expansion of $|A|$, the result is defined as a *cofactor* of A, denoted by A_{ij}. Thus $A_{11} = M_{11}$, but $A_{12} = -M_{12}$. Generally

$$A_{ij} = (-1)^{i+j} M_{ij} \tag{1.9.6}$$

1.10 COFACTOR EXPANSION OF A DETERMINANT

It would indeed be tedious if it were always necessary to write down all terms in the expansions of Eqs. (1.9.2) and (1.9.3), especially for matrices of high order. Fortunately the procedure utilizing cofactors, introduced in the last section, can be extended to determinants of any order. We now formally develop the cofactor expansion of a determinant.

In the expression for $|A|$ in Eq. (1.9.2), pick any number α between 1 and n and delete the element $a_{\alpha i_\alpha}$ from $|A|$. The right-hand side of Eq. (1.9.2) becomes

$$e_{i_1 i_2 \cdots i_\alpha \cdots i_n} a_{1i_1} a_{2i_2} \cdots a_{\alpha-1, i_{\alpha-1}} a_{\alpha+1, i_{\alpha+1}} \cdots a_{ni_n} \tag{1.10.1}$$

where we show i_α explicitly in the e index to stress the point that removing $a_{\alpha i_\alpha}$ from $|A|$ does not remove i_α from e. Rather, i_α is no longer a repeated index, and thus we must assign it a fixed value. Let $i_\alpha = k$, another number between 1 and n, and define the αkth cofactor of A by

$$A_{\alpha k} \equiv e_{i_1 i_2 \cdots k \cdots i_n} a_{1i_1} a_{2i_2} \cdots a_{\alpha-1, i_{\alpha-1}} a_{\alpha+1, i_{\alpha+1}} \cdots a_{ni_n} \tag{1.10.2}$$

$A_{\alpha k}$ differs from $|A|$ only in that it does not contain any of the elements of row α or column k of A. The elements $a_{\alpha 1}, a_{\alpha 2}, \ldots, a_{\alpha n}$ of row α are missing because the element $a_{\alpha i_\alpha}$ has been removed from $|A|$. The elements $a_{1k}, a_{2k}, \ldots, a_{nk}$ of column k are missing because when an i index (in addition to i_α) equals k, two indices in e are alike and that product vanishes from $A_{\alpha k}$.

Now let us substitute the definition of $A_{\alpha k}$ in Eq. (1.10.2) into Eq. (1.9.2) to obtain

$$|A| = a_{\alpha k} A_{\alpha k} \tag{1.10.3}$$

as the cofactor expansion of a determinant. Note that α is not a summation index but k is. This is in keeping with our previous practice. The expansion on k is across row α of A and yields

$$|A| = a_{\alpha 1} A_{\alpha 1} + a_{\alpha 2} A_{\alpha 2} + \cdots + a_{\alpha n} A_{\alpha n} \tag{1.10.4}$$

If the cofactor expansion is formulated using Eq. (1.9.3) instead of Eq. (1.9.2), the result is

$$|A| = a_{k\alpha} A_{k\alpha} \tag{1.10.5}$$

The expansion on k in Eq. (1.10.5) is down column α of A and yields

$$|A| = a_{1\alpha}A_{1\alpha} + a_{2\alpha}A_{2\alpha} + \cdots + a_{n\alpha}A_{n\alpha} \tag{1.10.6}$$

Example 1.10 Let us find the value of $|A|$ in Example 1.9 by a cofactor expansion across row 1 of A and thus verify the results obtained in our introduction to minors and cofactors following that example.

From Eq. (1.9.2),

$$|A| = e_{i_1 i_2 i_3} a_{1i_1} a_{2i_2} a_{3i_3}$$

We choose $\alpha = 1$ to expand across row 1 of A. Then Eq. (1.10.1) takes the form

$$e_{i_1 i_2 i_3} a_{2i_2} a_{3i_3}$$

Now arbitrarily choose $i_1 = 2$. Then Eq. (1.10.2) gives cofactor A_{12} as

$$\begin{aligned} A_{12} &= e_{2i_2 i_3} a_{2i_2} a_{3i_3} \\ &= e_{211}a_{21}a_{31} + e_{212}a_{21}a_{32} + e_{213}a_{21}a_{33} \\ &\quad + e_{221}a_{22}a_{31} + e_{222}a_{22}a_{32} + e_{223}a_{22}a_{33} \\ &\quad + e_{231}a_{23}a_{31} + e_{232}a_{23}a_{32} + e_{233}a_{23}a_{33} \\ &= a_{23}a_{31} - a_{21}a_{33} \end{aligned}$$

Terms 1, 5, and 9 vanish from A_{12} because the indices i_2 and i_3 have like values. These terms also vanish in the evaluation of $|A|$ by Eq. (1.9.2). Terms 2, 4, 6, and 8 vanish because either i_2 or i_3 equals 2, the fixed value of i_1. These terms do *not* vanish in the evaluation of $|A|$ by Eq. (1.9.2). It is the absence of these latter terms that distinguishes the cofactor from the determinant; i.e., in A_{12}, all elements from row 1 and column 2 of A are missing.

The expression for A_{12} is identical to the second cofactor in Eq. (1.9.5) (the negative of the second minor), as we intended to show. To continue and find $|A|$, we use Eq. (1.10.3) or its expansion in Eq. (1.10.4) with $\alpha = 1$ and obtain

$$|A| = a_{11}A_{11} + a_{12}A_{12} + a_{13}A_{13}$$

which, when each cofactor is expanded, is identical with Eq. (1.9.5).

1.11 PROPERTIES OF DETERMINANTS

Property 1 The expansion of a determinant of a matrix of order $n \times n$ contains $n!$ terms whose e indices do not vanish.

The e index in Eq. (1.9.2) or (1.9.3) has n subscripts. If any two are alike, that index vanishes. Since the number of permutations of n different numbers taken n at a time is $n!$, there are $n!$ nonvanishing indices.

Property 2 A statement involving row and determinant is equivalent to the same statement involving column and determinant.

This property is true because of the equivalent definitions of $|A|$ in Eqs. (1.9.2) and (1.9.3).

Property 3 The determinant of A^T is $|A|$.

This property is true because application of Eq. (1.9.2) to A is equivalent to applying Eq. (1.9.3) to A^T.

Property 4 If each element in any row of A is multiplied by r, the value of $|A|$ is multiplied by r.

In Eq. (1.9.2), multiply a_{2i_2}, for example, by r; this multiplies $|A|$ by r.

Property 5 If two rows of A are interchanged, the sign of $|A|$ is changed.

The proof of this property using index notation is outlined in Prob. 1.18, the reader being requested to fill in the details. If we let A' be the matrix resulting from the interchange of two rows of A and $|A'|$ be its determinant, then property 5 states that $|A'| = -|A|$.

Property 6 If two rows of A are identical, or if one is a multiple of the other, then $|A| = 0$.

We know from property 5 that the interchange of two rows of A changes the sign of the determinant. But if the two interchanged rows are identical, $A' = A$ and therefore $|A'| = |A|$. The only way $|A'|$ can equal $|A|$ and $-|A|$ at the same time is for $|A|$ to be zero.

Now let one of the two equal rows of A be multiplied by r. By property 4, the determinant becomes $r\,|A|$, which is zero because $|A|$ is zero.

Property 7 If r times the elements of a row of A are added to the corresponding elements of another row of A, the value of $|A|$ is unchanged.

In Eq. (1.9.2), add ra_{2i_2} to a_{1i_1}. Call the new matrix A'' and its determinant $|A''|$. Then $|A''|$ is given by

$$|A''| = e_{i_1i_2\cdots i_n}a_{1i_1}a_{2i_2}\cdots a_{ni_n} + e_{i_1i_2\cdots i_n}ra_{2i_2}a_{2i_2}\cdots a_{ni_n} \tag{1.11.1}$$

The second term in $|A''|$ vanishes because of property 6. The first term is $|A|$. Thus $|A''| = |A|$, and property 7 is proved.

Property 8 If A and B are two square matrices of equal order, the following relation holds:

$$|A|\,|B| = |AB| \tag{1.11.2}$$

A proof of Eq. (1.11.2) using index notation is outlined in Prob. 1.20. The reader is asked to fill in the details.

Property 9 If A is an $m \times n$ matrix and B is an $n \times m$ matrix, with $m < n$, the following relation holds:

$$|AB| = \Sigma \text{ (products of corresponding majors of } A \text{ and } B) \qquad (1.11.3)$$

This relation is known as the *Binet-Cauchy theorem*. A *major* of A is the determinant of any largest square submatrix of A, in this case any $m \times m$ submatrix of A. A major of B corresponds to a major of A if the rows of B and the columns of A included in the major are the same.

A proof of Eq. (1.11.3) is requested in Prob. 1.21.

Example 1.11 Find the determinant of the product of the matrices

$$A = \begin{bmatrix} 1 & 2 & -1 & 3 \\ 4 & 1 & 0 & 2 \end{bmatrix} \qquad B = \begin{bmatrix} 0 & 5 \\ 1 & 1 \\ 2 & -3 \\ 2 & 0 \end{bmatrix}$$

By conventional means, the matrix product and its determinant are

$$AB = \begin{bmatrix} 6 & 10 \\ 5 & 21 \end{bmatrix} \qquad |AB| = 6 \times 21 - 10 \times 5 = 76$$

By the Binet-Cauchy theorem,

$$|AB| = \begin{vmatrix} 1 & 2 \\ 4 & 1 \end{vmatrix}\begin{vmatrix} 0 & 5 \\ 1 & 1 \end{vmatrix} + \begin{vmatrix} 1 & -1 \\ 4 & 0 \end{vmatrix}\begin{vmatrix} 0 & 5 \\ 2 & -3 \end{vmatrix} + \begin{vmatrix} 1 & 3 \\ 4 & 2 \end{vmatrix}\begin{vmatrix} 0 & 5 \\ 2 & 0 \end{vmatrix}$$
$$+ \begin{vmatrix} 2 & -1 \\ 1 & 0 \end{vmatrix}\begin{vmatrix} 1 & 1 \\ 2 & -3 \end{vmatrix} + \begin{vmatrix} 2 & 3 \\ 1 & 2 \end{vmatrix}\begin{vmatrix} 1 & 1 \\ 2 & 0 \end{vmatrix} + \begin{vmatrix} -1 & 3 \\ 0 & 2 \end{vmatrix}\begin{vmatrix} 2 & -3 \\ 2 & 0 \end{vmatrix}$$
$$= -7 \times -5 + 4 \times -10 - 10 \times -10 + 1 \times -5 + 1 \times -2 - 2 \times 6 = 76$$

From Example 1.11, the Binet-Cauchy theorem does not appear to be notable for its laborsaving virtues. However, let us examine the situation more closely before accepting this conclusion. If the entries in A and B in Example 1.11 were letters instead of numbers, the product AB would have four elements, each the sum of four terms. A conventional determinant expansion would contain $2 \times 4^2 = 32$ terms before cancellation of like terms. On the other hand, A and B each have six majors, each major being the difference of two terms. Products of corresponding majors each contain four terms, and there are six such products. The total is 24 terms, eight less than by the conventional method. In general, the conventional method results in $n^m m!$ terms, whereas the Binet-Cauchy theorem yields

TABLE 1.1 Comparison of conventional and Binet-Cauchy determinant expansions

m	n	*Conventional expansion,* $n^m m!$ *terms*	*Binet-Cauchy expansion,* $\frac{(n-m+2)(n-m+1)m!^2}{2}$ *terms*
2	3	18	12
2	4	32	24
2	6	72	60
2	10	200	180
3	4	384	108
3	6	1,296	360
3	10	6,000	1,296
4	5	15,000	1,728
4	10	240,000	16,128
5	6	930,720	43,200
5	10	12,000,000	302,400

$[(n-m+2)(n-m+1)m!^2]/2$ terms. A comparison of the two methods for various values of m and n is given in Table 1.1.

In digital computation it is important to have programs that require minimum computer storage capacity and running time. Table 1.1 clearly shows the advantage of the Binet-Cauchy expansion over the conventional expansion in this respect.

Another important feature of the Binet-Cauchy theorem is its usefulness in the derivation of topological properties of networks, which will be considered in Chap. 6.

Property 10 Let A and B be square matrices of order $n \times n$. Then

$$|A + B| = A_0 + A_1 + \cdots + A_n \tag{1.11.4}$$

where A_k, $k = 0, \ldots, n$, is the summation of determinants obtained by replacing the columns (rows) of A by the corresponding columns (rows) of B, k at a time.

We shall prove this relation for $n = 3$ and extend the result by induction to matrices of higher order. From Eq. (1.9.3), we can write

$$|A + B| = e_{i_1 i_2 i_3}(a_{i_1 1} + b_{i_1 1})(a_{i_2 2} + b_{i_2 2})(a_{i_3 3} + b_{i_3 3}) \tag{1.11.5}$$

Expanding and grouping, we obtain

$$\begin{aligned} |A + B| = {} & e_{i_1 i_2 i_3}(a_{i_1 1} a_{i_2 2} a_{i_3 3}) \\ & + e_{i_1 i_2 i_3}(b_{i_1 1} a_{i_2 2} a_{i_3 3} + a_{i_1 1} b_{i_2 2} a_{i_3 3} + a_{i_1 1} a_{i_2 2} b_{i_3 3}) \\ & + e_{i_1 i_2 i_3}(b_{i_1 1} b_{i_2 2} a_{i_3 3} + b_{i_1 1} a_{i_2 2} b_{i_3 3} + a_{i_1 1} b_{i_2 2} b_{i_3 3}) \\ & + e_{i_1 i_2 i_3}(b_{i_1 1} b_{i_2 2} b_{i_3 3}) \end{aligned} \tag{1.11.6}$$

The first row in Eq. (1.11.6) is $|A|$, but, more important, it is A_0 of Eq. (1.11.4); i.e., it is the summation of determinants (one in this case) obtained by replacing the columns of A by the corresponding columns of B none at a time. The second row in Eq. (1.11.6) is A_1 in Eq. (1.11.4). Here the columns of A have been replaced by the corresponding columns of B one at a time. The third and fourth rows of Eq. (1.11.6) are A_2 and A_3 (A_n), respectively, in Eq. (1.11.4). The extension to the nth-order case follows immediately, and thus Eq. (1.11.4) is proved.

We shall use Eqs. (1.11.4) and (1.11.6) in the development of topological formulas for active and/or coupled netwnrks in Chap. 6.

1.12 CRAMER'S RULE

Consider a set of n linear algebraic equations in n unknowns given by

$$a_{ik}x_k = c_i \qquad i = 1, \ldots, n \tag{1.12.1}$$

We wish to solve the set for the x's. Let us premultiply both sides of Eq. (1.12.1) by the ilth cofactor of A.

$$A_{il}a_{ik}x_k = A_{il}c_i \tag{1.12.2}$$

There are n equations in Eq. (1.12.1) but only one equation in Eq. (1.12.2). In the latter we are summing on both i and k.

Let us examine the product $A_{il}a_{ik}$. Consider the cofactor expansion of a determinant by a column given in Eq. (1.10.5) and repeated below

$$|A| = a_{k\alpha}A_{k\alpha} \tag{1.12.3}$$

Recall that a summation on k (but not on α) is implied. In Eq. (1.12.3), let $a_{k\alpha}$ be replaced by $a_{k\beta}$. This means that column α is replaced by column β in A. Thus two columns are alike, and the determinant is 0 by property 6. We can now draw the conclusion that

$$a_{k\beta}A_{k\alpha} = \begin{cases} |A| & \alpha = \beta \\ 0 & \alpha \neq \beta \end{cases} \tag{1.12.4}$$

Equation (1.12.4) can be written more concisely using the Kronecker delta

$$a_{k\beta}A_{k\alpha} = \delta_{\beta\alpha}\,|A| \tag{1.12.5}$$

Substituting Eq. (1.12.5) (after changing the dummy indices) into Eq. (1.12.2) yields

$$\delta_{kl}\,|A|\,x_k = A_{il}c_i \tag{1.12.6}$$

The left-hand side of Eq. (1.12.6) is nonzero only when $k = l$. Thus Eq. (1.12.6) reduces to

$$|A|\, x_l = A_{il} c_i \tag{1.12.7}$$

which yields the Cramer's rule solution for the lth x as

$$x_l = \frac{A_{il} c_i}{|A|} \tag{1.12.8}$$

We observe that Cramer's rule is applicable if and only if $|A|$ is not zero. If $|A|$ is zero, we say that A is a *singular* matrix. The requirement here is that A be *nonsingular*. We note further that the numerator of Eq. (1.12.8) requires a summation from 1 to n on the index i. Thus

$$x_l = \frac{A_{1l} c_1 + A_{2l} c_2 + \cdots + A_{nl} c_n}{|A|} \tag{1.12.9}$$

If we delete the elements of column l from A and replace them by the c_i to obtain a new matrix, the numerator of Eq. (1.12.9) is seen to be the cofactor expansion down column l of the determinant of that new matrix. This yields the useful form

$$x_l = \frac{1}{|A|} \begin{vmatrix} a_{11} & a_{12} & \cdots & c_1 & \cdots & a_{1n} \\ a_{21} & a_{22} & \cdots & c_2 & \cdots & a_{2n} \\ \cdots & \cdots & \cdots & \cdots & \cdots & \cdots \\ a_{n1} & a_{n2} & \cdots & c_n & \cdots & a_{nn} \end{vmatrix} \tag{1.12.10}$$

where the c's are assumed to be in column l.

Example 1.12 Solve the following set of equations for x_2:

$$a_{11}x_1 + a_{12}x_2 = c_1$$

$$a_{21}x_1 + a_{22}x_2 = c_2$$

To illustrate the previous development, we shall proceed through the steps leading to Eq. (1.12.8), rather than using that equation [or Eq. (1.12.10)] directly. Since we are solving for x_2, we premultiply the given set of equations by A_{i2}. Thus the first equation is multiplied by A_{12} and the second by A_{22}, and the results are added to give

$$A_{12}a_{11}x_1 + A_{12}a_{12}x_2 + A_{22}a_{21}x_1 + A_{22}a_{22}x_2 = A_{12}c_1 + A_{22}c_2$$

By Eq. (1.12.5), the sum of the first and third terms on the left side vanishes. This leaves

$$(A_{12}a_{12} + A_{22}a_{22})x_2 = A_{12}c_1 + A_{22}c_2$$

The coefficient on the left side is $|A|$. The right-hand side is the determinant expansion down column 2 of the matrix formed by replacing a_{12} and a_{22} by c_1 and c_2, respectively, Thus

$$x_2 = \frac{a_{11}c_2 - a_{21}c_1}{a_{11}a_{22} - a_{21}a_{12}} = \frac{\begin{vmatrix} a_{11} & c_1 \\ a_{21} & c_2 \end{vmatrix}}{\begin{vmatrix} a_{11} & a_{12} \\ a_{21} & a_{22} \end{vmatrix}}$$

1.13 INVERSE MATRIX

Equation (1.12.8) lets us solve for any one of the x's. We now want to develop two alternative forms to Eq. (1.12.8) that display all the x's, namely, a closed matrix form, using a single capital letter for each matrix, and an expanded matrix form in which each matrix element is displayed.

Let us rewrite Eq. (1.12.1) in the form

$$AX = C \tag{1.13.1}$$

We can solve Eq. (1.13.1) symbolically to obtain

$$X = A^{-1}C \tag{1.13.2}$$

where A^{-1} is defined as the *inverse* matrix of A. Equation (1.13.2) is the closed matrix form of Cramer's rule. It is of considerable use in matrix manipulations but, without an explicit procedure for obtaining A^{-1}, is of no use in actually finding the x's in a given problem.

To obtain a form convenient for computation, we return to Eq. (1.12.8), solve for each x, and rewrite the result as an expanded matrix equation

$$\begin{bmatrix} x_1 \\ x_2 \\ \cdot \\ \cdot \\ \cdot \\ x_n \end{bmatrix} = \frac{1}{|A|} \begin{bmatrix} A_{11} & A_{21} & \cdots & A_{n1} \\ A_{12} & A_{22} & \cdots & A_{n2} \\ \cdots & \cdots & \cdots & \cdots \\ A_{1n} & A_{2n} & \cdots & A_{nn} \end{bmatrix} \begin{bmatrix} c_1 \\ c_2 \\ \cdot \\ \cdot \\ \cdot \\ c_n \end{bmatrix} \tag{1.13.3}$$

Comparing Eqs. (1.13.2) and (1.13.3), we see that

$$A^{-1} = \frac{1}{|A|} \begin{bmatrix} A_{11} & A_{21} & \cdots & A_{n1} \\ A_{12} & A_{22} & \cdots & A_{n2} \\ \cdots & \cdots & \cdots & \cdots \\ A_{1n} & A_{2n} & \cdots & A_{nn} \end{bmatrix} \tag{1.13.4}$$

Thus, to obtain A^{-1}, we replace each entry in A by its cofactor, transpose the resulting matrix, and divide by $|A|$. We note that A must be square and nonsingular for its inverse matrix to exist.

The product of a matrix and its inverse can be obtained by substituting Eq. (1.13.2) into Eq. (1.13.1) to obtain $AA^{-1}C = C$. It follows that

$$AA^{-1} = U \tag{1.13.5}$$

Performing the substitution in reverse order yields

$$A^{-1}A = U \tag{1.13.6}$$

Before proceeding to two examples, we summarize by saying that we now have three equivalent forms of Cramer's rule from which to choose, namely, the *index form* for each individual x_i in Eq. (1.12.8), the *closed matrix form* in Eq. (1.13.2), and the *expanded form* in Eq. (1.13.3).

Example 1.13 Find the inverse of

$$A = \begin{bmatrix} 2 & 1 & 3 \\ 0 & 2 & 1 \\ 3 & 1 & 2 \end{bmatrix}$$

Only the answer is given. The reader is urged to verify it.

$$A^{-1} = \frac{1}{-9}\begin{bmatrix} 3 & 1 & -5 \\ 3 & -5 & -2 \\ -6 & 1 & 4 \end{bmatrix}$$

Example 1.14 Given two square nonsingular matrices A and B of the same order, prove that

$$(AB)^{-1} = B^{-1}A^{-1}$$

We carry out this proof using the matrices as a whole rather than the individual elements and index notation. Premultiply both sides by B to obtain

$$B(AB)^{-1} = BB^{-1}A^{-1} = A^{-1}$$

Now premultiply both sides again, this time by A. The result is

$$AB(AB)^{-1} = U = AA^{-1} = U$$

Since both sides reduce to the identity matrix, the relation is proved.

1.14 SPECIAL MATRICES: II

In the last few years, considerable research has been done by circuit theorists in the area of n-port networks (see Ref. 1.4). Matrices have quite naturally played an important part in these studies. In this section we present certain special matrices commonly used in the classification, analysis, and synthesis of n-port networks.

PARAMOUNT MATRICES

A *paramount matrix* (also called an M matrix) is a real symmetric matrix in which any principal subdeterminant is greater than or equal to the absolute value of any other subdeterminant taken from the same rows (columns) of the matrix. A *principal subdeterminant* of a matrix A is the determinant of any submatrix of A whose main diagonal lies along that of A. Such a submatrix is appropriately termed a *principal submatrix* of A.

For a matrix to be paramount, all its main-diagonal elements must be positive. Off-diagonal terms may be negative.

The inverse of any nonsingular paramount matrix can be shown to be paramount. Any principal submatrix of a paramount matrix is also paramount.

Example 1.15 Determine whether the following matrix is paramount:

$$A = \begin{bmatrix} 3 & 1 & -2 \\ 1 & 4 & 3 \\ -2 & 3 & 6 \end{bmatrix}$$

We shall check A for paramountcy by tabulating its subdeterminants.

Subdeterminants of order 1 (vertical bars denote absolute values):

$3 > |-2| > |1|$ row 1

$4 > |3| > |1|$ row 2

$6 > |3| > |-2|$ row 3

Subdeterminants of order 2:

$12 - 1 = |9 + 2| = |3 + 8|$ rows 1 and 2

$18 - 4 > |6 + 6| > |9 + 2|$ rows 1 and 3

$24 - 9 > |6 + 6| > |3 + 8|$ rows 2 and 3

The matrix satisfies all the conditions for a paramount matrix.

DOMINANT MATRICES

A *dominant matrix* (also known as an R matrix) is a real symmetric matrix each of whose main-diagonal entries is greater than or equal to the sum of the absolute values of all other entries in the same row (column). Stated in index notation,

$$a_{ii} \geq \sum_{j=1}^{n} |a_{ij}| \qquad j \neq i \tag{1.14.1}$$

for every value of i. As in the case of a paramount matrix, main-diagonal terms must be positive, but off-diagonal terms may be negative.

Example 1.16 Find least values for a, b, and c such that the given matrix A is (*a*) paramount and (*b*) dominant.

$$A = \begin{bmatrix} a & -1 & -3 \\ -1 & b & 4 \\ -3 & 4 & c \end{bmatrix}$$

(*a*) For paramountcy, it is necessary that

$$a \geq 3 \qquad b \geq 4 \qquad c > 4$$

Let us choose the least values of a, b, and c and ascertain whether the conditions for paramountcy are satisfied by the subdeterminants of order 2.

$$12 - 1 > |12 - 3| > |-4 + 12| \qquad \text{rows 1 and 2}$$

$$12 - 9 \not> |12 - 3| > |-4 + 12| \qquad \text{rows 1 and 3}$$

$$16 - 16 > |-4 + 12| > |-4 + 9| \qquad \text{rows 2 and 3}$$

The matrix A is not paramount for these choices. A little study shows that changing a or b does not help unless c is changed. With $a = 3$ and $b = 4$, the minimum acceptable c is 6. Thus

$$A_p = \begin{bmatrix} 3 & -1 & -3 \\ -1 & 4 & 4 \\ -3 & 4 & 6 \end{bmatrix}$$

where the p subscript denotes paramount.

(*b*) To construct a dominant matrix we use Eq. (1.14.1) and find that

$$a \geq 4 \qquad b \geq 5 \qquad c \geq 7$$

Choosing the least values gives

$$A_d = \begin{bmatrix} 4 & -1 & -3 \\ -1 & 5 & 4 \\ -3 & 4 & 7 \end{bmatrix}$$

where the d subscript denotes dominant.

We note that the main-diagonal entries in A_d are larger than those in A_p. This is to be expected; the conditions for dominancy are more demanding on the main-diagonal entries than those for paramountcy.

UNIFORMLY TAPERED MATRICES

A real symmetric matrix is *uniformly tapered* if all its entries are nonnegative and

$$a_{ij} + a_{i-1,j+1} \geq a_{i-1,j} + a_{i,j+1} \qquad i \leq j \tag{1.14.2}$$

for all i and j, with the understanding that

$$a_{i,n+1} = a_{0,j} = 0 \tag{1.14.3}$$

For example, the matrix

$$A = \begin{bmatrix} a_{11} & a_{12} & a_{13} \\ a_{12} & a_{22} & a_{23} \\ a_{13} & a_{23} & a_{33} \end{bmatrix} \tag{1.14.4}$$

is uniformly tapered if

$$\begin{aligned} a_{11} &\geq a_{12} \\ a_{12} &\geq a_{13} \\ a_{22} + a_{13} &\geq a_{12} + a_{23} \\ a_{23} &\geq a_{13} \\ a_{33} &\geq a_{23} \end{aligned} \tag{1.14.5}$$

Example 1.17 Find least values for a, b, c, d, and e such that the matrix A is uniformly tapered

$$A = \begin{bmatrix} a & 3 & 2 & 1 \\ 3 & b & e & 2 \\ 2 & e & c & 3 \\ 1 & 2 & 3 & d \end{bmatrix}$$

From Eqs. (1.14.2) and (1.14.3), the constraints are

$$\begin{aligned} a &\geq 3 \\ b + 2 &\geq 3 + e \\ e + 1 &\geq 2 + 2 \\ c + 2 &\geq e + 3 \\ d &\geq 3 \end{aligned}$$

Using the equal signs, solutions are easily obtained. The result is

$$A = \begin{bmatrix} 3 & 3 & 2 & 1 \\ 3 & 4 & 3 & 2 \\ 2 & 3 & 4 & 3 \\ 1 & 2 & 3 & 3 \end{bmatrix}$$

UNIMODULAR MATRICES

A *unimodular* matrix (also called an E matrix) is a real rectangular matrix each of whose subdeterminants (of all orders) is ± 1 or 0.

If A is unimodular, it follows from the properties of determinants in Sec. 1.11 that:

1. Each element of A is ± 1 or 0.
2. A^T is unimodular.
3. If A is square and nonsingular, A^{-1} is unimodular and $|A|$ is ± 1.
4. Every submatrix of A is unimodular.
5. The matrix obtained by changing the signs of all elements of a row or column of A is unimodular.

Example 1.18 Find values for a, b, c, d, e, and f such that matrix A is unimodular

$$A = \begin{bmatrix} a & d & 1 & 0 & 0 \\ b & e & 0 & 1 & 0 \\ c & f & 0 & 0 & 1 \end{bmatrix}$$

Each of the unknown elements must be ± 1 or 0. In addition,

$$bf - ec = \pm 1 \text{ or } 0$$

$$af - dc = \pm 1 \text{ or } 0$$

$$ae - bd = \pm 1 \text{ or } 0$$

Many solutions exist, two of which are

$$A_1 = \begin{bmatrix} 1 & 0 & 1 & 0 & 0 \\ 1 & 1 & 0 & 1 & 0 \\ 0 & 1 & 0 & 0 & 1 \end{bmatrix} \qquad A_2 = \begin{bmatrix} -1 & 0 & 1 & 0 & 0 \\ 0 & 1 & 0 & 1 & 0 \\ -1 & -1 & 0 & 0 & 1 \end{bmatrix}$$

REFERENCES

1.1. Hildebrand, F. E.: "Methods of Applied Mathematics," Prentice-Hall, Inc., Englewood Cliffs, N.J., 1952. Chapter 1 contains a rigorous and yet readable discussion of matrices and determinants from an index-notation point of view.

1.2. Huelsman, L. P.: "Circuits, Matrices, and Linear Vector Spaces," McGraw-Hill Book Company, New York, 1963. A fine presentation of matrices and determinants is given in Chapter 2 at a level intermediate between Refs. 1.1 and 1.3.

1.3. Reza, F. M., and S. Seely: "Modern Network Analysis," McGraw-Hill Book Company, New York, 1959. Chapter 3 is an easy to read presentation of matrices and determinants, stressing their properties and manipulation.

1.4. N-port Networks without Transformers, *IRE Trans. Circuit Theory*, vol. CT-9, pp. 202–213, Sept., 1962. The special matrices of Sec. 1.14 are discussed from the point of view of their use in n-port networks. An extensive bibliography is given.

PROBLEMS

Drill

1.1. Where defined, find AB, BA, A^TB, AB^T, A^TB^T, A^2, B^2.

$$A = \begin{bmatrix} 1 & 2 & 1 \\ 1 & -1 & 0 \\ 2 & 0 & 1 \end{bmatrix} \qquad B = \begin{bmatrix} 1 & 0 & 1 \\ -1 & 1 & -1 \end{bmatrix}$$

1.2. Find $(A + B + C)^2$ and $A^2 + B^2 + C^2 + 2(AB + BC + CA)$ and compare.

$$A = \begin{bmatrix} 1 & -2 & 5 \\ -3 & 8 & -7 \\ 6 & -4 & 9 \end{bmatrix} \qquad B = \begin{bmatrix} 2 & 3 & -1 \\ 3 & -2 & 1 \\ 1 & 4 & 1 \end{bmatrix} \qquad C = \begin{bmatrix} -2 & -1 & -3 \\ 0 & -5 & 6 \\ -7 & 0 & -9 \end{bmatrix}$$

1.3. Partition the given matrices so as to minimize the labor in obtaining their product and carry out the multiplication in partitioned form.

$$\begin{bmatrix} 5 & 0 & 1 & 2 \\ 0 & 4 & -1 & 3 \\ 0 & 0 & 1 & 0 \\ 0 & 0 & 0 & 2 \end{bmatrix} \begin{bmatrix} 1 & 0 & 0 & 0 \\ 0 & 1 & 0 & 0 \\ 1 & -1 & 1 & 0 \\ 1 & 0 & 0 & 1 \end{bmatrix}$$

1.4. For what values of X does $|AB|$ vanish?

$$A = \begin{bmatrix} 1 & 0 & 5 & -2 \\ 2 & 3 & 2 & -1 \\ 4 & 1 & 3 & 2 \\ -3 & 0 & 6 & X \end{bmatrix} \qquad B = \begin{bmatrix} 2 & 0 & 1 & 6 \\ 1 & 2 & 0 & 4 \\ 3 & 0 & 1 & 5 \\ 1 & 2 & 0 & X \end{bmatrix}$$

1.5. Without expanding, show that the determinants of the given matrices are zero.

$$A = \begin{bmatrix} 1 & 2 & 7 \\ 1 & 3 & 6 \\ 1 & 4 & 5 \end{bmatrix} \qquad B = \begin{bmatrix} -7 & 1 & 1 & 3 \\ 1 & 4 & 2 & 1 \\ 4 & 3 & 1 & 2 \\ -3 & 5 & 3 & 1 \end{bmatrix}$$

1.6. Evaluate the determinant of the product of the given matrices by a conventional expansion and by a Binet-Cauchy expansion.

$$A = \begin{bmatrix} 1 & -3 & 1 & 2 & 1 \\ 2 & 1 & -1 & 1 & 0 \\ 4 & -3 & 0 & 3 & 2 \end{bmatrix} \qquad B = \begin{bmatrix} 1 & 0 & 0 \\ 2 & 1 & 0 \\ 4 & 2 & 1 \\ -2 & 3 & 1 \\ 1 & 2 & 3 \end{bmatrix}$$

1.7. Find solutions to the following sets of equations by Cramer's rule:

(*a*)
$$\begin{aligned} w + 2x - 3y + 3z &= 13 \\ 2w - x + y - 2z &= -7 \\ w + x + y + z &= 10 \\ 4w + 2x - y - 2z &= 0 \end{aligned}$$

(*b*)
$$\begin{aligned} \left(7 + \frac{1}{s}\right)e_1 - 3e_2 + 4e_3 &= 2 \\ -3e_1 + (3s + 3)e_2 + 2se_3 &= 1 \\ 4e_1 + 2se_2 + \left(2s + 4 + \frac{2}{s}\right)e_3 &= 0 \qquad \text{for } s = j1 \end{aligned}$$

1.8. (*a*) Find the inverses of the following matrices:

$$A = \begin{bmatrix} 1 & -3 & 1 \\ 2 & 1 & -1 \\ 4 & -3 & 0 \end{bmatrix} \qquad B = \begin{bmatrix} 1 & 0 & 0 & 0 \\ 2 & 1 & 0 & 0 \\ 4 & 2 & 1 & 0 \\ -2 & 3 & 1 & 1 \end{bmatrix}$$

(*b*) Find the inverse of the given impedance matrix directly and also by first partitioning and finding the individual inverses (see the result of Prob. 1.13).

$$\begin{bmatrix} 2s & s & 0 & 0 \\ s & 3s & 0 & 0 \\ 0 & 0 & 3 & 0 \\ 0 & 0 & 0 & \frac{1}{s} \end{bmatrix}$$

1.9. Let

$$\begin{bmatrix} y_1 \\ y_2 \\ y_3 \end{bmatrix} = \begin{bmatrix} 1 & 2 & 3 \\ 2 & -3 & -1 \\ -3 & 1 & 2 \end{bmatrix} \begin{bmatrix} x_1 \\ x_2 \\ x_3 \end{bmatrix} \quad \text{and} \quad \begin{bmatrix} z_1 \\ z_2 \\ z_3 \end{bmatrix} = \begin{bmatrix} 1 & -2 & 1 \\ -1 & 4 & -2 \\ 2 & 1 & 6 \end{bmatrix} \begin{bmatrix} y_1 \\ y_2 \\ y_3 \end{bmatrix}$$

Find a matrix C such that

$$X = CZ$$

Solve for X if $Z = \begin{bmatrix} 5 \\ 1 \\ -1 \end{bmatrix}$.

1.10. Are the given matrices paramount? Dominant? Uniformly tapered?

$$A = \begin{bmatrix} 2 & -1 & -2 \\ -1 & 3 & 3 \\ -2 & 3 & 3 \end{bmatrix} \qquad B = \begin{bmatrix} 7 & 6 & 5 \\ 6 & 9 & 7 \\ 5 & 7 & 9 \end{bmatrix}$$

If not, increase the main-diagonal elements just enough to make each matrix paramount; dominant.

Theory and proofs

1.11. The product of an $m \times p$ matrix A and a $p \times n$ matrix B is to be obtained. Let S be a $p \times 1$ column matrix such that s_i is the sum of the elements in the ith row of B. Show that the sum of the terms in the ith row of AB is equal to the term in the ith row of AS. Use this result to check the matrix product of the following matrices:

$$A = \begin{bmatrix} 11 & 6 & 3 \\ -2 & 1 & -4 \\ 3 & 5 & 1 \end{bmatrix} \qquad B = \begin{bmatrix} 4 & 7 \\ 3 & 2 \\ -6 & -1 \end{bmatrix}$$

1.12. If D is a diagonal matrix given by

$$D = [d_\alpha \delta_{\alpha j}]$$

show that

$$D^{-1} = \left[\frac{1}{d_\alpha}\,\delta_{\alpha j}\right]$$

1.13. Extend the result of Prob. 1.12 to the case where each main-diagonal element of D in partitioned form is itself a matrix, i.e., the αjth element of D is now given by

$$D_{\alpha j} = [A_\alpha]\,\delta_{\alpha j}$$

Prove that D^{-1} is given by

$$D^{-1} = [[A_\alpha]^{-1}\delta_{\alpha j}]$$

1.14. (*a*) Show that any square matrix A can be written as the sum of a symmetric matrix B and a skew-symmetric matrix C. Find B and C if A is given by

$$A = \begin{bmatrix} 1 & 6 & 3 \\ -2 & 1 & -4 \\ 3 & 5 & 1 \end{bmatrix}$$

(*b*) Repeat the proof of part (*a*) for a square matrix with complex entries. The matrices B and C are now hermitian and skew-hermitian, respectively. Find the two matrices for

$$A = \begin{bmatrix} 1 & 1+j & 3-j \\ -1-j & -j & 4 \\ 2 & 3+2j & 2+j \end{bmatrix}$$

1.15. Prove that premultiplication of a matrix A by a diagonal matrix D multiplies the ith row of A by d_i. Also prove that postmultiplication of A by D multiplies the jth column of A by d_j.

1.16. Let A and B be diagonal matrices of order n. Using index notation, prove that the matrix product AB is diagonal.

1.17. Prove that

$$e_{ijk}e_{klm} = \delta_{il}\delta_{jm} - \delta_{im}\delta_{jl}$$

118. (*a*) Consider the index $e_{i_1 \cdots i_k \cdots i_l \cdots i_n}$. Show that it always requires an odd number of interchanges of adjacent indices to interchange i_k in the kth position with i_l in the lth position.

(*b*) Rewrite Eq. (1.9.2) to include two arbitrary entries, the kth and the lth.

$$|A| = e_{i_1 \cdots i_k \cdots i_l \cdots i_n} a_{1i_1} \cdots a_{ki_k} \cdots a_{li_l} \cdots a_{ni_n}$$

Assume corresponding elements in rows k and l of A are interchanged to give a new matrix A', where

$$|A'| = e_{i_1 \cdots i_k \cdots i_l \cdots i_n} a_{1i_1} \cdots a_{li_k} \cdots a_{ki_l} \cdots a_{ni_n}$$

Now, using the result in part (*a*), prove that $|A'| = -|A|$ by showing that any product in $|A|$ for $i_k = p$ and $i = q$ is the negative of the corresponding product in $|A'|$ for $i_l = p$ and $i_k = q$.

1.19. Prove that the given matrix cannot be a unimodular matrix. The x's denote nonzero elements.

$$A = \begin{bmatrix} x & x & 0 & x \\ x & 0 & x & x \\ 0 & x & x & x \end{bmatrix}$$

1.20. (*a*) Prove that

$$e_{i_1 i_2} a_{i_1 1} a_{i_2 2} e_{k_1 k_2} = e_{i_1 i_2} a_{i_1 k_1} a_{i_2 k_2}$$

(*b*) A and B are two square matrices of the same order. Let $|A|$ be given by Eq. (1.9.3), and let $|B|$ be given by

$$|B| = e_{k_1 k_2 \cdots k_n} b_{k_1 1} b_{k_2 2} \cdots b_{k_n n}$$

Using the result of part (*a*), prove that $|A|\,|B| = |AB|$.

1.21. Let A and B be given by

$$A = \begin{bmatrix} a_{11} & a_{12} & a_{13} \\ a_{21} & a_{22} & a_{23} \end{bmatrix} \qquad B = \begin{bmatrix} b_{11} & b_{12} \\ b_{21} & b_{22} \\ b_{31} & b_{32} \end{bmatrix}$$

Show that $|AB|$ is given by

$$|AB| = e_{i_1 i_2} a_{i_1 k_1} b_{k_1 1} a_{i_2 k_2} b_{k_2 2}$$

where each i has the value 1 or 2 and each k has the value 1, 2, or 3. Convert this expression into the sum of three expressions and thus justify the Binet-Cauchy theorem for $m = 2$ and $n = 3$. Extend this result to the general case.

1.22. Let A be a unimodular matrix of order $m \times n$, with $m < n$. Let D be a diagonal matrix of order $n \times n$ whose main-diagonal elements are $d_1, d_2, \ldots, d_n$. Prove that all elements of the matrix product ADA^T are of the form $\pm(d_r + d_s + \cdots)$, where $r, s, \ldots$ are some of the indices $1, 2, \ldots, n$.

Application

1.23.(*a*) The voltage-current matrix relations of two 2-port networks are

$$\begin{bmatrix} e_1 \\ e_2 \end{bmatrix} = \begin{bmatrix} z_{11} & z_{12} \\ z_{21} & z_{22} \end{bmatrix} \begin{bmatrix} i_1 \\ i_2 \end{bmatrix} \qquad \begin{bmatrix} e_2 \\ e_3 \end{bmatrix} = \begin{bmatrix} -z_{22} & z_{21} \\ -z_{12} & z_{11} \end{bmatrix} \begin{bmatrix} i_2 \\ i_3 \end{bmatrix}$$

Find the unknown elements in the matrix equation

$$\begin{bmatrix} e_1 \\ e_3 \end{bmatrix} = \begin{bmatrix} Z_{11} & Z_{12} \\ Z_{21} & Z_{22} \end{bmatrix} \begin{bmatrix} i_1 \\ i_3 \end{bmatrix}$$

What significance can be attached to the quantities $Z_{11} + Z_{12}$ and $Z_{11} - Z_{12}$?

(*b*) Repeat part (*a*) when the second matrix equation is

$$\begin{bmatrix} e_2 \\ e_3 \end{bmatrix} = \begin{bmatrix} -R & R \\ -R & R \end{bmatrix} \begin{bmatrix} i_2 \\ i_3 \end{bmatrix}$$

where R is a positive constant. What are Z_{11} and Z_{12}?

1.24.

$$A = \begin{bmatrix} 1 & 0 & -1 & 1 & 0 & 0 \\ -1 & 1 & 0 & 0 & 1 & 0 \\ 0 & -1 & 1 & 0 & 0 & 1 \end{bmatrix} \qquad D = \begin{bmatrix} 4 & & & & & \\ & 3 & & & \mathbf{0} & \\ & & 5 & & & \\ & \mathbf{0} & & 2 & & \\ & & & & 6 & \\ & & & & & 2 \end{bmatrix}$$

(*a*) Show that A is unimodular and write down the product AD by inspection.

(*b*) Compute the product ADA^T by suitably partitioning A and D and show that ADA^T is paramount and dominant.

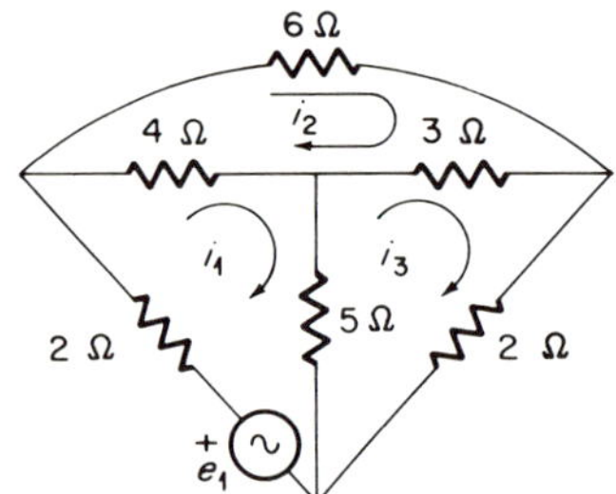

Fig. P1.24

(*c*) Write loop equations in matrix form for the given network to show that the loop-impedance matrix is ADA^T. Give an interpretation of dominancy and paramountcy in terms of the network configuration.

1.25. The networks of Fig. 1.1 are to be cascaded in reverse order. What constraints must be imposed on the individual network matrices if the matrix of the combined network is to be the same as that obtained in Example 1.2? In other words, what are the conditions for two 2×2 matrices to be commutative?

1.26. A matrix product Z is defined by

$$Z = ABC^T$$

If B is symmetric, find conditions on A and C such that Z is symmetric. Find suitable B and C matrices when

$$A = \begin{bmatrix} 1 & 0 & 1 \\ 0 & 1 & 1 \end{bmatrix} \qquad \text{and} \qquad Z = \begin{bmatrix} 2 & 1 \\ 1 & 1 \end{bmatrix}$$

1.27. Matrices can be used to perform phasor manipulations.

(*a*) Three phasors, $P = A\underline{/\alpha + \beta + \gamma}$, $Q = B\underline{/\beta + \gamma}$, and $R = C\underline{/\gamma}$, are shown in Fig. P1.27*a*. Let the components of P be P_x and P_y and those of Q be Q_x and Q_y. Find a 2×2 matrix M_1 such that

$$\begin{bmatrix} P_x \\ P_y \end{bmatrix} = [M_1] \begin{bmatrix} Q_x \\ Q_y \end{bmatrix}$$

(*b*) Now let the components of R be R_x and R_y and find a second 2×2 matrix M_2 such that

$$\begin{bmatrix} Q_x \\ Q_y \end{bmatrix} = [M_2] \begin{bmatrix} R_x \\ R_y \end{bmatrix}$$

(*c*) Show that M_1 and M_2 commute and find matrices to express P in terms of R and R in terms of P.

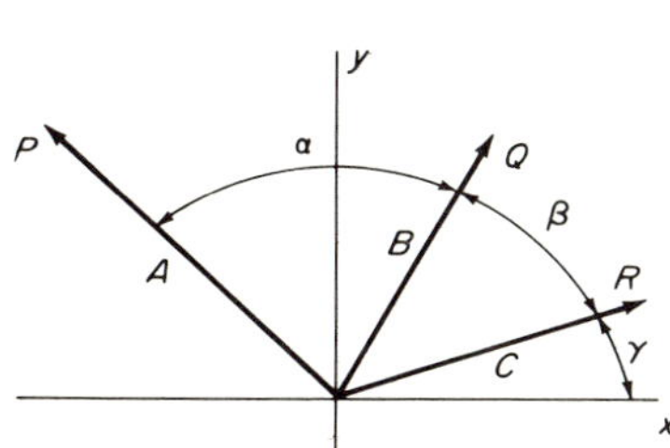

Fig. P1.27a

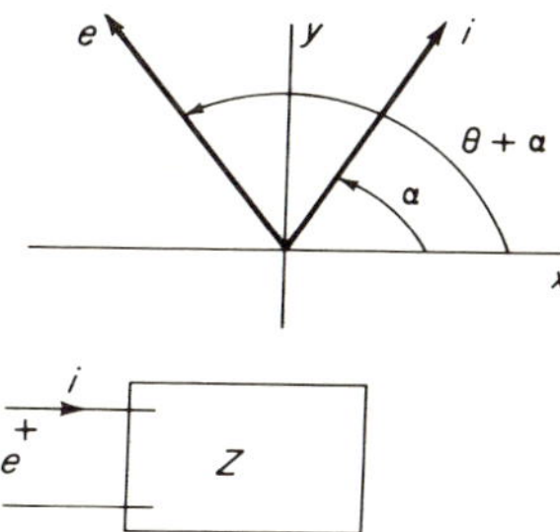

Fig. P1.27b

(*d*) Express e in terms of i in Fig. P1.27*b* through a transformation matrix whose entries are the real and imaginary parts of Z and whose order is 2×2.

1.28. Matrices can be used to perform manipulations of complex quantities. Define

$$\begin{bmatrix} a & -b \\ b & a \end{bmatrix} \equiv a + jb$$

Based on this definition, perform the following addition of complex quantities using matrices:

$$(a + jb) + (c + jd)$$

Repeat for the division

$$\frac{a + jb}{c + jd}$$

1.29. Matrices can be used to prove vector identities. Let two vectors be given by

$$\mathbf{A} = A_1\mathbf{u}_1 + A_2\mathbf{u}_2 + A_3\mathbf{u}_3$$

$$\mathbf{B} = B_1\mathbf{u}_1 + B_2\mathbf{u}_2 + B_3\mathbf{u}_3$$

where $\mathbf{u}_1$, $\mathbf{u}_2$, and $\mathbf{u}_3$ are unit vectors in, say, the x, y, z cartesian coordinate system. Show that the dot and cross products of $\mathbf{A}$ and $\mathbf{B}$ can be expressed in index notation by the equations

$$\mathbf{A} \cdot \mathbf{B} = A_k B_k$$

$$(\mathbf{A} \times \mathbf{B})_i = e_{ijk} A_j B_k$$

Use these results to find the scalar product

$$\mathbf{A} \cdot \mathbf{B} \times \mathbf{C}$$

where

$$\mathbf{C} = C_1\mathbf{u}_1 + C_2\mathbf{u}_2 + C_3\mathbf{u}_3$$

1.30. Use the results of Probs. 1.29 and 1.17 to prove the following vector identities by index notation:

(*a*) $\mathbf{A} \times \mathbf{B} \times \mathbf{C} = (\mathbf{A} \cdot \mathbf{C})\mathbf{B} - (\mathbf{A} \cdot \mathbf{B})\mathbf{C}$
(*b*) $(\mathbf{A} \cdot \mathbf{A})(\mathbf{B} \cdot \mathbf{B}) - (\mathbf{A} \cdot \mathbf{B})^2 = (\mathbf{A} \times \mathbf{B}) \cdot (\mathbf{A} \times \mathbf{B})$

1.31. The Y parameters of the network in Fig. P1.31 are give by

$$Y_{11} = G_1 + \frac{(G_a + G_b)(G_c + G_d)}{G_a + G_b + G_c + G_d}$$

$$Y_{22} = G_2 + \frac{(G_a + G_b)(G_c + G_d)}{G_a + G_b + G_c + G_d}$$

$$Y_{12} = Y_{21} = \frac{G_a G_d - G_b G_c}{G_a + G_b + G_c + G_d}$$

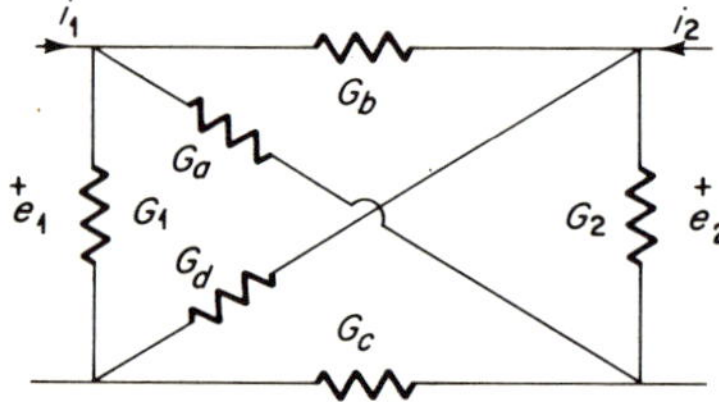

Fig. P1.31

For each of the following cases, determine whether the Y-parameter matrix is dominant, paramount, and/or uniformly tapered. All conductance values are in mhos.

(*a*)	$G_1 = 1$	$G_2 = 3$	$G_a = 1$	$G_b = 2$	$G_c = 4$	$G_d = 4$
(*b*)	$G_1 = 1$	$G_2 = 3$	$G_a = 2$	$G_b = 1$	$G_c = 4$	$G_d = 4$
(*c*)	$G_1 = 0$	$G_2 = 0$	$G_a = 1$	$G_b = 2$	$G_c = 4$	$G_d = 4$
(*d*)	$G_1 = 0$	$G_2 = 0$	$G_a = 0$	$G_b = 2$	$G_c = 4$	$G_d = 0$
(*e*)	$G_1 = 0$	$G_2 = 0$	$G_a = 1$	$G_b = 0$	$G_c = 0$	$G_d = 4$

What conclusions can you draw from these results regarding the conditions that must be satisfied by a 2×2 matrix if it is to be the Y-parameter matrix of the network in Fig. P1.31? Can the network be simplified in any of these cases? Explain.

2 SETS OF LINEAR ALGEBRAIC EQUATIONS

2.1 INTRODUCTION

In Chap. 1, matrices and determinants were presented as two important mathematical tools of network theory. In this chapter, these tools will be used in the analysis of sets of linear algebraic equations. We shall develop methods for determining whether a given set of equations possesses a solution and if so, how many, and we shall obtain the solution(s).

First, however, we should establish how sets of linear algebraic equations arise in the analysis of electric networks. This will be accomplished through a review of loop, node-to-datum, and node-pair analysis. At the same time, we shall present a consistent notation for use throughout the remainder of the text.

2.2 NOTATION AND REFERENCE CONDITIONS

Consider Fig. 2.1, which is a general representation of a network branch and its associated independent sources. The term branch has a broad meaning

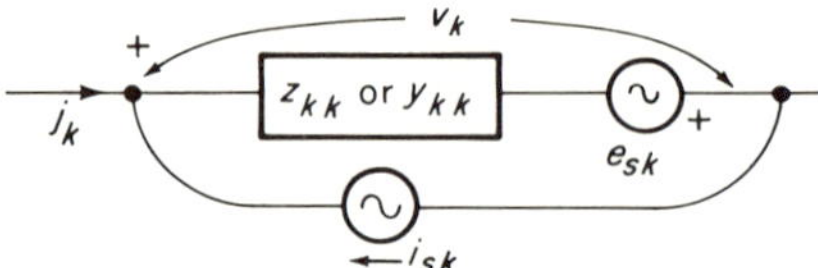

Fig. 2.1 The kth branch of a network

in network theory. As we use it here, a branch consists of one immittance box (z_{kk} or y_{kk}), which may contain one or more network elements.

Lowercase letters are used to represent branch quantities. The symbols e_s and i_s are reserved for voltage and current sources, respectively, and the symbols v and j are used to denote branch voltages and currents, respectively. Note that the branch voltage includes the voltage source as well as the drop across the branch impedance and that the branch current includes the current source as well as the current through the branch impedance.

If there is no coupling with other branches, the Ohm's law relations for the branch are

$$v_k + e_{sk} = z_{kk}(j_k + i_{sk}) \tag{2.2.1}$$

or

$$j_k + i_{sk} = y_{kk}(v_k + e_{sk}) \tag{2.2.2}$$

Note that the reference directions in Fig. 2.1 have been chosen such that a positive branch current produces a positive branch-voltage drop and such that the source voltage and current aid the branch voltage and current, respectively. This choice yields all plus signs in Eqs. (2.2.1) and (2.2.2).

To include coupling, Eqs. (2.2.1) and (2.2.2) are modified to

$$v_k + e_{sk} = \sum_{l=1}^{B} z_{kl}(j_l + i_{sl}) \tag{2.2.3}$$

$$j_k + i_{sk} = \sum_{l=1}^{B} y_{kl}(v_l + e_{sl}) \tag{2.2.4}$$

where B is the number of branches in the network. The double subscripts permit us to distinguish between the self-impedance of branch k (z_{kk}) and the mutual impedance between branches k and l (z_{kl}).

2.3 LOOP, NODE-TO-DATUM, AND NODE-PAIR ANALYSIS

Consider the network of Fig. 2.2, where single subscripts are used on the impedances because there is no coupling. We choose this network because it is simple and yet general. It has both a current and a voltage source, the former an *ideal current source* in shunt with z_3, the latter an *ideal voltage source* in series with z_1. There are five branches (the generators are not

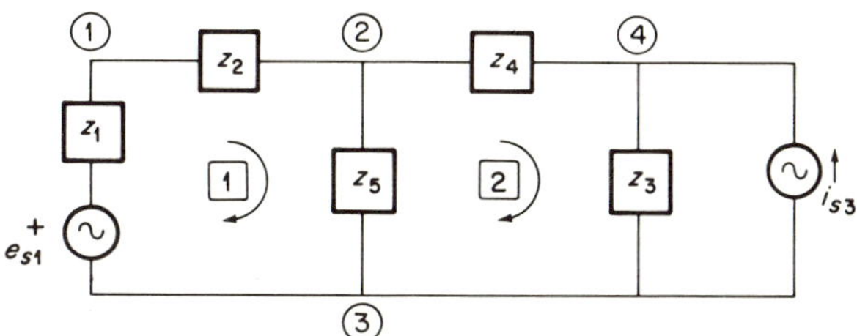

Fig. 2.2 Illustrative network.

counted as separate branches, but their internal impedances are). The network has four simple nodes, indicated by the circled numbers, and two simple loops, indicated by the boxed numbers. The junction of e_{s1} and z_1 is not considered a node, and the closed circuit formed by i_{s3} and z_3 is not considered a loop, since both are included within their respective branches, as shown in Fig. 2.1.

To perform a *loop* analysis of the network in Fig. 2.2, a Kirchhoff voltage law (KVL) equation is written for each loop

$$
\begin{aligned}
&\text{Loop 1: } e_{s1} = i_1 z_1 + i_1 z_2 + (i_1 - i_2) z_5 \\
&\text{Loop 2: } -z_3 i_{s3} = (i_2 - i_1) z_5 + i_2 z_4 + i_2 z_3
\end{aligned}
\tag{2.3.1}
$$

Here we adopt the convention that a small i represents a loop current. In each of Eqs. (2.3.1), the voltage rise of a source is equated to the summation of voltage drops around a loop containing that source, all rises and drops being taken in the direction of that loop current. Special attention should be given to the equation for loop 2. That portion of the circuit has been redrawn in Fig. 2.3, with the current source transformed to an equivalent voltage source. It is often convenient to transform all current sources to equivalent voltage sources before writing the KVL equations, but it is not necessary to do so.

Equation (2.3.1) can be rewritten in matrix form as

$$
\begin{bmatrix} e_{s1} \\ -z_3 i_{s3} \end{bmatrix} = \begin{bmatrix} z_1 + z_2 + z_5 & -z_5 \\ -z_5 & z_3 + z_4 + z_5 \end{bmatrix} \begin{bmatrix} i_1 \\ i_2 \end{bmatrix}
\tag{2.3.2}
$$

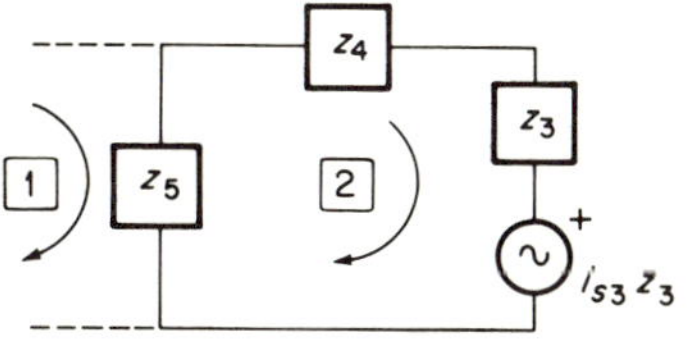

Fig. 2.3 Current-source transformation.

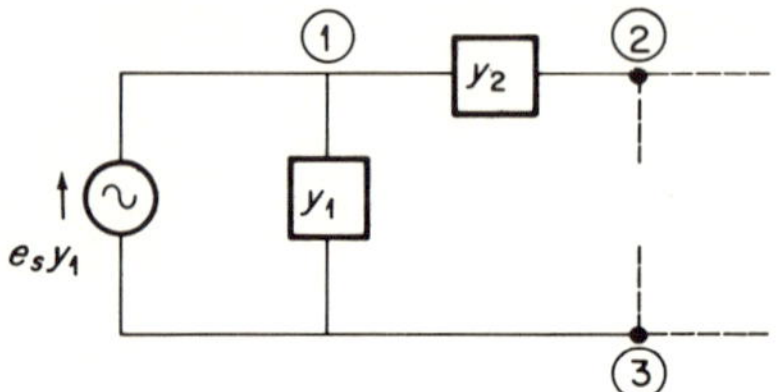

Fig. 2.4 Voltage-source transformation.

and solved for the currents i_1 and i_2 by Cramer's rule. When written in this form, the main-diagonal entries in the impedance matrix are the self-impedances of the loops, whereas the off-diagonal entries are the negatives of the impedances common to two loops.

To carry out a *node-to-datum* analysis of the network of Fig. 2.2, a Kirchhoff current law (KCL) equation is written at each node except one, the latter being called a *datum node*. Let node 4 be the datum node.† Then the KCL equations are:

$$
\begin{aligned}
&\text{Node 1:} \quad & y_1 e_{s1} &= (e_1 - e_3)y_1 + (e_1 - e_2)y_2 \\
&\text{Node 2:} \quad & 0 &= (e_2 - e_1)y_2 + (e_2 - e_3)y_5 + e_2 y_4 \\
&\text{Node 3:} \quad & -y_1 e_{s1} - i_{s3} &= (e_3 - e_1)y_1 + (e_3 - e_2)y_5 + e_3 y_3
\end{aligned}
\tag{2.3.3}
$$

Here we use branch admittances instead of branch impedances in each equation and assign a small e to represent a node-to-datum voltage, with the nondatum node having the positive reference sense. In each of Eqs. (2.3.3), we have equated the source currents at a node (assumed positive entering) to the currents in the admittances touching this same node (assumed positive leaving). Special attention should be given to the equations at nodes 1 and 3. Figure 2.4 shows this portion of the network redrawn with the voltage source transformed to a current source. The KCL equations at nodes 1 and 3 are of course the same as they were before the change. It is often convenient to convert all voltage sources to current sources *before* writing the KCL equations, but it is not necessary to do so.

In matrix form, Eqs. (2.3.3) become

$$
\begin{bmatrix} y_1 e_{s1} \\ 0 \\ -y_1 e_{s1} - i_{s3} \end{bmatrix} = \begin{bmatrix} y_1 + y_2 & -y_2 & -y_1 \\ -y_2 & y_2 + y_4 + y_5 & -y_5 \\ -y_1 & -y_5 & y_1 + y_3 + y_5 \end{bmatrix} \begin{bmatrix} e_1 \\ e_2 \\ e_3 \end{bmatrix}
\tag{2.3.4}
$$

† Node 3 is actually a preferable datum node from a common-ground point of view. We choose node 4 because it permits a better comparison of node-to-datum and node-pair analysis.

The main-diagonal entries are self-admittances at a node; off-diagonal entries are the negatives of admittances common to two nodes. Cramer's rule can be applied to Eq. (2.3.4) to obtain the node-to-datum voltages.

Finally, let us perform a *node-pair* analysis of the network in Fig. 2.2. We define a node-pair voltage as the voltage between a pair of nodes at the ends of a single branch (rather than at the ends of two or more branches in series). Just as we needed three independent node-to-datum voltages in the analysis above, so we need three independent node-pair voltages for the present analysis. The three voltages chosen cannot form a closed loop or they will not be independent. We choose the following voltages, illustrated diagrammatically in Fig. 2.5*a*:

e_a voltage across z_2 with node 1 positive

e_b voltage across z_5 with node 2 positive

e_c voltage across z_3 with node 3 positive

Then the KCL equations at nodes 1, 2, and 3 are:

$$\begin{aligned} &\text{Node 1:} & y_1e_{s1} &= y_1(e_a + e_b) + y_2e_a \\ &\text{Node 2:} & 0 &= -y_2e_a + y_5e_b + y_4(e_b + e_c) \\ &\text{Node 3:}\ -i_{s3} - y_1e_{s1} & &= -y_1(e_a + e_b) - y_5e_b + y_3e_c \end{aligned} \tag{2.3.5}$$

which can be written in matrix form as

$$\begin{bmatrix} y_1e_{s1} \\ 0 \\ -y_1e_{s1} - i_{s3} \end{bmatrix} = \begin{bmatrix} y_1 + y_2 & y_1 & 0 \\ -y_2 & y_4 + y_5 & y_4 \\ -y_1 & -(y_1 + y_5) & y_3 \end{bmatrix} \begin{bmatrix} e_a \\ e_b \\ e_c \end{bmatrix} \tag{2.3.6}$$

We note that whereas the admittance matrix in Eq. (2.3.4) is symmetric, that in Eq. (2.3.6) is not. The reason for this lies in our choice of the nodes at which we wrote Eqs. (2.3.5). The concept of a supernode must be introduced in order to convert Eqs. (2.3.6) to symmetric form.

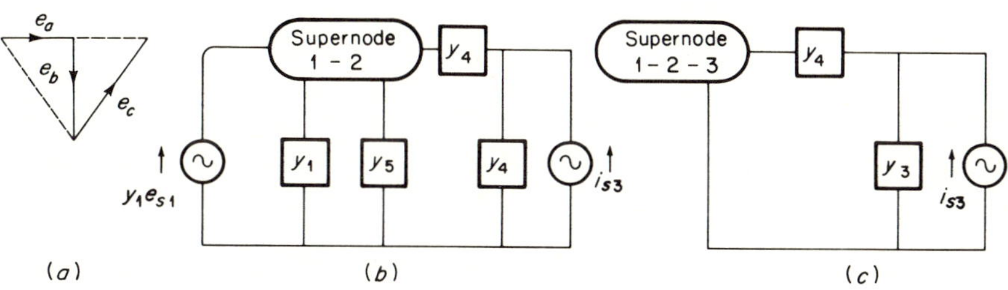

Fig. 2.5 *Supernodes.*

Define a *supernode* as a combination of one or more simple nodes. Supernode 1-2 (the combination of nodes 1 and 2) is shown diagrammatically in Fig. 2.5*b*. One source current enters this supernode, and three branch currents leave it. Supernode 1-2-3 is shown in Fig. 2.5*c*. One source current and two branch currents leave it.

Now, using the node-pair voltages previously defined, write KCL equations at the nodes 1, 1-2, and 1-2-3:

$$\begin{aligned} &\text{Node 1:} && y_1 e_{s1} = y_1(e_a + e_b) + y_2 e_a \\ &\text{Node 1-2} && y_1 e_{s1} = y_1(e_a + e_b) + y_5 e_b + y_4(e_b + e_c) \\ &\text{Node 1-2-3:} && -i_{s3} = y_4(e_b + e_c) + y_3 e_c \end{aligned} \tag{2.3.7}$$

Putting Eq. (2.3.7) in matrix form yields

$$\begin{bmatrix} y_1 e_{s1} \\ y_1 e_{s1} \\ -i_{s3} \end{bmatrix} = \begin{bmatrix} y_1 + y_2 & y_1 & 0 \\ y_1 & y_1 + y_4 + y_5 & y_4 \\ 0 & y_4 & y_3 + y_4 \end{bmatrix} \begin{bmatrix} e_a \\ e_b \\ e_c \end{bmatrix} \tag{2.3.8}$$

Note that the new choice of nodes has resulted in a symmetric admittance matrix. Since the question of how we knew what choice of nodes to make is difficult to answer without the concept of a cut set, we defer this question until Sec. 5.7.

Equations (2.3.6) and (2.3.8) can be solved by Cramer's rule. The main- and off-diagonal elements in the admittance matrices in Eq. (2.3.8) can be interpreted in the same way as they were in Eq. (2.3.4). This is not true in Eq. (2.3.6). For example, in Eq. (2.3.8), $y_1 + y_4 + y_5$ is the self-admittance of supernode 1-2, $y_3 + y_4$ is the self-admittance of supernode 1-2-3, and y_4 is the admittance common to supernodes 1-2 and 1-2-3. In Eq. (2.3.6), $y_4 + y_5$ is not the self-admittance of node 2, and y_1 is not the admittance common to nodes 1 and 2.

Let us introduce an additional point regarding voltage and current sources before closing this section. Situations are often encountered in which an

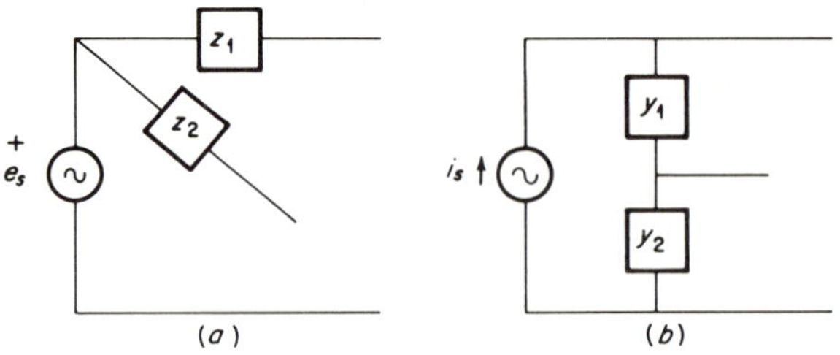

Fig. 2.6 Sources not in or across branches.

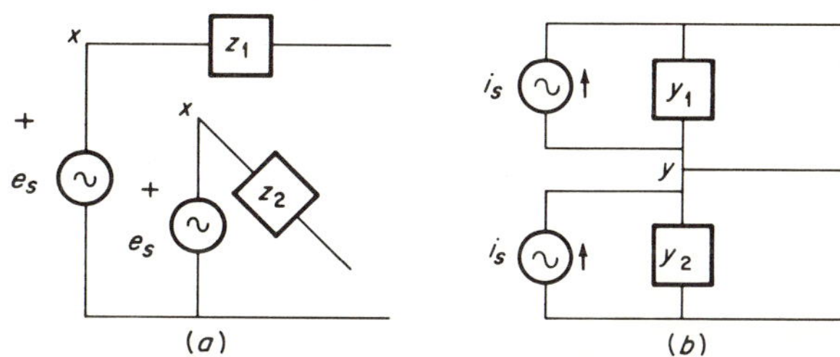

Fig. 2.7 Transformations of Fig. 2.6.

ideal voltage source is not in series with any one branch or an ideal current source is not in parallel with any one branch. These cases are depicted in Fig. 2.6 and can easily be remedied, as shown in Fig. 2.7. To verify that the networks in Fig. 2.7 are identical to the corresponding ones in Fig. 2.6, observe that the voltages at point x and the currents at point y are unchanged in the transformations.

We have shown through Eqs. (2.3.1), (2.3.3), (2.3.5), and (2.3.7) that loop and node analyses lead to sets of linear algebraic equations involving the network currents and voltages. In the remainder of this chapter the solvability of such sets of equations will be examined.

2.4 GAUSS–JORDAN REDUCTION

Consider the problem of obtaining solutions of a set of m linear algebraic equations in n unknown x variables of the form

$$\begin{aligned} a_{11}x_1 + a_{12}x_2 + a_{13}x_3 + \cdots + a_{1n}x_n &= c_1 \\ a_{21}x_1 + a_{22}x_2 + a_{23}x_3 + \cdots + a_{2n}x_n &= c_2 \\ \cdots\cdots\cdots\cdots\cdots\cdots\cdots\cdots\cdots\cdots \\ a_{m1}x_1 + a_{m2}x_2 + a_{m3}x_3 + \cdots + a_{mn}x_n &= c_m \end{aligned} \tag{2.4.1}$$

We already know from Sec. 1.12 how to solve Eqs. (2.4.1) by Cramer's rule if $m = n$ and $|A| \neq 0$. There is a systematic procedure, known as the *Gauss-Jordan reduction*, for solving Eqs. (2.4.1) in the general case when $m \neq n$ or $|A| = 0$.

FIRST STEP

Assume $a_{11} \neq 0$ (otherwise reorder the equations so that the first equation has a nonzero x_1 coefficient). Multiply the first of Eqs. (2.4.1) by $1/a_{11}$ and let the result be the new first equation. Now multiply the first of Eqs.

(2.4.1) by a_{21}/a_{11} and subtract the second of Eqs. (2.4.1) from it (or vice versa). Let the result be the new second equation. Next multiply the first of Eqs. (2.4.1) by a_{31}/a_{11} and subtract the third of Eqs. (2.4.1) from it (or, again, vice versa), letting the result be the new third equation. Repeat the process for all m equations. The new equations are

$$\begin{aligned} x_1 + a'_{12}x_2 + a'_{13}x_3 + \cdots + a'_{1n}x_n &= c'_1 \\ a'_{22}x_2 + a'_{23}x_3 + \cdots + a'_{2n}x_n &= c'_2 \\ \cdots\cdots\cdots\cdots\cdots\cdots \\ a'_{m2}x_2 + a'_{m3}x_3 + \cdots + a'_{mn}x_n &= c_m \end{aligned} \tag{2.4.2}$$

Notice that the variable x_1 has been eliminated from all equations except the first.

SECOND STEP

Assume $a'_{22} \neq 0$ (otherwise interchange the second equation with an equation below it so that the new second equation has a nonzero x_2 coefficient).† Repeat the first step so as to eliminate x_2 from all equations except the second. The result is

$$\begin{aligned} x_1 \qquad + a''_{13}x_3 + \cdots + a''_{1n}x_n &= c''_1 \\ x_2 + a''_{23}x_3 + \cdots + a''_{2n}x_n &= c''_2 \\ \cdots\cdots\cdots\cdots\cdots\cdots \\ a''_{m3}x_3 + \cdots + a''_{mn}x_n &= c''_m \end{aligned} \tag{2.4.3}$$

THIRD STEP

Repeat the process of the first and second steps r times, until either $r = m$ or all coefficients of variables in equations $r + 1$ through m ($m - r$ equations) are zero. These $m - r$ equations are known as *residual* equations. Assuming $r < m$, the final equations are (using b's and k's for the final coefficients)

$$\begin{aligned} x_1 \qquad + b_{1,r+1}x_{r+1} + \cdots + b_{1n}x_n &= k_1 \\ x_2 \qquad + b_{2,r+1}x_{r+1} + \cdots + b_{2n}x_n &= k_2 \\ \cdots\cdots\cdots\cdots\cdots\cdots \\ 0x_{r+1} + \cdots + \ 0x_n &= k_{r+1} \\ \cdots\cdots\cdots\cdots\cdots\cdots \\ 0x_{r+1} + \cdots + \ 0x_n &= k_m \end{aligned} \tag{2.4.4}$$

† It is possible after the first step that $a'_{22}, a'_{23}, \ldots, a'_{2m}$ will all be zero. If this happens, reorder the variables in Eqs. (2.4.2) so that x_2 appears after x_n in each equation.

Example 2.1 Solve the following set of linear equations by a Gauss-Jordan reduction.

$$\begin{aligned}
v_1 + v_2 \qquad\qquad\quad + v_5 \qquad &= 3\\
v_1 + v_2 \qquad\quad - v_4 \qquad + v_6 &= 5\\
v_1 \qquad + v_3 + v_4 + v_5 \qquad &= 0\\
- v_2 + v_3 \qquad\quad - v_5 + v_6 &= -1\\
- v_2 + v_3 + v_4 \qquad\qquad\quad &= -3
\end{aligned}$$

The solution proceeds as follows:

First step

$$\begin{aligned}
v_1 + v_2 \qquad\qquad\quad + v_5 \qquad &= 3\\
v_4 + v_5 - v_6 &= -2\\
v_2 - v_3 - v_4 \qquad\qquad\quad &= 3\\
- v_2 + v_3 \qquad\quad - v_5 + v_6 &= -1\\
- v_2 + v_3 + v_4 \qquad\qquad\quad &= -3
\end{aligned}$$

Reorder equations

$$\begin{aligned}
v_1 + v_2 \qquad\qquad\quad + v_5 \qquad &= 3\\
v_2 - v_3 - v_4 \qquad\qquad\quad &= 3\\
v_4 + v_5 - v_6 &= -2\\
- v_2 + v_3 \qquad\quad - v_5 + v_6 &= -1\\
- v_2 + v_3 + v_4 \qquad\qquad\quad &= -3
\end{aligned}$$

Second step

$$\begin{aligned}
v_1 \qquad + v_3 + v_4 + v_5 \qquad &= 0\\
v_2 - v_3 - v_4 \qquad\qquad\quad &= 3\\
v_4 + v_5 - v_6 &= -2\\
- v_4 - v_5 + v_6 &= 2\\
0 &= 0
\end{aligned}$$

Reorder variables

$$\begin{aligned}
v_1 \qquad + v_4 + v_5 \qquad + v_3 &= 0\\
v_2 - v_4 \qquad\qquad\quad - v_3 &= 3\\
v_4 + v_5 - v_6 \qquad\quad &= -2\\
- v_4 - v_5 + v_6 \qquad\quad &= 2\\
0 &= 0
\end{aligned}$$

Third step

$$\begin{aligned}
v_1 \qquad\qquad\qquad + v_6 + v_3 &= 2\\
v_2 \qquad + v_5 - v_6 - v_3 &= 1\\
v_4 + v_5 - v_6 \qquad\quad &= -2\\
0 &= 0\\
0 &= 0
\end{aligned}$$

Note that a reordering of the equations was necessary after the first step and that a reordering of the variables was necessary after the second step. Also note that $n = 6$, $m = 5$, and $r = 3$.

The Gauss-Jordan reduction is more easily accomplished in tabular form. We append the C matrix to the A matrix as an extra column and carry out the reduction as follows:

$$
\begin{array}{ccccccc}
1 & 1 & 0 & 0 & 1 & 0 & 3 \\
1 & 1 & 0 & -1 & 0 & 1 & 5 \\
1 & 0 & 1 & 1 & 1 & 0 & 0 \\
0 & -1 & 1 & 0 & -1 & 1 & -1 \\
0 & -1 & 1 & 1 & 0 & 0 & -3
\end{array}
\longrightarrow
\begin{array}{ccccccc}
1 & 1 & 0 & 0 & 1 & 0 & 3 \\
0 & 0 & 0 & 1 & 1 & -1 & -2 \\
0 & 1 & -1 & -1 & 0 & 0 & 3 \\
0 & -1 & 1 & 0 & -1 & 1 & -1 \\
0 & -1 & 1 & 1 & 0 & 0 & -3
\end{array}
\longrightarrow
$$

$$
\begin{array}{ccccccc}
1 & 0 & 1 & 1 & 1 & 0 & 0 \\
0 & 0 & 0 & 1 & 1 & -1 & -2 \\
0 & 1 & -1 & -1 & 0 & 0 & 3 \\
0 & 0 & 0 & -1 & -1 & 1 & 2 \\
0 & 0 & 0 & 0 & 0 & 0 & 0
\end{array}
\longrightarrow
\begin{array}{ccccccc}
1 & 0 & 1 & 0 & 0 & 1 & 2 \\
0 & 0 & 0 & 1 & 1 & -1 & -2 \\
0 & 1 & -1 & 0 & 1 & -1 & 1 \\
0 & 0 & 0 & 0 & 0 & 0 & 0 \\
0 & 0 & 0 & 0 & 0 & 0 & 0
\end{array}
\longrightarrow
$$

$$
\begin{aligned}
v_1 + v_3 + v_6 &= 2 \\
v_4 + v_5 - v_6 &= -2 \\
v_2 - v_3 + v_5 - v_6 &= 2 \\
0 &= 0 \\
0 &= 0
\end{aligned}
$$

In this streamlined procedure, the reordering of either the equations or the variables is avoided. In the second tabulation, x_1 is eliminated from all equations except the first. In the third tabulation, x_2 is eliminated from all equations except the *third.* In the fourth tabulation, x_4 is eliminated from all equations except the *second.*

2.5 SOLVABILITY

Several possibilities exist in the final set of equations given in Eqs. (2.4.4).

1. If k_{r+1} through k_m are not all zero, at least one of the $m - r$ residual equations takes the form $0 = k_i$, where $k_i \neq 0$. Thus no solution exists, and the set of Eqs. (2.4.4) is said to be *inconsistent.* The equations in Example 2.1 are *consistent.*
2. If Eqs. (2.4.4) are consistent, and if $r < n$, then r of the variables may be expressed in terms of the remaining $n - r$ variables. (These r variables are not necessarily x_1 through x_r since reordering of the variables may have been necessary at one or more steps in the reduction.) In Eqs. (2.4.4) all terms containing these $n - r$ variables in the first r equations may be moved to the right of the equal signs. The set of Eqs. (2.4.4) is said to be of *defect* $n - r$ and has an $(n - r)$-fold *infinity of solutions.* By this we mean that each of the $n - r$ variables moved to the right side of the equations may be asigned an infinite number of values. Each set of values of these $n - r$ variables yields a solution for the other r variables. In Example 2.1 the defect is 3, and there is a threefold infinity of solutions.

3. If Eqs. (2.4.4) are consistent, and if $r < m$, then $m - r$ residual equations become *ignorable* equations. Again they are not necessarily equations $r + 1$ through m since reordering of the equations may have been necessary at one or more steps in the reduction. The set of Eqs. (2.4.4) is then said to be *linearly dependent* in that each of the $m - r$ ignorable equations of the set can be written as a linear combination of the other r equations of the set. If $r = m$, the set of Eq. (2.4.4) is linearly independent. The equations in Example 2.1 are linearly dependent, and there are two ignorable equations.
4. If Eqs. (2.4.4) are consistent, and if $r = n$, a *unique* solution for the n variables results; i.e., each x variable has one and only one value. The defect is zero, and if $n < m$, there are $m - n$ ignorable equations. As we have noted, there is no unique solution for the equations in Example 2.2.

Example 2.2 Consider the network of Fig. 2.8. It is obvious that two equations are sufficient to solve for the unknowns i_1 and i_2. However, to illustrate the ideas of Secs. 2.4 and 2.5, let the following three loop equations be written:

$$\begin{aligned} 4i_1 - 3i_2 &= 11 \\ -3i_1 + 5i_2 &= 0 \\ i_1 + 2i_2 &= 11 \end{aligned}$$

Performing a Gauss-Jordan reduction yields:

Step 1:

$$\begin{aligned} i_1 - \tfrac{3}{4}i_2 &= \tfrac{11}{4} \\ \tfrac{11}{4}i_2 &= \tfrac{33}{4} \\ \tfrac{11}{4}i_2 &= \tfrac{33}{4} \end{aligned}$$

Step 2:

$$\begin{aligned} i_1 \quad &= 5 \\ i_2 &= 3 \\ 0 &= 0 \end{aligned}$$

In this example $r = n = 2$ and $m = 3$. Thus the equations are consistent, the defect is zero, a unique solution exists, and there is $m - r = 1$ ignorable equation.

Suppose now that the third equation had been incorrectly written $i_1 + i_2 = 11$. Then the second step in the Gauss-Jordan reduction would yield

$$\begin{aligned} i_1 \quad &= 5 \\ i_2 &= 3 \\ 0 &= 3 \end{aligned}$$

The set of equations is now inconsistent, and no solution exists.

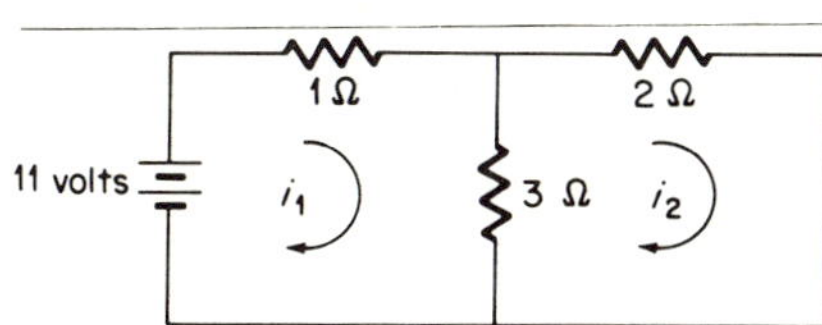

Fig. 2.8

Suppose next that an unknown current source i_0 is placed across the 3-ohm resistor, its direction being such as to send current down through that resistor. The network equations become

$$\begin{aligned} 4i_1 - 3i_2 + 3i_0 &= 11 \\ -3i_1 + 5i_2 - 3i_0 &= 0 \\ i_1 + 2i_2 \qquad &= 11 \end{aligned}$$

The Gauss-Jordan reduction yields

$$\begin{aligned} i_1 \qquad + \tfrac{6}{11}i_0 &= 5 \\ i_2 - \tfrac{3}{11}i_0 &= 3 \\ 0 &= 0 \end{aligned}$$

In this case, $r = 2$, $n = 3$, and $m = 3$. Thus the defect is 1, and there is $m - r = 1$ ignorable equation. The equations are consistent. Any one of the variables (i_1, i_2, or i_0) may be assigned a value and the remaining two variables determined.

2.6 RANK OF A MATRIX

The Gauss-Jordan reduction is a systematic means of determining whether a set of linear equations possesses a solution, whether a solution is unique, and whether there are ignorable equations. It is pertinent to inquire whether an alternate method using matrices can be devised for obtaining this information.

As was done in Chap. 1, Eqs. (2.4.1) can be written in matrix notation in the following three equivalent forms:

$$\begin{bmatrix} a_{11} & a_{12} & \cdots & a_{1n} \\ a_{21} & a_{22} & \cdots & a_{2n} \\ \cdot & \cdot & \cdots & \cdot \\ a_{m1} & a_{m2} & \cdots & a_{mn} \end{bmatrix} \begin{bmatrix} x_1 \\ x_1 \\ \cdot \\ \cdot \\ \cdot \\ x_n \end{bmatrix} = \begin{bmatrix} c_1 \\ c_2 \\ \cdot \\ \cdot \\ \cdot \\ c_m \end{bmatrix} \tag{2.6.1}$$

$$[a_{ik}][x_k] = [c_i] \tag{2.6.2}$$

$$AX = C \tag{2.6.3}$$

We shall call A the *coefficient* matrix. Its order is $m \times n$. The matrix obtained by appending the C matrix as an extra $(n + 1)$st column to the A matrix will be called the *augmented* matrix. The augmented matrix is of order $m \times (n + 1)$.

We now define the *rank* of a matrix as the order of the largest square submatrix of the matrix whose determinant does not vanish.

Example 2.3 Determine the rank of the coefficient and augmented matrices associated with the following set of equations:

$$2x_1 + x_3 = 4$$
$$x_1 - 2x_2 + 2x_3 = 7$$
$$3x_1 + 2x_2 = 1$$

The coefficient and augmented matrices are, respectively,

$$\begin{bmatrix} 2 & 0 & 1 \\ 1 & -2 & 2 \\ 3 & 2 & 0 \end{bmatrix} \qquad \begin{bmatrix} 2 & 0 & 1 & 4 \\ 1 & -2 & 2 & 7 \\ 3 & 2 & 0 & 1 \end{bmatrix}$$

The determinant of the 3×3 coefficient matrix vanishes. However, there are several 2×2 submatrices whose determinants do not vanish. For example,

$$\begin{vmatrix} 2 & 0 \\ 1 & -2 \end{vmatrix} = -4 \qquad \begin{vmatrix} 2 & 1 \\ 1 & 2 \end{vmatrix} = 3$$

Hence the rank of the coefficient matrix is 2. The augmented matrix has no 3×3 submatrices with nonzero determinants. Hence its rank is also 2.

Performing a Gauss-Jordan reduction does not change the rank of either the coefficient or augmented matrices; i.e., the rank of the coefficient matrix of Eqs. (2.4.4) is the same as the rank of the coefficient matrix of Eqs. (2.4.1), and likewise for the augmented matrices. This conclusion is true because the Gauss-Jordan reduction involves only the elementary matrix transformations of (1) interchange of two rows or columns; (2) multiplication of the elements of a row by a constant; and (3) addition to the elements of one row of k times the elements of another row. None of these operations causes the determinant of a matrix to vanish, as was shown in Sec. 1.11.†

Now refer to Eqs. (2.4.4). The rank of the coefficient matrix is r since an $r \times r$ submatrix is present whose main diagonal has all nonzero terms. If k_{r+1} through k_m are all zero, the rank of the augmented matrix is also r; otherwise its rank is $r + 1$. From this information we can draw the following conclusions about a set of linear algebraic equations:

1. The equations are consistent if and only if the ranks of the coefficient and augmented matrices are the same.
2. The defect of the set of equations is obtained by subtracting the rank of the coefficient matrix from the total number of unknowns.
3. The number of ignorable equations in the set is obtained by subtracting the rank of the coefficient matrix from the total number of equations.

† A further proof is required when operation 3 is applied and one of the rows involved is outside the nonvanishing determinant. This proof is given in Ref. 2.2.

Example 2.4 Let us illustrate the conclusions just stated by reconsidering the equations of Example 2.3. The coefficient and augmented matrices are, respectively,

$$\begin{bmatrix} 2 & 0 & 1 \\ 1 & -2 & 2 \\ 3 & 2 & 0 \end{bmatrix} \qquad \begin{bmatrix} 2 & 0 & 1 & 4 \\ 1 & -2 & 2 & 7 \\ 3 & 2 & 0 & 1 \end{bmatrix}$$

The rank of each is 2. Now performing a Gauss-Jordan reduction, the coefficient and augmented matrices become

$$\begin{bmatrix} 1 & 0 & \frac{1}{2} \\ 0 & 1 & -\frac{3}{4} \\ 0 & 0 & 0 \end{bmatrix} \qquad \begin{bmatrix} 1 & 0 & \frac{1}{2} & 2 \\ 0 & 1 & -\frac{3}{4} & -\frac{5}{2} \\ 0 & 0 & 0 & 0 \end{bmatrix}$$

The rank of the coefficient and augmented matrices is still 2. The set is consistent, the defect is 1, and there is one ignorable equation.

2.7 EXTENSION OF CRAMER'S RULE

In Sec. 1.12, Cramer's rule was developed. Application of this rule permitted us to obtain a unique solution of a set of n equations in n unknowns when the coefficient determinant was nonvanishing, i.e., when $r = m = n$. We now use the results of Secs. 2.4 to 2.6 to extend Cramer's rule to the general case when $r \neq m \neq n$.

Let us discuss the various possibilities, using Eqs. (2.4.1), in terms of r, m, and n.

Case 1: $r = n < m$ In this case there are more equations than unknowns, but the defect is zero. First a check of the augmented matrix is made to ensure that the equations are consistent. Next, there is at least one $r \times r$ submatrix of the coefficient matrix whose determinant does not vanish. The $m - r$ equations not associated with this submatrix are ignored. Cramer's rule is then applied to the remaining r equations to obtain the unique solution.

Case 2: $r = m < n$ In this case there are more unknowns than equations, and the defect is $n - r$. Again the coefficient matrix contains at least one submatrix of order $r \times r$ whose determinant does not vanish. The $n - r$ unknowns not associated with this submatrix may be considered as independent variables and lumped with the constants on the right sides of the equations. The remaining r unknowns (the dependent variables) can now be found in terms of these independent variables by Cramer's rule.

Case 3: $r < m \leq n$ *or* $r < n \leq m$ These cases are merely combinations of cases 1 and 2; that is, before Cramer's rule is applied, a consistency check must be made, $m - r$ equations must be ignored, and $n - r$ variables must be moved to the right sides of the equations.

Example 2.5 Consider the set of equations of Example 2.1. Here $r < m < n$, which is the situation of case 3. The coefficient matrix is

$$\begin{bmatrix} 1 & 1 & 0 & 0 & 1 & 0 \\ 1 & 1 & 0 & -1 & 0 & 1 \\ 1 & 0 & 1 & 1 & 1 & 0 \\ 0 & -1 & 1 & 0 & -1 & 1 \\ 0 & -1 & 1 & 1 & 0 & 0 \end{bmatrix}$$

No 5×5 or 4×4 submatrices can be found whose determinants do not vanish. There are several 3×3 submatrices with nonvanishing determinants. One of these is the submatrix composed of rows 1, 2, 3 and columns 1, 2, 4, whose determinant is

$$\begin{vmatrix} 1 & 1 & 0 \\ 1 & 1 & -1 \\ 1 & 0 & 1 \end{vmatrix} = -1$$

If we choose this submatrix, we then ignore the fourth and fifth equations and move the terms containing v_3, v_5, and v_6 to the right sides of the equations. This gives

$$\begin{aligned} v_1 + v_2 &= 3 - v_5 \\ v_1 + v_2 - v_4 &= 5 - v_6 \\ v_1 + v_4 &= -v_3 - v_5 \end{aligned}$$

Now v_3, v_5, and v_6 may be assigned arbitrary values and Cramer's rule applied to obtain v_1, v_2, and v_4.

2.8 SOLUTIONS IN MATRIX FORM

The set of Eqs. (2.4.1) is written in closed matrix form in Eq. (2.6.3). A is of order $m \times n$, and we shall assume that its rank is $r < m$ and $r < n$. Let us further assume that we have reordered the rows and columns of A, and thus of X and C, so that an $r \times r$ submatrix whose determinant does not vanish is contained in the first r rows and r columns of A. Now let Eq. (2.6.3) be partitioned to give

$$\begin{bmatrix} A_1 & A_2 \\ A_3 & A_4 \end{bmatrix} \begin{bmatrix} X_1 \\ X_2 \end{bmatrix} = \begin{bmatrix} C_1 \\ C_2 \end{bmatrix} \tag{2.8.1}$$

A_1 is the $r \times r$ submatrix just discussed. A_2 is an $r \times (n - r)$ submatrix associated with the independent variables. A_3 and A_4 are $(m - r) \times r$ and $(m - r) \times (n - r)$ submatrices, respectively, that define the ignorable equations. Finally, X_1 and C_1 are $r \times 1$ matrices, x_2 is an $(n - r) \times 1$ matrix, and C_2 is an $(m - r) \times 1$ matrix.

Expanding Eq. (2.8.1) yields

$$A_1X_1 + A_2X_2 = C_1$$
$$A_3X_1 + A_4X_2 = C_2 \qquad (2.8.2)$$

The second of Eqs. (2.8.2) is ignorable. In the first of Eqs. (2.8.2), the independent variables are moved to the right side, yielding

$$A_1X_1 = C_1 - A_2X_2 \qquad (2.8.3)$$

Since A_1 is nonsingular, X_1 can be obtained by premultiplying each side of Eq. (2.8.3) by A_1^{-1}

$$X_1 = A_1^{-1}(C_1 - A_2X_2) \qquad (2.8.4)$$

For a *homogeneous* set of m equations in n unknowns, C (and therefore C_1) is zero. Such a set is always consistent with the trivial solution $X = 0$. This is the only solution unless $r < n$ because if $r = n$, none of the x variables appears on the right side of Eq. (2.8.4).

Example 2.6 The following matrix equation represents the application of Kirchhoff's current law (KCL) at the nodes of a certain network:

$$\begin{bmatrix} -1 & 1 & 0 & 0 & 0 \\ 0 & -1 & 0 & -1 & 1 \\ 0 & 0 & 1 & 1 & 0 \\ 1 & 0 & -1 & 0 & -1 \end{bmatrix} \begin{bmatrix} j_1 \\ j_2 \\ j_3 \\ j_4 \\ j_5 \end{bmatrix} = \begin{bmatrix} 0 \\ 0 \\ 0 \\ 0 \end{bmatrix}$$

We wish to answer the following questions without performing a Gauss-Jordan reduction:

Are the equations consistent?
Are the equations linearly independent?
What is the defect of the set?
Which currents may be chosen as independent variables?

The equations are homogeneous and thus consistent. An examination of the coefficient matrix reveals that it has a rank of 3. There is an $m - r = 1$ ignorable equation; hence the set of equations is not linearly independent. The defect is $n - r = 2$. Thus two of the current variables may be arbitrarily assigned values (assumed as independent variables), but not any two since it is necessary that the remaining coefficient determinant not vanish. For example, j_4 and j_5 are an acceptable set, but j_1 and j_2 are not, since in the first case the remaining coefficient determinant does not vanish whereas in the second case it does

$$\begin{vmatrix} -1 & 1 & 0 \\ 0 & -1 & 0 \\ 0 & 0 & 1 \end{vmatrix} = +1 \qquad \begin{vmatrix} 0 & 0 & 0 \\ 0 & -1 & 1 \\ 1 & 1 & 0 \end{vmatrix} = 0$$

Choosing j_4 and j_5 as independent variables, a solution for j_1, j_2, and j_3 as functions of j_4 and j_5 proceeds as follows:

$$\left[\begin{array}{ccc:cc} -1 & 1 & 0 & 0 & 0 \\ 0 & -1 & 0 & -1 & 1 \\ 0 & 0 & 1 & 1 & 0 \\ \hdashline 1 & 0 & -1 & 0 & -1 \end{array}\right] \begin{bmatrix} j_1 \\ j_2 \\ j_3 \\ - \\ j_4 \\ j_5 \end{bmatrix} = \begin{bmatrix} 0 \\ 0 \\ 0 \\ - \\ 0 \end{bmatrix}$$

$$\begin{bmatrix} -1 & 1 & 0 \\ 0 & -1 & 0 \\ 0 & 0 & 1 \end{bmatrix} \begin{bmatrix} j_1 \\ j_2 \\ j_3 \end{bmatrix} + \begin{bmatrix} 0 & 0 \\ -1 & 1 \\ 1 & 0 \end{bmatrix} \begin{bmatrix} j_4 \\ j_5 \end{bmatrix} = \begin{bmatrix} 0 \\ 0 \\ 0 \end{bmatrix}$$

$$[1 \quad 0 \quad -1] \begin{bmatrix} j_1 \\ j_2 \\ j_3 \end{bmatrix} + [0 \quad -1] \begin{bmatrix} j_4 \\ j_5 \end{bmatrix} = [0]$$

The second matrix equation is ignorable. The first matrix equation can be written as

$$\begin{bmatrix} j_1 \\ j_2 \\ j_3 \end{bmatrix} = - \begin{bmatrix} -1 & 1 & 0 \\ 0 & -1 & 0 \\ 0 & 0 & 1 \end{bmatrix}^{-1} \begin{bmatrix} 0 & 0 \\ -1 & 1 \\ 1 & 0 \end{bmatrix} \begin{bmatrix} j_4 \\ j_5 \end{bmatrix} = \begin{bmatrix} -1 & 1 \\ -1 & 1 \\ -1 & 0 \end{bmatrix} \begin{bmatrix} j_4 \\ j_5 \end{bmatrix}$$

This equation expresses the dependent variables (j_1, j_2, j_3) in terms of the chosen independent variables (j_4, j_5).

2.9 MIXED EQUATIONS

Consider a set of n equations in n unknowns given by

$$AX = Y \tag{2.9.1}$$

Assume that the first k of the y's and the last l of the x's are prescribed, where $k + l = n$. Partition Eq. (2.9.1) as follows:

$$\begin{bmatrix} A_{kk} & A_{kl} \\ A_{lk} & A_{ll} \end{bmatrix} \begin{bmatrix} X_k \\ X_l \end{bmatrix} = \begin{bmatrix} Y_k \\ Y_l \end{bmatrix} \tag{2.9.2}$$

where the subscripts denote the orders of the various submatrices, X_l and Y_k contain the prescribed (independent) variables, and X_k and Y_l contain the unknown (dependent) variables. Our problem is to find solutions for X_k and Y_l, if indeed they exist.

Equation (2.9.2) can be expanded to yield

$$A_{kk}X_k + A_{kl}X_l = Y_k \tag{2.9.3}$$

and

$$A_{lk}X_k + A_{ll}X_l = Y_l \tag{2.9.4}$$

All variables in Eq. (2.9.3) are known except those contained in X_k. If A_{kk} is nonsingular, we can write

$$X_k = A_{kk}^{-1}(Y_k - A_{kl}X_l) \tag{2.9.5}$$

Then Eq. (2.9.5) can be substituted into Eq. (2.9.4) to give

$$\begin{aligned} Y_l &= A_{lk}A_{kk}^{-1}(Y_k - A_{kl}X_l) + A_{ll}X_l \\ &= A_{lk}A_{kk}^{-1}Y_k + (A_{ll} - A_{lk}A_{kk}^{-1}A_{kl})X_l \end{aligned} \tag{2.9.6}$$

The solution for X_k, and therefore Y_l, is unique in this case.

Example 2.7 In the network in Fig. 2.9, assume that e_1, e_2, and i_3 are prescribed. Solutions are desired for i_1, i_2, and e_3. The KVL equations in matrix form are

$$\begin{bmatrix} 7 & 4 & 1 \\ 4 & 8 & -3 \\ 1 & -3 & 4 \end{bmatrix} \begin{bmatrix} i_1 \\ i_2 \\ i_3 \end{bmatrix} = \begin{bmatrix} e_1 \\ e_2 \\ e_3 \end{bmatrix}$$

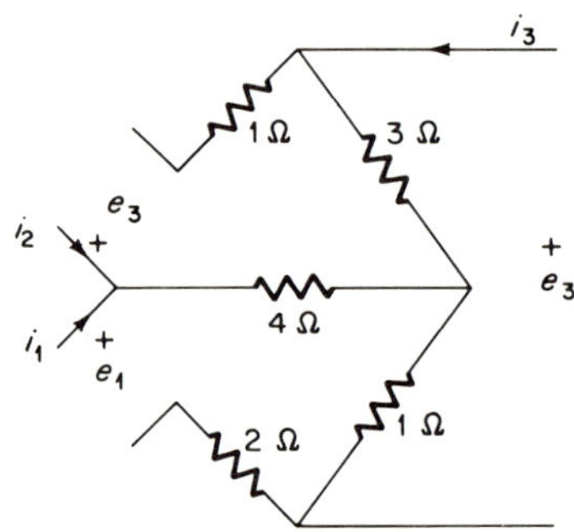

Fig. 2.9

Noting that $k = 2$ and $l = 1$, we partition this equation to obtain

$$\begin{bmatrix} 7 & 4 \\ 4 & 8 \end{bmatrix} \begin{bmatrix} i_1 \\ i_2 \end{bmatrix} + \begin{bmatrix} 1 \\ -3 \end{bmatrix} [i_3] = \begin{bmatrix} e_1 \\ e_2 \end{bmatrix}$$

and

$$[1 \quad -3] \begin{bmatrix} i_1 \\ i_2 \end{bmatrix} + 4i_3 = e_3$$

Since the determinant of A_{kk} does not vanish, we can write

$$\begin{bmatrix} i_1 \\ i_2 \end{bmatrix} = \begin{bmatrix} 7 & 4 \\ 4 & 8 \end{bmatrix}^{-1} \left(\begin{bmatrix} e_1 \\ e_2 \end{bmatrix} - \begin{bmatrix} 1 \\ -3 \end{bmatrix} i_3 \right)$$

and

$$e_3 = [1 \quad -3]\begin{bmatrix} 7 & 4 \\ 4 & 8 \end{bmatrix}^{-1}\begin{bmatrix} e_1 \\ e_2 \end{bmatrix} + 4\left(-[1 \quad -3]\begin{bmatrix} 7 & 4 \\ 4 & 8 \end{bmatrix}^{-1}\begin{bmatrix} 1 \\ -3 \end{bmatrix} i_3\right)$$

Simplifying these equations and recombining in matrix form yields

$$\begin{bmatrix} i_1 \\ i_2 \\ e_3 \end{bmatrix} = \frac{1}{40}\begin{bmatrix} 8 & -4 & -20 \\ -4 & 7 & 25 \\ 20 & -25 & 65 \end{bmatrix}\begin{bmatrix} e_1 \\ e_2 \\ i_3 \end{bmatrix}$$

Let us now return to Eqs. (2.9.3) and (2.9.4) and consider the situation when A_{kk} is singular. We can rewrite Eq. (2.9.3) in the form

$$A_{kk}X_k = Y_k - A_{kl}X_l \tag{2.9.7}$$

Since $|A_{kk}| = 0$, A_{kk} can be at most of rank $k - 1$. Assume this is its rank. Then, if a Gauss-Jordan reduction is carried out on Eq. (2.9.7), the process will terminate after $k - 1$ steps. There will be one residual equation whose right-hand side must vanish if the set of Eq. (2.9.7) is to be consistent. The final form of the reduction will be

$$\begin{bmatrix} 1 & 0 & \cdots & 0 & k_1 \\ 0 & 1 & \cdots & \cdots & \cdots \\ \cdots & \cdots & \cdots & \cdots & \cdots \\ 0 & 0 & \cdots & 1 & k_{k-1} \\ 0 & 0 & \cdots & 0 & 0 \end{bmatrix}\begin{bmatrix} x_1 \\ x_2 \\ \vdots \\ x_{k-1} \\ x_k \end{bmatrix} = \begin{bmatrix} f_1(Y_k,X_l) \\ f_2(Y_k,X_l) \\ \cdots\cdots \\ f_{k-1}(Y_k,X_l) \\ f_k(Y_k,X_l) \end{bmatrix} \tag{2.9.8}$$

where the k's are constants and $f_k(Y_k,X_l)$ means a function of y_1 through y_k and x_{k+1} through x_{k+l}. In other words, each of the right-hand members in Eq. (2.9.8) is, in general, a function of *all* the entries in Y_k and *all* the entries in X_l.

There are two cases to consider.

Only X_l specified In this case, one way (and perhaps the easiest way) to ensure that Eq. (2.9.7) is consistent is to choose Y_k such that

$$A_{kl}X_l = Y_k \tag{2.9.9}$$

This can always be done when only X_l is specified, and it eliminates the need for going through the steps of the Gauss-Jordan reduction. The required Y_k matrix is obtained simply by substituting the known X_l matrix into Eq. (2.9.9). The disadvantage of the method is that Eq. (2.9.9) fixes *each* of the y_k's whereas to make $f_k(Y_k,X_l)$ vanish in the residual equation in Eq. (2.9.8) in order to ensure consistency, it is necessary that only *one* of the y_k's be prescribed. The other $k-1$ y_k's may have any desired values. Thus the Gauss-Jordan reduction affords greater flexibility than Eq. (2.9.9) in the choice of the y_k's.

If Eq. (2.9.9) is satisfied, nonunique solutions for X_k are obtained from

$$A_{kk}X_k = 0 \tag{2.9.10}$$

and solutions for Y_l follow from Eq. (2.9.4).

Only Y_k specified In this case Eq. (2.9.9) is not always a consistent set of equations. Consistency is guaranteed only if A_{kl} is of rank k because only then are we sure that there will be no residual equation in the Gauss-Jordan reduction of Eq. (2.9.9). Assuming that A_{kl} is of rank k, the solution for X_l is unique only if $l = k$.

If, on the other hand, a Gauss-Jordan reduction of Eq. (2.9.7) is carried out to obtain Eq. (2.9.8), we must again require that $f_k(Y_k,X_l)$ vanish. This is assured if A_{kl} is of rank k, for then we can always choose X_l to guarantee consistency. If A_{kl} is less than k in rank, it is possible for X_l to disappear completely from $f_k(Y_k,X_l)$. This happens if the Gauss-Jordan reduction of A_{kk} causes all zero coefficients to appear in the kth row of A_{kl}. We are then left with the consistency requirement $f_k(Y_k) = 0$, which, for a prescribed Y_k, is not, in general, satisfied.

Example 2.8 For the set of equations

$$\begin{aligned}
2x_1 + 4x_2 - 6x_3 - 2x_4 &= y_1\\
3x_1 + 6x_2 + 3x_3 + x_4 &= y_2\\
x_1 + 3x_2 + 2x_3 - x_4 &= y_3\\
4x_1 + x_2 - x_3 + 3x_4 &= y_4
\end{aligned}$$

find solutions when (*a*) x_3 and x_4 are specified; (*b*) y_1 and y_2 are specified.

(*a*) With x_3 and x_4 specified, $k = 2$, $l = 2$, and $|A_{kk}| = 0$. The situation is case 1, and y_1 and y_2 can be found from Eq. (2.9.9)

$$\begin{bmatrix} y_1 \\ y_2 \end{bmatrix} = \begin{bmatrix} -6 & -2 \\ 3 & 1 \end{bmatrix} \begin{bmatrix} x_3 \\ x_4 \end{bmatrix}$$

$$y_1 = -6x_3 - 2x_4$$

$$y_2 = 3x_3 + x_4$$

Note that there is no choice in the values of y_1 and y_2. Next, nonunique solutions for x_1 and x_2 are obtained from Eq. (2.9.10)

$$\begin{bmatrix} 2 & 4 \\ 3 & 6 \end{bmatrix} \begin{bmatrix} x_1 \\ x_2 \end{bmatrix} = \begin{bmatrix} 0 \\ 0 \end{bmatrix}$$

$$x_1 = -2c \qquad x_2 = c \qquad c = \text{arbitrary const}$$

Finally, y_3 and y_4 are found from Eq. (2.9.4)

$$\begin{bmatrix} y_3 \\ y_4 \end{bmatrix} = \begin{bmatrix} 1 & 3 \\ 4 & 1 \end{bmatrix} \begin{bmatrix} -2c \\ c \end{bmatrix} + \begin{bmatrix} 2 & -1 \\ -1 & 3 \end{bmatrix} \begin{bmatrix} x_3 \\ x_4 \end{bmatrix}$$

$$y_3 = c + 2x_3 - x_4$$

$$y_4 = -7c + x_3 + 3x_4$$

If rather than using Eq. (2.9.9) a Gauss-Jordan reduction is carried out, the result is

$$\begin{bmatrix} 1 & 2 \\ 0 & 0 \end{bmatrix} \begin{bmatrix} x_1 \\ x_2 \end{bmatrix} = \begin{bmatrix} \tfrac{1}{2}y_1 + 3x_3 + x_4 \\ -\tfrac{3}{2}y_1 + y_2 - 12x_3 - 4x_4 \end{bmatrix}$$

Note that $f_k(Y_k, X_l)$ includes all entries of both Y_k and X_l. To ensure consistency,

$$-\tfrac{3}{2}y_1 + y_2 = 12x_3 + 4x_4$$

As predicted, there is a choice in the Y_k values. Choose $y_1 = 0$. Then

$$y_2 = 12x_3 + 4x_4 \qquad \text{and} \qquad x_1 + 2x_2 = 3x_3 + x_4$$

Let $x_2 = c$ as before. Then

$$x_1 = 3x_3 + x_4 - 2c$$

Finally, y_3 and y_4 are found from Eq. (2.9.4)

$$\begin{bmatrix} y_3 \\ y_4 \end{bmatrix} = \begin{bmatrix} 1 & 3 \\ 4 & 1 \end{bmatrix} \begin{bmatrix} 3x_3 + x_4 - 2c \\ c \end{bmatrix} + \begin{bmatrix} 2 & -1 \\ -1 & 3 \end{bmatrix} \begin{bmatrix} x_3 \\ x_4 \end{bmatrix}$$

$$y_3 = c + x_3$$

$$y_4 = -7c + 11x_3 + 7x_4$$

(*b*) We next consider the situation where only y_1 and y_2 are specified. As before, $k = 2$, $l = 2$, and $|A_{kk}| = 0$. This is case 2, and we note that l is not less than k but that A_{kk} is of rank $k - 1$. Because of the latter, we cannot use Eq. (2.9.9).

If a Gauss-Jordan reduction is carried out, the resulting equations are the same as before, only this time y_1 and y_2 are the prescribed quantities. For consistency

$$12x_3 + 4x_4 = -\tfrac{3}{2}y_1 + y_2$$

There is a choice in the X_l values. Choose $x_3 = 0$. Then

$$x_4 = -\tfrac{3}{8}y_1 + \tfrac{1}{4}y_2$$

Let $x_2 = c$ as before. Then

$$x_1 = -2c + \tfrac{1}{8}y_1 + \tfrac{1}{4}y_2$$

Finally, y_3 and y_4 are found from Eq. (2.9.4)

$$\begin{bmatrix} y_3 \\ y_4 \end{bmatrix} = \begin{bmatrix} 1 & 3 \\ 4 & 1 \end{bmatrix} \begin{bmatrix} -2c + \tfrac{1}{8}y_1 + \tfrac{1}{4}y_2 \\ c \end{bmatrix} + \begin{bmatrix} 2 & -1 \\ -1 & 3 \end{bmatrix} \begin{bmatrix} 0 \\ -\tfrac{3}{8}y_1 + \tfrac{1}{4}y_2 \end{bmatrix}$$

$$y_3 = c + \tfrac{1}{2}y_1$$

$$y_4 = -7c - \tfrac{5}{8}y_1 + \tfrac{7}{4}y_2$$

As a final point, consider what would happen if the coefficients of x_3 and x_4 in the second equation of the original set were -9 and -3, respectively. A_{kl} is still of rank $k-1$, but now the Gauss-Jordan reduction of A_{kk} yields

$$\begin{bmatrix} 1 & 2 \\ 0 & 0 \end{bmatrix} \begin{bmatrix} x_1 \\ x_2 \end{bmatrix} = \begin{bmatrix} \frac{1}{2}y_1 + 3x_3 + x_4 \\ -\frac{3}{2}y_1 + y_2 \end{bmatrix}$$

X_l is missing from the consistency equation. If y_1 and y_2 have been prescribed, then unless $y_1 = \frac{2}{3}y_2$, the equations are inconsistent and no solution is possible.

2.10 GAUSS-SEIDEL PROCEDURE

As an alternative to the Gauss-Jordan reduction, which is a direct procedure for solving a set of n linear algebraic equations in n unknowns, there is an iterative method known as the *Gauss-Seidel procedure* that is well suited to digital-computer programming.

Consider the following set of equations:

$$\begin{aligned} x + 4y - z &= 5 \\ 3x - y + 5z &= 18 \\ 3x + y + z &= 12 \end{aligned}$$

We first rearrange the equations so that, as nearly as possible, the largest coefficient in each equation appears on the main diagonal of the coefficient matrix

$$\begin{aligned} 3x + y + z &= 12 \\ x + 4y - z &= 5 \\ 3x - y + 5z &= 18 \end{aligned}$$

Each equation is then solved for its main-diagonal variable

$$x = \frac{12 - y - z}{3}$$

$$y = \frac{5 - x + z}{4}$$

$$z = \frac{18 - 3x + y}{5}$$

We now choose a solution for all variables except the first. The most common choice is all zeros. With $y = z = 0$, a new x value of 4.00† is

† We use two decimal places here. A computer would, of course, permit greater accuracy than this.

found from the first equation. Using $x = 4.00$ and $z = 0$, the new y value from the second equation is 0.25. With $x = 4.00$ and $y = 0.25$, the third equation gives $z = 1.25$. We continue the iteration, finding new values for each variable from the most current values of the other variables. The second cycle is

$$x = \frac{12 - 0.25 - 1.25}{3} = 3.50$$

$$y = \frac{5 - 3.50 + 1.25}{4} = 0.69$$

$$z = \frac{18 - 12.00 + 0.25}{5} = 1.25$$

Subsequent cycles yield

$$\begin{array}{lll}
x = 3.23 & y = 0.85 & z = 1.83 \\
\ \ = 3.11 & \ \ = 0.93 & \ \ = 1.92 \\
\ \ = 3.05 & \ \ = 0.97 & \ \ = 1.96 \\
\ \ = 3.02 & \ \ = 0.98 & \ \ = 1.98 \\
\ \ = 3.01 & \ \ = 0.99 & \ \ = 1.99 \\
\cdots & \cdots & \cdots \\
\ \ = 3.00 & \ \ = 1.00 & \ \ = 2.00
\end{array}$$

In this example, convergence to the values $x = 3$, $y = 1$, and $z = 2$ is reasonably fast. In general, convergence is assured only if the main-diagonal coefficients satisfy the condition

$$|a_{ii}| > \sum_{j=1}^{n} |a_{ij}| \qquad j \neq i \tag{2.10.1}$$

which is very similar to the condition in Eq. (1.14.1) for a matrix to be dominant.

REFERENCES

2.1. Hamming, R. W.: "Numerical Methods for Scientists and Engineers," McGraw-Hill Book Company, New York, 1962. The Gauss-Jordan and Gauss-Seidel reduction procedures are discussed from the point of view of machine computation.

2.2. Hildebrand, F. E.: "Methods of Applied Mathematics," Prentice-Hall, Inc., Englewood Cliffs, N.J., 1952. Chapter 1 includes an excellent discussion of the Gauss-Jordan reduction, the rank of a matrix, and solutions of m equations in n unknowns.

2.3. Guillemin, E. A.: "Theory of Linear Physical Systems," John Wiley & Sons, Inc., New York, 1963. Chapter 1 considers systems of mixed equations.

2.4. Seshu, S., and N. Balabanian: "Linear Network Analysis," John Wiley & Sons, Inc.. New York, 1959. The matrix solution of systems of linear algebraic equations is discussed in Chapter 3. Chapter 4 contains a thorough review of loop and node analysis.

PROBLEMS

Drill

2.1. Find the current in the 1-ohm resistor by loop, node-to-datum, and node-pair analysis.

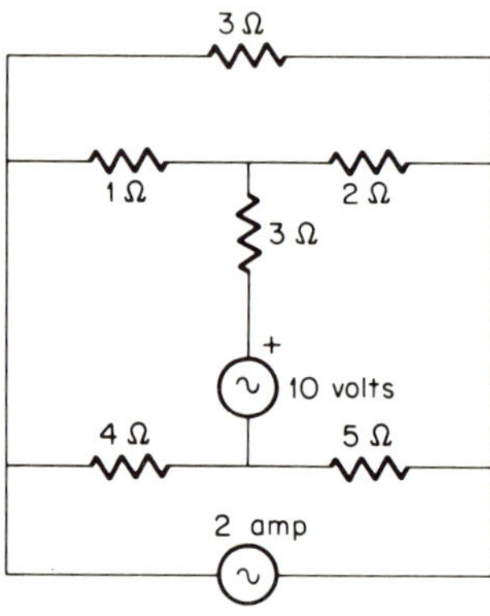

Fig. P2.1

2.2 Find the voltage across the ¼-ohm resistor by loop, node-to-datum, and node-pair analysis.

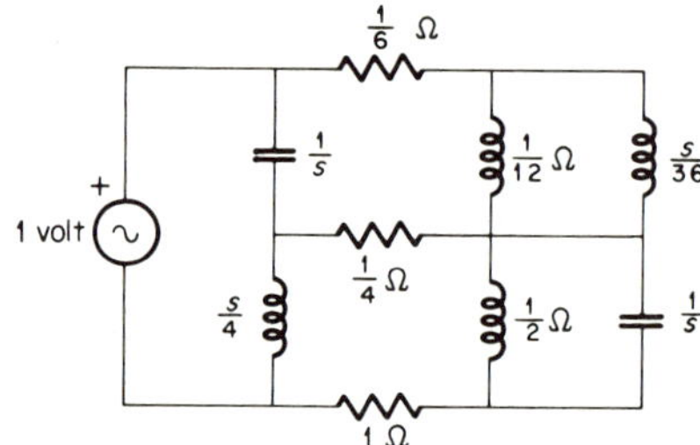

Fig. P2.2

2.3. Find the voltages across the four 2-ohm resistors by node-pair analysis.

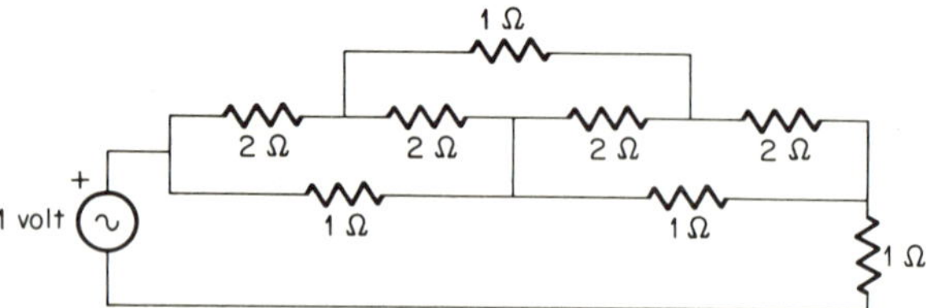

Fig. P2.3

2.4. Find the value of I_s that will make the voltage across the 1-ohm load resistor zero.

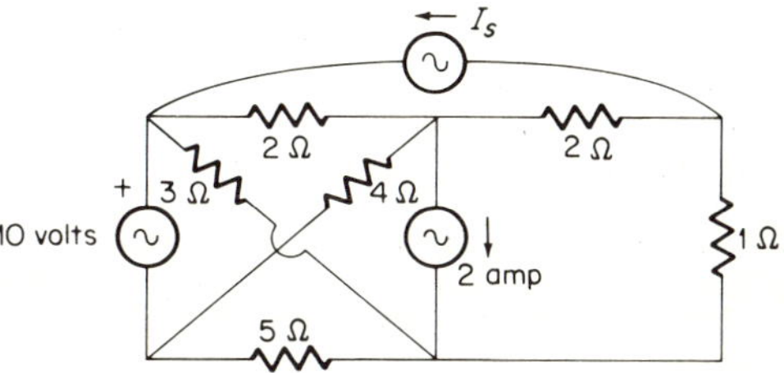

Fig. P2.4

2.5. Solve for the current i by node-to-datum and node-pair analysis.

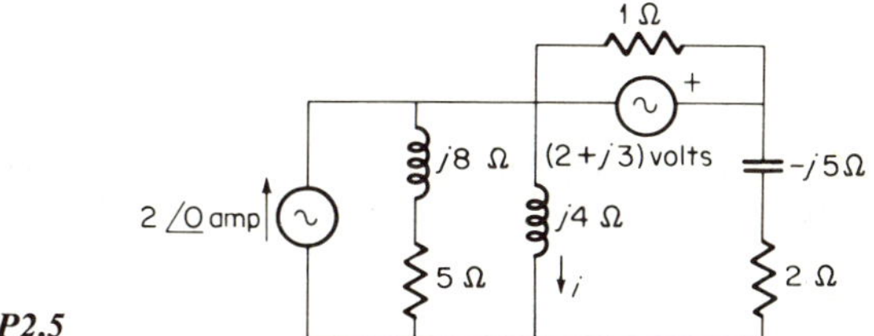

Fig. P2.5

2.6. In the given network, find the ratio e_2/e_1.

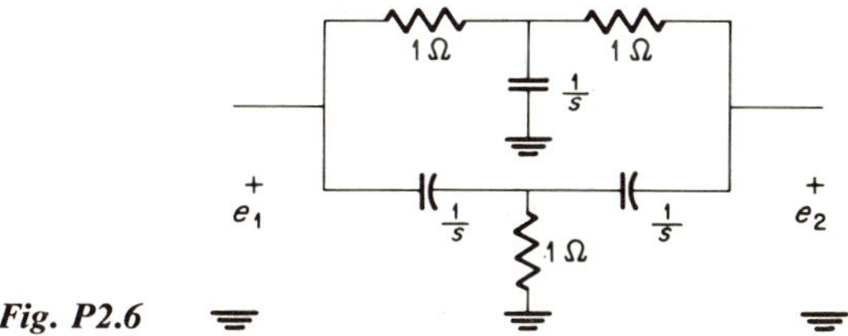

Fig. P2.6

2.7. The following set of equations represents the application of Kirchhoff's voltage law around the loops of a network.

$$v_1 + v_2 + v_3 = 0$$

$$v_1 + v_4 + v_5 = 0$$

$$v_2 - v_4 - v_6 = 0$$

$$v_3 - v_5 + v_6 = 0$$

$$v_1 + v_2 + v_5 - v_6 = 0$$

$$v_1 + v_3 + v_4 + v_0 = 0$$

$$v_2 + v_3 - v_4 - v_5 = 0$$

Perform a Gauss-Jordan reduction and answer the following questions:

(*a*) Are the equations consistent?

(*b*) Are the equations linearly independent?

(*c*) What is the defect of the system?

Select a set of independent voltage variables. Express the remaining dependent voltage variables in terms of this set and solve. Is this the only set?

2.8. Find all solutions to each of the following sets of equations:

(*a*)
$$\begin{aligned} 2x_1 + 3x_2 + 4x_3 &= 18 \\ x_1 + x_2 + 6x_3 &= 9 \\ 3x_1 + x_2 + 2x_3 &= 11 \\ 4x_1 + 2x_2 + 2x_3 &= 17 \end{aligned}$$

(*b*)
$$\begin{aligned} 2x_1 + 3x_2 + 4x_3 &= 18 \\ x_1 + x_2 + 6x_3 &= 9 \\ 3x_1 + 5x_2 + 2x_3 &= 27 \\ 4x_1 + 5x_2 + 16x_3 &= 36 \end{aligned}$$

2.9. Without performing a Gauss-Jordan reduction, determine whether the following set of equations is consistent and linearly independent. Find the defect.

$$\begin{aligned} x_1 + 2x_2 + x_3 + 5x_4 &= 6 \\ 2x_1 + x_2 + x_3 + 6x_4 &= 5 \\ x_1 - x_2 - x_3 &= -2 \\ x_1 - x_2 - 2x_3 - x_4 &= -3 \end{aligned}$$

Select a set of independent variables and express the dependent variables in terms of these. Set up the solution in matrix form and solve.

2.10. A network has equations on a loop basis given by

$$\begin{aligned} j_1 + 2j_2 - 3j_3 + j_4 + j_5 &= e_{s1} \\ 3j_1 - j_2 + 5j_3 - 4j_4 + j_5 &= e_{s2} \\ 4j_1 - j_2 + 6j_3 - 5j_4 + 2j_5 &= e_{s3} \\ j_1 + j_2 - 4j_3 + 5j_4 - j_5 &= e_{s4} \\ j_1 + 3j_2 - 2j_3 - 3j_4 + 2j_5 &= e_{s5} \end{aligned}$$

(*a*) If $j_4 = 4$ amp and $j_5 = 5$ amp, solve for the voltages and the remaining currents. Can the voltages e_{s1}, e_{s2}, and e_{s3} have more than one set of values while still ensuring that nontrivial solutions for j_1, j_2, and j_3 exist?

(*b*) If $e_{s1} = 5$ volts and $e_{s2} = 1$ volt, specify e_{s3} such that a solution exists. Find the currents and the remaining voltages. Note that j_4 and j_5 are not specified in this case.

2.11. Repeat Prob. 2.10 for the following set of equations:

$$\begin{aligned} 2j_1 + 2j_2 - 3j_3 + j_4 + j_5 &= e_{s1} \\ -j_1 - j_2 + 5j_3 - 4j_4 + j_5 &= e_{s2} \\ -j_1 - j_2 + 6j_3 - 5j_4 + 2j_5 &= e_{s3} \\ j_1 + j_2 - 4j_3 + 5j_4 - j_5 &= e_{s4} \\ j_1 + 3j_2 - 2j_3 - 3j_4 + 2j_5 &= e_{s5} \end{aligned}$$

2.12. Solve the following sets of equations by the Gauss-Seidel procedure:

(*a*)
$$\begin{aligned} w - x + y + 4z &= 10 \\ 2w + 4x + y - z &= 16 \\ 2w - x + 4y \quad &= 15 \\ 4w + x - y + z &= 8 \end{aligned}$$

(*b*)
$$\begin{aligned} x + 3y - z &= 4 \\ 4x - y + 5z &= 21 \\ 2x + y + z &= 9 \end{aligned}$$

Theory and proofs

2.13. (*a*) Given two $n \times n$ matrices B and C, find the conditions for which a nontrivial $n \times n$ matrix A exists such that $AB = AC$.

(*b*) Find a nontrivial A matrix (if one exists) when

$$B = \begin{bmatrix} 1 & 2 & 5 \\ 3 & 4 & -3 \\ 1 & 3 & -4 \end{bmatrix} \qquad C = \begin{bmatrix} 3 & 4 & 1 \\ 4 & 3 & 3 \\ 1 & 2 & 0 \end{bmatrix}$$

(*c*) Compute the products BA and CA. What relationship exists between the columns of these product matrices?

2.14. Two matrices A and C are given, where C is symmetric

$$A = \begin{bmatrix} 1 & 0 & 1 \\ 0 & 1 & 1 \end{bmatrix} \qquad C = \begin{bmatrix} 2 & 1 \\ 1 & 1 \end{bmatrix}$$

Find a symmetric matrix B having nonnegative entries such that

$$C = ABA^T$$

Is your answer unique?

2.15. We are given a set of n linear algebraic equations in n variables in x which is homogeneous and whose coefficient matrix A is of rank $r = n - 1$. Show that the solution for the x's takes the form

$$x_l = cA_{il}$$

where c is an arbitrary constant and A_{il} is the ilth cofactor of A. This shows that the unknowns are proportional to the cofactors in any row of the coefficient matrix.

Use this result to obtain all nontrivial solutions of the following set of equations

$$\begin{aligned} 2kx + ky + z &= 0 \\ 3x + y + 2z &= 0 \\ 3x + 2y + kz &= 0 \end{aligned}$$

2.16. Consider a set of n equations in n unknowns given by

$$AX = C \qquad |A| \neq 0$$

By a procedure similar to the Gauss-Jordan reduction, show that this set of equations can be converted to the form

$$TX = K$$

where T is a *triangular* matrix given by

$$T = [a_{ij}] \qquad a_{ij} = \begin{cases} 0 & i > j \quad \text{upper triangular} \\ 0 & i < j \quad \text{lower triangular} \end{cases}$$

Convert the following set of equations to triangular form and solve

$$\begin{aligned} x_1 + 5x_2 + x_3 + 2x_4 &= 16 \\ 2x_1 - 3x_2 + 2x_3 - x_4 &= 9 \\ -3x_1 + x_2 + x_3 - 2x_4 &= 0 \\ x_1 + 4x_2 - 3x_3 + 5x_4 &= 5 \end{aligned}$$

2.17. Consider an $n \times n$ nonsingular matrix A. Append an $n \times n$ identity matrix to A to obtain

$$\begin{bmatrix} a_{11} & a_{12} & \cdots & a_{1n} & 1 & 0 & \cdots & 0 \\ a_{21} & a_{22} & \cdots & a_{2n} & 0 & 1 & \cdots & 0 \\ \cdot & \cdot & \cdot & \cdot & \cdot & \cdot & \cdot & \cdot \\ a_{n1} & a_{n2} & \cdots & a_{nn} & 0 & 0 & \cdots & 1 \end{bmatrix}$$

Carry out a Gauss-Jordan reduction of this combined matrix. The result has the form

$$\begin{bmatrix} 1 & 0 & \cdots & 0 & b_{11} & b_{12} & \cdots & b_{1n} \\ 0 & 1 & \cdots & 0 & b_{21} & b_{22} & \cdots & b_{2n} \\ \cdot & \cdot & \cdot & \cdot & \cdot & \cdot & \cdot & \cdot \\ 0 & 0 & \cdots & 1 & b_{n1} & b_{n2} & \cdots & b_{nn} \end{bmatrix}$$

Call the right-hand portion of this matrix B. Prove that $B = A^{-1}$. Illustrate the procedure by obtaining the inverse of

$$A = \begin{bmatrix} 1 & 3 & 2 \\ 2 & -1 & 0 \\ -1 & -2 & 1 \end{bmatrix}$$

This method of obtaining the inverse of a matrix is well suited to the digital computer.

Application

2.18. Find j_1, j_2, j_3, and j_4 in terms of i_1, i_2, and i_3. Establish whether or not your solution is unique.

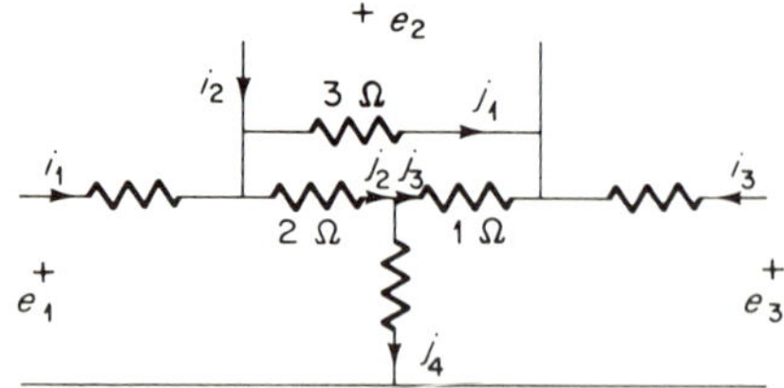

Fig. P2.18

2.19. For the given network, find all values of complex frequency s that yield nonzero values for i_1 and i_2 when e becomes zero. Find all such values of i_1 and i_2.

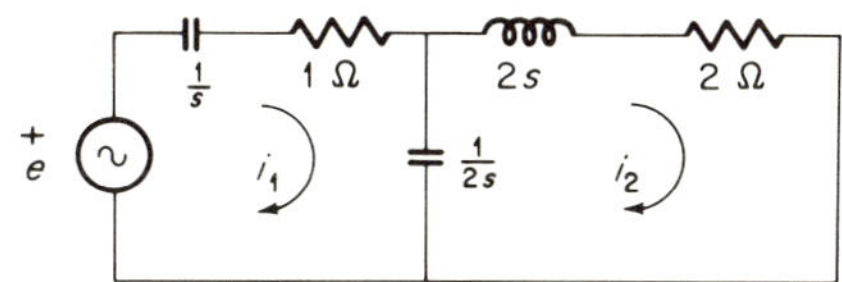

Fig. P2.19

2.20. Assume that e_1, e_2, i_3, and i_4 are specified sources. Find e_3, e_4, i_1, and i_2. Are your solutions unique?

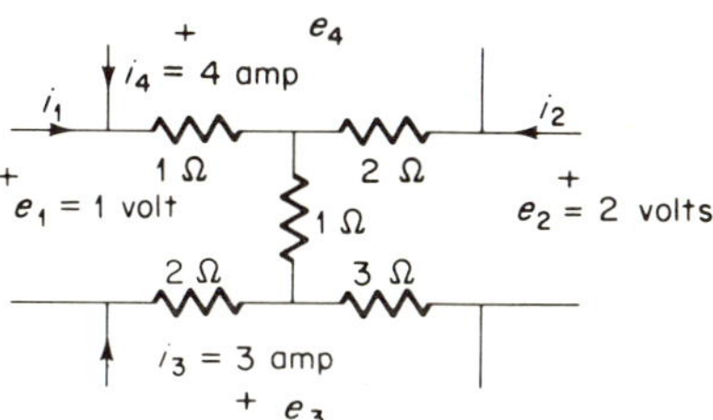

Fig. P2.20

2.21. We wish 2-port equations of the form

$$e_1 = z_{11}i_1 + z_{12}i_2$$

$$e_2 = z_{21}i_1 + z_{22}i_2$$

for the given network. Compose a matrix equation relating i_a, i_b, i_c, and i_d to i_1 and i_2 and use it to find numerical values for the z parameters.

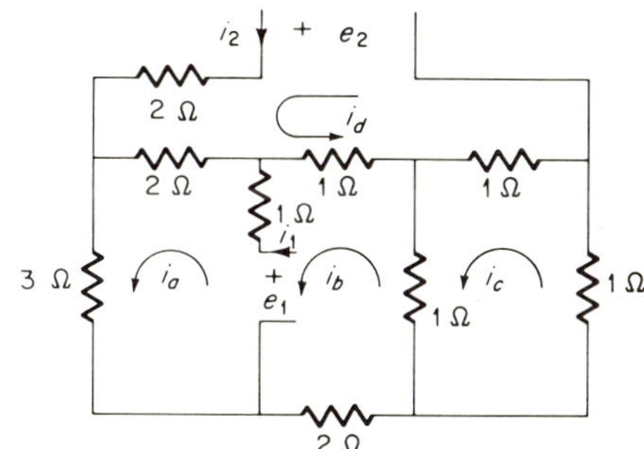

Fig. P2.21

3 NETWORK PARAMETERS AND MATRICES

3.1 INTRODUCTION

In this chapter we derive sets of parameters that describe the external-terminal and port behavior of linear networks. Interconnections of networks are discussed. We review coupled-circuit theory and extend it to include resistive coupling. The properties of ideal and perfect transformers, ideal gyrators, and negative-impedance converters are developed. We show that the gyrator, vacuum triode, and transistor may be considered as resistively coupled 3-terminal networks. The parameter matrices of these network elements are presented, and a systematic matrix method of passive- and active-network analysis is developed.

3.2 TERMINALS AND PORTS

Figure 3.1 depicts a linear network from which n leads have been brought. Such a network is termed an *n-terminal network*. The network itself may

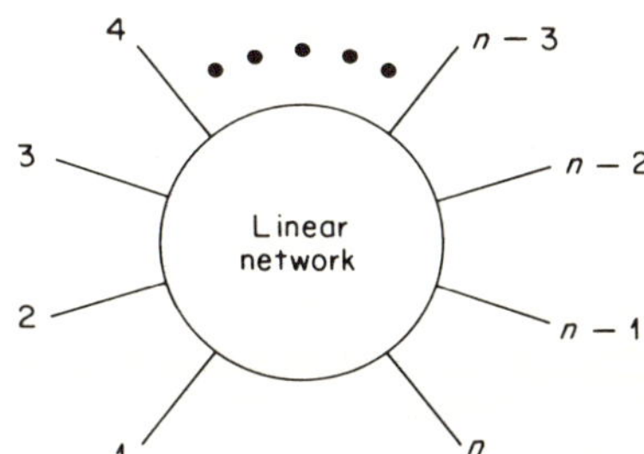

Fig. 3.1 Terminal and port designations.

contain any number of internal loops and nodes. This does not concern us. What we care about are its external characteristics at the n terminals, i.e., the voltage-current relations that exist there.

In Fig. 3.1, we can define $n - 1$ independent currents, one entering each of the first $n - 1$ terminals.† The current at the nth terminal is not an independent variable because the summation of currents entering the entire network, considered as one supernode, must be zero. The voltages at the $n - 1$ terminals with respect to the nth terminal are functions of the $n - 1$ currents. We require $n - 1$ equations in the $n - 1$ independent currents to specify the $n - 1$ dependent voltages. In the time domain these are linear differential equations with constant coefficients. In the frequency domain, they are linear algebraic equations in the variable s and can be written in closed matrix form as

$$E(s) = Z(s)I(s) \tag{3.2.1}$$

where E is an $(n - 1) \times 1$ column matrix of voltages, I is an $(n - 1) \times 1$ column matrix of currents, and Z is an $(n - 1) \times (n - 1)$ matrix of network impedance parameters.

Very often it is possible to represent the external behavior of a network on a port (rather than a terminal) basis and achieve a considerable reduction in the number of equations required for its characterization. By definition, a *port* consists of any pair of terminals such that the current into one of the terminals is identical to the current out of the other terminal. To create a port, we attach some external device to the two terminals in question. This device may be a simple voltage or current source, an impedance, or a complicated passive or active circuit, provided that it has only these two connections to the network. For example, let us place resistors across terminals 1 and 2, 3 and 4, . . . , $n - 3$ and $n - 2$, $n - 1$ and n in Fig. 3.1, where we assume that n is even. The result is an $(n/2)$-*port network*.

In the n-terminal case, there were n currents, $n - 1$ independent and 1

† It is, of course, not necessary to choose $n - 1$ currents as independent variables; $n - 1$ voltages or a mixture of $n - 1$ voltages and currents may be chosen. These choices are explored in Sec. 3.4.

dependent (the latter assumed at the nth terminal). When the resistors are added to form $n/2$ ports, $n/2 - 1$ of these $n - 1$ independent currents become dependent currents. The current at the nth terminal remains a dependent current. Thus we have a total of $n/2 - 1 + 1 = n/2$ dependent currents, and these are functions of $n - n/2 = n/2$ independent currents. E and I in Eq. (3.2.1) become $(n/2) \times 1$ column matrices, and Z is an $(n/2) \times (n/2)$ matrix. For large n, the number of equations required is reduced by approximately 50 percent.

Suppose that we now move the resistors so that they are across terminals 1 and n, 2 and $n, \ldots, n - 1$ and n (we now need $n - 1$, rather than $n/2$, resistors). The resulting configuration is an $(n - 1)$-port network in which all the ports have a common (ground) terminal. We call such a network a *grounded* $(n - 1)$*-port*. Is there any difference between this configuration and the original n-terminal representation? The answer is "no." We are back to the case of $n - 1$ independent currents and 1 dependent current, the latter at the nth (ground) terminal. It makes no difference whether we consider the network as an n-terminal network or as a grounded $(n - 1)$-port. The voltage-current equations are the same, and $n - 1$ of them are required.

Example 3.1 We wish to write equations in the form of Eq. (3.2.1) for the network in Fig. 3.2.

We consider the network on a terminal basis first. There are four terminals; hence we need three equations. If we choose terminal 4 as the reference (ground) terminal, they are given by

$$\begin{bmatrix} e_{14} \\ e_{24} \\ e_{34} \end{bmatrix} = \begin{bmatrix} 9 & 9 & 6 \\ 9 & 13 & 6 \\ 6 & 6 & 6 \end{bmatrix} \begin{bmatrix} i_1 \\ i_2 \\ i_3 \end{bmatrix} \qquad \text{4-terminal or grounded 3-port basis}$$

If we now consider terminals 1 and 2 as a port, and also terminals 3 and 4, we obtain

$$\begin{bmatrix} e_{21} \\ e_{34} \end{bmatrix} = \begin{bmatrix} 4 & 0 \\ 0 & 6 \end{bmatrix} \begin{bmatrix} i_2 \\ i_3 \end{bmatrix} \qquad \text{2-port basis}$$

We see that $n - 1 = 3$ equations are required in the 4-terminal representation but that only $n/2 = 2$ equations are necessary in the 2-port representation. In the 4-terminal representation, i_4 is a dependent current. When we consider the network as a

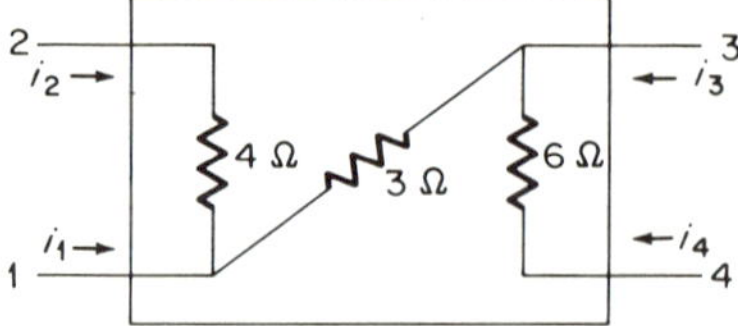

Fig. 3.2

2-port, i_4 is still a dependent current and i_1 becomes one. We can also observe that the 3-ohm resistor does not influence the 2-port equations but is included in the 4-terminal equations. This result is expected, since there is no closed path linking the 2-ports in the network of Fig. 3.2.

3.3 THE TWO 1-PORT PARAMETERS

If, in Fig. 3.1, we delete terminals 3 through n and attach an external device to terminals 1 and 2 (renumbered 1′ and 1), we arrive at a 1-port network, as shown in Fig. 3.3. In this case $n = 2$, we need $n/2 = 1$ equation of the matrix set in Eq. (3.2.1), and there is one independent current. The result is

$$e_1 = z_1 i_1 \tag{3.3.1}$$

where z_1 is the driving-point impedance at terminals 1-1′.

From the theory of combinations, if we have p different things and we wish to use them q at a time, there are

$$pCq = \frac{p(p-1)\cdots(p-q+1)}{q!} \tag{3.3.2}$$

possible combinations from which to choose.† In our present case $p = 2$ (one voltage and one current) and $q = 1$ (one independent variable). Thus $pCq = 2$, and we should be able to find another companion equation for Eq. (3.3.1). The second equation is, of course,

$$i_1 = y_1 e_1 \tag{3.3.3}$$

where y_1 is the driving-point admittance at terminals 1-1′.

3.4 THE SIX SETS OF 2-PORT PARAMETERS

A 2-port network‡ is shown in Fig. 3.4. Again, it may contain any number of internal loops and nodes, but it has only two sets of leads brought out. We

† For a derivation of Eq. (3.3.2), see F. M. Morgan, "College Algebra," American Book Company, New York, 1943, or similar text.

‡ If terminals 1′ and 2′ are connected together, the network becomes a 3-terminal network or grounded 2-port with a common input-output lead. No changes are necessary in the following development if this is so.

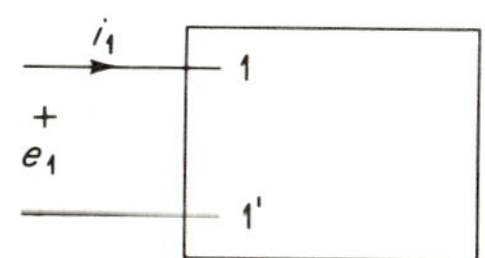

Fig. 3.3 A 1-port.

TABLE 3.1

Number of variables, $p = n$	*Number of independent variables,* $q = \frac{n}{2}$	*Sets of equations,* pCq	*Parameters per set,* q^2
2	1	2	1
4	2	6	4
6	3	20	9
8	4	70	16
10	5	252	25

wish to characterize this network through certain parameters, as we did in the 1-port case. First, current and voltage reference directions are chosen such that all currents enter unprimed terminals and all voltages are positive there. With this choice, either set of terminals may be considered as the input. Our next step is to find how many different sets of equations can be written to describe the network. There are four variables, two voltages and two currents. If values are assigned to any two of these, the other two are uniquely determined. Thus there are two independent variables. In Eq. (3.3.2), $p = 4$ and $q = 2$, which gives $pCq = 6$ possible sets of equations. Our third step is to determine, for any one of the six choices of independent variables, how many parameters are needed to describe the network. We require $n/2 = 2$ equations, and each equation contains $n/2 = 2$ independent variables. We thus require $(n/2)^2 = 4$ parameters.

We can use Eq. (3.2.1) to obtain the first set of 2-port equations and parameters. These are

$$\begin{aligned} e_1 &= z_{11}i_1 + z_{12}i_2 \\ e_2 &= z_{21}i_1 + z_{22}i_2 \end{aligned} \tag{3.4.1}$$

where the *z parameters* are given by

$$z_{11} = \left.\frac{e_1}{i_1}\right|_{i_2=0} \qquad z_{12} = \left.\frac{e_1}{i_2}\right|_{i_1=0}$$
$$z_{21} = \left.\frac{e_2}{i_1}\right|_{i_2=0} \qquad z_{22} = \left.\frac{e_2}{i_2}\right|_{i_1=0} \tag{3.4.2}$$

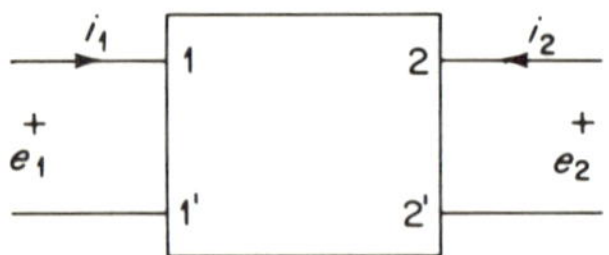

Fig. 3.4 A 2-port.

The set of equations (3.4.1) is called the *open-circuit* equations of the network. Thus z_{11} and z_{22} are *driving-point impedances*, and z_{12} and z_{21} are *transfer impedances*, each obtained with one terminal pair open.

Five other choices of independent variables are possible. Choosing the two currents yields a set of *y parameters*. The equations are

$$\begin{aligned} i_1 &= y_{11}e_{11} + y_{12}e_2 \\ i_2 &= y_{21}e_1 + y_{22}e_2 \end{aligned} \tag{3.4.3}$$

where y_{11} and y_{22} are *driving-point admittances* and y_{12} and y_{21} are *transfer admittances*. The set of equations (3.4.3) is called the *short-circuit* equations of the network, and the resulting y parameters are given by

$$\begin{aligned} y_{11} &= \left.\frac{i_1}{e_1}\right|_{e_2=0} \\ y_{12} &= \left.\frac{i_1}{e_2}\right|_{e_1=0} \\ y_{21} &= \left.\frac{i_2}{e_1}\right|_{e_2=0} \\ y_{22} &= \left.\frac{i_2}{e_2}\right|_{e_1=0} \end{aligned} \tag{3.4.4}$$

In dealing with transistors it is often convenient to express the input voltage and output current in terms of the input current and output voltage through a set of *hybrid parameters*. The resulting equations are

$$\begin{aligned} e_1 &= h_{11}i_1 + h_{12}e_2 \\ i_2 &= h_{21}i_1 + h_{22}e_2 \end{aligned} \tag{3.4.5}$$

where

$$\begin{aligned} h_{11} &= \left.\frac{e_1}{i_1}\right|_{e_2=0} \\ h_{12} &= \left.\frac{e_1}{e_2}\right|_{i_1=0} \\ h_{21} &= \left.\frac{i_2}{i_1}\right|_{e_2=0} \\ h_{22} &= \left.\frac{i_2}{e_2}\right|_{i_1=0} \end{aligned} \tag{3.4.6}$$

Note that h_{11} is a driving-point impedance whereas h_{22} is a driving-point admittance. However, these are not equal to z_{11} and y_{22}, respectively, because different quantities are zero during their evaluation. They are equal to $1/y_{11}$ and $1/z_{22}$, respectively. Note further that h_{12} is a voltage transfer ratio whereas h_{21} is a current transfer ratio.

A fourth very useful set of parameters is the *abcd*, or *chain*, parameters, used to express the quantities at the input-terminal pair in terms of those at the output-terminal pair through the set of equations

$$\begin{aligned} e_1 &= ae_2 - bi_2 \\ i_1 &= ce_2 - di_2 \end{aligned} \tag{3.4.7}$$

where

$$\begin{aligned} a &= \left.\frac{e_1}{e_2}\right|_{i_2=0} \\ b &= \left.\frac{e_1}{i_2}\right|_{e_2=0} \\ c &= \left.\frac{i_1}{e_2}\right|_{e_2=0} \\ d &= \left.\frac{i_1}{i_2}\right|_{e_2=0} \end{aligned} \tag{3.4.8}$$

Here a is the voltage transfer ratio under open-circuited output conditions (the reciprocal of the voltage gain of the network with infinite load impedance), and d is the negative of the current transfer ratio under short-circuited output conditions (the negative reciprocal of the current gain of the network with zero load impedance); b is a transfer impedance, and c is a transfer admittance. The minus signs in Eqs. (3.4.7) should be associated with i_2 rather than with b and d. Thus $-i_2$ is the current leaving the output reference terminal. The usefulness of the *abcd* parameters in problems involving cascaded networks should be evident.

Two other sets of four parameters each should be included to complete the development. These are the *g parameters* and the *a′b′c′d′ parameters*

$$\begin{aligned} i_1 &= g_{11}e_1 + g_{12}i_2 \\ e_2 &= g_{21}e_1 + g_{22}i_2 \end{aligned} \tag{3.4.9}$$

and

$$\begin{aligned} e_2 &= a'e_1 - b'i_1 \\ i_2 &= c'e_1 - d'i_1 \end{aligned} \tag{3.4.10}$$

These two sets are less often used. If needed, they can always be obtained by inverting Eqs. (3.4.5) and (3.4.7), respectively.

Example 3.2 Obtain the z, y, h and $abcd$ parameters of the T network shown in Fig. 3.5. The z parameters are obtained by inspection.

$$z_{11} = z_1 + z_3$$

$$z_{22} = z_2 + z_3$$

$$z_{12} = z_{21} = z_3$$

The other sets of parameters can be found as functions of the z parameters, or they can be obtained directly from the network. Only the results are given. The reader is urged to verify them by both methods.

$$y_{11} = \frac{z_2 + z_3}{|z|} \qquad y_{22} = \frac{z_1 + z_3}{|z|}$$

$$y_{12} = y_{21} = \frac{-z_3}{|z|} \qquad |z| = z_1 z_2 + z_2 z_3 + z_3 z_1$$

$$h_{11} = z_1 + \frac{z_2 z_3}{z_2 + z_3} \qquad h_{12} = \frac{z_3}{z_2 + z_3}$$

$$h_{21} = \frac{-z_3}{z_2 + z_3} \qquad h_{22} = \frac{1}{z_2 + z_3}$$

$$a = 1 + \frac{z_1}{z_3} \qquad b = z_1 + z_2 + \frac{z_1 z_2}{z_3}$$

$$c = \frac{1}{z_3} \qquad d = 1 + \frac{z_2}{z_3}$$

Several points can be brought out through Example 3.2. First note that

$$z_{12} = z_{21} \qquad y_{12} = y_{21} \qquad h_{12} = -h_{21} \qquad ad - bc = 1 \tag{3.4.11}$$

These four relationships result because the network in Fig. 3.5 is *reciprocal.* Reciprocity is discussed in detail in Sec. 7.3. Suffice it to say here that reciprocity means that the network is electrically symmetric; i.e., it conducts equally well in each direction (is bilateral). Because reciprocity holds, only

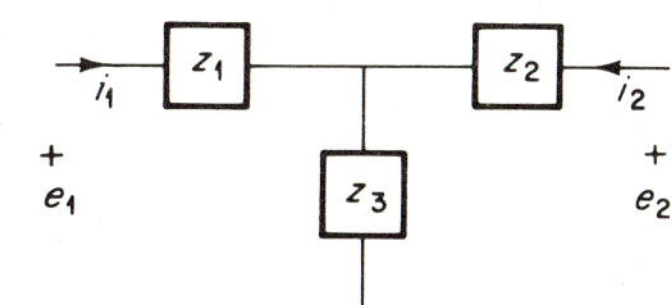

Fig. 3.5

TABLE 3.2 Parameter conversion relationships

z_{11}	$\frac{y_{22}}{\lvert y\rvert}$	$\frac{\lvert h\rvert}{h_{22}}$	$\frac{a}{c}$
z_{12}	$\frac{-y_{12}}{\lvert y\rvert}$	$\frac{h_{12}}{h_{22}}$	$\frac{ad - bc}{c}$
z_{21}	$\frac{-y_{21}}{\lvert y\rvert}$	$\frac{-h_{21}}{h_{22}}$	$\frac{1}{c}$
z_{12}	$\frac{y_{11}}{\lvert y\rvert}$	$\frac{1}{h_{22}}$	$\frac{d}{c}$
y_{11}	$\frac{z_{22}}{\lvert z\rvert}$	$\frac{1}{h_{11}}$	$\frac{d}{b}$
y_{12}	$\frac{-z_{12}}{\lvert z\rvert}$	$\frac{-h_{12}}{h_{11}}$	$\frac{-(ad - bc)}{b}$
y_{21}	$\frac{-z_{21}}{\lvert z\rvert}$	$\frac{h_{21}}{h_{22}}$	$\frac{-1}{b}$
y_{22}	$\frac{z_{11}}{\lvert z\rvert}$	$\frac{\lvert h\rvert}{h_{11}}$	$\frac{a}{b}$
h_{11}	$\frac{\lvert z\rvert}{z_{22}}$	$\frac{1}{y_{11}}$	$\frac{b}{d}$
h_{12}	$\frac{z_{12}}{z_{22}}$	$\frac{-y_{12}}{y_{11}}$	$\frac{ad - bc}{d}$
h_{21}	$\frac{-z_{21}}{z_{22}}$	$\frac{y_{21}}{y_{11}}$	$\frac{-1}{d}$
h_{22}	$\frac{1}{z_{22}}$	$\frac{\lvert y\rvert}{y_{11}}$	$\frac{c}{d}$
a	$\frac{z_{11}}{z_{21}}$	$\frac{-y_{22}}{y_{21}}$	$\frac{-\lvert h\rvert}{h_{21}}$
b	$\frac{\lvert z\rvert}{z_{21}}$	$\frac{-1}{y_{21}}$	$\frac{-h_{11}}{h_{21}}$
c	$\frac{1}{z_{21}}$	$\frac{-\lvert y\rvert}{y_{21}}$	$\frac{-h_{22}}{h_{21}}$
d	$\frac{z_{22}}{z_{21}}$	$\frac{-y_{11}}{y_{21}}$	$\frac{-1}{h_{21}}$

three parameters are necessary to characterize the network, the fourth parameter being prescribed by Eqs. (3.4.11).

As a second point, suppose that $z_1 = z_2$ in Fig. 3.5. Then

$$z_{11} = z_{22} \qquad y_{11} = y_{22} \qquad h_{11}h_{22} - h_{12}h_{21} = 1 \qquad a = d \tag{3.4.12}$$

The network is now physically symmetric, or simply *symmetrical.* It can be rotated about a vertical centerline with no change in any network parameter. The reader should note carefully the difference between electrical and physical symmetry. To be (physically) symmetrical, a network must also be reciprocal, but the reverse is not true. If a network is reciprocal and symmetrical, two parameters are sufficient to characterize it, the other two coming from Eqs. (3.4.11) and (3.4.12).

A third point to note is that simple relationships exist between the various sets of network parameters. To save labor in converting from one set of parameters to another, the conversion relationships for the four most common sets of parameters are assembled in Table 3.2.

3.5 TWO-PORTS WITH SOURCE AND LOAD IMPEDANCES

If we add load and source impedances to the network in Fig. 3.4, the network in Fig. 3.6 results. Noting that

$$e_2 = -i_2 z_L \tag{3.5.1}$$

we can include the load impedance in the z-parameter relations in Eqs. (3.4.1) by rewriting them in the form

$$\begin{aligned} e_1 &= z_{11}i_1 + z_{12}i_2 \\ -i_2 z_L &= z_{21}i_1 + z_{22}i_2 \end{aligned} \tag{3.5.2}$$

To include the source impedance as well, we note that

$$e_s = e_1 + i_1 z_s \tag{3.5.3}$$

Substituting Eq. (3.5.3) into the first of Eqs. (3.5.2) and rearranging the second of Eqs. (3.5.2), we obtain

$$\begin{aligned} e_s &= (z_s + z_{11})i_1 + z_{12}i_2 \\ 0 &= z_{21}i_1 + (z_L + z_{22})i_2 \end{aligned} \tag{3.5.4}$$

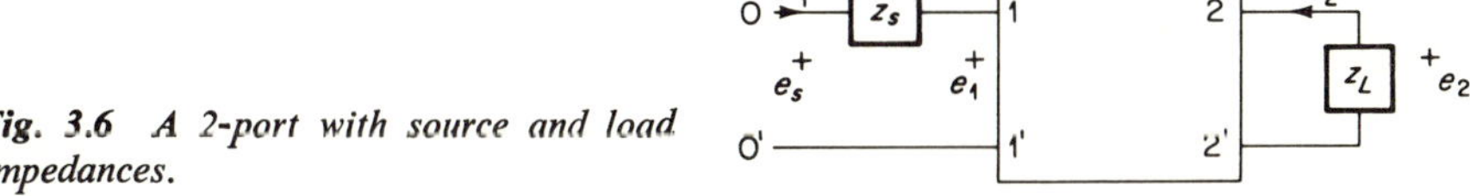

Fig. 3.6 A 2-port with source and load impedances.

We now wish to use Eqs. (3.5.1) to (3.5.4) to develop several driving-point and transfer relationships which will be of use in network analysis and in subsequent work in network synthesis.

Consider the driving-point relations first. The driving-point impedance at terminals 0-0′ is the sum of z_s and the driving-point impedance at terminals 1-1′. The latter is e_1/i_1 and can be obtained by solving the second of Eqs. (3.5.2) for i_2 in terms of i_1 and inserting this quantity in the first of Eqs. (3.5.2). The result is

$$Z_{1\text{-}1'} = \frac{z_{11}z_L + |z|}{z_{22} + z_L} \tag{3.5.5}$$

Consider next the current transfer ratio i_2/i_1, which can be found directly from the second of Eqs. (3.5.2)

$$I_{21} \equiv \frac{i_2}{i_1} = -\frac{z_{21}}{z_{22} + z_L} \tag{3.5.6}$$

The transfer impedance Z_{21} is defined by

$$Z_{21} \equiv \frac{e_2}{i_1} = \frac{i_2}{i_1}\frac{e_2}{i_2} = \frac{z_{21}z_L}{z_{22} + z_L} \tag{3.5.7}$$

and the transfer admittance by

$$Y_{21} \equiv \frac{i_2}{e_1} = \frac{i_2}{i_1}\frac{i_1}{e_1} = -\frac{z_{21}}{z_{11}z_L + |z|} \tag{3.5.8}$$

Finally, consider the two voltage transfer ratios, e_2/e_1 and e_2/e_s. The former is given by

$$E_{21} \equiv \frac{e_2}{e_1} = \frac{i_2}{e_1}\frac{e_2}{i_2} = \frac{z_{21}z_L}{z_{11}z_L + |z|} \tag{3.5.9}$$

Then e_2/e_s is formed as

$$E_{2s} \equiv \frac{e_2}{e_s} = \frac{e_2}{e_1}\frac{e_1}{e_s} = \frac{E_{21}Z_{1-1'}}{z_s + Z_{1-1'}} = \frac{z_{21}z_L}{z_s(z_{22} + z_L) + z_{11}z_L + |z|} \tag{3.5.10}$$

A similar set of relations can be derived on a y-parameter basis. In Fig. 3.6 the series z_s is replaced by a shunt y_s across terminals 1 and 1′, and the source voltage e_s is replaced by a source current i_s. The resulting network is shown in Fig. 3.7. The derivations are requested in a problem at the end of the chapter.

The technique of expressing a ratio as the product of two other ratios, used in Eqs. (3.5.7) to (3.5.10), can result in a considerable saving in labor in finding a given transfer function. The reader is invited to test this statement by finding the transfer functions in these equations by conventional circuit-analysis methods.

The various transfer functions for a 2-port with source and load impedances are assembled in Table 3.3 on both a z- and a y-parameter basis.

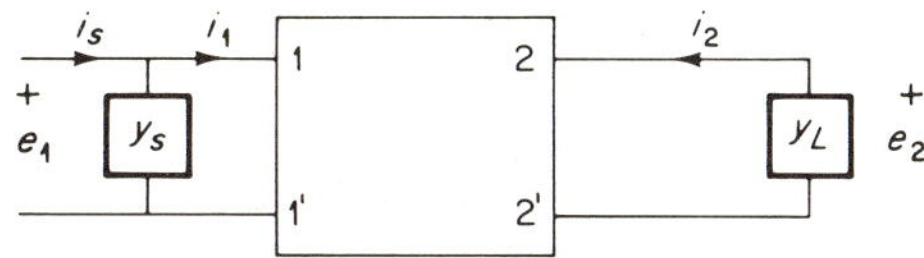

Fig. 3.7 Counterpart of Fig. 3.6.

The transfer functions given in the upper part of the table are independent of z_s and y_s, and thus the formulas given are equally valid for either the configuration of Fig. 3.6 or that of Fig. 3.7. This is not true in the lower part of the table, where each formula applies either to Fig. 3.6 or to Fig. 3.7 but not to both.

The use of the formulas in Table 3.3 in the analysis of a given network is rather obvious, but let us mention one point that can result in a saving in computational labor. Suppose that the current ratio I_{21} of a network is desired. We have the choice of computing it on a z- or a y-parameter basis. Which do we choose? A glance at the two formulas indicates that the z-parameter basis is preferable because only two z parameters need to be found whereas all three y parameters are included in the y-parameter formula. A similar situation exists for many of the other transfer functions in Table 3.3. Also, in network synthesis, if we are given an I_{21} to synthesize, it is best to do it on a z-parameter basis, since only two of the z parameters are tied down and the third can be chosen to achieve the most desirable network.

3.6 *n*-PORT PARAMETERS

It is of general and practical interest to extend the results of Sec. 3.4 to networks having three or more ports. Consider, for example, the triode amplifier with an unbypassed cathode resistor shown in Fig. 3.8. This is a 4-terminal network, or a grounded 3-port. Generally, an input voltage is applied at terminal 1, an output voltage is taken from terminal 2, and a portion of this output voltage, after being processed by additional stages, is *fed back* to terminal 3. All three ports are of interest. This means that we need a 3×3 matrix of parameters to relate the voltages at the ports to the currents there. In the general case, for an n-port network, we need an $n \times n$ parameter matrix that relates a set of n independent variables (voltages and/or currents) to a set of n dependent variables (again voltages and/or currents). The general form of the desired relationship is

$$\begin{bmatrix} w_1 \\ w_2 \\ \cdot \\ \cdot \\ \cdot \\ w_n \end{bmatrix} = \begin{bmatrix} k_{11} & k_{12} & \cdots & k_{1n} \\ k_{21} & k_{22} & \cdots & k_{2n} \\ \cdot & \cdot & \cdots & \cdot \\ k_{n1} & k_{n2} & \cdots & k_{nn} \end{bmatrix} \begin{bmatrix} x_1 \\ x_2 \\ \cdot \\ \cdot \\ \cdot \\ x_n \end{bmatrix} \tag{3.6.1}$$

TABLE 3.3

Relationship	*Condition*	*Symbol*	*Formula Figs. 3.6 and 3.7* z basis	y basis
$\frac{e_1}{i_1}$		$Z_{1\text{-}1'}$	$\frac{z_{11}z_L + \|z\|}{z_{22} + z_L}$	
$\frac{i_1}{e_1}$		$Y_{1\text{-}1'}$		$\frac{y_{11}y_L + \|y\|}{y_{22} + y_L}$
$\frac{i_2}{i_1}$		I_{21}	$\frac{-z_{21}}{z_{22} + z_L}$	$\frac{y_{21}y_L}{y_{11}y_L + \|y\|}$
$\frac{i_2}{i_1}$	2-2 shorted	$I_{21}{}^s$	$-\frac{z_{21}}{z_{22}}$	$\frac{y_{21}}{y_{11}}$
$\frac{e_2}{i_1}$		Z_{21}	$\frac{z_{21}z_L}{z_{22} + z_L}$	$\frac{-y_{21}}{y_{11}y_L + \|y\|}$
$\frac{e_2}{i_1}$	2-2 open	$Z_{21}{}^0$	z_{21}	$-\frac{y_{21}}{\|y\|}$
$\frac{i_2}{e_1}$		Y_{21}	$\frac{-z_{21}}{z_{11}z_L + \|z\|}$	$\frac{y_{21}y_L}{y_{22} + y_L}$
$\frac{i_2}{e_1}$	2-2 shorted	$Y_{21}{}^s$	$-\frac{z_{21}}{\|z\|}$	y_{21}
$\frac{e_2}{e_1}$		E_{21}	$\frac{z_{21}z_L}{z_{11}z_L + \|z\|}$	$\frac{-y_{21}}{y_{22} + y_L}$
$\frac{e_2}{e_1}$	2-2 open	$E_{21}{}^0$	$\frac{z_{21}}{z_{11}}$	$-\frac{y_{21}}{y_{22}}$
			Fig. 3.6, z basis	*Fig. 3.7,* y basis
$\frac{i_2}{i_s}$		I_{2s}		$\frac{I_{21}Y_{1\text{-}1'}}{y_s + Y_{1\text{-}1'}}$
$\frac{i_2}{i_s}$	2-2′ shorted	$I_{2s}{}^s$		$\frac{I_{21}{}^s y_{11}}{y_s + y_{11}}$
$\frac{e_s}{e_s}$		E_{2s}	$\frac{E_{21}Z_{1\text{-}1'}}{z_s + Z_{1\text{-}1'}}$	
$\frac{e_2}{e_s}$	2-2′ open	$E_{2s}{}^0$	$\frac{E_{21}{}^0 z_{11}}{z_s + z_{11}}$	
$\frac{e_s}{i_1}$		$Z_{0\text{-}0'}$	$z_s + z_{1\text{-}1'}$	
$\frac{i_s}{e_1}$		$Y_{0\text{-}0'}$		$y_s + Y_{1\text{-}1'}$

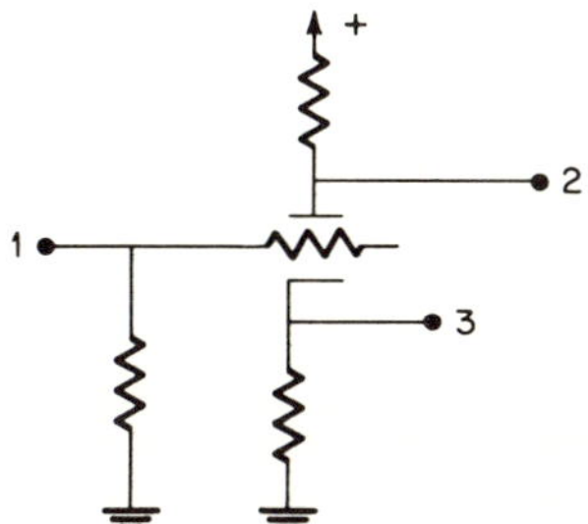

Fig. 3.8 A 3-port.

If each of the w's in Eq. (3.6.1) is a voltage at one of the ports, then each of the x's is the current at the corresponding port and each of the k's is an impedance function. Conversely, if each of the w's is a current, each of the x's is a voltage and each of the k's is an admittance function. A third possibility is for W to contain a mixture of voltages and currents, and likewise X. Then each k is either an impedance, admittance, voltage ratio, or current ratio. There are two situations to consider: (1) when the W (and thus the X) matrix contains only one variable from each of the ports and (2) when W (and thus X) contains both variables from at least one port.

For a 3-port, the four possibilities just considered are illustrated by the following equations:

$$\begin{bmatrix} e_1 \\ e_2 \\ e_3 \end{bmatrix} = \begin{bmatrix} z_{11} & z_{12} & z_{13} \\ z_{21} & z_{22} & z_{23} \\ z_{31} & z_{32} & z_{33} \end{bmatrix} \begin{bmatrix} i_1 \\ i_2 \\ i_3 \end{bmatrix} \tag{3.6.2}$$

$$\begin{bmatrix} i_1 \\ i_2 \\ i_3 \end{bmatrix} = \begin{bmatrix} y_{11} & y_{12} & y_{13} \\ y_{21} & y_{22} & y_{23} \\ y_{31} & y_{32} & y_{33} \end{bmatrix} \begin{bmatrix} e_1 \\ e_3 \\ e_3 \end{bmatrix} \tag{3.6.3}$$

$$\begin{bmatrix} e_1 \\ e_2 \\ i_3 \end{bmatrix} = \begin{bmatrix} m_{11} & m_{12} & m_{13} \\ m_{21} & m_{22} & m_{23} \\ m_{31} & m_{32} & m_{33} \end{bmatrix} \begin{bmatrix} i_1 \\ i_2 \\ e_3 \end{bmatrix} \tag{3.6.4}$$

$$\begin{bmatrix} e_1 \\ e_2 \\ i_2 \end{bmatrix} = \begin{bmatrix} n_{11} & n_{12} & n_{13} \\ n_{21} & n_{22} & n_{23} \\ n_{31} & n_{32} & n_{33} \end{bmatrix} \begin{bmatrix} i_1 \\ e_3 \\ i_3 \end{bmatrix} \tag{3.6.5}$$

Equations (3.6.2) and (3.6.3) are conventional z- and y-parameter formulations and are merely extensions of Eqs. (3.4.1) and (3.4.3) to the 3-port case. The z and y parameters are defined by, and can be obtained from, relations similar to Eqs. (3.4.2) and (3.4.4). In general form,

$$z_{ij} = \frac{e_i}{i_j}\bigg|_{\text{all other port currents zero}} \tag{3.6.6}$$

and

$$y_{ij} = \frac{i_i}{e_j}\bigg|_{\text{all other port voltages zero}} \tag{3.6.7}$$

Note that in obtaining z_{ij} we apply a *current* at the jth port and *open* all other ports; to find y_{ij}, we apply a *voltage* at the jth port and *short* all other ports.

Let us next consider Eq. (3.6.4). Here each column matrix contains mixed variables, but only one from a given port. Each m_{ij} parameter can be found in a manner analogous to Eqs. (3.6.6) and (3.6.7), except that some ports must be opened and others shorted to obtain a given parameter. For example, $m_{11} = e_1/i_1$ with i_2 and e_3 zero, i.e., port 2 opened and port 3 shorted.

Equation (3.6.5) poses a new problem. Suppose we try to obtain $n_{11} = e_1/i_1$ with i_3 and e_3 zero. How do we force the current and voltage at the same port to be zero simultaneously? The answer is that we cannot by the same direct method of opening and shorting ports we have used heretofore. We need to find an indirect method of obtaining n_{11}, if indeed this parameter exists.

In searching for such a method suppose we try to transform one of the sets of equations whose parameters are easily determined into the form of Eq. (3.6.5). Consider the z-parameter set in Eq. (3.6.2). If we can interchange e_3 and i_2, the column matrices will match those of Eq. (3.6.5). Then the n_{ij} can be obtained in terms of the known z parameters.

Are there any constraints on this procedure? Consider the interchange of any two variables, w_i and x_j, in Eq. (3.6.1). To effect this interchange, we would (1) solve the ith equation for x_j, (2) insert the expression for x_j into each of the other equations, and (3) reassemble the resulting set of equations in matrix form. The only constraint imposed is $k_{ij} \neq 0$. For if $k_{ij} = 0$, it is impossible to perform step 1 and solve the ith equation in Eq. (3.6.1) for x_j. In terms of Eqs. (3.6.2) and (3.6.5), this means that each of the n_{ij} in Eq. (3.6.5) exists (is zero or a finite number) if $z_{32} \neq 0$ in Eq. (3.6.2).

We could also find the n_{ij}, if they exist, through Eq. (3.6.4). This time we would interchange the variables i_3 and i_2 [in Eq. (3.6.5) we would also have to interchange columns 2 and 3 in N and rows 2 and 3 in the right-side column matrix to match the variables in the two equations]. In this case the condition for the n_{ij} to exist is $m_{32} \neq 0$.

As one further illustration, consider the 2-port conversion relations in Table 3.2. Suppose we are given the z parameters and we wish to find the h parameters. Comparing Eq. (3.4.1) with Eq. (3.4.5), we see that e_2 and i_2 have traded places. Thus the h parameters exist if and only if $z_{22} \neq 0$. A glance at Table 3.2 confirms this conclusion.

Example 3.3 We wish, if possible, to write the set of Eqs. (3.6.5) for the 3-port network inside the dashed box in Fig. 3.9. Our approach will be to write the z-parameter equations of Eq. (3.6.2) and, if possible, convert them to the form of Eq. (3.6.5).

From conventional loop analysis,

$$\begin{bmatrix} e_1 \\ e_2 \\ e_3 \end{bmatrix} = \begin{bmatrix} 7/3 & 1 & 1/3 \\ 1 & 7/2 & -1/2 \\ 1/3 & -1/2 & 23/6 \end{bmatrix} \begin{bmatrix} i_1 \\ i_2 \\ i_3 \end{bmatrix}$$

To obtain Eq. (3.6.5), we must interchange e_3 and i_2. This is possible since $z_{32} \neq 0$. To perform the interchange, we solve the third equation for i_2 and obtain

$$i_2 = 2/3 i_1 - 2e_3 + 23/3 i_3$$

Substituting this equation into the first two equations yields

$$e_1 = 3i_1 - 2e_3 + 8i_3 \qquad \text{and} \qquad e_2 = 10/3 i_1 - 7e_3 + 79/3 i_3$$

The desired mixed-variable matrix equation is

$$\begin{bmatrix} e_1 \\ e_2 \\ i_2 \end{bmatrix} = \begin{bmatrix} 3 & -2 & 8 \\ 10/3 & -7 & 79/3 \\ 2/3 & -2 & 23/3 \end{bmatrix} \begin{bmatrix} i_1 \\ e_3 \\ i_3 \end{bmatrix}$$

It was pointed out earlier in this section that the parameter n_{11} $(= 3)$ could not be obtained directly by opening and shorting ports because it was impossible to make e_3 and i_3 zero simultaneously by this means. Let us examine this situation for the network of Fig. 3.9. If we short port 3, e_3 is zero but i_3 is not. To make i_3 zero also, the voltage across the right-hand 1-ohm resistor must be zero. Choosing $i_2 = 2/3 i_1$ accomplishes this and results in $e_2 = 5i_2 = 10/3 i_1$. These two relations are the last two equations in the mixed-variable matrix equation if we set i_3 and e_3 zero. Thus to measure n_{11}, we short port 3, apply a current i_1 at port 1, and apply a current i_2, adjusted to be $2/3 i_1$, at port 2. Then $i_3 = 0$, and we measure e_1 and obtain n_{11}. Note that port 2 is neither opened nor shorted in this procedure.

Suppose now that the lower 2-ohm resistor in Fig. 3.9 is opened. Is it still possible to write the set of equations (3.6.5) for the remaining network?

The z-parameter equations become

$$\begin{bmatrix} e_1 \\ e_1 \\ e_2 \end{bmatrix} = \begin{bmatrix} 5 & 3 & 1 \\ 3 & 5 & 0 \\ 1 & 0 & 4 \end{bmatrix} \begin{bmatrix} i_1 \\ i_2 \\ i_3 \end{bmatrix}$$

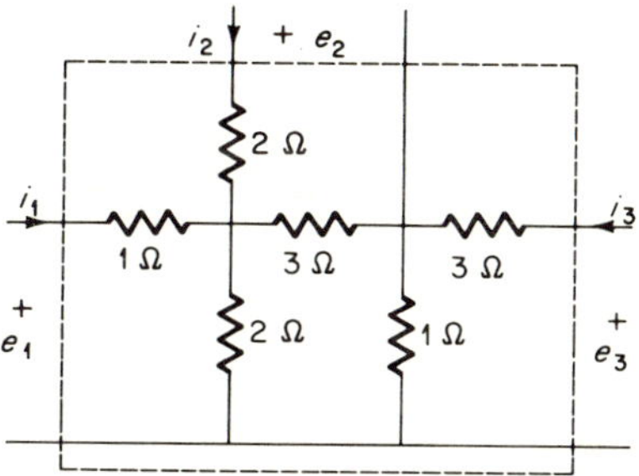

Fig. 3.9

Now $z_{32} = 0$, and it is impossible to interchange e_3 and i_2 and formulate Eq. (3.6.5). Physically this means that there is no coupling between ports 2 and 3; that is, the voltage at port 3 does not depend on the current at port 2, and likewise the voltage at port 2 does not depend on the current at port 3.

3.7 COUPLED ELEMENTS AND CIRCUITS

Certain coupled elements occur frequently in network analysis and synthesis problems. Among these are the transformer, the gyrator, the vacuum triode, the transistor, and, more recently, the negative-impedance converter (NIC). Each of these may be represented as a 2-port that has nonzero off-diagonal, as well as main-diagonal, entries in its matrix. We now investigate the properties of these special elements, showing that each involves coupling of one kind or another and developing their matrices so that we can include them, along with the customary R, L, and C elements, in our discussion of linear graph theory in the next chapter.

Inductive coupling is undoubtedly the most familiar. To review the formulation of equations of coupled circuits, consider the network of Fig. 3.10. Assume that the coils are coupled in such a way that if a current enters the marked terminal of one coil, a voltage is induced in the second coil that is positive at its marked terminal. With this convention, the loop equations for the network are

$$\begin{aligned} e_1 &= sL_1i_1 + sL_2(i_1 + i_2) + sM_1(i_1 + i_2) + sM_1i_1 - sM_2i_2 \\ 0 &= sL_2(i_1 + i_2) + sL_3i_2 + z_Li_2 + sM_1i_1 - sM_2(i_1 + i_2) - sM_2i_2 \end{aligned} \tag{3.7.1}$$

which simplify to

$$\begin{aligned} e_1 &= s(L_1 + L_2 + 2M_1)i_1 + s(L_2 + M_1 - M_2)i_2 \\ 0 &= s(L_2 + M_1 - M_2)i_1 + s(L_2 + L_3 - 2M_2)i_2 + z_Li_2 \end{aligned} \tag{3.7.2}$$

Noting that $e_2 = -z_Li_2$, we can consider the network as 2-port terminated

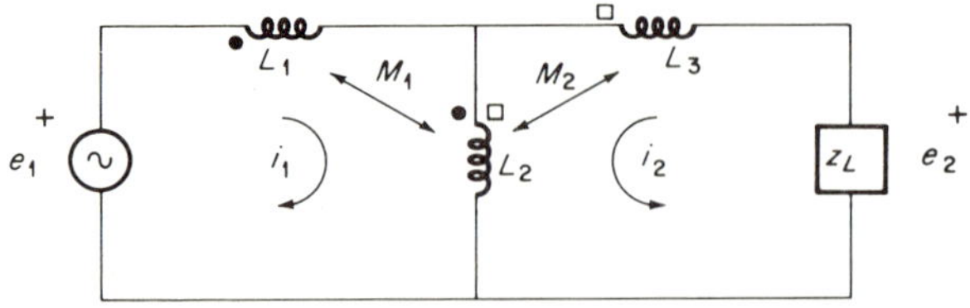

Fig. 3.10 An inductively coupled circuit.

in z_L and rewrite Eqs. (3.7.2) in the matrix form

$$\begin{bmatrix} e_1 \\ e_2 \end{bmatrix} = \begin{bmatrix} s(L_1 + L_2 + 2M_1) & s(L_2 + M_1 - M_2) \\ s(L_2 + M_1 - M_2) & s(L_2 + L_3 - 2M_2) \end{bmatrix} \begin{bmatrix} i_1 \\ i_2 \end{bmatrix} \tag{3.7.3}$$

3.8 PERFECT AND IDEAL TRANSFORMERS

A transformer and its corresponding 2-port representation appear in Fig. 3.11. The dots in Fig. 3.11*b* have the same meaning as in Sec. 3.7. Thus currents entering dotted terminals induce voltages which aid each other in a given winding. In terms of Fig. 3.11*a*, this means that currents entering dotted terminals produce fluxes which add in the core.

Five constraints will now be imposed on the transformer of Fig. 3.11 in order to derive the 2-port equations of a special device known as a *perfect transformer*. Then a sixth constraint will be imposed to convert the perfect transformer into another special device known as an *ideal transformer*.

Constraint 1 The transformer is a linear non-time-varying device. This means that the transformer's inductance and resistance parameters are constants.

Constraint 2 The self-inductances are related by

$$\sqrt{\frac{L_{11}}{L_{22}}} = \frac{N_1}{N_2} = a \tag{3.8.1}$$

where N_1 and N_2 are the numbers of turns on the primary and secondary, respectively, and a is the turns ratio. The self- and mutual inductances of a transformer can be expressed as

$$L_{ii} = \frac{N_i^2}{\mathscr{R}_{ii}} \qquad L_{ij} = \frac{N_i N_j}{\mathscr{R}_{ij}} \tag{3.8.2}$$

where $\mathscr{R}_{ii}$ and $\mathscr{R}_{ij}$ are the self- and mutual reluctances, respectively. From Eq. (3.8.2), the constraint in Eq. (3.8.1) simply means that the self-reluctances $\mathscr{R}_{11}$ and $\mathscr{R}_{22}$ must be equal.

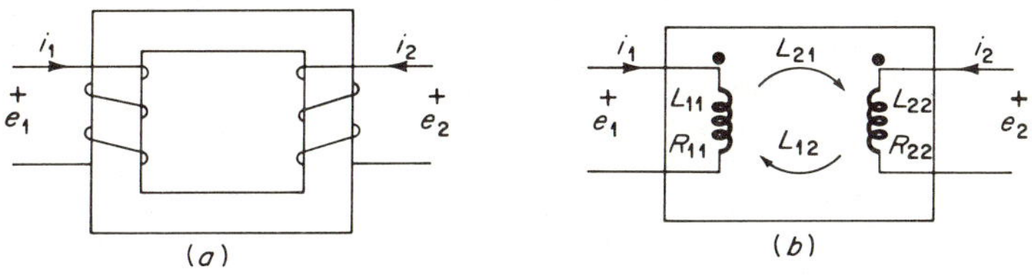

Fig. 3.11 (*a*) *Transformer and* (*b*) *2-port representation.*

Constraint 3 The mutual inductances are related by

$$L_{12} = L_{21} \equiv M \tag{3.8.3}$$

This constraint, from the second of Eqs. (3.8.2), requires the mutual reluctances $\mathscr{R}_{12}$ and $\mathscr{R}_{21}$ to be equal and means that the transformer is reciprocal.

Constraint 4 The winding resistances are negligible.

With constraints 1 to 4 applied, the equations for the transformer in the s domain take the simplified form

$$e_1 = sL_{11}i_1 + sMi_2 \qquad e_2 = sMi_1 + sL_{22}i_2 \tag{3.8.4}$$

The voltage ratio, current ratio, and input impedance of a transformer satisfying constraints 1 to 4 and terminated in a load impedance z_L are of interest. Noting that $e_2 = -i_2z_L$, these ratios can be obtained from Eqs. (3.8.4) and are

$$\frac{e_1}{e_2} = \frac{s(L_{11}L_{22} - M^2) + L_{11}z_L}{Mz_L} \tag{3.8.5}$$

$$\frac{i_2}{i_1} = -\frac{sM}{sL_{22} + z_L} \tag{3.8.6}$$

$$Z_{\text{in}} = \frac{e_1}{i_1} = \frac{s^2(L_{11}L_{22} - M^2) + sL_{11}z_L}{sL_{22} + z_L} \tag{3.8.7}$$

Constraint 5 The coefficient of coupling, defined by

$$k = \frac{M}{\sqrt{L_{11}L_{22}}} \tag{3.8.8}$$

is unity.

With this additional constraint, Eqs. (3.8.5) to (3.8.7) become

$$\frac{e_1}{e_2} = a \tag{3.8.9}$$

$$\frac{i_2}{i_1} = -\frac{asL_{22}}{sL_{22} + z_L} \tag{3.8.10}$$

$$Z_{\text{in}} = \frac{a^2sL_{22}z_L}{sL_{22} + z_L} \tag{3.8.11}$$

where we have made use of

$$\frac{L_{11}}{M} = \frac{M}{L_{22}} = \sqrt{\frac{L_{11}}{L_{22}}} = a \tag{3.8.12}$$

which is obtained from Eqs. (3.8.1) and (3.8.8), with the latter equated to unity.

Also the transformer equations (3.8.4) can be written in concise chain-parameter form as

$$\begin{bmatrix} e_1 \\ i_1 \end{bmatrix} = \begin{bmatrix} a & 0 \\ \dfrac{1}{saL_{22}} & -\dfrac{1}{a} \end{bmatrix} \begin{bmatrix} e_2 \\ i_2 \end{bmatrix} \tag{3.8.13}$$

The requirement of unity coupling has forced the voltage ratio of the transformer to be a constant equal to the turns ratio. However, the current ratio and the input impedance still depend on frequency. The form of Z_{in} in Eq. (3.8.11) lends itself nicely to physical interpretation. The output circuit of the transformer contains two impedances, z_L and sL_{22}. The impedance seen at the input terminals of the transformer is simply the parallel combination of these multiplied by the square of the turns ratio.

The volt-ampere input of the perfect transformer loaded with z_L is given by

$$e_1 i_1 = ae_2 \frac{(sL_{22} + z_L)i_2}{asL_{22}} = -e_2 i_2 \left(1 + \frac{z_L}{sL_{22}}\right) \tag{3.8.14}$$

The voltages e_1 and e_2 are in phase, but for $s = j\omega$, i_1 and i_2 have a phase relation dependent on z_L. Thus Eq. (3.8.14) is not a power equation. However, for any z_L, the power out must equal the power in since the perfect transformer is lossless.

To summarize, a transformer that satisfies constraints 1 to 5 is called a perfect transformer. Thus a perfect transformer is one that fulfills the following conditions:

$$\begin{aligned} &L_{11}, L_{22}, L_{12}, L_{21} = \text{const} \\ &L_{11} = a^2 L_{22} \\ &L_{12} = L_{21} = M \\ &R_{11} = R_{22} = 0 \\ &k = 1.0 \end{aligned} \tag{3.8.15}$$

Let us now proceed to the ideal transformer, which is a perfect transformer with one additional constraint imposed on the self-inductances.

Constraint 6 L_{11} and L_{22} approach infinity, but their ratio remains finite and constant.

The ratio involved is a^2. Thus L_{11} and L_{22} are required to become very large in such a way that their ratio continues to be a^2.

Let us examine the effect of this additional constraint on the current ratio,

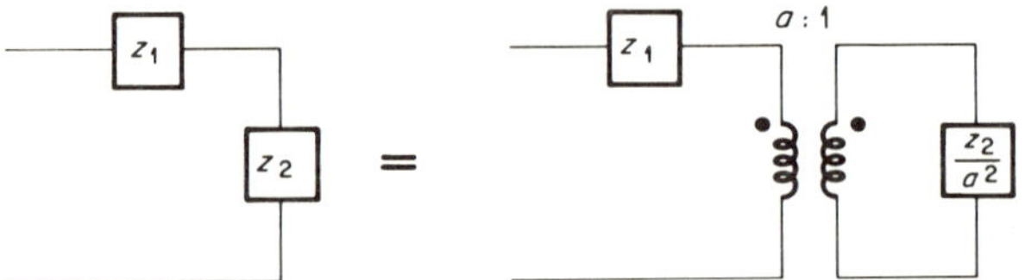

Fig. 3.12 Level changing and isolation of an impedance.

input impedance, and volt-ampere relations developed for the perfect transformer.

In Eq. (3.8.10), if L_{22} approaches infinity, the current ratio becomes

$$\frac{i_2}{i_1} = -a \tag{3.8.16}$$

Thus the chain-parameter relations for the ideal transformer are simply

$$\begin{bmatrix} e_1 \\ i_1 \end{bmatrix} = \begin{bmatrix} a & 0 \\ 0 & -\frac{1}{a} \end{bmatrix} \begin{bmatrix} e_2 \\ i_2 \end{bmatrix} \tag{3.8.17}$$

The input impedance of the ideal transformer terminated in a load impedance z_L is obtained directly from Eq. (3.8.11) by letting L_{22} approach infinity and is

$$Z_{\text{in}} = a^2 z_L \tag{3.8.18}$$

The impedance-transforming and isolation properties of the ideal transformer are used a great deal in network theory. The ideal transformer acts as an impedance-level changer but does not otherwise affect the circuit in which it is used. This property is illustrated in Fig. 3.12, where two equivalent representations of the same network are shown. To make $Z_{\text{in}} \approx a^2 z_L$, it is not necessary, practically, for L_{11} and L_{22} to become infinite. All that is necessary is for $sL_{22} \gg z_L$ over the range of s of interest, as Eq. (3.8.11) shows.

The volt-ampere relation in Eq. (3.8.14) reduces to

$$e_1 i_1 = -e_2 i_2 \tag{3.8.19}$$

for the ideal transformer. Since e_1 and e_2 are in phase and i_1 and i_2 are 180° out of phase, Eq. (3.8.19) is also a power equation and demonstrates that the ideal transformation is lossless.

Example 3.4 In the network of Fig. 3.13, let $L_{11} = 4$ henrys, $L_{22} = 1$ henry, and $z_L = 3000$ ohms (resistive). If $\omega = 4000$ rad/sec, find Z_{in} for $M = 1$, 1.9, and 2 henrys. How large must L_{22} become for the transformer to act as an ideal transformer?

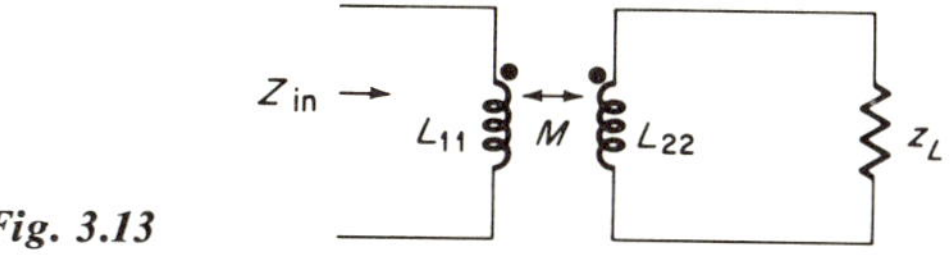

Fig. 3.13

For $M = 1$ henry, $k = 0.5$; for $M = 1.9$ henrys, $k = 0.95$. In either case the transformer is not perfect. For $M = 2$ henrys, $k = 1.0$ and the transformer is perfect. If the transformer were ideal, Z_{in} from Eq. (3.8.18) would be 12,000 ohms resistive. Using Eq. (3.8.7) with s replaced by $j\omega$, Z_{in} for $M = 1$ henry is

$$Z_{in} = j(4000)(4) + \frac{(4000^2)(1)}{j(4000)(1) + 3000} = 1920 + j13{,}440 \text{ ohms}$$

The transformer is seen to be far from ideal. For $M = 1.9$ henrys,

$$Z_{in} = j(4000)(4) + \frac{(4000^2)(3.61)}{j(4000)(1) + 3000} = 6940 + j6750 \text{ ohms}$$

The reactive portion has been reduced, and the resistive portion is closer to the ideal value of 12,000 ohms, but the discrepancy is still large. Consider next the case where $M = 2$ henrys. Either Eq. (3.8.7) or Eq. (3.8.11) may be used, the latter giving

$$Z_{in} = \frac{j(4)(4000)(1)(3000)}{j(4000)(1) + 3000} = 7700 + j5760 \text{ ohms}$$

This is some improvement but not really significant. *The problem is that sL_{22} is not enough larger than z_L.* Suppose we try to approach an ideal transformer by letting L_{11} and L_{22} increase by a factor of 10 (their ratio is still 4:1). This change also forces M to increase by a factor of 10. If we consider the $k = 1.0$ case, the new parameter values are $L_{11} = 40$ henrys, $L_{22} = 10$ henrys, $M = 20$ henrys. Then, again using Eq. (3.8.11),

$$Z_{in} = \frac{j(4)(4000)(10)(3000)}{j(4000)(10) + 3000} = 11{,}900 + j896 \text{ ohms}$$

This result is much better. The resistive portion of Z_{in} is almost the ideal-transformer value of 12,000 ohms. The reactive portion of Z_{in} is perhaps small enough to be neglected.

Example 3.5 Consider again the 2-port network in Fig. 3.4. Convert this network into a 1-port by using an ideal transformer, as shown in Fig. 3.14. Compute the driving-point impedances Z_1 and Z_{in} in terms of the 2-port z parameters.

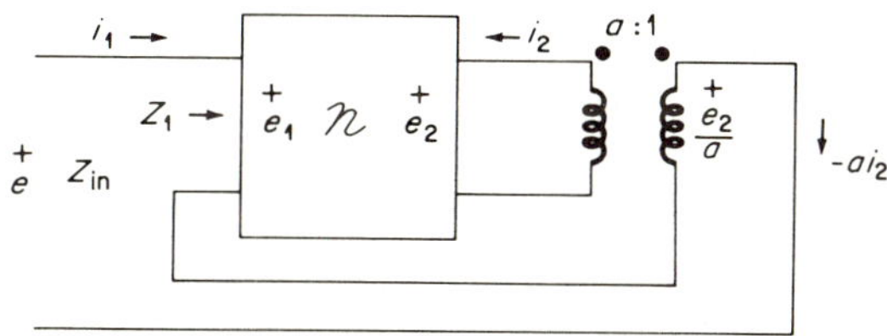

Fig. 3.14 Converting a 2-port into a 1-port.

The z-parameter equations of a 2-port are given in Eqs. (3.4.1). Since Eq. (3.8.16) holds, the voltages e_1 and e_2 can be expressed in terms of i_1 only by

$$e_1 = z_{11}i_1 - \frac{1}{a}z_{12}i_1 \qquad e_2 = z_{21}i_1 - \frac{1}{a}z_{22}i_1$$

The desired driving-point impedances are

$$Z_1 = \frac{e_1}{i_1} \qquad \text{and} \qquad Z_{\text{in}} = \frac{e}{i_1} = \frac{e_1 - e_2/a}{i_1}$$

Substituting the z-parameter equations into these relations gives

$$Z_1 = z_{11} - \frac{1}{a}z_{12}$$

$$Z_{\text{in}} = z_{11} - \frac{1}{a}(z_{12} + z_{21}) + \frac{1}{a^2}z_{22}$$

If the 2-port were as shown in Fig. 3.4, Z_1 would equal z_{11}. Why does Z_1 change when the ideal transformer is added? The answer is that there is a finite load impedance at the 2-port output terminals in Fig. 3.14, since i_2 is not zero. This load impedance is given by

$$z_L = -\frac{e_2}{i_2} = az_{21} - z_{22}$$

3.9 NEGATIVE IMPEDANCE CONVERTER

An ideal negative-impedance converter (NIC) is shown diagrammatically in Fig. 3.15. It is a 2-port having the property that the driving-point impedance at one of the ports is the negative of the load impedance connected to the other port. Such a device is of considerable interest in network theory because of its ability to generate negative R, L, or C simply by loading it with positive R, L, or C. For example, losses in the positive resistances of a network can be canceled by including resistively loaded NICs in the network.

Let us first concern ourselves with the external behavior of the NIC, i.e., the current-voltage relationships at its ports. For reasons that will be discussed momentarily, it is convenient to use h (rather than z or y) parameters in describing an NIC. For the NIC in Fig. 3.15 we can write

$$e_1 = h_{11}i_1 + h_{12}e_2 \qquad i_2 = h_{21}i_1 + h_{22}e_2 \tag{3.9.1}$$

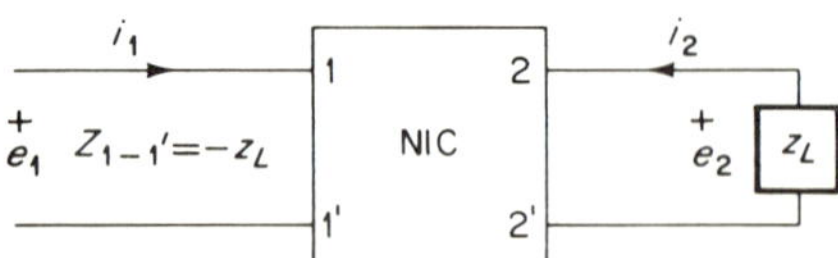

Fig. 3.15 An ideal NIC.

Solving for the driving-point impedance at terminals 1-1′ yields

$$Z_{1\text{-}1'} \equiv \frac{e_1}{i_1} = h_{11} - \frac{h_{12}h_{21}}{1/z_L + h_{22}} \tag{3.9.2}$$

For the device to be an ideal NIC, it is required that

$$Z_{1\text{-}1'} = -z_L \tag{3.9.3}$$

Combining Eqs. (3.9.2) and (3.9.3), we obtain the following necessary and sufficient conditions:

$$h_{11} = h_{22} = 0 \qquad h_{12}h_{21} = 1 \tag{3.9.4}$$

If the constraints in Eqs. (3.9.4) are inserted into Eqs. (3.9.1), the latter reduce to

$$e_1 = \frac{1}{h_{21}}\,e_2 \qquad i_2 = h_{21}i_1 \tag{3.9.5}$$

which can be written in chain-parameter matrix form as

$$\begin{bmatrix} e_1 \\ i_1 \end{bmatrix} = \begin{bmatrix} \dfrac{1}{h_{21}} & 0 \\ 0 & \dfrac{1}{h_{21}} \end{bmatrix} \begin{bmatrix} e_2 \\ i_2 \end{bmatrix} \tag{3.9.6}$$

The chain matrix in Eq. (3.9.6) should be compared with that of the ideal transformer in Eq. (3.8.17).

If we had tried to derive the constraints in Eq. (3.9.4) in terms of z parameters, we would have used Eq. (3.5.5) and obtained

$$z_{1\text{-}1'} = \frac{z_{11}z_L + |z|}{z_{22} + z_L} = -z_L \tag{3.9.7}$$

Careful study shows the constraints to be

$$\begin{aligned} z_{22} &= \pm\infty \\ \frac{z_{11}}{z_{22}} &= -1 \\ \frac{z_{12}z_{21}}{z_{22}^{\,2}} &= -1 \end{aligned} \tag{3.9.8}$$

The constraints in Eqs. (3.9.8) are not as easy to apply as those in Eqs. (3.9.4), which explains our preference for h parameters.

Next connect an impedance z_s across terminals 1 and 1′ in Fig. 3.15 and determine the driving-point impedance at terminals 2-2′ from Eqs. (3.9.1)

$$Z_{2\text{-}2'} = \frac{e_2}{i_2} - \frac{1}{h_{22} - h_{12}h_{21}/(h_{11} + z_s)} \tag{3.9.9}$$

Substituting the constraints of Eqs. (3.9.4) into Eq. (3.9.9) yields

$$Z_{2\text{-}2'} = -z_s \tag{3.9.10}$$

Thus the device in Fig. 3.15 performs as an ideal NIC in both directions.

It is informative to consider the volt-ampere and power relations in an ideal NIC. From Eqs. (3.9.5) we obtain

$$e_1 i_1 = \frac{1}{h_{21}^2} e_2 i_2 \tag{3.9.11}$$

as the volt-ampere equation. e_1 and e_2 are either in phase or 180° out of phase, from the first of Eqs. (3.9.5). This is also true of i_1 and i_2 from the second of Eqs. (3.9.5). Furthermore if e_1 and e_2 are in phase, so too are i_1 and i_2, and likewise for 180° out of phase. Thus we can convert Eq. (3.9.11) into a power equation by multiplying each side by the load power factor (denote it by $\cos\theta$)

$$P_{\text{in}} = e_1 i_1 \cos\theta = \frac{1}{h_{21}^2} e_2 i_2 \cos\theta \tag{3.9.12}$$

Equation (3.9.12) gives the power *into* the NIC. The power to the load is given by

$$P_L = -e_2 i_2 \cos\theta \tag{3.9.13}$$

We see that the power into the NIC and the power into the load differ in magnitude unless $h_{21} = 1$. We also note that the power into the NIC is negative with respect to the load power. This means that power is flowing simultaneously out of (or into) both ports of the NIC, and thus the NIC is an active 2-port.

Thus far we have considered only ideal NICs. In the nonideal case, the parameters h_{11} and h_{22} would not be identically zero. Consider first the situation in Fig. 3.16, where the solid box represents an ideal NIC. The dashed box is a nonideal NIC whose $h_{11} = z_1$. The purpose of z_2 will become apparent if we compute the driving-point impedance at terminals 1-1′. From Eq. (3.9.2),

$$Z_{1\text{-}1'} = z_1 - \frac{1}{1/(z_L + z_2) + 0} = z_1 - z_2 - z_L \tag{3.9.14}$$

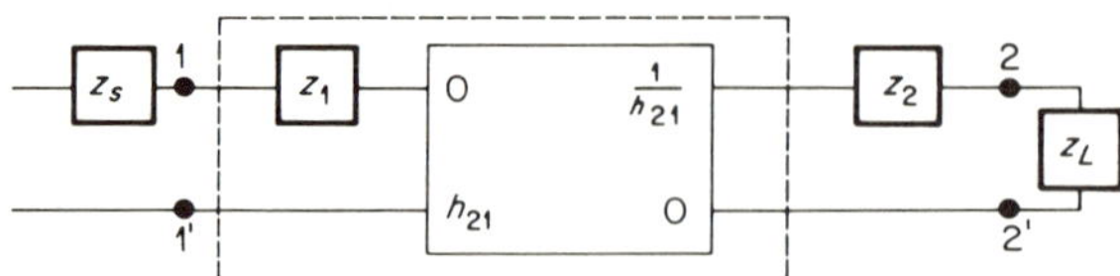

Fig. 3.16 A nonideal NIC; $h_{11} \neq 0$.

If we choose $z_2 = z_1$, the driving-point impedance in Eq. (3.9.14) becomes identical to that in Eq. (3.9.3), and we again have ideal NIC action at terminals 1-1′. To check that the action is also ideal at terminals 2-2′, Eq. (3.9.9) is used

$$Z_{2\text{-}2'} = \frac{1}{0 - 1/(z_1 + z_s)} + z_2 = -z_1 + z_2 - z_s \tag{3.9.15}$$

Once again ideal NIC action results if $z_1 = z_2$.

It is left as a problem for the reader to show that a nonideal NIC with $h_{22} \neq 0$ can easily be converted to an ideal NIC in a similar way.

This concludes our discussion of the NIC. Readers interested in the internal circuitry of these devices are urged to read the appropriate references listed at the end of the chapter.

3.10 THE GYRATOR

A gyrator† is a nonreciprocal 2-port network element described by the equations

$$\begin{aligned} e_1 &= r_{11}i_1 + ki_2 \\ e_2 &= -ki_1 + r_{22}i_2 \end{aligned} \tag{3.10.1}$$

where r_{11} and r_{22} are its self-resistances and k is its gyration constant. The conventional circuit symbol for a gyrator appears in Fig. 3.17. If the gyrator is ideal, the self-resistances are negligible and Eqs. (3.10.1) can be written simply in chain- and z-parameter form as

$$\begin{bmatrix} e_1 \\ i_1 \end{bmatrix} = \begin{bmatrix} 0 & k \\ -\dfrac{1}{k} & 0 \end{bmatrix} \begin{bmatrix} e_2 \\ i_2 \end{bmatrix} \qquad \begin{bmatrix} e_1 \\ e_2 \end{bmatrix} = \begin{bmatrix} 0 & k \\ -k & 0 \end{bmatrix} \begin{bmatrix} i_1 \\ i_2 \end{bmatrix} \tag{3.10.2}$$

The first of Eqs. (3.10.2) should be compared with Eq. (3.8.17) for the ideal transformer and with Eq. (3.9.6) for the ideal NIC.

The gyrator takes its name from its gyroscopic counterpart in mechanics. Consider a set of forces acting on a coupled two-mass system. We may

† This device was introduced by B. H. H. Tellegen in The Gyrator: A New Electric Network Element, *Philips Res. Rept.*, Apr., 1948.

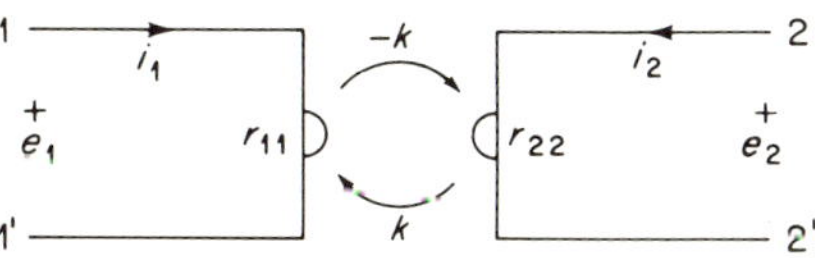

Fig. 3.17 Gyrator z-parameter notation.

write

$$F_1 = k_{11}v_1 + k_{12}v_2$$
$$F_2 = k_{21}v_1 + k_{22}v_2 \tag{3.10.3}$$

where F denotes force, v denotes velocity, and the k's are constants. If $k_{ij} = k_{ji}$, we have a reciprocal dissipative system. If $k_{ij} = -k_{ji}$, then k_{11} and k_{22} are zero and we have a nonreciprocal, nondissipative system. Forces in this latter type of system are gyroscopic forces, i.e., forces which are perpendicular to the velocities and which do no work. With $k_{ij} = -k_{ji}$, Eqs. (3.10.3) match the second set of Eqs. (3.10.2).

To develop some of the properties of the gyrator, connect an impedance z_L across terminals 2 and 2′ so that $e_2 = -i_2 z_L$. If the gyrator is nonideal, the driving-point impedance at terminals 1-1′ from Eqs. (3.10.1) is

$$Z_{1\text{-}1'} = r_{11} + \frac{k^2}{r_{22}z_L} \tag{3.10.4}$$

which, for the ideal gyrator, becomes

$$Z_{1\text{-}1'} = \frac{k^2}{z_L} \tag{3.10.5}$$

Thus the ideal gyrator, unlike the ideal transformer, is an impedance inverter as well as an impedance-level changer. Like the ideal transformer, the ideal gyrator does not otherwise affect the circuit into which it is inserted. These properties are illustrated in Fig. 3.18, where two circuits equivalent with respect to their input impedances are shown. The ideal-gyrator volt-ampere relation can be obtained from Eqs. (3.10.2) as

$$e_1 i_1 = k i_2 \frac{-e_2}{k} = -e_2 i_2 \tag{3.10.6}$$

Equation (3.10.6) illustrates the important fact that the ideal gyrator is lossless.

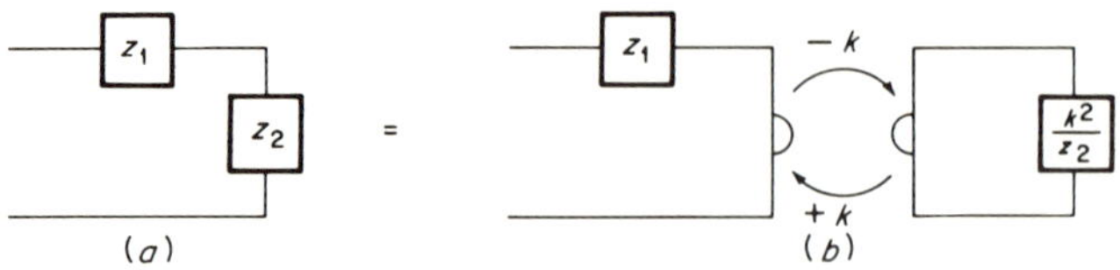

Fig. 3.18 The ideal gyrator as an impedance inverter.

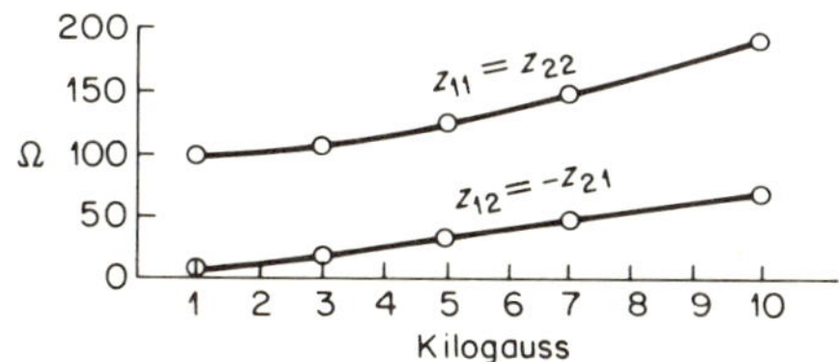

Fig. 3.19 z parameters of an experimental Hall effect generator.

An example of a nonideal gyrator is the Hall effect generator. The z parameters of an experimental model of such a device as a function of applied magnetic field are shown in Fig. 3.19. Note that the best ratio of transfer to self-resistance, obtained at 10 kilogauss, is only 70:195, far from ideal.

Ideal gyrators can be obtained by using circuits containing NICs.† An example of such a circuit using two ideal NICs and three resistors is shown in Fig. 3.20*a*. To show that this circuit produces ideal-gyrator action, note first that the right NIC can be replaced by a resistor of $-R$. Next the remaining two resistors can be combined with the left NIC to yield a nonideal NIC. These changes are shown in Fig. 3.20*b*.

For the network in Fig. 3.20*b* the following h-parameter equations can be written:

$$e_1 = Ri_1 + \frac{1}{h_{21}}(i_2R + e_2)$$

$$i_2 = h_{21}i_1 + \frac{1}{R}(i_2R + e_2) \tag{3.10.7}$$

† See T. J. Harrison, A Gyrator Realization, *IEEE Trans. Circuit Theory*, vol. C.T. 10, no. 2, June, 1963.

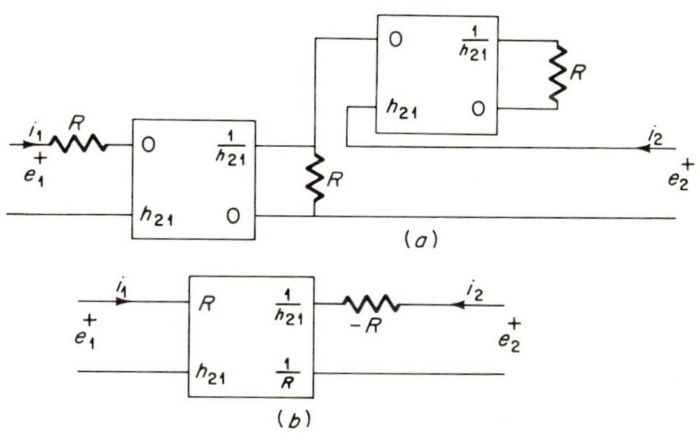

Fig. 3.20 An ideal gyrator built from NICs.

Equations (3.10.6) reduce to

$$\begin{bmatrix} e_1 \\ i_1 \end{bmatrix} = \begin{bmatrix} 0 & \dfrac{R}{h_{21}} \\ -\dfrac{1}{Rh_{21}} & 0 \end{bmatrix} \begin{bmatrix} e_2 \\ i_2 \end{bmatrix} \tag{3.10.8}$$

Comparison with Eq. (3.10.2) shows the requirement for ideal-gyrator action to be $h_{21} = 1$.

Example 3.6 Compute the driving-point impedance of the network in Fig. 3.21.

Loop equations can be written and solved for the ratio $Z = e_1/i_1$, or the z parameters of the network can be found and Eq. (3.4.5) used. Pursuing the latter course, we find

$$z_{11} = z_{22} = \frac{1}{sC} + R$$

$$z_{12} = k + \frac{1}{sC} \qquad z_{21} = -k + \frac{1}{sC}$$

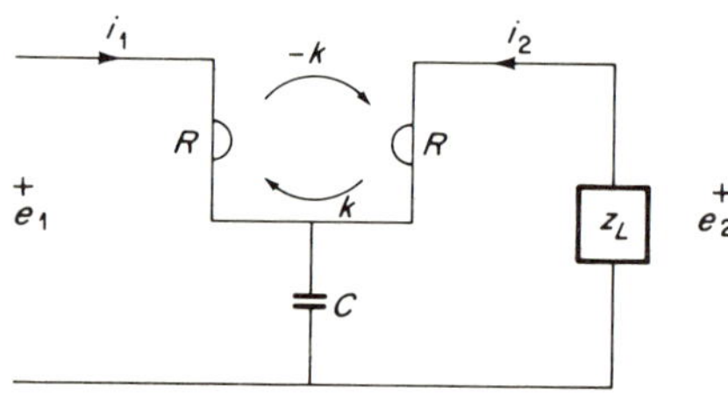

Fig. 3.21

and

$$Z = \frac{(R + 1/sC)z_L + R^2 + k^2 + 2R/sC}{R + 1/sC + z_L}$$

For an ideal gyrator, $R = 0$ and

$$Z = \frac{(1/sC)z_L + k^2}{1/sC + z_L}$$

3.11 INTERCONNECTED 2-PORTS

In Sec. 3.4 the parameter matrices of a single 2-port were developed. In this section we examine how the parameter matrices of two or more 2-ports can be combined to yield an overall 2-port parameter matrix. We introduced this problem in Chap. 1. Reconsider Example 1.2, in which a pair of 2-ports are connected in cascade. We found that chain parameters were the most convenient set to use in representing the overall network. The overall chain matrix was obtained by multiplying the individual chain matrices together in the order of cascading. Note that this cascading procedure is valid regardless of the network structures within the boxes of Fig. 1.1 and may readily be extended to the cascading of more than two 2-ports.

A second interconnection of 2-ports was discussed in Example 1.3, where the inputs of the 2-ports were connected in series, as were the outputs. This is known as a *series-series connection*. We found that z parameters were the preferred set and that each overall z parameter was obtained as the sum of the respective individual z parameters.

In all interconnections of 2-ports, except the cascade one, we must make sure that the parameters of the individual 2-ports do not change when the interconnection is made. Consider the proposed series-series connection in Fig. 3.22*a*. If the dotted connections are made, the lower series arm of the lattice is shorted. This alters the z parameters of the upper network and invalidates our plan to sum the individual z parameters of the two 2-ports.

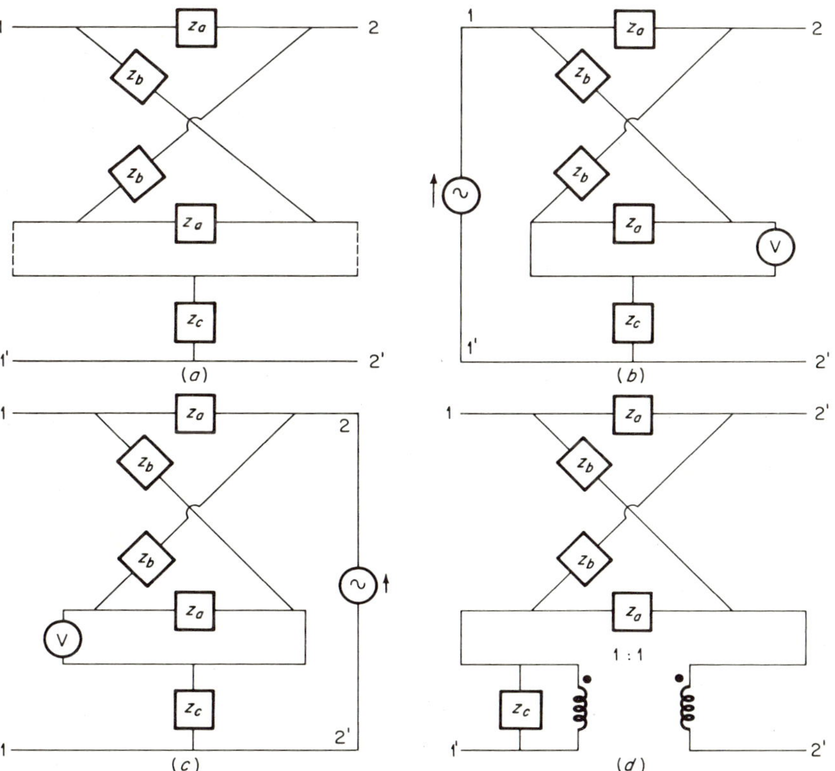

Fig. 3.22 A series-series connection.

We need a simple external test to determine whether the parameters of either 2-port will change when they are interconnected, since we often cannot peek inside to see whether an element becomes shorted in the process. The test must establish whether each terminal pair still satisfies the definition of a port after the interconnection is made, i.e., whether the current entering one terminal of a port is still the same as the current leaving the other terminal of the same port. In Fig. 3.22*b* we have energized the input port of the combination with a current source and inserted a voltmeter between the interconnection points of the output port. If the voltmeter reads other than zero volts, we know that a current will flow when the output-port interconnection is made, even though the current at terminals 2-2′ is zero. We see in Fig. 3.22*b* that $V \neq 0$ because of the voltage across the lower series arm of lattice.

If V were zero in Fig. 3.22*b*, we would have to make a second test with the output energized and the voltmeter at the input interconnection, as shown in Fig. 3.22*c*. For the given networks, this test also shows $V \neq 0$.

In Fig. 3.22*d* we show how a pair of networks not satisfying the interconnection tests can be altered so that they do. Note that a 1:1 ideal transformer has been inserted in the lower 2-port. The lower series arm of the lattice is no longer shorted when the interconnection is made, and thus the individual z parameters can be summed to give the overall z parameters.

In Fig. 3.23 we consider a proposed *parallel-parallel* connection of two 2-ports, of interest when we wish to sum y parameters. In Fig. 3.23*a* we note that if the dotted connections are made, the lower series arm of the lattice is shorted, altering the y parameters of the upper network.

As in the series-series case, simple tests are needed to determine whether the y parameters of either 2-port will change when a parallel-parallel interconnection is made. In Fig. 3.23*b*, the input is energized by a voltage source, and each output port is shorted. A voltmeter is inserted between the two shorted ports. If the voltmeter reads other than zero, we know that the currents at points *a* and *b* will differ when the output ports are paralleled, and likewise for the currents at points *c* and *d*. For the given networks in Fig. 3.23*b*, there is a net voltage because of the lower series arm of the lattice, and therefore $V \neq 0$.

If V were zero in Fig. 3.23*b*, a second test, indicated in Fig. 3.23*c*, would have to be made. Once again, for the given networks, the voltmeter does not read zero.

The use of a 1:1 ideal transformer to permit the parallel-parallel interconnection is shown in Fig. 3.23*d*.

Two other types of interconnections are possible, the series-parallel arrangement, applicable when h parameters are to be added, and the parallel-series hookup for summing g parameters. In testing these interconnections, one each of the tests of Figs. 3.22 and 3.23 would be required.

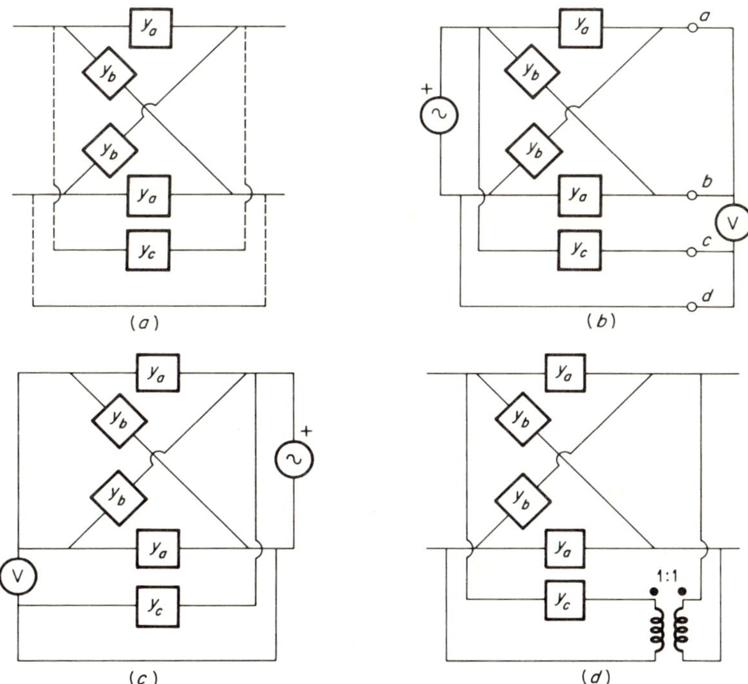

Fig. 3.23 A parallel-parallel connection.

Example 3.7 In Fig. 3.22*b*, let $z_a = 4$ ohms, $z_b = 6$ ohms, and $z_c = 3$ ohms, all resistances.

(*a*) Find the z parameters of each network and their sums.

(*b*) Find the z parameters of the overall network after the series-series interconnection is made.

(*c*) For a current source I_s in Fig. 3.22*b*, what will the voltmeter read?

(*a*) The z parameters of the lattice are given by

$$z_{11} = z_{22} = \tfrac{1}{2}(z_b + z_a) = 5 \text{ ohms}$$

$$z_{12} = z_{21} = \tfrac{1}{2}(z_b - z_a) = 1 \text{ ohm}$$

The z parameters of the 3-ohm resistor are each 3 ohms. Summing the z parameters gives

$$\begin{bmatrix} 5 & 1 \\ 1 & 5 \end{bmatrix} + \begin{bmatrix} 3 & 3 \\ 3 & 3 \end{bmatrix} = \begin{bmatrix} 8 & 4 \\ 4 & 8 \end{bmatrix}$$

(*b*) When the networks are interconnected, the bottom 4-ohm resistor is shorted out, leaving a π network in place of the lattice. Conventional circuit analysis yields the overall z parameters as

$$\begin{bmatrix} 15/4 & 9/4 \\ 9/4 & 15/4 \end{bmatrix} + \begin{bmatrix} 3 & 3 \\ 3 & 3 \end{bmatrix} = \begin{bmatrix} 27/4 & 21/4 \\ 21/4 & 27/4 \end{bmatrix}$$

The result differs from that in part (*a*) because the interconnection altered the upper network.

(*c*) One-half I_s flows through the bottom 4-ohm resistor, giving a voltage drop of $2I_s$ volts.

3.12 THREE-TERMINAL COUPLED ELEMENTS

In Figs. 3.11 and 3.17, the transformer and gyrator are represented as 2-terminal-pair networks. If we now join the lower input and output leads, each device becomes a 3-terminal network, as shown in Fig. 3.24 (the self-resistances are assumed negligible.

Two types of coupling are indicated in Fig. 3.24. The transformer, of course, has inductive coupling; the off-diagonal terms in its 2-port z-parameter matrix are inductive. The gyrator, on the other hand, may be thought of as having resistive coupling, since the off-diagonal terms in its z-parameter matrix are constants rather than functions of s. The representation in Fig. 3.24*b* has been redrawn in Fig. 3.25 to illustrate this point.

Let us now extend the concept of resistive coupling to the vacuum triode. Representation of the triode as a 3-terminal element appears in Fig. 3.26*a*, and its low-frequency current equivalent circuit is shown in Fig. 3.26*b*. The circuit equations applicable to Fig. 3.26*b* are

$$i_g = 0 \qquad i_p = g_m e_g + g_p e_p \tag{3.12.1}$$

or in matrix form

$$\begin{bmatrix} i_g \\ i_p \end{bmatrix} = \begin{bmatrix} 0 & 0 \\ g_m & g_p \end{bmatrix} \begin{bmatrix} e_g \\ e_p \end{bmatrix} \tag{3.12.2}$$

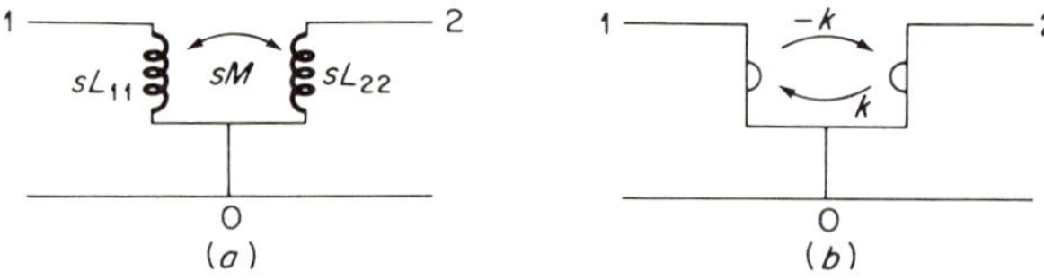

Fig. 3.24 Three-terminal coupled elements.

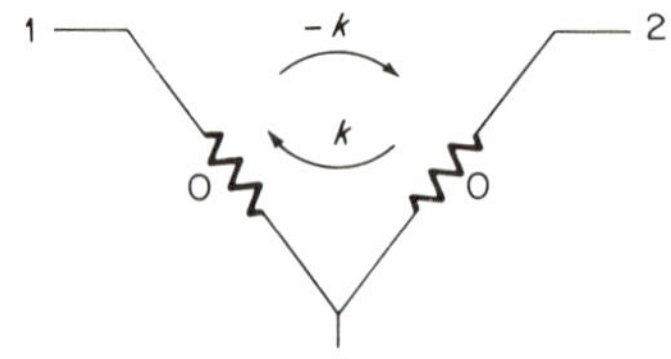

Fig. 3.25 Coupled-resistor representation of a gyrator (z-parameter notation).

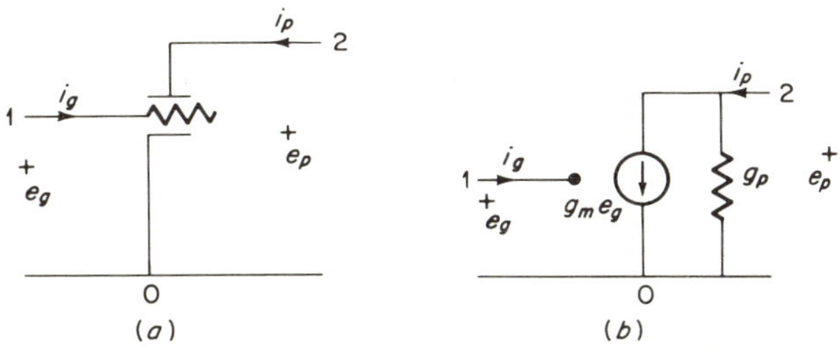

Fig. 3.26 The triode and its equivalent circuit.

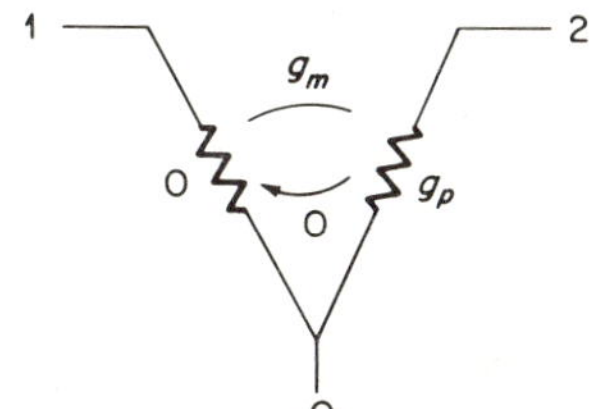

Fig. 3.27 Coupled-resistor representation of a triode (y-parameter notation).

The parameters in Eq. (3.12.2) are y parameters, and the equation suggests that the triode, like the gyrator, can be thought of as a resistively coupled element (conductively coupled in this instance). This representation is shown in Fig. 3.27, which should be compared with Fig. 3.25. Note that the absence of feedback from the output to the input is clearly indicated in Fig. 3.27.

Consider next the transistor. Its representation as a 3-terminal element and its 2-generator low-frequency equivalent circuit appear in Fig. 3.28, where we have chosen to consider the common-emitter case. The equations from which the circuit of Fig. 3.28*b* is derived are

$$\begin{aligned} v_{be} &= h_{11}i_b + h_{12}v_{ce} \\ i_c &= h_{21}i_b + h_{22}v_{ce} \end{aligned} \tag{3.12.3}$$

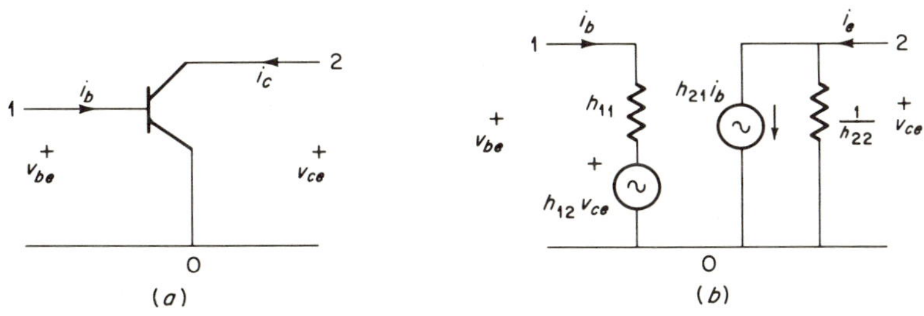

Fig. 3.28 The transistor and its equivalent circuit.

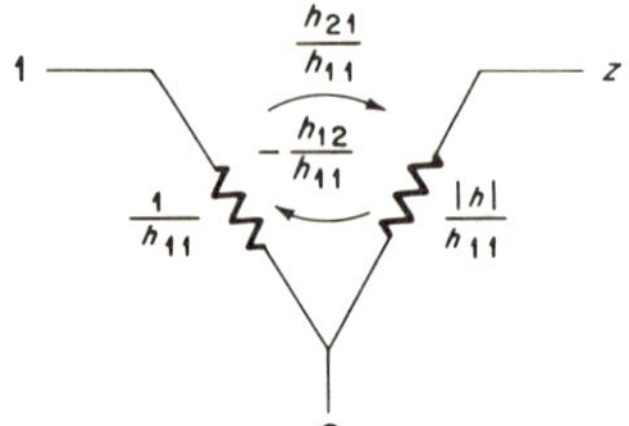

Fig. 3.29 Coupled-resistor representation of a transistor (y-parameter notation).

These are h-parameter equations. If we convert them to y-parameter form and obtain

$$\begin{bmatrix} i_b \\ i_c \end{bmatrix} = \begin{bmatrix} \dfrac{1}{h_{11}} & -\dfrac{h_{12}}{h_{11}} \\ \dfrac{h_{21}}{h_{11}} & \dfrac{|h|}{h_{11}} \end{bmatrix} \begin{bmatrix} v_{be} \\ v_{ce} \end{bmatrix} \tag{3.12.4}$$

we can interpret the matrix equation as being that of a conductively coupled element (see Fig. 3.29). In the transistor there *is* feedback from output to input, as is clearly shown in Fig. 3.29.

3.13 MATRIX ANALYSIS OF NETWORKS

What is the advantage of representing the elements in Sec. 3.12 as conductively coupled 3-terminal networks? The answer is that if we represent them in this way, it becomes quite easy to carry out a systematic matrix analysis of complicated networks in which they are present without requiring the use of dependent generators. Let us develop such an analysis through two examples.

Example 3.8 A transistor amplifier is shown in Fig. 3.30*a*, and its small-signal ac equivalent circuit appears in Fig. 3.30*b*. We want to obtain an overall 3-terminal network

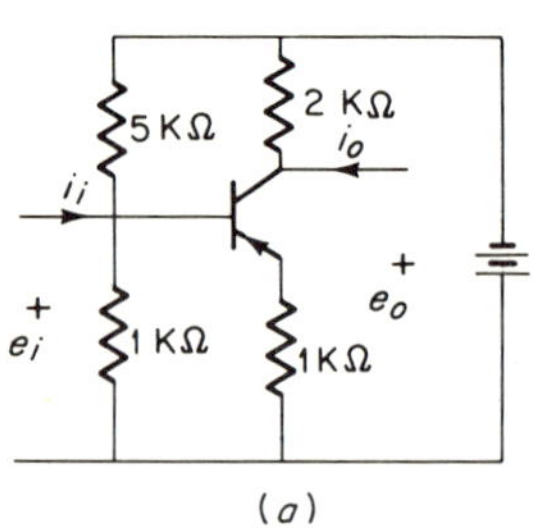

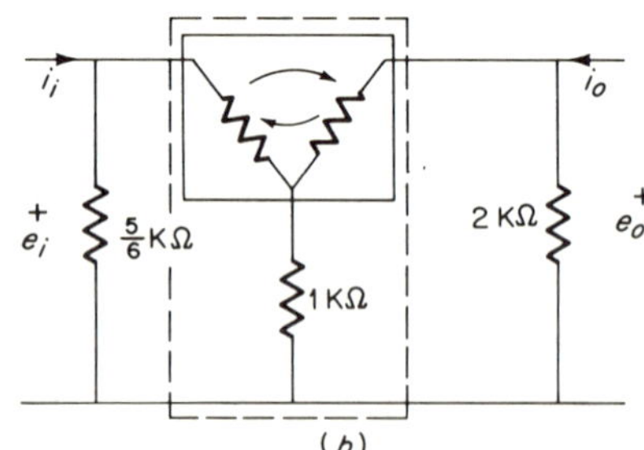

Fig. 3.30 A transistor amplifier and its ac equivalent circuit.

matrix equation of the form

$$\begin{bmatrix} i_i \\ i_o \end{bmatrix} = \begin{bmatrix} Y_{11} & Y_{12} \\ Y_{21} & Y_{22} \end{bmatrix} \begin{bmatrix} e_i \\ e_o \end{bmatrix}$$

The y parameters of the transistor are given as

$$y_{11} = 5 \times 10^{-4} \text{ mho} \qquad y_{12} = -3 \times 10^{-7} \text{ mho}$$

$$y_{21} = 2 \times 10^{-2} \text{ mho} \qquad y_{22} = 1.8 \times 10^{-5} \text{ mho}$$

Our procedure is as follows. If we change the transistor y parameters to z parameters, we can combine the transistor and the 1-kilohm resistor. This will give us the z parameters of the dashed box in Fig. 3.30*b*. We convert these back to a new set of y parameters. Now we combine the ⅚- and 2-kilohm resistors with this new set of y parameters to obtain the overall y parameters of the amplifier. The calculations follow. For *z parameters of dotted box*

$$z_{11} = \frac{y_{22}}{|y|} + 10^3 = 2.20 \times 10^3 \text{ ohms}$$

$$z_{12} = \frac{-y_{12}}{|y|} + 10^3 = 1.02 \times 10^3 \text{ ohms}$$

$$z_{21} = \frac{-y_{21}}{|y|} + 10^3 = -1.33 \times 10^6 \text{ ohms}$$

$$z_{22} = \frac{y_{11}}{|y|} + 10^3 = 3.43 \times 10^4 \text{ ohms}$$

For *overall y parameters*

$$Y_{11} = \frac{z_{22}}{|z|} + 1.20 \times 10^{-3} = 1.22 \times 10^{-3} \text{ mho}$$

$$Y_{12} = \frac{-z_{12}}{|z|} = -0.71 \times 10^{-6} \text{ mho}$$

$$Y_{21} = \frac{-z_{21}}{|z|} = 0.93 \times 10^{-3} \text{ mho}$$

$$Y_{22} = \frac{z_{11}}{|z|} + 0.50 \times 10^{-3} = 0.50 \times 10^{-3} \text{ mho}$$

The final matrix equation is

$$\begin{bmatrix} i_i \\ i_o \end{bmatrix} = \begin{bmatrix} 1.22 \times 10^{-3} & -0.71 \times 10^{-6} \\ 0.93 \times 10^{-3} & 0.50 \times 10^{-3} \end{bmatrix} \begin{bmatrix} e_i \\ e_o \end{bmatrix}$$

What if the 1-kilohm resistor in Fig. 3.30*b* is bypassed by a capacitor whose reactance is negligible over the frequency range of interest? Our problem becomes considerably simplified; the overall y parameters can be written by inspection as

$$Y_{11} = y_{11} + 1.20 \times 10^{-3} = 1.70 \times 10^{-3} \text{ mho}$$

$$Y_{12} = y_{12} = -3 \times 10^{-7} \text{ mho}$$

$$Y_{21} = y_{21} = 2 \times 10^{-2} \text{ mho}$$

$$Y_{22} = y_{22} + 0.50 \times 10^{-3} = 5.18 \times 10^{-4} \text{ mho}$$

Fig. 3.31 The replacement of the transistor in Fig. 3.30 by a triode.

Next let us suppose that the 1-kilohm resistor is again unbypassed and that the transistor is replaced by a triode (bias voltages and resistor values would have to be changed), as indicated schematically in Fig. 3.31. It appears that the analysis just concluded still applies, i.e., that only the numbers have changed. There is a subtle difference though in the two problems. Consider the matrix representation of the triode given in Eq. (3.12.2). Recall that we wish to change the y parameters of the active element to z parameters so that we can add the series (1-kilohm) resistor. But to perform this inversion we need a nonvanishing $|y|$, which is not present in the case of the triode. To circumvent this problem we rewrite Eq. (3.12.2) in the form

$$\begin{bmatrix} i_g \\ i_p \end{bmatrix} = \begin{bmatrix} \epsilon & \epsilon \\ g_m & g_p \end{bmatrix} \begin{bmatrix} e_g \\ e_p \end{bmatrix} \tag{3.13.1}$$

where ϵ is any small number. We then proceed through the steps of the previous analysis, retaining the ϵ until the end, where we let it become zero.

Example 3.9 A triode amplifier with feedback is shown in Fig. 3.32*a*. Its small-signal ac equivalent circuit is drawn in Fig. 3.32*b* in such a way as to display the amplifier and the feedback network as two interconnected 2-ports. We again want to obtain an overall matrix equation for the network

$$\begin{bmatrix} i_i \\ i_o \end{bmatrix} = \begin{bmatrix} Y_{11} & Y_{12} \\ Y_{21} & Y_{22} \end{bmatrix} \begin{bmatrix} e_i \\ e_o \end{bmatrix}$$

To begin with we note that the two 2-ports are interconnected so that their inputs are in series and their outputs in parallel. From the discussion in connection with Fig. 3.23, this suggests that we obtain the overall matrix in terms of h parameters.

The y-parameter matrix of the amplifier 2-port is obtained by adding G_4 and G_1 to the y-parameter matrix of the triode

$$Y_a = \begin{bmatrix} G_4 & 0 \\ g_m & g_p + G_1 \end{bmatrix}$$

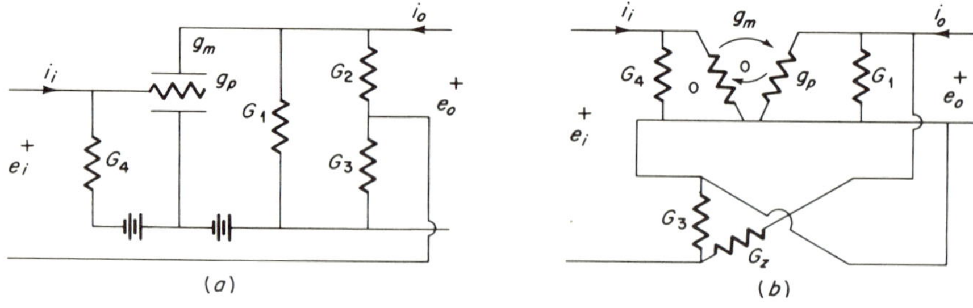

Fig. 3.32 A triode amplifier and its equivalent circuit.

Converting this matrix to h parameters with the aid of Table 3.2 yields

$$H_a = \begin{bmatrix} \dfrac{1}{G_4} & 0 \\ \dfrac{g_m}{G_4} & g_p + G_1 \end{bmatrix}$$

The h-parameter matrix of the feedback network can be obtained directly from the defining relations in Eqs. (3.4.6)

$$H_f = \begin{bmatrix} \dfrac{1}{G_2 + G_3} & -\dfrac{G_2}{G_2 + G_3} \\ \dfrac{G_2}{G_2 + G_3} & \dfrac{G_2 G_3}{G_2 + G_3} \end{bmatrix}$$

Now H_a and H_f can be added together to obtain the h-parameter matrix of the overall network. (The reader should verify that it is permissible to combine them.)

$$H = \begin{bmatrix} \dfrac{1}{G_4} + \dfrac{1}{G_2 + G_3} & \dfrac{-G_2}{G_2 + G_3} \\ \dfrac{g_m}{G_4} + \dfrac{G_2}{G_2 + G_3} & g_p + G_1 + \dfrac{G_2 G_3}{G_2 + G_3} \end{bmatrix}$$

The reader may feel that this procedure is unduly lengthy. If so, he is invited to derive H by other means.

From these two examples, a general procedure to follow in the matrix analysis of active networks can be tabulated.

1. Replace each active device, transformer, and gyrator by its 3-terminal coupled-element equivalent circuit.
2. Using the techniques of Sec. 3.11, redraw the network as interconnections of subnetworks—series-series, parallel-parallel, series-parallel, parallel-series, and cascade.
3. Write the appropriate parameter matrix (z, y, h, g, $abcd$) for each subnetwork.
4. Combine the subnetwork parameter matrices, using Table 3.2, so as to obtain an overall 2×2 matrix equation for the network.

The usefulness of the final overall 2×2 matrix should be discussed. Representing the network in terms of its input and output variables permits us to find the driving-point and transfer properties of the network directly, as discussed in Sec. 3.5 and tabulated in Table 3.3. For example, from the y-parameter matrix equation in Example 3.8 and with a load admittance y_L of 10^{-4} mho, the voltage gain of the network can be found directly as

$$\frac{e_o}{e_i} = \frac{-Y_{21}}{Y_{22} + y_L} = -\frac{0.93 \times 10^{-3}}{0.50 \times 10^{-3} + 0.1 \times 10^{-3}} = -1.55$$

REFERENCES

3.1. Chen, W. H.: "The Analysis of Linear Systems," McGraw-Hill Book Company, New York, 1963. Chapter 4 discusses network parameters and methods of interconnecting networks.

3.2. Cote, A.J., and J. B. Oakes: "Linear Vacuum Tube and Transistor Circuits," McGraw-Hill Book Company, New York, 1961. Contains a thorough treatment of the use of 2-port network theory and matrices in the analysis of active networks.

3.3. Guillemin, E. A.: "Synthesis of Passive Networks," John Wiley & Sons, Inc., New York, 1957. Discusses 2-port parameters and interconnections of 2-ports.

3.4. Hazony, D.: "Elements of Network Synthesis, Reinhold Publishing Coporation, New York, 1963. Discusses perfect and ideal transformers and gyrators.

3.5. Huelsman, L. P.: "Circuits, Matrices, and Linear Vector Spaces," McGraw-Hill Book Company, New York, 1963. Includes an excellent chapter on 2- to- n-port parameters and considerable material on NICs.

3.6. Larky, A. I.: Negative Impedance Converter Design, *Stanford Electro Lab. Tech. Rept.* 11, Oct. 30, 1956. Presents a clear description of the NIC and its internal circuitry.

PROBLEMS

Drill

3.1. For the network shown, find the z, y, h, and $abcd$ parameters.

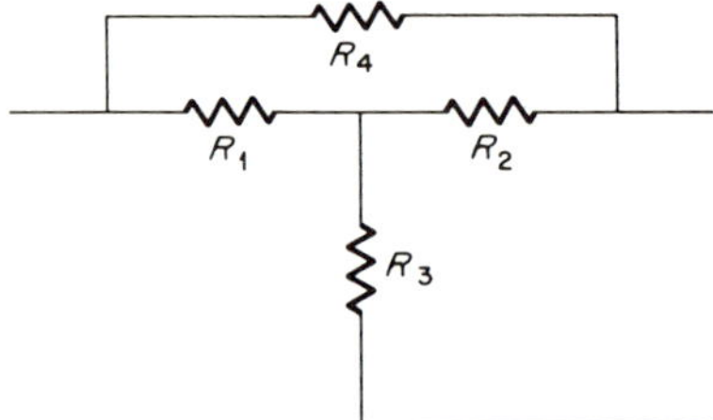

Fig. P3.1

3.2. For the network in Fig. 3.7, verify each of the transfer-function formulas given on a y basis in Table 3.3.

3.3. Write loop equations for the given coupled circuit. Do not solve.

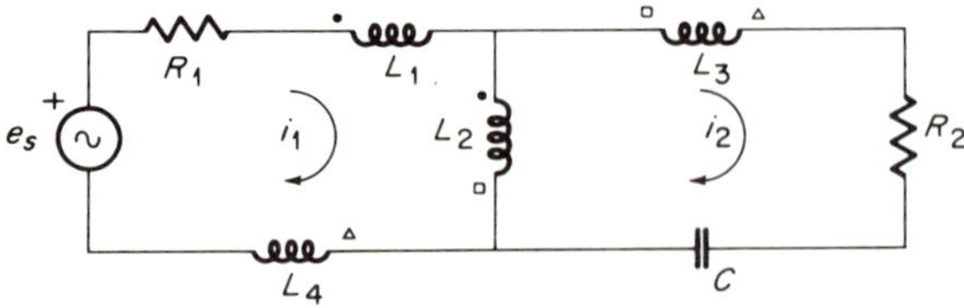

Fig. P3.3

3.4. If possible, write the set of Eqs. (3.6.5) for the 3-port shown.

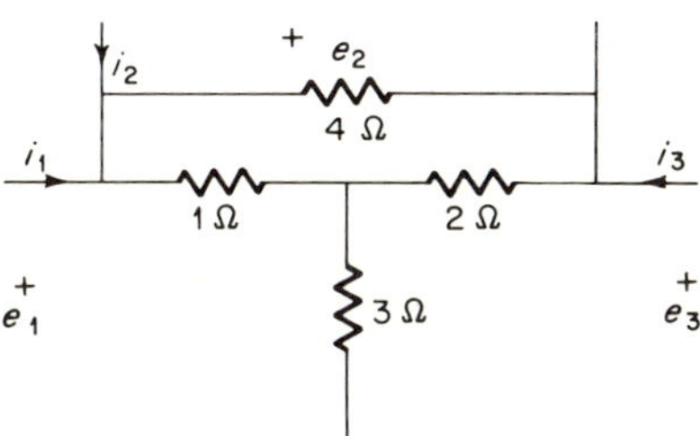

Fig. P3.4

3.5. Calculate the driving-point impedance Z for the network in Fig. P3.5 assuming that the transformer is perfect. Show that the final result can be put in the form

$$Z = \frac{M_1 z_L + N_1}{M_2 + N_2 z_L}$$

where M_1 and M_2 are even polynomials in s (contain terms in s^{2k}, $k = 0, 1, 2, \ldots$) and N_1 and N_2 are odd polynomials in s (contain terms in s^{2l+1}, $l = 0, 1, 2, \ldots$).

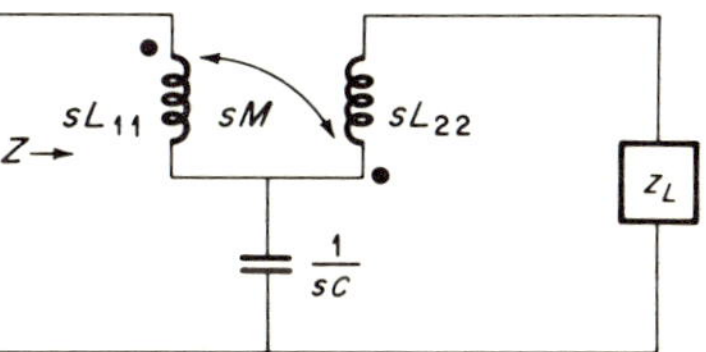

Fig. P3.5

3.6. Find the driving-point impedances of the given networks. Assume a perfect transformer and an ideal gyrator. Can you find a simple equivalent network for each of these circuits?

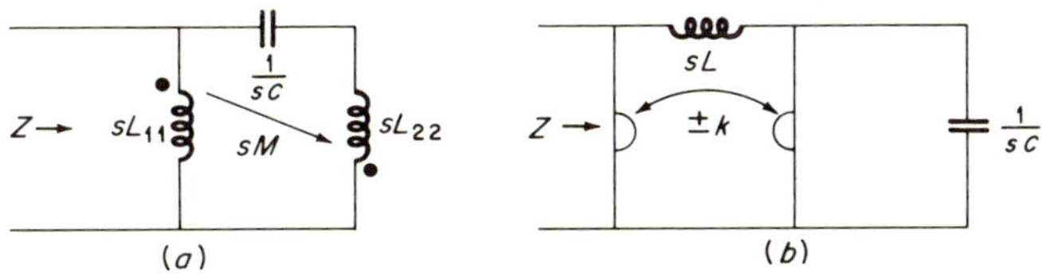

Fig. P3.6

3.7. The network within the solid box has z parameters z_{11}, z_{12}, z_{21}, and z_{22}, as shown. Find the z parameters of the network within the dashed box.

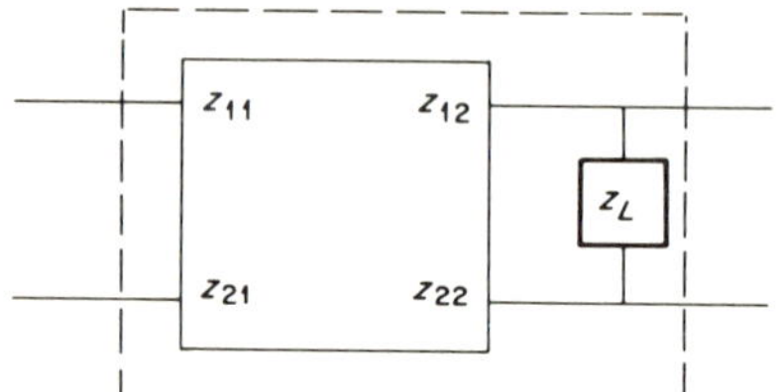

Fig. P3.7

3.8. Find the z parameters of the network within the dashed box.

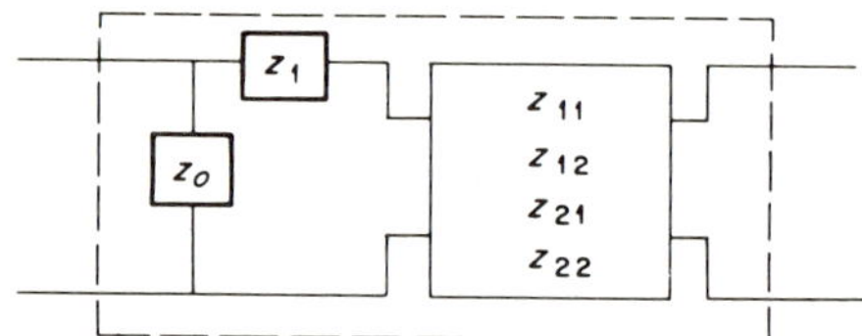

Fig. P3.8

3.9. Find the driving-point impedance of the given NIC-gyrator network.

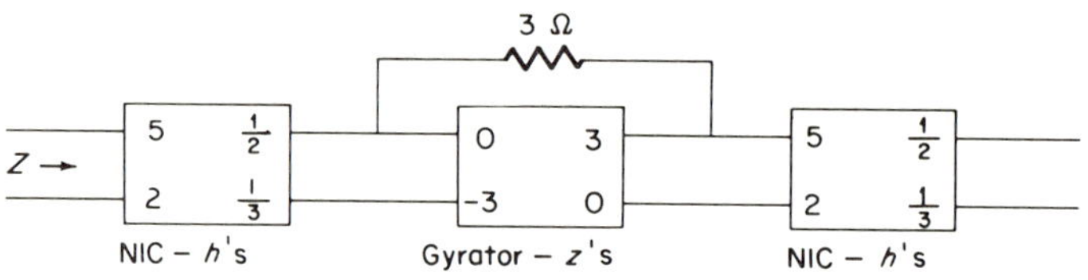

Fig. P3.9

3.10. Connect a resistance of 100 ohms across the output terminals of the network in Fig. 3.30. Find the following quantities:

(*a*) Voltage gain $E_{21} = \dfrac{e_o}{e_i}$

(*b*) Current gain $I_{21} = \dfrac{i_o}{i_i}$

(*c*) Input impedance $Z_{\text{in}} = \dfrac{e_i}{i_i}$

(*d*) Output impedance $Z_{\text{out}} = \left.\dfrac{e_o}{i_o}\right|_{e_i=0}$

(*e*) Output admittance $Y_{\text{out}} = \left.\dfrac{i_o}{e_o}\right|_{i_i=0}$

3.11. Perform a matrix analysis of the triode amplifier for small-signal ac conditions. Specifically obtain y parameters in the matrix equation

$$\begin{bmatrix} i_i \\ i_o \end{bmatrix} = \begin{bmatrix} y_{11} & y_{12} \\ y_{21} & y_{22} \end{bmatrix} \begin{bmatrix} e_i \\ e_o \end{bmatrix}$$

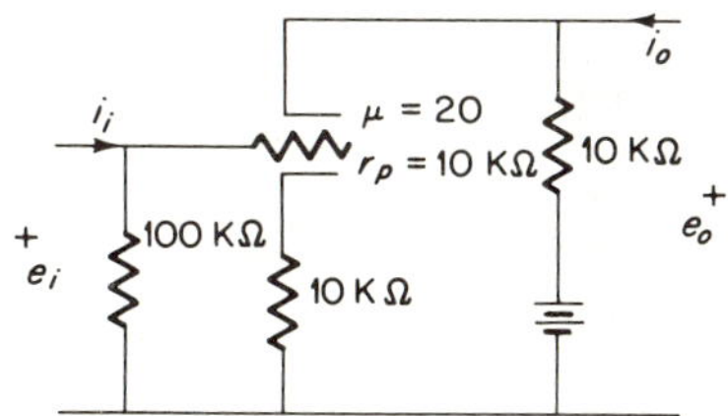

Fig. P3.11

3.12. For the two-stage triode amplifier, find the overall chain parameters of the ac equivalent circuit by finding the individual chain parameters of each stage and combining. Assume all reactances negligible.

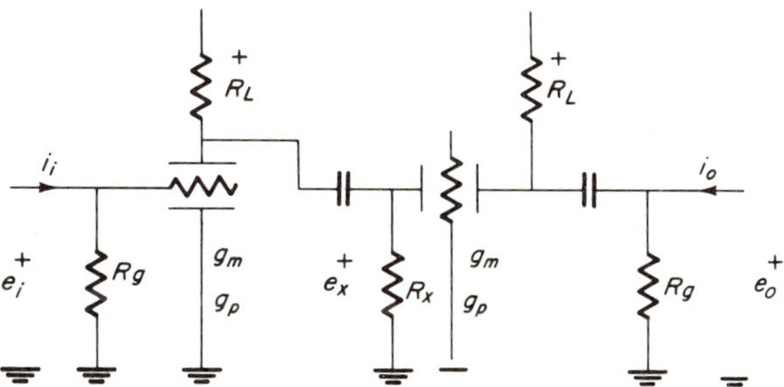

Fig. P3.12 $g_p = 0.1 \times 10^{-3}$ *mho,* $g_m = 1.9 \times 10^{-3}$ *mho,* $R_L = 20$ *kilohms,* $R_g = 100$ *kilohms, and* $R_x = 500$ *kilohms.*

Theory and proofs

3.13. The network within the dashed box is a nonideal NIC with $h_{22} \neq 0$. Show that the addition of $z_1 = z_2$ as shown will make the network within the outer solid box an ideal NIC.

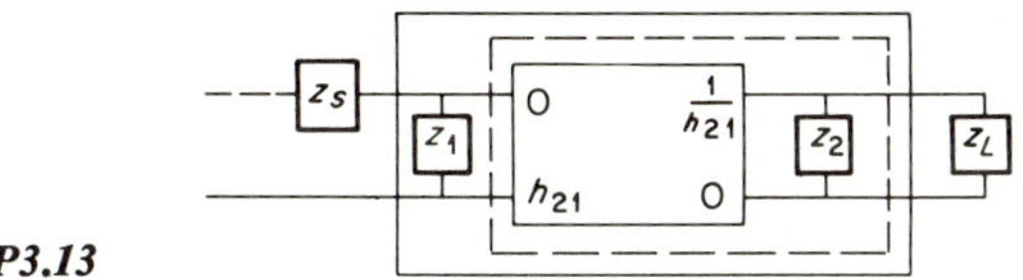

Fig. P3.13

3.14. For the ideal transformer shown, the turns ratio is 2:1 between the primary and each half of the secondary. Find the driving-point impedance at the input terminals for each load connection. How are the results affected if $z_{L1} = z_{L2}$?

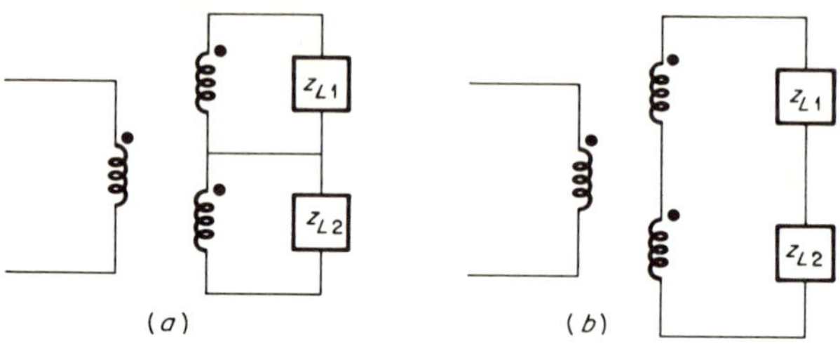

Fig. P3.14

3.15. A constant-resistance network is a 2-port which, when terminated in R ohms, has an input impedance of R ohms. Show that the necessary and sufficient condition for a symmetrical passive reciprocal 2-port to be a constant-resistance network for $R = 1$ ohm is $z_{11} = y_{11}$.

3.16. (*a*) Write y-parameter equations for the given 4-port. All voltages are to ground.

(*b*) What conditions must the y parameters satisfy for it to be possible to write equations of the form

$$\begin{bmatrix} i_1 \\ e_1 \\ i_3 \\ e_3 \end{bmatrix} = \begin{bmatrix} n_{11} & n_{12} & n_{13} & n_{14} \\ n_{21} & n_{22} & n_{23} & n_{24} \\ n_{31} & n_{32} & n_{33} & n_{34} \\ n_{41} & n_{42} & n_{43} & n_{44} \end{bmatrix} \begin{bmatrix} i_2 \\ e_2 \\ i_4 \\ e_4 \end{bmatrix}$$

(*c*) Find the n parameters.

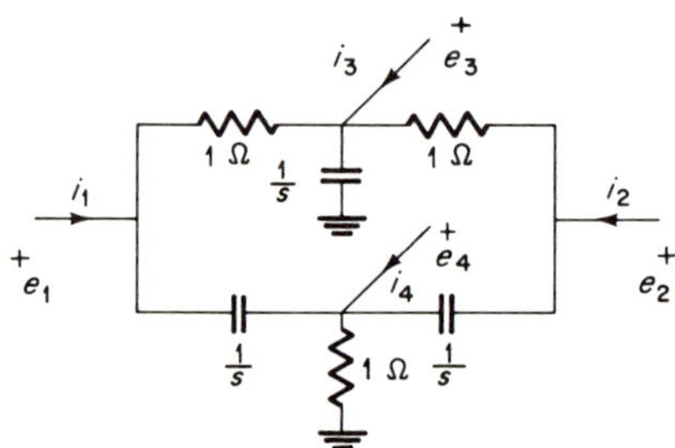

Fig. P3.16

3.17. Find the quantities

$$Y_{12} = \frac{i_1}{e_3}\bigg|_{e_1=0} \qquad \text{and} \qquad Y_{22} = \frac{i_3}{e_3}\bigg|_{e_1=0}$$

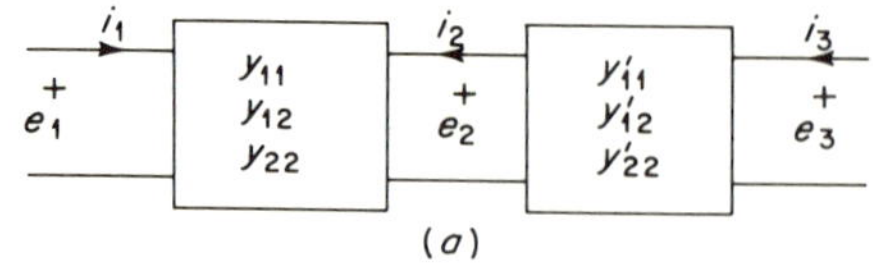

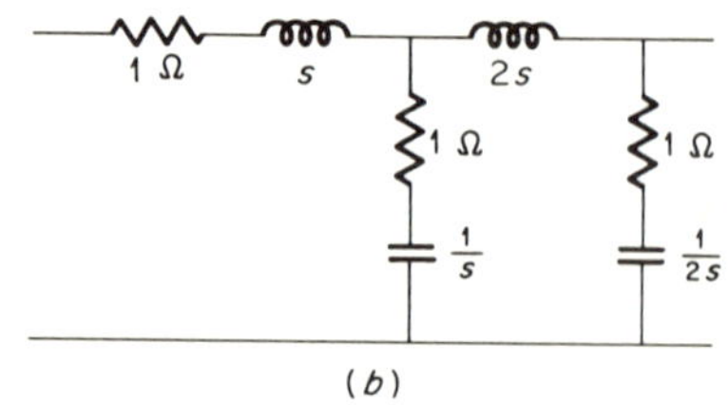

Fig. P3.17

for the network of Fig. P3.17*a* in terms of the y and y' parameters. Use the results to obtain Y_{12} and Y_{22} of the network in Fig. P3.17*b*.

3.18. Find the open-circuit voltage transfer ratio

$$E_{21}{}^0 = \left.\frac{e_2}{e_1}\right|_{i_2=0}$$

What relationship between y_{22a} and y_{22b} will make the overall y_{22} of the network equal to 1 mho?

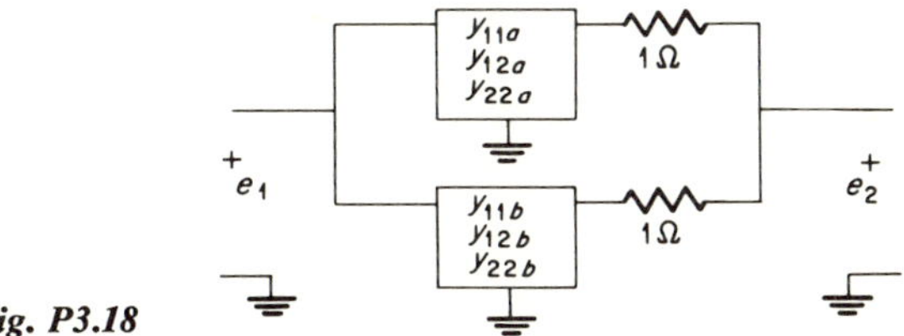

Fig. P3.18

3.19. The network is entirely resistive. Each box has the same y_{22}. The three transfer admittances are in the ratio $y_{12a}/y_{12b}/y_{12c} = 1:2:3$. The resistors in each box are to be scaled so that this ratio becomes 4:5:6 with the overall $Y_{22} = y_{22}$. By what factor must each box be scaled?

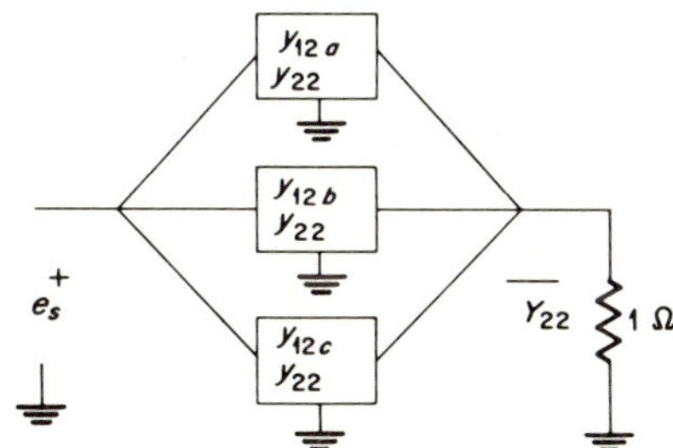

Fig. P3.19

Application

3.20. Consider the 5-terminal network shown. Its open-circuit voltage transfer equations can be written in the geneal form

$$\begin{bmatrix} e_3 \\ e_4 \end{bmatrix} = \begin{bmatrix} t_{31} & t_{32} \\ t_{41} & t_{42} \end{bmatrix} \begin{bmatrix} e_1 \\ e_2 \end{bmatrix}$$

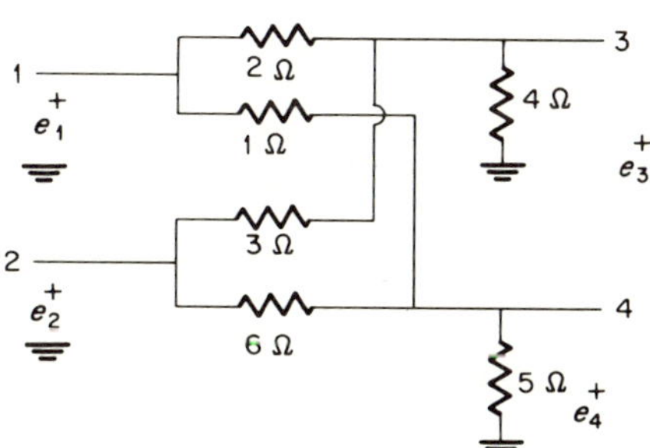

Fig. P3.20

The entry t_{31}, for example, is e_3/e_1 with i_3 and i_4 zero and with e_2 zero. Find the components of the T matrix for the given network.

3.21. A two-stage transistor amplifier is shown (dc excitations are omitted). Find the y parameters of each dashed box. From these find the current gain of each stage and the overall current gain. Repeat for the voltage gains.

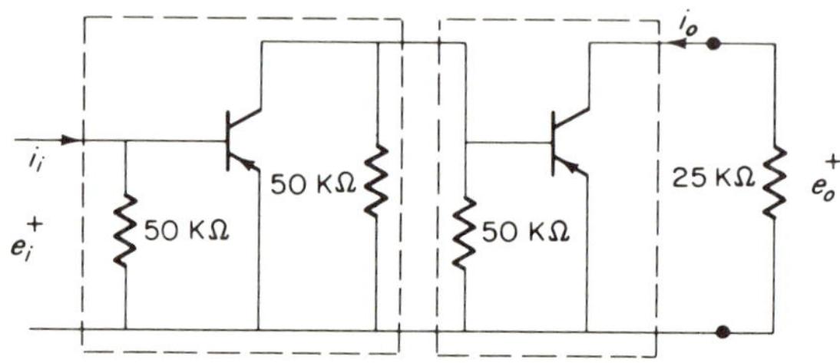

Fig. P3.21 $n_{11} = 2360$ *ohms,* $n_{12} = 1.04 \times 10^{-3}$, $n_{21} = 39$, *and* $n_{22} = 40 \times 10^{-6}$ *mho.*

3.22. Determine the elements of a T network which would yield the following set of data.

dc		*1000 Hz*	
Measurement	*Condition*	*Measurement (rms)*	*Condition*
$e_1 = 2$ volts	2-2′ open	$e_1 = 5\underline{/0^\circ}$ volts	2-2′ open
$i_1 = 10$ ma		$i_1 = 13.4\underline{/-57.5^\circ}$ ma	
$e_2 = 1$ volt		$e_2 = 4.4\underline{/14.8^\circ}$ volts	
$e_1 = 0$ volts	1-1′ open	$e_1 = 12.4\underline{/46.6^\circ}$	1-1′ open
$i_2 = 0$ ma		$i_2 = 37.6\underline{/-25.7^\circ}$ ma	
$e_2 = 2$ volts		$e_2 = 5\underline{/0^\circ}$ volts	

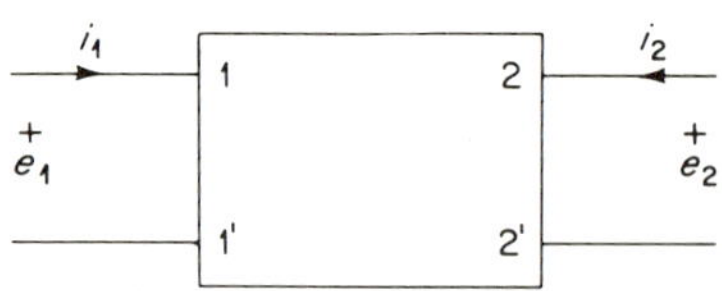

Fig. P3.22

3.23. The following questions refer to the network in Example 3.4.

(*a*) With $L_{11} = 40$ henrys, $L_{22} = 10$ henrys, $k = 1.0$, and $z_L = 3000$ ohms, what is the least value of ω that will keep the resistive portion of Z_{in} within 10 percent of the ideal-transformer value? How large is the reactive portion of Z_{in} at this value of ω?

(*b*) With $L_{11} = 40$ henrys, $L_{22} = 10$ henrys, $Z_L = 3000$ ohms, and $\omega = 4000$ rad/sec, what is the least value of k that will keep the resistive portion of Z_{in} within 10 percent of the ideal transformer value? How large is the reactive portion of Z_{in} at this value of k?

3.24. Assuming the capacitors have negligible reactances, find the overall y-parameter matrix for the small-signal ac equivalent circuit of the given network.

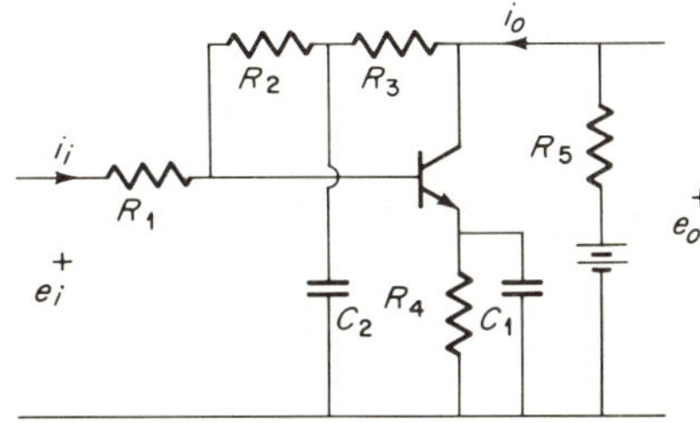

Fig. P3.24 $R_1 = 5$ *kilohms*, $R_2 = 3$ *kilohms*, $R_3 = 2$ *kilohms*, $R_4 = 5$ *kilohms*, $R_5 = 10$ *kilohms. The transistor common-emitter parameters are* $y_{11} = 5 \times 10^{-4}$ *mho*, $y_{12} = -3 \times 10^{-7}$ *mho*, $y_{21} = 2 \times 10^{-2}$ *mho, and* $y_{22} = 1.8 \times 10^{-5}$ *mho.*

3.25. Obtain an appropriate matrix to describe the behavior of the network at its ports.

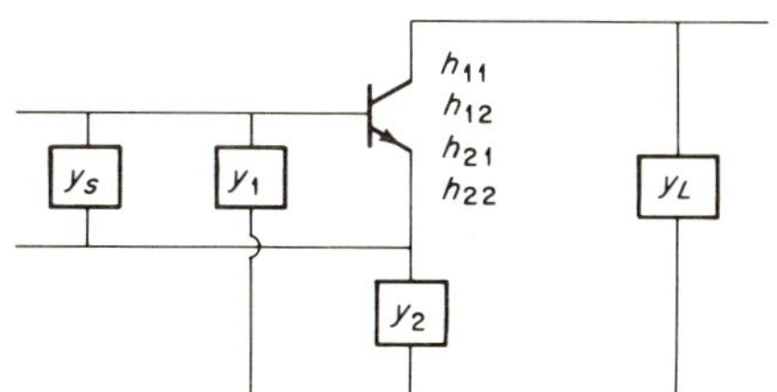

Fig. P3.25

3.26. For the triode amplifier, derive expressions for the parameters in the matrix equation (3.6.4).

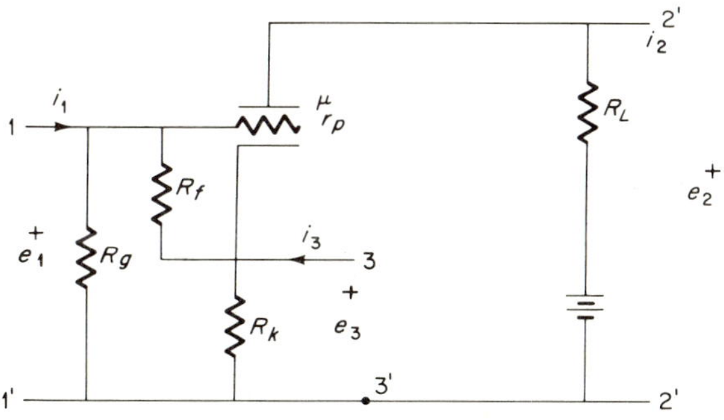

Fig. P3.26

3.27. C is a short circuit to ac. The y-parameter matrix representation of the transistor on a common-emitter basis is given as

$$\begin{bmatrix} i_b \\ i_c \end{bmatrix} = \begin{bmatrix} y_i & y_r \\ y_f & y_o \end{bmatrix} \begin{bmatrix} v_{be} \\ v_{ce} \end{bmatrix}$$

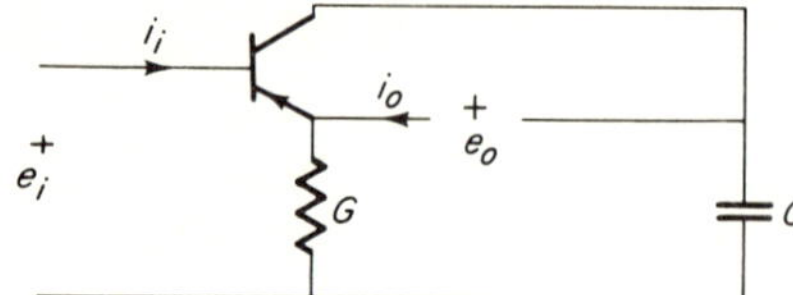

Fig. P3.27

Find the overall 2-port parameters in the equation

$$\begin{bmatrix} i_i \\ i_o \end{bmatrix} = \begin{bmatrix} Y_i & Y_r \\ Y_f & Y_o \end{bmatrix} \begin{bmatrix} e_i \\ e_o \end{bmatrix}$$

4 LINEAR GRAPH THEORY

4.1 INTRODUCTION

In this chapter we use the mathematical tools and network concepts of the first three chapters to develop a systematic procedure for obtaining and using sets of independent loop-current and node-pair voltage variables in a given network. We shall find it convenient to use certain concepts from *linear graph theory* that relate to the geometrical structure of networks. Linear graph theory permits a systematic analysis of a network from a consideration of its geometry only. This geometrical structure permits a unique determination of how many current or voltage variables we need and how to define them. It also facilitates formulation of the pertinent KCL and KVL equations relating the branch currents and voltages to the loop-current and node-pair voltage variables, respectively.

The relationship between current and voltage in a given branch is not obtained from geometrical considerations. Nor are sources taken into account. In Chap. 5 we consider both these matters through application of

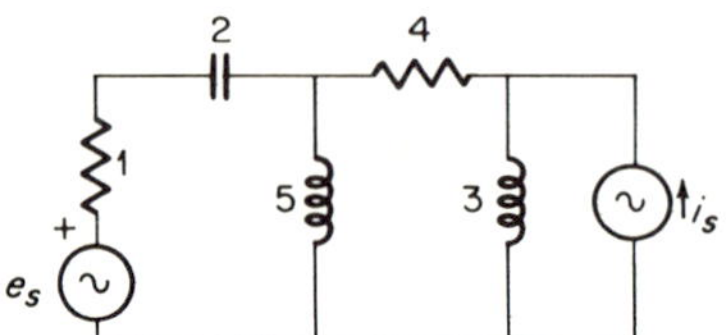

Fig. 4.1

Ohm's law to each branch, leading directly to the formulation of loop-current and node-pair voltage equilibrium equations for the overall network on a topological basis.

4.2 LINEAR AND DIRECTED LINEAR GRAPHS

We assume the network in Fig. 4.1 to be composed of linear sources and elements. The *linear graph* of this network is formed by replacing each branch by a line and each node by a dot. Recall from Sec. 2.2 that a branch is a 1-port and includes its associated ideal current and/or voltage sources. Thus sources do not appear explicitly in a linear graph since they are parts of branches.† The linear graph of the network in Fig. 4.1 is given in Fig. 4.2, assuming only one network element per branch.

To form a *directed linear graph*, it is necessary to assign positive reference directions to all branch currents and voltages. The convention will be adopted that the positive current reference is in the direction of the arrow and the voltage reference is such that the tail of the arrow is positive. Figure 4.2 is a directed linear graph.

4.3 CONNECTED AND NONCONNECTED GRAPHS

We define a *connected graph* as one having a path between every pair of nodes (there may be intervening nodes). The graph in Fig. 4.2 is a connected

† An alternate approach is to consider an ideal voltage source as a shorted branch, since it has zero internal impedance and only the geometry of the source is being represented. Similarly, an ideal current source can be considered as an open branch since it has infinite short impedance.

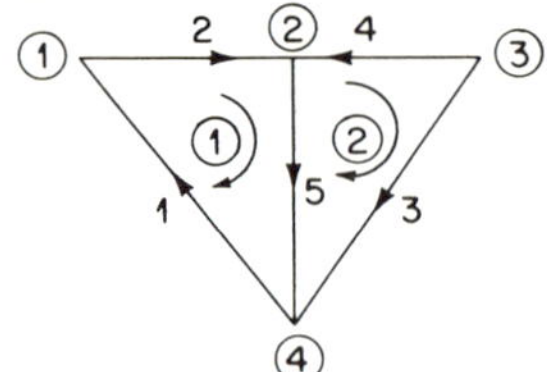

Fig. 4.2 Graph of the network in Fig. 4.1.

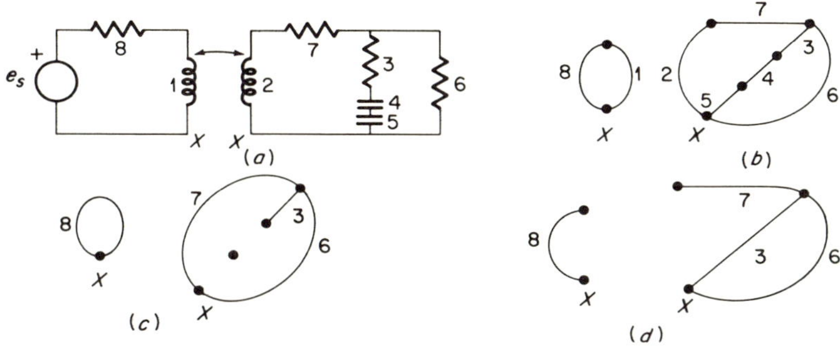

Fig. 4.3 Graphs of a network having separate parts.

graph. Now consider the network of Fig. 4.3*a* and its linear graph in Fig. 4.3*b*. This graph is nonconnected; it is said to have two *separate parts*. If we join the two nodes labeled x in Fig. 4.3*a*, the two nodes labeled x in Fig. 4.3*b* coalesce into one node. The graph becomes connected and has one separate part. Coalescing of nodes in this way is permissible if and only if no new loops are formed.

Three degenerate cases—the *self-loop*, the *isolated node*, and the *dangling branch*—require special interpretation. Consider the network in Fig. 4.3*a* under very low-frequency conditions. The capacitors in branches 4 and 5 become open, whereas the inductors in branches 1 and 2 become shorted. The graph degenerates to that of Fig. 4.3*c*. Branch 8 forms a self-loop; the node between branches 4 and 5 becomes an isolated node; branch 3 becomes a dangling branch. The graph now has three separate parts.

Now consider the network of Fig. 4.3*a* under very high-frequency conditions. The two capacitors are now shorted branches, whereas the two inductors are open branches. The graph becomes that of Fig. 4.3*d*. Branches 7 and 8 are dangling branches; there are no isolated nodes or self-loops. The graph now has two separate parts.

To summarize: we consider a self-loop as having one node and an isolated node as being a separate part.

4.4 TERMINOLOGY

The following definitions and symbols will be useful in the subsequent development. Some of them were presented in Chap. 2, and the others will be explained when they first appear in the discussion to follow.

Simple node intersection point of two or more branches
Simple loop a closed branch contour having only one opening (window)

Supernode combination of one or more simple nodes
Superloop combination of one or more simple loops
B number of branches
N_s number of simple nodes
L_s number of simple loops
N number of tree branches, also *rank* of graph
L number of tree links, also *nullity* of graph
P number of separate parts
v_k kth-branch voltage
j_k kth-branch current
e_k kth independent voltage variable
i_k kth independent current variable
e_{sk} summation of potential sources in branch k
i_{sk} summation of current sources in branch k
V_B $B \times 1$ branch-voltage matrix
J_B $B \times 1$ branch-current matrix
E_N $N \times 1$ voltage-variable matrix
I_L $L \times 1$ current-variable matrix

In these definitions, small letters are used for branch quantities and capital letters for matrices. Note that the subscripts on the voltage and current matrices indicate the number of rows in each (one column being understood).

4.5 VERTEX AND CIRCUIT MATRICES

Consider again the network of Fig. 4.1 and its associated graph in Fig. 4.2. The four simple nodes are labeled. Writing the four KCL equations in matrix form yields

$$\begin{matrix} \text{Node 1} \\ \text{Node 2} \\ \text{Node 3} \\ \text{Node 4} \end{matrix} \begin{bmatrix} -1 & 1 & 0 & 0 & 0 \\ 0 & -1 & 0 & -1 & 1 \\ 0 & 0 & 1 & 1 & 0 \\ 1 & 0 & -1 & 0 & -1 \end{bmatrix} \begin{bmatrix} j_1 \\ j_2 \\ j_3 \\ j_4 \\ j_5 \end{bmatrix} = \begin{bmatrix} 0 \\ 0 \\ 0 \\ 0 \end{bmatrix} \tag{4.5.1}$$

In closed matrix form, Eq. (4.5.1) becomes

$$[a_{ik} \quad j_k] = [0] \tag{4.5.2}$$

or

$$AJ_B = 0 \tag{4.5.3}$$

where A is called the *vertex matrix* (also known as the *vertex-incidence matrix* or, simply, *incidence matrix*). Its order is $N_S \times B$ and

$$a_{ik} = \begin{cases} +1 & \text{branch } k \text{ is connected to node } i \text{ and branch-current reference direction is away from node } i \\ -1 & \text{branch } k \text{ is connected to node } i \text{ and branch-current reference direction is into node } i \\ 0 & \text{branch } k \text{ is not connected to node } i \end{cases} \tag{4.5.4}$$

The equations in (4.5.1) are not linearly independent. Using the notation of Chap. 2, $m = 4$, $n = 5$, and $r = 3$. Thus there is one ignorable equation, and the defect is 2. Two branch currents may be chosen as independent variables and moved to the right side of the equations, subject to the constraint that the coefficient determinant of the remaining variables does not vanish. It is left to the reader to show that there are eight such sets of currents in Eq. (4.5.1).

Proceeding to the KVL equations for Fig. 4.1, we find two simple loops and the superloop around the circuit exterior. Thus $L_S = 2$, and the three KVL equations are

$$\begin{matrix} \text{Loop 1} \\ \text{Loop 2} \\ \text{Loop 12} \end{matrix} \begin{bmatrix} 1 & 1 & 0 & 0 & 1 \\ 0 & 0 & 1 & -1 & -1 \\ 1 & 1 & 1 & -1 & 0 \end{bmatrix} \begin{bmatrix} v_1 \\ v_2 \\ v_3 \\ v_4 \\ v_5 \end{bmatrix} = \begin{bmatrix} 0 \\ 0 \\ 0 \end{bmatrix} \tag{4.5.5}$$

In closed-matrix form, we have

$$[c_{ik} \quad v_k] = [0] \tag{4.5.6}$$

or

$$CV_B = 0 \tag{4.5.7}$$

C is known as the *circuit matrix* of the given network and is of order $(L_S + 1) \times B$. We note that

$$b_{ik} = \begin{cases} +1 & \text{branch } k \text{ is in loop } i \text{ and branch-voltage reference direction is clockwise in loop } i \\ -1 & \text{branch } k \text{ is in loop } i \text{ and branch-voltage reference direction is counterclockwise in loop } i \\ 0 & \text{branch } k \text{ is not in loop } i \end{cases} \tag{4.5.8}$$

In this case, $m = 3$, $n = 5$, and $r = 2$. There is one ignorable equation and a defect of 3. Three branch voltages may be moved to the right side of the equations as independent variables, again subject to the constraint that the coefficient determinant of the remaining variables not vanish. There are eight such sets of independent voltages, the same number as in the KCL case. The reader should verify this fact.

4.6 TREES, TREE BRANCHES, AND TREE LINKS

Consider a connected graph G. Define a *subgraph* of G as a graph that remains after deleting any K branches from G, where $1 \leq K \leq B$. Now define a *tree* of G as a connected subgraph of G that includes all the nodes of G but contains no closed loops. Further, define a *tree link* (or simply *link*) as a branch of G that does not belong to the particular tree under consideration and a *tree branch* as one that does.

For the graph of Fig. 4.2 there are eight trees, as shown in Fig. 4.4. Thus the number of trees, sets of independent currents, and sets of independent voltages is the same for the network of Fig. 4.1. This is a general result, but we defer the proof to Sec. 6.3, where an expression for the number of trees in a graph will be derived.

For a connected graph, the numbers of tree branches, tree links, simple nodes, simple loops, and branches are related by

$$\begin{aligned} B &= N + L \\ N &= N_S - 1 \\ L &= L_S \end{aligned} \tag{4.6.1}$$

When a graph contains more than one separate part, we interpret the definition of a tree as applying to each separate part individually. Then Eqs. (4.6.1) become

$$\begin{aligned} B &= N + L \\ N &= N_S - P \\ L &= L_S \end{aligned} \tag{4.6.2}$$

Thus, for each separate part, we require one less tree branch than the number

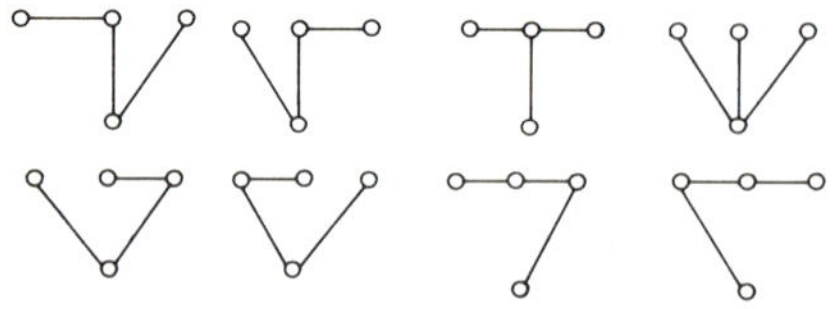

Fig. 4.4 The eight trees of the graph of Fig. 4.2.

of simple nodes in that part; i.e., the first tree branch consumes two nodes and each additional tree branch one node. Also, the number of links in each separate part is equal to the number of simple loops in that part, since for a proper choice of tree, e.g., the first tree in Fig. 4.4, each link completes one simple loop not completed by any other link.

For the graph of Fig. 4.2, $B = 5$, $P = 1$, $N_S = 4$, $N = 3$, and $L = 2$. For the graph of Fig. 4.3*b*, $B = 8$, $P = 2$, $N_S = 7$, $N = 5$, and $L = 3$. In Fig. 4.3*c*, $B = 4$, $P = 3$, $N_S = 5$, $N = 2$, and $L = 2$; for Fig. 4.3*d*, $B = 4$, $P = 2$, $N_S = 5$, $N = 3$, and $L = 1$.

Two theorems permit an exact determination of the number of independent loop currents and node-pair voltages in a given network.

Theorem 1 The number of independent loop currents in a network is L, and the number of independent KCL equations is $B - L = N$.

Theorem 2 The number of independent node-pair voltages in a network is N, and the number of independent KVL equations is $B - N = L$.

A heuristic justification of these theorems follows. For Theorem 1, consider the first tree in Fig. 4.4, repeated in more detail in Fig. 4.5, where it should be noted that there are two link currents (links are shown by dashed lines) and three tree-branch currents.

If the $L(2)$ link currents are chosen zero simultaneously, all $N(3)$ tree-branch currents must also be zero. Choosing $L + 1$ branch currents zero does not change the result. Hence there are at most L independent currents. Choosing $L - 1$ branch currents zero provides one loop where the currents need not be zero. Hence there are at least L independent currents. It follows that there are exactly L independent currents. But not any L will do. For example, j_3 and j_4 in Fig. 4.5 are not an independent set. A set of link currents is always an independent set.

Since there are L independent branch currents, there must be $B - L = N$ dependent branch currents. Thus exactly N independent KCL equations are required to express these N dependent branch currents uniquely in terms of the L independent branch currents.

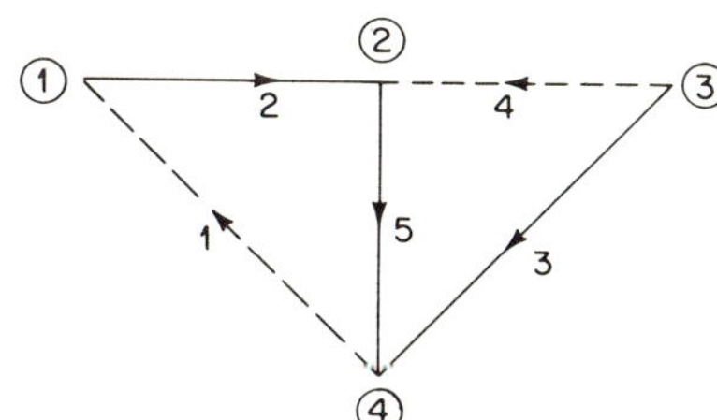

Fig. 4.5 Tree used to illustrate Theorems 1 and 2.

Theorem 2 is justified in an analogous manner. Thus if the $N(3)$ tree-branch voltages are chosen zero, all $L(2)$ tree-link voltages must also be zero. Choosing $N + 1$ branch voltages zero does not change the result, and choosing $N - 1$ branch voltages zero provides one node whose voltages with respect to the other nodes are not determined. Hence there are exactly N independent voltages but not any N. For example, v_3, v_4, and v_5 in Fig. 4.5 do not form an independent set. The tree-branch voltages are always an independent set. There are N independent branch voltages, L dependent branch voltages, and thus exactly L independent KVL equations.

4.7 FUNDAMENTAL-TIE-SET MATRIX

Define a *fundamental tie set* as a set of branches energized when one link current flows and the other link currents are zero.

Let us try to establish a "physical feel" for a fundamental tie set. The link currents are effectively loop currents, since the addition of a link to a tree permits that link current to circulate around the completed loop composed of that link and one or more tree branches (no tree branches in the case of a self-loop). It was stated in Sec. 4.6 that the link currents (and therefore the resulting loop currents) are always an *independent* set of current variables. It is now clear that this is so since each loop formed by adding a link to the tree has at least one element (the link element) not contained in any other loop. The branches appearing in each completed loop constitute a fundamental tie set. The fundamental tie sets for the tree of Fig. 4.5 are shown in Fig. 4.6.

Expressing the branch currents in terms of these independent link-current variables is accomplished through the *tie-set schedule*. To prepare such a schedule, a tree is selected from the directed linear graph of the network. The link currents are interpreted as being independent loop currents. Each such loop current flows through one link and one or more tree branches. These contributions of the independent loop currents to the branch currents make up the tie-set schedule. The tie-set schedule for Fig. 4.5 is used to illustrate the construction.

	j_1	j_2	j_3	j_4	j_5
i_1 (j_1)	1	1	0	0	1
i_2 (j_4)	0	0	-1	1	1

Note that link current j_1 becomes loop current i_1 circulating around loop 125. Thus i_1 contributes to j_1, j_2, and j_5, all in the positive sense, as indicated by the first row of the tie-set schedule. Similarly, link current j_4 becomes

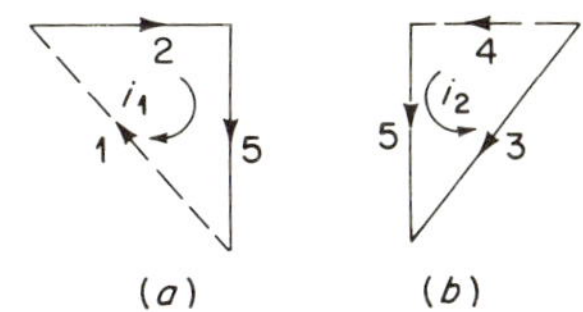

Fig. 4.6 Tie sets of Fig. 4.5.

loop current i_2, and the latter contributes to j_4 and j_5 in the positive sense and j_3 in the negative sense, as indicated by the second row in the tie-set schedule. The schedule can also be formulated by columns. For example, j_5 is composed of i_1 and i_2, both in the positive sense, whereas j_3 is composed of i_2 in the negative sense.

The tie-set schedule is more useful in matrix form. For the above example

$$\beta_F = \begin{bmatrix} 1 & 1 & 0 & 0 & 1 \\ 0 & 0 & -1 & 1 & 1 \end{bmatrix} \tag{4.7.1}$$

where β_F is the *fundamental-tie-set matrix*. It is always an $L \times B$ matrix. The fundamental-tie-set matrix can be used to relate the branch currents to the independent current variables by the matrix equation

$$J_B = \beta_F{}^T I_L \tag{4.7.2}$$

For example, application of Eq. (4.7.2) to Fig. 4.5 yields

$$\begin{bmatrix} j_1 \\ j_2 \\ j_3 \\ j_4 \\ j_5 \end{bmatrix} = \begin{bmatrix} 1 & 0 \\ 1 & 0 \\ 0 & -1 \\ 0 & 1 \\ 1 & 1 \end{bmatrix} \begin{bmatrix} i_1 \\ i_2 \end{bmatrix} \tag{4.7.3}$$

In Eq. (4.7.3), $m = B = 5$, $n = L = 2$, $r = L = 2$. Thus there is no defect, but there are three ignorable equations, which means that i_1 and i_2 can be expressed uniquely in terms of only two of the branch currents, but not any two, since the determinant of the coefficient matrix of the two equations chosen must not vanish.

Equation (4.7.2) is the first of several topological matrix equations we shall derive. A further consideration of the rows of β_F yields a second equation. Recall that each row is formed by determining the contribution of a particular loop current to the branch currents in that loop. By replacing each branch current by its branch voltage, each row of β_F becomes the KVL equation around the loop in question. Thus the summation of branch voltages in any fundamental tie set is zero, or

$$\beta_F V_B = 0 \tag{4.7.4}$$

For the given example,

$$\begin{bmatrix} 1 & 1 & 0 & 0 & 1 \\ 0 & 0 & -1 & 1 & 1 \end{bmatrix} \begin{bmatrix} v_1 \\ v_2 \\ v_3 \\ v_4 \\ v_5 \end{bmatrix} = \begin{bmatrix} 0 \\ 0 \end{bmatrix} \tag{4.7.5}$$

Here $m = L = 2$, $n = B = 5$, $r = L = 2$. There are no ignorable equations, but a defect of 3 is present. This confirms that $N = 3$ of the branch voltages may be chosen as independent variables, but not any three since the coefficient determinant of the remaining two variables must not vanish.

As a final point we note that each tree of a directed graph has one and only one fundamental-tie-set matrix.

4.8 FUNDAMENTAL-CUT-SET MATRIX

Define a *fundamental cut set* as a set of branches energized when one tree-branch voltage is applied and the other tree-branch voltages are zero.

The tree-branch voltages are node-pair voltages. Let one tree-branch voltage in a given tree be energized with all other tree-branch voltages zero. Since the tree branches represent conducting elements, the absence of a voltage across a given tree branch may be interpreted as meaning that the two nodes at the ends of that tree branch coalesce. Thus when one tree-branch voltage is energized with the others zero, all nodes of the tree coalesce to the two nodes at the ends of the energized tree branch. In this process, the various links of the tree are either shorted out or placed in parallel with the energized tree branch. Thus a given link voltage is either equal in magnitude to the energized tree-branch voltage or zero, depending on whether the given link does or does not bridge the energized tree branch.

It was stated in Sec. 4.6 that the tree-branch voltages are always an independent set of voltage variables. This must be true since each of the node pairs formed by the tree branch and the energized tree links contains one element (the tree-branch element) not found in any other node pair. The branches appearing in each such node pair constitute a fundamental cut set. The fundamental cut sets for the tree of Fig. 4.5 appear in Fig. 4.7*b* to *d*, curved arrows indicating the coalescing of the nodes.

To motivate the name cut set, a dashed line is drawn through the energized tree branch and links in Fig. 4.7*a*. This line cuts the graph into two parts, one of which may be an isolated node.

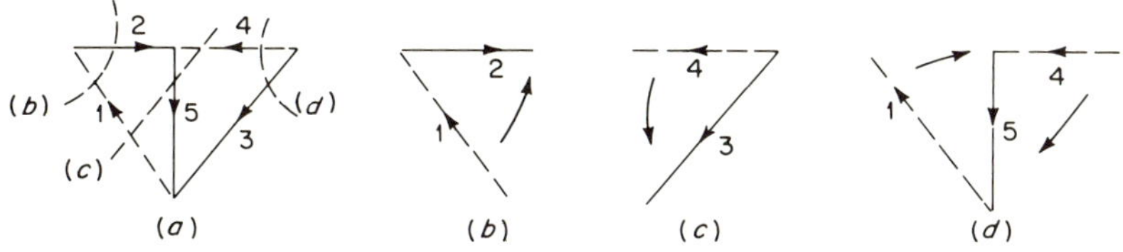

Fig. 4.7 Cut sets of Fig. 4.5.

To express the branch voltages in terms of the independent tree-branch voltage variables, a cut-set schedule is formed. A tree is selected from the directed linear graph. The tree-branch voltages are considered as node-pair voltages. Each independent node-pair voltage energizes one tree branch and one or more links (no links in the case of a dangling branch). These contributions of the independent node-pair voltages to the branch voltages make up the cut-set schedule. The cut-set schedule for Fig. 4.5 is

	v_1	v_2	v_3	v_4	v_5
e_1 (v_2)	-1	1	0	0	0
e_2 (v_3)	0	0	1	1	0
e_3 (v_5)	-1	0	0	-1	1

As in the case of the tie-set schedule, the cut-set schedule can be formulated either by rows or by columns. For example, e_1 contributes to v_2 in the positive sense and v_1 in the negative sense. Or v_1 receives contributions from e_1 and e_3, both in the negative sense.

The matrix representation of the cut-set schedule is

$$\alpha_F = \begin{bmatrix} -1 & 1 & 0 & 0 & 0 \\ 0 & 0 & 1 & 1 & 0 \\ -1 & 0 & 0 & -1 & 1 \end{bmatrix} \tag{4.8.1}$$

where α_F is the *fundamental-cut-set matrix.* Its order is $N \times B$. Now the matrix equation relating the branch voltages to the independent voltage variables can be written as

$$V_B = \alpha_F{}^T E_N \tag{4.8.2}$$

For the given example, Eq. (4.8.2) gives

$$\begin{bmatrix} v_1 \\ v_2 \\ v_3 \\ v_4 \\ v_5 \end{bmatrix} = \begin{bmatrix} -1 & 0 & -1 \\ 1 & 0 & 0 \\ 0 & 1 & 0 \\ 0 & 1 & -1 \\ 0 & 0 & 1 \end{bmatrix} \begin{bmatrix} e_1 \\ e_2 \\ e_3 \end{bmatrix} \tag{4.8.3}$$

In this case $m = B = 5, n = N = 3, r = N = 3$. Thus there is no defect, but there are $L = 2$ ignorable equations. This means that e_1, e_2, and e_3 can be expressed uniquely in terms of only three of the branch voltages, but not any three, since the coefficient determinant of the equations chosen must not vanish.

If each branch voltage in the cut-set schedule is now replaced by its branch current, each row of α_F becomes a KCL equation at a node. It follows that the summation of branch currents in any fundamental cut set is zero, or

$$\alpha_F J_B = 0 \tag{4.8.4}$$

For the given example, Eq. (4.8.4) yields

$$\begin{bmatrix} -1 & 1 & 0 & 0 & 0 \\ 0 & 0 & 1 & 1 & 0 \\ -1 & 0 & 0 & -1 & 1 \end{bmatrix} \begin{bmatrix} j_1 \\ j_2 \\ j_3 \\ j_4 \\ j_5 \end{bmatrix} = \begin{bmatrix} 0 \\ 0 \\ 0 \end{bmatrix} \tag{4.8.5}$$

Here $m = N = 3$, $n = B = 5$, $r = N = 3$. There are no ignorable equations, but a defect of 2 is present. This confirms that $L = 2$ of the branch currents may be chosen as independent variables, but not any two, since the remaining coefficient determinant must not vanish.

Finally, each tree of a directed graph has one and only one fundamental cut-set matrix.

Example 4.1 Draw a directed linear graph, select a tree and links, and find the fundamental-tie-set and cut-set matrices for the network of Fig. 4.8*a*.

The directed linear graph appears in Fig. 4.8*b*. Note that it contains two common degeneracies, namely, a self-loop and a dangling branch. The graph has five branches, four nodes, and one separate part. Thus $N = 3$ and $L = 2$. We choose the tree 234 and thus define the link currents j_1 and j_5 as independent current variables and the tree-branch voltages v_2, v_3, and v_4 as independent voltage variables. The fundamental tie sets are branches 1, 2, and 3 and branch 5; the fundamental cut sets are branches

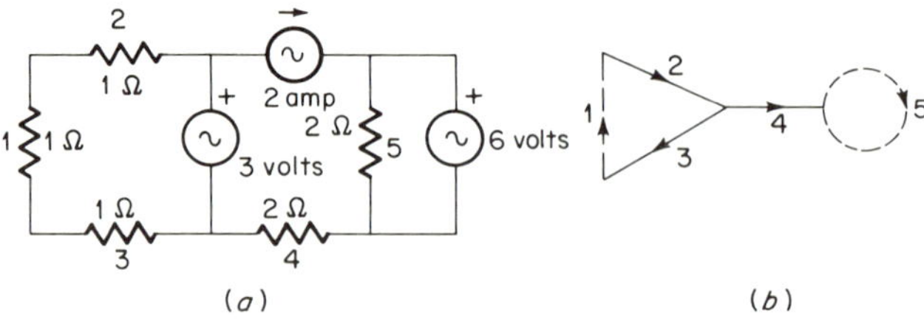

Fig. 4.8 Network for Example 4.1.

1 and 2, branches 1 and 3, and branch 4. β_F and α_F are given by

$$\beta_F = \begin{bmatrix} 1 & 1 & 1 & 0 & 0 \\ 0 & 0 & 0 & 0 & 1 \end{bmatrix} \qquad \alpha_F = \begin{bmatrix} -1 & 1 & 0 & 0 & 0 \\ -1 & 0 & 1 & 0 & 0 \\ 0 & 0 & 0 & 1 & 0 \end{bmatrix}$$

Note that there is only one entry in the second row of β_F because of the self-loop and only one entry in the third row of α_F because of the dangling branch.

4.9 INVERSE RELATIONS

In Eqs. (4.7.2) and (4.8.2), the branch currents and branch voltages are expressed in terms of the independent current and voltage variables, respectively. We now wish to formulate inverse relations, i.e., expressions giving the independent variables in terms of the branch quantities. These inversions cannot, of course, be performed directly since β_F and α_F are not square matrices. Let us define the following relations:

$$I_L = \gamma_F J_B \tag{4.9.1}$$

and

$$E_N = \delta_F V_B \tag{4.9.2}$$

Each tree of a network has only one β_F and one α_F matrix. Are γ_F and δ_F also unique in this respect? Substituting Eq. (4.7.2) into Eq. (4.9.1), and Eq. (4.8.2) into Eq. (4.9.2) yields

$$I_L = \gamma_F \beta_F{}^T I_L \tag{4.9.3}$$

and

$$E_N = \delta_F \alpha_F \; E_N \tag{4.9.4}$$

Thus γ_F and δ_F must satisfy the matrix equations

$$\gamma_F \beta_F{}^T = U_L \tag{4.9.5}$$

and

$$\delta_F \alpha_F{}^T = U_N \tag{4.9.6}$$

where U_L and U_N are identity matrices of order L and N, respectively. Equations (4.9.5) and (4.9.6) do not yield unique solutions for γ_F and δ_F for specified β_F and α_F.† For example, in Fig. 4.5, two γ_F matrices are

$$\gamma_{F_1} = \begin{bmatrix} 1 & 0 & 0 & 0 & 0 \\ 0 & 0 & 0 & 1 & 0 \end{bmatrix} \qquad \gamma_{F_2} = \begin{bmatrix} 0 & 1 & 0 & 0 & 0 \\ 0 & -1 & 0 & 0 & 1 \end{bmatrix}$$

† The solution of Prob. 2.13 may be helpful in understanding this point.

4.10 NONFUNDAMENTAL TIE SETS AND CUT SETS

The definitions of fundamental tie set and fundamental cut set given in Secs. 4.7 and 4.8 pertained to a particular tree and links of a graph. We defined the L link currents as independent current variables and the N tree-branch voltages as independent voltage variables. One link current defined each fundamental tie set and one tree-branch voltage each fundamental cut set.

In this section we consider less restrictive definitions of the independent current and voltage variables, which, in turn, will lead us to definitions of nonfundamental tie sets and cut sets.

Consider a connected graph of B branches. Instead of defining each independent current variable as a single link current, let us try to define each one as a linear combination of the B branch currents as follows:

$$\begin{aligned} i_1 &= \gamma_{11}j_1 + \gamma_{12}j_2 + \cdots + \gamma_{1B}j_B \\ i_2 &= \gamma_{21}j_1 + \gamma_{22}j_2 + \cdots + \gamma_{2B}j_B \\ &\cdots\cdots\cdots\cdots\cdots \\ i_L &= \gamma_{L1}j_1 + \gamma_{L2}j_2 + \cdots + \gamma_{LB}j_B \end{aligned} \tag{4.10.1}$$

In matrix form, Eqs. (4.10.1) are written

$$I = \gamma J_B \tag{4.10.2}$$

Compare Eqs. (4.10.2) and (4.9.1). They have the same form but we have omitted the subscripts from I and γ in Eq. (4.10.2) to stress the fact that the current variables are no longer single-link currents.

We can obtain similar relations on a voltage basis. Thus

$$\begin{aligned} e_1 &= \delta_{11}v_1 + \delta_{12}v_2 + \cdots + \delta_{1B}v_B \\ e_2 &= \delta_{21}v_1 + \delta_{22}v_2 + \cdots + \delta_{2B}v_B \\ &\cdots\cdots\cdots\cdots\cdots \\ e_N &= \delta_{N1}v_1 + \delta_{N2}v_2 + \cdots + \delta_{NB}v_B \end{aligned} \tag{4.10.3}$$

and

$$E = \delta V_B \tag{4.10.4}$$

What constraints must be imposed on γ and δ in Eqs. (4.10.2) and (4.10.4) if I and E are to be acceptable sets of current and voltage variables for a given graph? To answer this question, we choose any tree of the graph and form the fundamental-tie-set and cut-set matrices for that tree. We denote the variables thus defined by I_L and E_N. They are given by Eqs. (4.7.2) and (4.8.2), respectively. Substituting Eq. (4.7.2) into Eq. (4.10.2) and Eq. (4.8.2) into Eq. (4.10.4) yields

$$I = \gamma\beta_F{}^T I_L \equiv P I_L \tag{4.10.5}$$

and

$$E = \delta \alpha_F{}^T E_N \equiv Q E_N \tag{4.10.6}$$

Equations (4.10.5) and (4.10.6) express the proposed variables in terms of known acceptable sets of link-current and tree-branch voltage variables through the new matrices P and Q. The only requirement for the proposed variables I and E to be acceptable is that Eqs. (4.10.5) and (4.10.6) each be a linearly independent set of equations. This will be true if P is of rank L and Q is of rank N, that is, if $|P|$ and $|Q|$ do not vanish.

Assuming that the proposed new variables are acceptable, we would like to express the branch currents and branch voltages in terms of them through new tie-set and cut-set matrices. We do this through Eqs. (4.7.2), (4.8.2), (4.10.5), and (4.10.6) as follows:

$$J_B = \beta_F{}^T I_L = \beta_F{}^T P^{-1} I \equiv \beta^T I \tag{4.10.7}$$

and

$$V_B = \alpha_F{}^T E_N = \alpha_F{}^T Q^{-1} E \equiv \alpha^T E \tag{4.10.8}$$

where

$$\beta = (P^{-1})^T \beta_F \tag{4.10.9}$$

and

$$\alpha = (Q^{-1})^T \alpha_F \tag{4.10.10}$$

Equations (4.10.7) and (4.10.8) suggest general definitions of the terms tie set and cut set. Thus define a *tie set* as a set of branches energized when one independent current flows and the other independent currents are zero, and define a *cut set* as a set of branches energized when one independent voltage is applied and the other independent voltages are zero. Note that these definitions differ from the previous ones only in that "link current" and "tree-branch voltage" are replaced by "independent current" and "independent voltage," respectively.

In subsequent discussions we shall wish to continue to distinguish those cases where link currents and tree-branch voltages are the independent variables from cases where they are not. We thus adopt the convention of subscripting quantities which pertain to fundamental cut sets and tie sets and leaving unsubscripted those which do not. For example, β_F will denote a fundamental tie-set matrix, whereas β will represent one that is not fundamental. Also E_N will represent a tree-branch voltage matrix, whereas E will denote one that contains other independent voltage variables. With this subscripting convention, Eqs. (4.7.2), (4.7.4), (4.8.2), and (4.8.4) become, for the nonfundamental case,

$$J_B = \beta^T I \tag{4.10.11}$$

$$\beta V_B = 0 \tag{4.10.12}$$

$$V_B = \alpha^T E \tag{4.10.13}$$

$$\alpha J_B = 0 \tag{4.10.14}$$

Example 4.2 Consider the graph of Fig. 4.2, which is redrawn in Fig. 4.9. Let us propose the following set of voltage variables:

$$e_1 = -v_1 - v_4 - v_5$$

$$e_2 = -v_4 - v_5$$

$$e_3 = -v_1 - v_2 - v_3$$

We wish to determine whether they form an acceptable set. A δ matrix can be formed by inspection of the equations

$$\delta = \begin{bmatrix} -1 & 0 & 0 & -1 & -1 \\ 0 & 0 & 0 & -1 & -1 \\ -1 & -1 & -1 & 0 & 0 \end{bmatrix}$$

Choose any tree, say 235 as was done previously. Then α_F is given by Eq. (4.8.1) as

$$\alpha_F = \begin{bmatrix} -1 & 1 & 0 & 0 & 0 \\ 0 & 0 & 1 & 1 & 0 \\ -1 & 0 & 0 & -1 & 1 \end{bmatrix}$$

From Eq. (4.10.6),

$$Q = \delta\alpha_F{}^T = \begin{bmatrix} 1 & -1 & 1 \\ 0 & -1 & 0 \\ 0 & -1 & 1 \end{bmatrix}$$

We note that $|Q| = -1$ and thus the proposed voltage variables are an acceptable set. The new cut-set matrix is, from Eq. (4.10.10),

$$\alpha = (Q^{-1})^T\alpha_F = \begin{bmatrix} -1 & 1 & 0 & 0 & 0 \\ 1 & 0 & -1 & 0 & -1 \\ 0 & -1 & 0 & -1 & 1 \end{bmatrix}$$

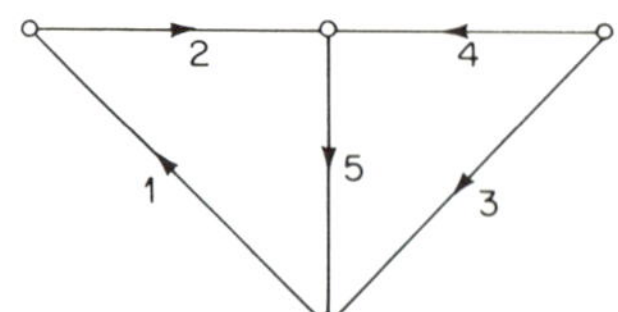

Fig. 4.9 Graph of the network in Fig. 4.1.

The reader should note that the proposed voltage variables are node voltages with respect to the extreme right-hand node as a datum node in Fig. 4.9. Thus this example illustrates that node-to-datum voltages are acceptable as independent voltage variables.

4.11 MORE ABOUT TIE-SET AND CUT-SET MATRICES

In Sec. 4.10 we were given a proposed set of current (voltage) variables in terms of the branch currents (voltages), and we derived expressions for the latter in terms of the former, thus defining nonfundamental tie sets (cut sets). The process involved the intermediate step of finding a fundamental tie-set (cut-set) matrix from the network graph.

Suppose we are given Eq. (4.10.7) [Eq. (4.10.8)] but not the network graph. Can we proceed backward to a fundamental tie-set (cut-set) matrix?

Assume β is given and thus Eq. (4.10.7) is known. Reorder the rows of β^T (and thus of J_B) in Eq. (4.10.7) so that the first L rows have an $L \times L$ nonvanishing determinant, and partition Eq. (4.10.7) as follows:

$$\begin{bmatrix} J_L \\ J_N \end{bmatrix} = \begin{bmatrix} \beta_{LL}{}^T \\ \beta_{NL}{}^T \end{bmatrix} [I] \tag{4.11.1}$$

where the subscripts denote the orders of the submatrices. Since β_{LL} is nonsingular, the upper portion of Eq. (4.11.1) can be solved for the current variables

$$I = (\beta_{LL}{}^T)^{-1} J_L \tag{4.11.2}$$

Equation (4.11.2) is identical in form to Eq. (4.10.5) because J_L in Eq. (4.11.2) [like I_L in Eq. (4.10.5)] must be a set of link currents, since β_{LL} is nonsingular. The distinction between these two developments is that in Sec. 4.10 we were given Eq. (4.10.2) and the graph, whereas here only Eq. (4.10.7) is available.

To obtain the fundamental-tie-set matrix associated with J_L in Eq. (4.11.2), we substitute that equation into Eq. (4.10.7)

$$J_B = \beta^T I = \beta^T (\beta_{LL}{}^T)^{-1} J_L \tag{4.11.3}$$

Consider Eq. (4.11.3) carefully. It relates all the branch currents to L independent branch currents. But, as we have noted, the L independent branch currents are L link currents. Thus J_L is I_L, and $\beta^T(\beta_{LL}{}^T)^{-1}$ is the transpose of the fundamental-tie-set matrix for that set of links. Thus

$$\beta_F = [\beta^T (B_{LL}{}^T)^{-1}]^T = \beta_{LL}{}^{-1} \beta \tag{4.11.4}$$

The result in Eq. (4.11.4) is most significant, for it permits us to transform a given tie-set matrix of a graph into a fundamental-tie-set matrix of that same graph. We shall use this result in Sec. 4.13 as the first step in obtaining a graph from a given tie-set matrix.

Example 4.3 Assume that the following tie-set matrix is given without its graph:

$$\beta = \begin{bmatrix} 1 & 1 & 2 & -2 & -1 \\ 2 & 2 & 1 & -1 & 1 \end{bmatrix}$$

We wish to define the independent variables in terms of L independent branch currents and relate the other $B - L$ branch currents to them.

β is obviously not a fundamental-tie-set matrix since it contains entries other than $+1$, $-1, 0$. However, we do know that the network to which it applies has five branches and two links. Thus we can define the current variables in terms of two independent branch currents. The governing relation is Eq. (4.10.7), which yields in this case

$$\begin{bmatrix} j_1 \\ j_2 \\ j_3 \\ j_4 \\ j_5 \end{bmatrix} = \begin{bmatrix} 1 & 2 \\ 1 & 2 \\ 2 & 1 \\ -2 & -1 \\ -1 & 1 \end{bmatrix} \begin{bmatrix} i_1 \\ i_2 \end{bmatrix} \tag{4.11.5}$$

There are eight nonvanishing 2×2 determinants in β. Choosing j_1 and j_4 as independent branch currents and reordering rows according to Eq. (4.11.1), we obtain

$$\begin{bmatrix} j_1 \\ j_4 \\ j_2 \\ j_3 \\ j_5 \end{bmatrix} = \begin{bmatrix} 1 & 2 \\ -2 & -1 \\ 1 & 2 \\ 2 & 1 \\ -1 & 1 \end{bmatrix} \begin{bmatrix} i_1 \\ i_2 \end{bmatrix} \tag{4.11.6}$$

Performing the inversion indicated in Eq. (4.11.2) gives the current variables in terms of the chosen independent branch currents

$$\begin{bmatrix} i_1 \\ i_2 \end{bmatrix} = \begin{bmatrix} -\frac{1}{3} & -\frac{2}{3} \\ \frac{2}{3} & \frac{1}{3} \end{bmatrix} \begin{bmatrix} j_1 \\ j_4 \end{bmatrix} \tag{4.11.7}$$

To relate all the branch currents to the two independent ones, Eq. (4.11.3) is used. The final result is

$$\begin{bmatrix} j_1 \\ j_4 \\ j_2 \\ j_3 \\ j_5 \end{bmatrix} = \begin{bmatrix} 1 & 2 \\ -2 & -1 \\ 1 & 2 \\ 2 & 1 \\ -1 & 1 \end{bmatrix} \begin{bmatrix} -\frac{1}{3} & -\frac{2}{3} \\ \frac{2}{3} & \frac{1}{3} \end{bmatrix} \begin{bmatrix} j_1 \\ j_4 \end{bmatrix} = \begin{bmatrix} 1 & 0 \\ 0 & 1 \\ 1 & 0 \\ 0 & -1 \\ 1 & 1 \end{bmatrix} \begin{bmatrix} j_1 \\ j_4 \end{bmatrix} \tag{4.11.8}$$

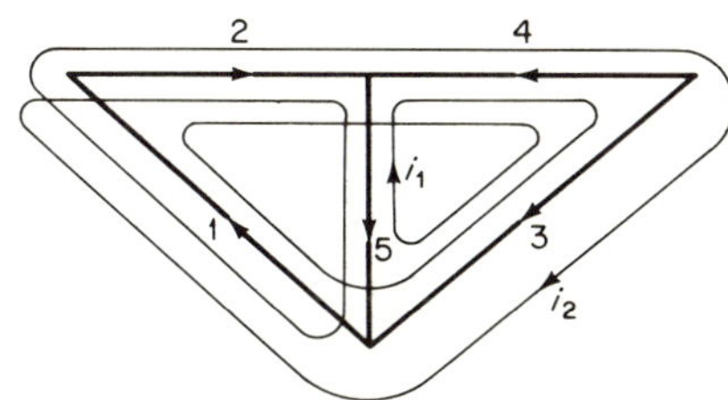

Fig. 4.10 A new choice of current variables.

It is interesting to note that if a Gauss-Jordan reduction is performed on Eq. (4.11.5) so as to obtain i_1 and i_2 in terms of j_1 and j_4, the result will be

$$\begin{bmatrix} 1 & 0 \\ 0 & 1 \\ 0 & 0 \\ 0 & 0 \\ 0 & 0 \end{bmatrix} \begin{bmatrix} i_1 \\ i_2 \end{bmatrix} = \begin{bmatrix} -\tfrac{1}{3}j_1 - \tfrac{2}{3}j_4 \\ \tfrac{2}{3}j_1 + \tfrac{1}{3}j_4 \\ j_2 - j_1 \\ j_3 + j_4 \\ j_5 - j_1 - j_4 \end{bmatrix} \tag{4.11.9}$$

For Eq. (4.11.9) to be consistent, it is necessary and sufficient that the three current expressions in rows, 3, 4, and 5 be identically zero. These are the same as the last three expressions in Eq. (4.11.8). The equations of the first two rows in Eq. (4.11.9) are identical to the two equations obtained in Eq. (4.11.7). Thus our partitioned-matrix formulation is really just an artful (and often laborsaving) way of carrying out a Gauss-Jordan reduction.

Two questions raised by Example 4.3 remain to be answered: (1) What is the physical significance of those entries in a tie-set matrix which are not $+1$, -1, or 0? (2) What is the physical significance of the three branch-current equations in Eq. (4.11.8) that must be identically zero for consistency? To answer these questions, consider the following example.

Example 4.4 A set of current variables is defined in Fig. 4.10. Show that the tie-set matrix is that of Example 4.3.

We note from the graph that

$$\begin{aligned} j_1 &= i_1 + 2i_2 \\ j_2 &= i_1 + 2i_2 \\ j_3 &= 2i_1 + i_2 \\ j_4 &= -2i_1 - i_2 \\ j_5 &= -i_1 + i_2 \end{aligned}$$

Thus the tie-set matrix is, by inspection, that of Example 4.3

$$\beta = \begin{bmatrix} 1 & 1 & 2 & -2 & -1 \\ 2 & 2 & 1 & -1 & 1 \end{bmatrix}$$

It is easy to see from the graph that the value of each entry in β is the net number of times a current variable passes through a branch. For example, i_1 passes through branch

4 no times in the positive direction and two times in the negative direction; hence that entry in β is -2.

The three branch-current constraint relations of Eq. (4.11.8) can also be given a physical interpretation in terms of the graph in Fig. 4.10. These equations are repeated for ease of reference

$$j_2 - j_1 = 0$$
$$j_3 + j_4 = 0$$
$$j_5 - j_1 - j_4 = 0$$

Each of these equations is a KCL equation at a node (two simple and one super) of the graph. Is there any significance to the particular nodes used? Recall that the current variables in Example 4.3 were defined in terms of j_1 and j_4 by

$$i_1 = -\tfrac{1}{3}j_1 - \tfrac{2}{3}j_4$$
$$i_2 = \tfrac{2}{3}j_1 + \tfrac{1}{3}j_4$$

As we have noted, these current variables are not single link currents. But, if we choose the tree 235, these currents are defined in terms of a linear combination of link currents. Then each of the three KCL equations is the summation of the branch currents in the fundamental cut sets of the tree 235 in Fig. 4.9. These summations have previously been given in Eq. (4.8.5).

4.12 RELATING THE TIE-SET AND CUT-SET MATRICES

Equations (4.7.4) and (4.8.2) can be combined to yield a relation between the fundamental-tie-set and cut-set matrices

$$\beta_F \alpha_F{}^T E_N = 0 \tag{4.12.1}$$

Since E_N is arbitrary, it follows that

$$\beta_F \alpha_F{}^T = 0 \tag{4.12.2}$$

There is no stipulation in Eqs. (4.7.4) and (4.8.2) that β_F and α_F must come from the same tree. Thus Eq. (4.12.2) may be generalized to

$$\beta_{F1} \alpha_{F2}{}^T = 0 \tag{4.12.3}$$

We have shown that it is not necessary to use fundamental-tie-set and cut-set matrices in relating the branch currents and voltages to the independent current and voltage variables, respectively. Thus Eq. (4.12.3) can be further generalized to

$$\beta_1 \alpha_2{}^T = 0 \tag{4.12.4}$$

Equation (4.12.4) states that, for a given network, the matrix product of any tie-set matrix and the transpose of any cut-set matrix is an $L \times N$ null matrix.

The reader should verify that a similar development using Eqs. (4.7.2) and (4.8.4) yields

$$\alpha_F \beta_F{}^T = 0 \tag{4.12.5}$$

$$\alpha_{F1} \beta_{F2}{}^T = 0 \tag{4.12.6}$$

$$\alpha_1 \beta_1{}^T = 0 \tag{4.12.7}$$

where the null matrices are now $N \times L$.

Let us consider Eq. (4.12.2) and ask whether we can use it to find β_F if α_F is known or vice versa. If we number the graph so that the links are 1 through L and the tree branches $L + 1$ through B, the fundamental-tie-set and cut-set matrices will take the partitioned forms

$$\beta_F = [U_{LL} \quad \beta_{LN}] \tag{4.12.8}$$

and

$$\alpha_F = [\alpha_{NL} \quad U_{NN}] \tag{4.12.9}$$

where the subscripts denote the orders of the submatrices. Inserting Eqs. (4.12.8) and (4.12.9) into Eq. (4.12.2) yields

$$[U_{LL} \quad \beta_{LN}] \begin{bmatrix} \alpha_{NL}{}^T \\ U_{NN} \end{bmatrix} = \alpha_{NL}{}^T + \beta_{LN} = 0 \tag{4.12.10}$$

Thus

$$\beta_{LN} = -\alpha_{NL}{}^T \tag{4.12.11}$$

Knowing α_F, we of course know α_{NL}. We merely transpose α_{NL} and negate each entry to obtain β_{LN}. Knowing β_{LN}, we can assemble β_F. The process works equally well in reverse to find α_F knowing β_F.

4.13 FROM TIE-SET OR CUT-SET MATRIX TO GRAPH†

In this section we develop a method for deriving a graph from a fundamental-cut-set matrix. No dual development for a tie-set matrix is given since it is always possible, by the method of Sec. 4.11, to convert a given tie-set matrix into a fundamental-tie-set matrix and then, through the relationships in Sec. 4.12, to convert the latter into a fundamental-cut-set matrix. It is also always possible, by a method paralleling that of Sec. 4.11, to convert a cut-set matrix into a fundamental one.

The reader will soon recognize that we are embarking on a synthesis. Since the theme of the book is analysis, why include a synthesis topic? The reason is that this particular procedure provides considerable insight into the

† The development in this section is based on Ref. 4.1.

construction of a graph and its relation to its cut-set matrix. The reader is invited to test this statement after studying this section.

STEP 1 DERIVATION OF α_{NL} FROM α_F

Our first step is to reorder the columns of the fundamental-cut-set matrix so that the last N columns are the tree-branch voltage columns and form an $N \times N$ identity matrix. Thus

$$\alpha_F = [\alpha_{NL} \quad U_{NN}] \tag{4.13.1}$$

If we agree always to place α_F in the form of Eq. (4.13.1), we gain nothing, in terms of constructing the graph, by retaining the U_{NN} portion of α_F in the subsequent discussion. The submatrix α_{NL} contains all the information necessary to construct the graph.

Example 4.5 The essential idea in the method to be developed is that a tree and its associated links can be assembled directly from an inspection of the rows of α_{NL}. To illustrate, let

$$\alpha_{NL} = \begin{array}{c} \\ 4 \\ 5 \\ 6 \\ 7 \end{array} \begin{array}{c} \begin{array}{ccc} 1 & 2 & 3 \end{array} \\ \begin{bmatrix} 0 & 0 & -1 \\ 1 & 1 & 0 \\ 0 & -1 & -1 \\ 1 & 0 & -1 \end{bmatrix} \end{array}$$

The numbers at the top and to the left of α_{NL} refer to the links and tree branches, respectively. Let us try to develop the graph by considering the rows (cut sets) of α_{NL} one at a time, starting with the bottom row.

Branch 7 and its associated links (the links included in its fundamental cut set) are shown in Fig. 4.11*a*. Now branch 6 is added. Its cut set includes links 2 and 3. Thus, in Fig. 4.11*b*, link 3 must span both tree branches, whereas link 1 is across only tree branch 7 and link 2 is across only tree branch 6. Consider next the cut set for tree branch 5. It contains links 1 and 2 and thus must be inserted as shown in Fig. 4.11*c*. Finally,

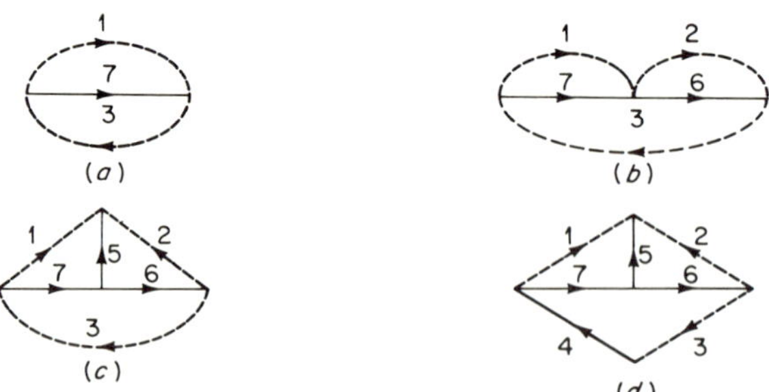

Fig. 4.11 How to grow a tree.

the cut set for branch 4 contains only link 3. Branch 4 is placed in series with link 3 at either end of the graph. The final construction appears in Fig. 4.11*d*. The reader should verify that it yields the given α_{NL}.

It might appear that Example 4.5 is in itself a complete procedure for deriving a graph from a fundamental-cut-set matrix. In this sense the example is deceptively simple because the order in which we considered the tree branches was not important. In the general case this order is important, for it determines the structure of the tree. We thus devote the next two steps of our procedure to dissecting and reordering α_{NL} so as to identify the growth pattern of the tree, i.e., the arrangement of the tree branches from the "trunk" of the tree out to its "tips."

STEP 2 PRELIMINARY REDUCTIONS

Let us first identify entries in α_{NL} that pertain to:

1. A link in parallel with a single tree branch
2. A link in series with a tree branch at a tip of the tree, hereafter denoted as a *tree-tip branch*
3. Links in parallel with each other

Having identified these entries, we shall delete them from α_{NL} for the present, repeating the process until none of the steps can be repeated again.

To discuss the significance of each of these special entries consider the graph in Fig. 4.12. The α_{NL} portion of its fundamental-cut-set matrix is a 6×8 submatrix given by

$$\alpha_{68} = \begin{array}{c} \\ 9 \\ 10 \\ 11 \\ 12 \\ 13 \\ 14 \end{array} \begin{array}{c} \begin{array}{cccccccc} 1 & 2 & 3 & 4 & 5 & 6 & 7 & 8 \end{array} \\ \begin{bmatrix} -1 & 1 & 0 & 0 & 0 & 0 & 0 & 0 \\ -1 & 0 & -1 & 1 & 0 & 0 & 0 & 0 \\ -1 & 0 & -1 & 0 & -1 & 0 & 0 & 0 \\ 0 & 0 & 0 & 1 & 1 & 1 & 0 & 0 \\ 0 & 0 & 0 & 0 & 1 & 0 & 1 & 1 \\ 0 & 0 & 0 & 0 & 0 & 1 & -1 & -1 \end{bmatrix} \end{array} \qquad (4.13.2)$$

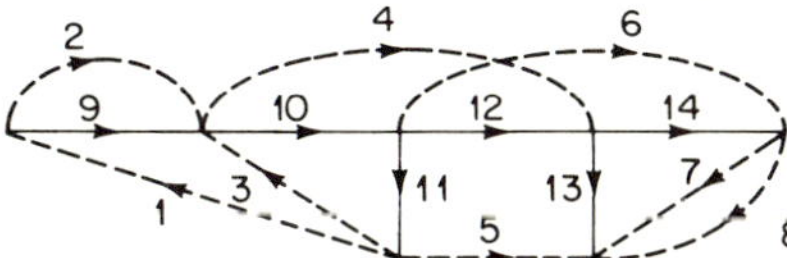

Fig. 4.12 Graph to illustrate reduction techniques.

where the link numbers are indicated at the top and the tree-branch numbers to the left.

First note that columns 7 and 8 are identical. This means that links 7 and 8 belong to the same fundamental cut sets. Referring to the graph, we see that this is possible if and only if links 7 and 8 are in parallel. We thus delete all but one of these columns (delete column 8) from α_{68} and open these links (link 8) in the graph.

Our next observation is that column 2 has only one nonzero entry. This means that link 2 belongs to only one fundamental cut set (cut set 9). Again referring to the graph, we reason that this is possible if and only if link 2 is in parallel with a single tree branch (tree branch 9). We eliminate column 2 from α_{68} and remove link 2 from the graph. Our reduced matrix is now given by

$$\alpha_{66} = \begin{array}{c} \\ 9 \\ 10 \\ 11 \\ 12 \\ 13 \\ 14 \end{array} \begin{array}{c} \begin{array}{cccccc} 1 & 3 & 4 & 5 & 6 & 7 \end{array} \\ \begin{bmatrix} -1 & 0 & 0 & 0 & 0 & 0 \\ -1 & -1 & 1 & 0 & 0 & 0 \\ -1 & -1 & 0 & -1 & 0 & 0 \\ 0 & 0 & 1 & 1 & 1 & 0 \\ 0 & 0 & 0 & 1 & 0 & 1 \\ 0 & 0 & 0 & 0 & 1 & -1 \end{bmatrix} \end{array} \tag{4.13.3}$$

We next turn our attention to the first row in Eq. (4.13.3). Note that it has only one entry. This means that link 1 is the only link receiving a voltage contribution from tree-branch voltage 9. Recalling that link 2 has been removed from the graph, we note that one way in which a row can have a single entry is for a link and a tree-tip branch to be in series. This is not the only way, but it is the easiest way. Thus in Fig. 4.12 we short tree branch 9 and eliminate the first row from Eq. (4.13.3). This places links 1 and 3 in parallel in Fig. 4.12 and makes the first two columns alike in the remaining α_{56} matrix. We open link 1 and remove the first column from α_{56}. Our graph now appears in the reduced form of Fig. 4.13. Its

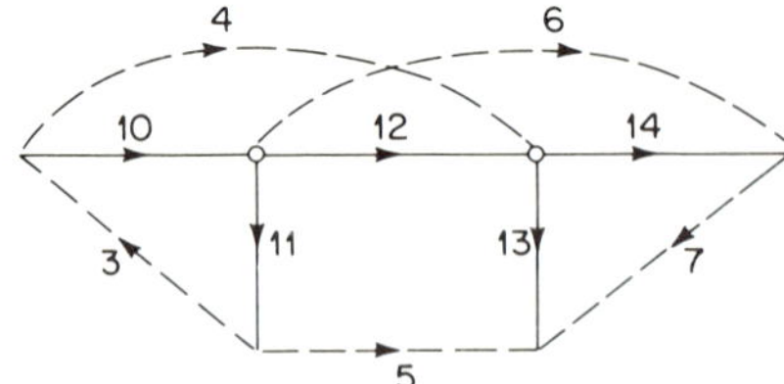

***Fig. 4.13** Graph of Fig. 4.12 after preliminary reductions.*

fundamental-cut-set matrix is given by

$$\alpha_{55} = \begin{array}{c} \\ 10 \\ 11 \\ 12 \\ 13 \\ 14 \end{array} \begin{array}{c} \begin{array}{ccccc} 3 & 4 & 5 & 6 & 7 \end{array} \\ \begin{bmatrix} -1 & 1 & 0 & 0 & 0 \\ -1 & 0 & -1 & 0 & 0 \\ 0 & 1 & 1 & 1 & 0 \\ 0 & 0 & 1 & 0 & 1 \\ 0 & 0 & 0 & 1 & -1 \end{bmatrix} \end{array} \tag{4.13.4}$$

The matrix α_{55} has no two columns alike and no rows or columns with single entries. Hence no further preliminary reductions can be made.

STEP 3 IDENTIFICATION OF SIMPLE-NODE CUT SETS

This step forms the core of the method. In it we consider, in turn, each of the fundamental cut sets in the reduced cut-set matrix and establish whether a given fundamental cut set relates to a simple node or a supernode. A simple-node cut set contains a tree-tip branch whereas a supernode cut set does not. Identification of the tree-tip branches by this means permits us to establish the growth pattern of the tree and to reorder the rows of the fundamental-cut-set matrix so that we can grow the tree in a step-by-step manner from a set of trunk tree branches connected to a simple seed node.

Consider the fundamental cut set of branch 12 in Fig. 4.13. Branch 12 is not a tree-tip branch. Its fundamental cut set consists of branches 4, 5, 6, and 12. Deleting these branches from the graph yields a subgraph with two separate parts, neither of which is a simple isolated node, as shown in Fig. 4.14*a*. The fundamental cut set of branch 12 is a supernode cut set. How is the cut-set matrix affected by the removal of these branches? Deleting the third row and the second, third, and fourth columns from α_{55} leaves

$$\alpha_{42}{}^{12} = \begin{array}{c} \\ 10 \\ 11 \\ \\ 13 \\ 14 \end{array} \begin{array}{c} \begin{array}{cc} 3 & 7 \end{array} \\ \left[\begin{array}{c:c} -1 & 0 \\ -1 & 0 \\ \hdashline 0 & 1 \\ 0 & -1 \end{array}\right] \end{array} \tag{4.13.5}$$

where the superscript identifies the tree branch of the cut set. Note that $\alpha_{42}{}^{12}$ is effectively split into two parts, as indicated by the dashed partition lines. Branches 3, 10, and 11 form one part; branches 7, 13, and 14 the other.

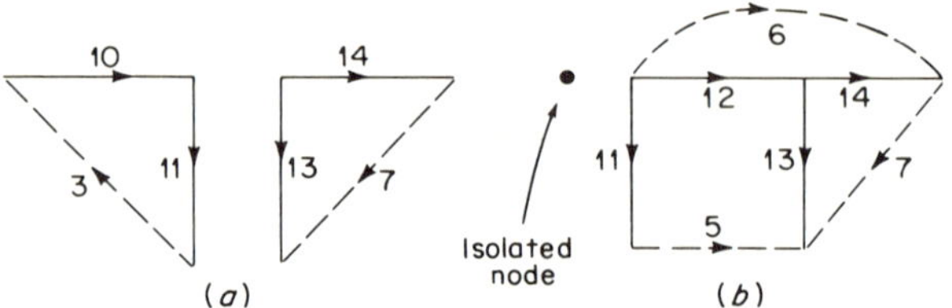

Fig. 4.14 A supernode and a simple-node cut set.

Next consider the fundamental cut set of branch 10 in Fig. 4.13. Branch 10 is a tree-tip branch. Its fundamental cut set contains branches 3, 4, and 10. Deleting these branches from the graph in Fig. 4.13 again yields a subgraph with two separate parts, but this time one of them is an isolated node, as shown in Fig. 4.14*b*. Thus branch 10 forms a simple-node cut set. The remaining submatrix, after deleting the first row and the first two columns from α_{55}, is

$$\alpha_{43}{}^{10} = \begin{array}{c} \\ 11 \\ 12 \\ 13 \\ 14 \end{array} \begin{array}{c} \begin{array}{ccc} 5 & 6 & 7 \end{array} \\ \begin{bmatrix} -1 & 0 & 0 \\ 1 & 1 & 0 \\ 1 & 0 & 1 \\ 0 & 1 & -1 \end{bmatrix} \end{array} \tag{4.13.6}$$

It is not possible, even by rearranging rows and columns, to put $\alpha_{43}{}^{10}$ in the partitioned form of $\alpha_{42}{}^{12}$ in Eq. (4.13.5). And it should not be possible, for an isolated node contributes nothing to a cut-set matrix. Thus $\alpha_{43}{}^{10}$ should, and does, retain the form of a fundamental-cut-set matrix of a graph with one separate part.

Herein lies the philosophy of step 3. We examine, by the procedure just outlined, each fundamental cut set of the cut-set matrix remaining after completion of the preliminary reductions in step 2. This examination permits us to determine whether a given cut set is of the simple-node or supernode variety. If it is the former, we know that the tree branch involved is a tree-tip branch. After examining all the cut sets in this way, we remove all rows associated with the tree-tip branches from the fundamental-cut-set matrix. We then return to step 2 and once again exhaust the preliminary reduction possibilities. Next we return to step 3 and identify the next set of tree-tip branches, and so on. The ultimate result is the reduction of the fundamental-cut-set matrix to a set of trunk tree branches connected to a simple seed node.

For the graph in Fig. 4.13 and its fundamental-cut-set matrix in Eq. (4.13.4), we identify branches 10, 11, 13, and 14 as tree-tip branches. Deleting these (and links 3 and 7) leaves a single trunk branch 12 with links 4, 5, and 6 across it.

STEP 4 REORDERING ROWS IN α_{NL}

We return now to the original α_{68} in Eq. (4.13.2) and reorder its rows so that trunk tree branch 12 forms the bottom row. Next those rows containing the last set of removed tree-tip branches (branches 10, 11, 13, and 14) are added in any order. Finally the row containing the second last set of tree-tip branches is added (branch 9). Now the graph of α_{68} can be realized by the technique of Example 4.5.

REFERENCES

4.1. Guillemin, E. A.: How to Grow Your Own Trees from Given Cut-set or Tie-set Matrices, *IRE Trans. Circuit Theory*, vol. CT-6, pp. 110–126, May, 1959. In addition to developing the procedure presented in Sec. 4.13, the article provides an invaluable physical feel for the topological structure of networks.

4.2. *IRE Trans. Circuit Theory*, Special Issue on Topology, vol. CT-8, March, 1961. This issue will acquaint the interested student with the wide variety of research activity in linear graph theory.

4.3. Standards on Circuits, *Proc. IRE*, vol 39, Jan., 1951. A compilation of definitions of terms in network topology.

4.4. Reza, F. M., and S. Seely: "Modern Network Analysis," McGraw-Hill Book Company, New York, 1959. Chapter 4 is an excellent introduction to linear graphs and fundamental tie sets and cut sets. The terminology and mathematical background of topology are presented in the appendix.

4.5. Seshu, S., and N. Balabanian: "Linear Network Analysis," John Wiley & Sons, Inc., New York, 1959. An introduction to topology is presented in Chapter 3.

4.6. Seshu, S., and M. B. Reed: "Linear Graphs and Electrical Networks," Addison-Wesley Publishing Company, Inc., Reading, Mass., 1961. Linear graph theory and its applications are discussed in depth.

4.7. Weinberg, L.: "Network Analysis and Synthesis," McGraw-Hill Book Company, New York, 1962. Chapter 3 includes discussions of network geometry and vertex (incidence), cut-set, and tie-set matrices.

PROBLEMS

Drill

4.1. For the given network with branches numbered as shown:
(*a*) Draw a directed linear graph.

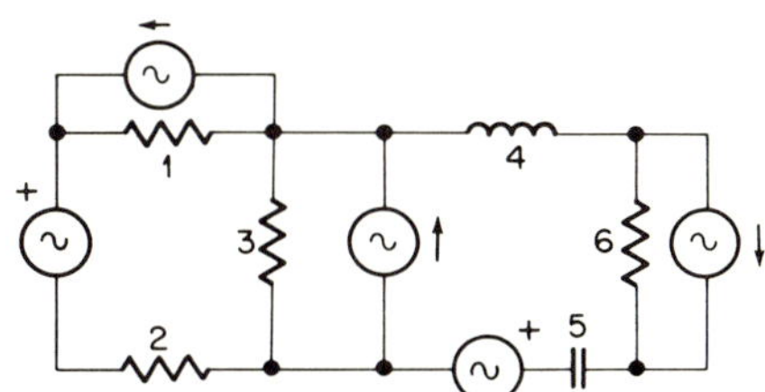

Fig. P4.1

(*b*) Draw all possible trees.

(*c*) Select a tree and define a set of independent loop currents and a set of independent node-pair voltages.

4.2. Repeat Prob. 4.1 for the network in Fig. P4.2.

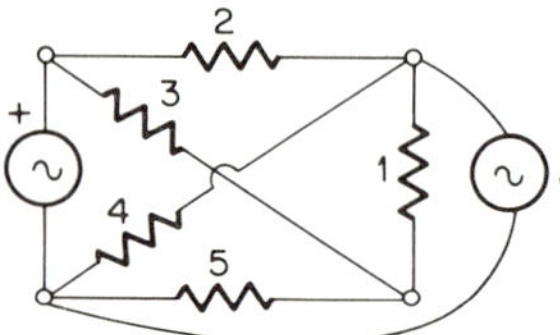

Fig. P4.2

4.3. For the network of Fig. P4.1, select a tree and determine the fundamental-tie-set and cut-set matrices.

4.4. Repeat Prob. 4.3 for the network of Fig. P4.2.

4.5. For the network shown:

(*a*) Draw a directed linear graph.

(*b*) Draw all trees.

(*c*) Choose a tree and find the fundamental-tie-set and cut-set matrices.

(*d*) Verify that the number of trees is given by $|\alpha_F \alpha_F^T| = |\beta_F \beta_F^T|$.

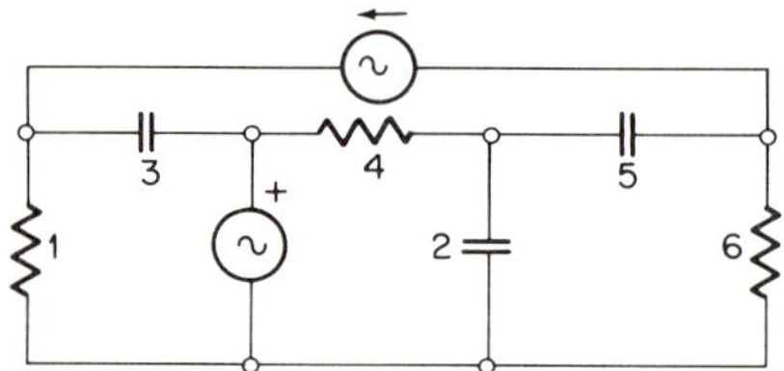

Fig. P4.5

4.6. Repeat Prob. 4.5 for the given networks. In each network find the voltages across current generators and the currents through voltage generators.

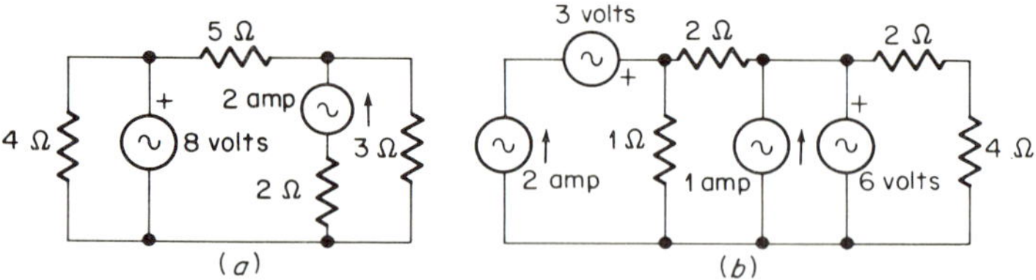

Fig. P4.6

4.7. For each of the networks draw a directed linear graph, choose a tree, and write the fundamental-tie-set and cut-set matrices. In each case the tree should be chosen so that voltage sources appear in links and current sources across tree branches.

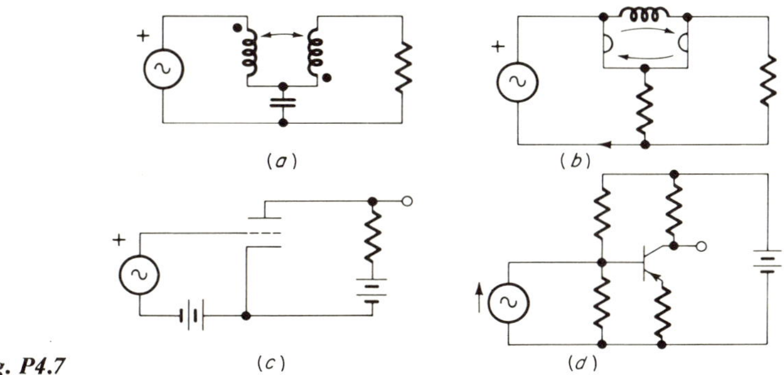

Fig. P4.7

4.8. For the given graph define the independent current variables by

$$i_1 = j_1 + j_2 + j_3 - j_4$$

$$i_2 = j_1 + 2j_2 + j_5$$

$$i_3 = j_1 - j_3 + j_4 - j_6$$

Verify that these currents form an independent set and determine the tie-set matrix associated with them *(a)* by solving the *j*'s in terms of the *i*'s directly and *(b)* by using the procedure of Sec. 4.10. Show the path of each current variable on the graph.

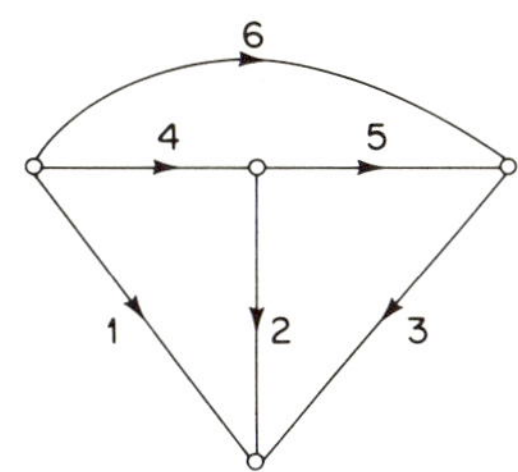

Fig. P4.8

4.9. For the given graph, the currents indicated by the dashed lines are proposed as an independent set of variables.

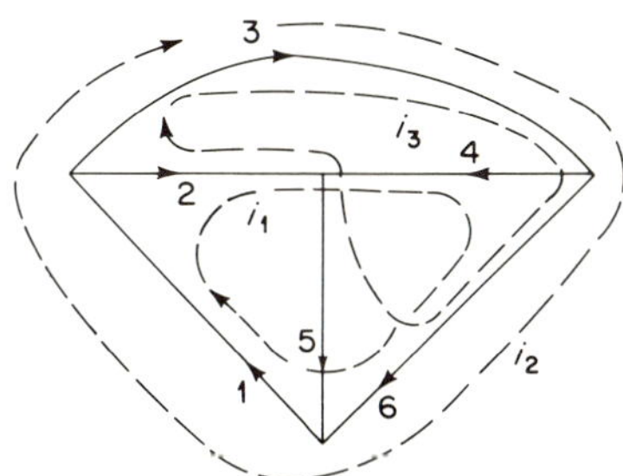

Fig. P4.9

(*a*) Determine whether or not the proposed set is acceptable.

(*b*) If the set is not acceptable, change one of the current variables so that the set becomes so.

(*c*) If the set is acceptable, determine the tie-set matrix. State which branches contribute to each tie set.

4.10. A tie-set matrix of a network is given by

$$\beta = \begin{bmatrix} 2 & -2 & -3 & 0 & 2 & -4 & -1 & 3 \\ -1 & 2 & 3 & 1 & -2 & 3 & 1 & -2 \\ 1 & -1 & -2 & -1 & 2 & -2 & -1 & 1 \\ -1 & 1 & 2 & 0 & -1 & 2 & 1 & -2 \end{bmatrix}$$

Define the independent current variables in terms of a minimum number of branch currents. Find fundamental-tie-set and cut-set matrices.

4.11. A proposed tie-set matrix is given by

$$\beta = \begin{bmatrix} 0 & -1 & 1 & 0 & -1 & -1 \\ 1 & 1 & 0 & -1 & -1 & 0 \\ 1 & 0 & 1 & 1 & 0 & 1 \end{bmatrix}$$

Find a graph if one exists.

4.12. A proposed cut-set matrix is given by

$$\alpha = \begin{bmatrix} 0 & 1 & 0 & -1 & 0 & 0 & 0 & 0 & 1 & 0 & 0 & 0 \\ 0 & 0 & 1 & 0 & 1 & 1 & 1 & 0 & 0 & 0 & 0 & 0 \\ 1 & 0 & 1 & 1 & 0 & 0 & 0 & 0 & 0 & 0 & 1 & 0 \\ 0 & 1 & 1 & 0 & 1 & 0 & 0 & 0 & 0 & 0 & 0 & 1 \\ 0 & 0 & 1 & 0 & 1 & 1 & 0 & 0 & 0 & 1 & 0 & 0 \\ 1 & 0 & 1 & 0 & 0 & 0 & 0 & 1 & 0 & 0 & 0 & 0 \end{bmatrix}$$

Find a graph if one exists.

4.13. The α_{NL} portion of a fundamental-cut-set matrix is given by

$$\alpha_{NL} = \begin{bmatrix} 1 & 1 & -1 & 0 & 0 & 0 & 1 & 1 & 0 \\ 0 & 1 & -1 & 0 & 0 & 0 & 1 & 1 & 1 \\ 1 & 0 & -1 & 0 & 0 & 0 & 1 & 1 & 0 \\ 0 & 0 & -1 & 1 & 1 & 1 & 0 & 0 & 0 \\ 0 & 0 & 0 & 0 & 0 & -1 & 0 & 1 & 0 \\ 0 & 0 & -1 & 0 & 1 & 1 & 0 & 0 & 0 \end{bmatrix}$$

Find a graph if one exists.

4.14. Repeat Prob. 4.13 for the following α_{NL}:

$$\alpha_{NL} = \begin{bmatrix} -1 & 0 & 0 & 0 & 0 & 0 & -1 & 0 & 0 & -1 \\ 0 & 0 & 0 & 0 & 0 & 1 & 0 & 0 & 0 & 0 \\ 0 & 0 & 0 & 0 & 0 & -1 & 1 & 0 & 0 & 0 \\ 0 & 1 & 1 & 0 & 0 & 0 & 1 & 1 & 0 & 0 \\ 0 & 0 & 1 & 1 & 1 & 0 & 0 & 0 & 0 & 0 \\ 0 & 0 & 0 & -1 & -1 & 0 & 1 & 1 & 0 & 0 \\ 0 & 0 & 0 & 0 & 0 & 0 & 0 & 0 & -1 & 1 \\ 0 & 0 & 0 & 0 & 0 & 0 & 1 & 1 & 1 & 0 \end{bmatrix}$$

Theory and proofs

4.15. For the given graph, two partially completed cut-set matrices are given, where the ellipses represent missing columns. Complete each cut-set matrix if it is *uniquely possible* to do so.

$$\alpha_1 = \begin{array}{c} \\ 1 \\ 2 \\ 3 \\ 4 \end{array} \begin{array}{c} \begin{array}{cccccccc} 1 & 2 & 3 & 4 & 5 & 6 & 7 & 8 \end{array} \\ \begin{bmatrix} 1 & 0 & \cdot & \cdot & 0 & 0 & \cdot & \cdot \\ -1 & 1 & \cdot & \cdot & -1 & 0 & \cdot & \cdot \\ 0 & 0 & \cdot & \cdot & 0 & 0 & \cdot & \cdot \\ 0 & -1 & \cdot & \cdot & 0 & -1 & \cdot & \cdot \end{bmatrix} \end{array}$$

$$\alpha_2 = \begin{array}{c} \\ 1 \\ 2 \\ 3 \\ 4 \end{array} \begin{array}{c} \begin{array}{cccccccc} 1 & 2 & 3 & 4 & 5 & 6 & 7 & 8 \end{array} \\ \begin{bmatrix} -1 & 1 & 0 & \cdot & \cdot & 0 & \cdot & \cdot \\ 0 & 0 & 0 & \cdot & \cdot & 1 & \cdot & \cdot \\ 1 & 0 & 0 & \cdot & \cdot & 0 & \cdot & \cdot \\ 0 & 0 & -1 & \cdot & \cdot & 0 & \cdot & \cdot \end{bmatrix} \end{array}$$

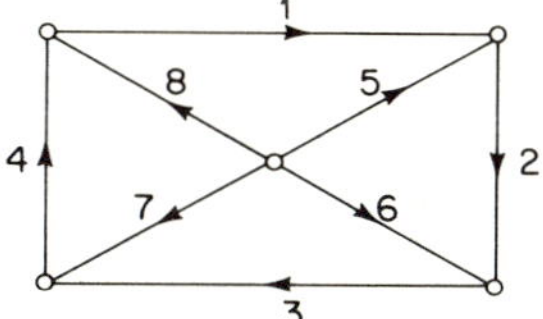

Fig. P4.15

4.16. Obtain the cut-set matrix α in Example 4.2 directly from the graph in Fig. 4.9 by applying the techniques of Sec. 4.8 and Fig. 4.7. What sets of branches define the cut sets in α?

4.17. A cut-set and a tie-set matrix are given. Prove whether or not they represent the same graph. If they do, find the graph. If they do not, state exactly how the graphs differ. Include any necessary assumptions made.

$$\alpha_1 = \begin{bmatrix} 1 & 0 & -1 & 0 & 0 & -1 & -1 \\ 0 & \frac{1}{2} & -\frac{1}{2} & 0 & 0 & 0 & -\frac{1}{2} \\ 0 & 0 & 0 & \frac{1}{4} & 0 & \frac{1}{4} & \frac{1}{4} \\ 0 & 0 & 0 & 0 & \frac{1}{5} & \frac{1}{5} & \frac{1}{5} \end{bmatrix}$$

$$\beta_2 = \begin{bmatrix} 0 & 0 & 0 & 0 & 2 & 2 & -2 \\ 0 & 0 & 1 & 1 & 1 & 0 & -1 \\ 1 & 1 & 2 & 1 & 0 & -1 & 0 \end{bmatrix}$$

4.18. Find the missing columns in the given nonfundamental-cut-set matrix. Is your answer unique?

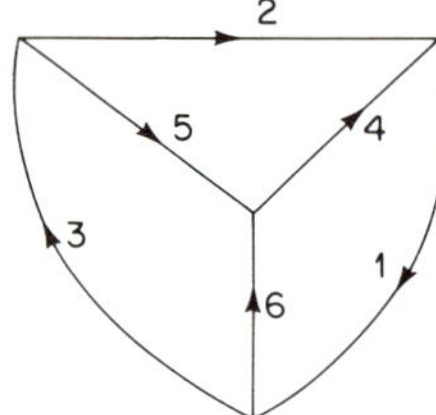

Fig. P4.18

4.19. (*a*) Define a *branch train* as a sequence of connected branches traversed in such a way that each branch is passed through once and only once, although any node may be passed more than once. Define a *closed branch train* as a branch train whose first and last nodes are the same and an *open branch train* as one where they are not. Define the degree of a node as the number of branches incident at the node. Prove (1) that in any connected graph there is an even number of nodes of odd degree and (2) that a connected graph is an open branch train if and only if it contains two nodes of odd degree and a closed branch train if and only if it contains no nodes of odd degree.

(*b*) In a large city built on a river and its tributaries a system of bridges connects every part of the city, including the islands, with every other part (Fig. P4.19). Can a motorist

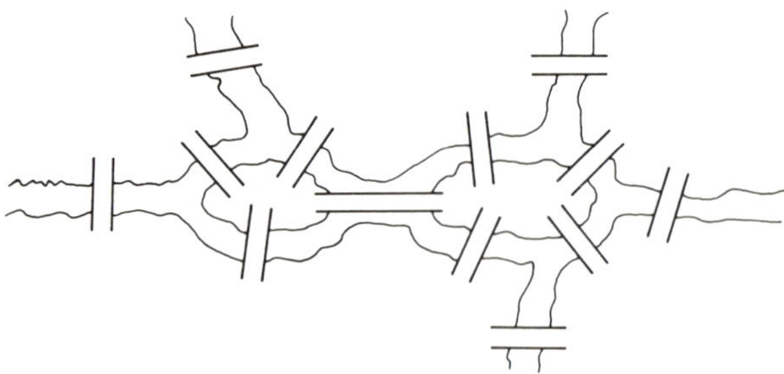

Fig. P4.19

cross over all 13 bridges while crossing over each bridge only once? Use the results of part (*a*) to answer the question. Explain your reasoning.

Application

4.20. It is desired to standardize the analysis of the given *RC* ladder network by finding fundamental-tie-set and cut-set matrices in which the resistors are tree branches and the capacitors are links. The network graph is to be numbered and oriented so that these matrices take the partitioned forms

$$\beta_F = [U_{44} \quad T_{l44}] \qquad \alpha_F = [T_{u44} \quad U_{44}]$$

where the numbered subscripts indicate the orders of the submatrices, and where T_{l44} is a lower triangular matrix and T_{u44} is an upper triangular matrix (see Prob. 2.16). Furthermore, all main-diagonal entries in T_{l44} are to be negative and all nonzero off-diagonal entries positive. For T_{u44}, all main-diagonal entries are to be positive and all nonzero off-diagonal entries negative. Find the matrices independently and then show that one can be derived from the other.

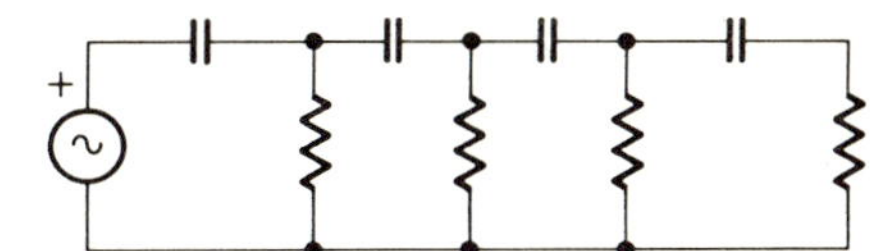

Fig. P4.20

4.21. Draw the graphs of each of the two common transistor networks shown. For each, choose a tree such that the voltages from terminals 1, 2, 3, and 4 to ground appear across tree branches. Find the fundamental-tie-set and cut-set matrices for the trees chosen.

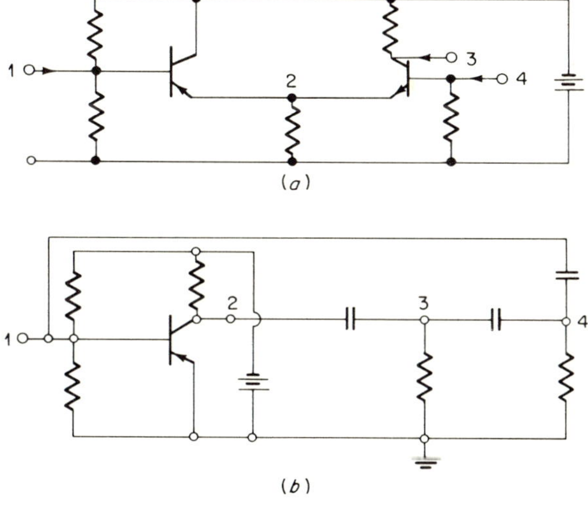

Fig. P4.21

4.22. The network is an *RC* twin-T filter with a load impedance z_L. Draw its graph and choose branch directions and a tree such that (*a*) α_F contains no negative entries and (*b*) α_F contains no more than three nonzero entries per row. Find β_F and examine its structure. Must it contain negative entries if α_F is to have none?

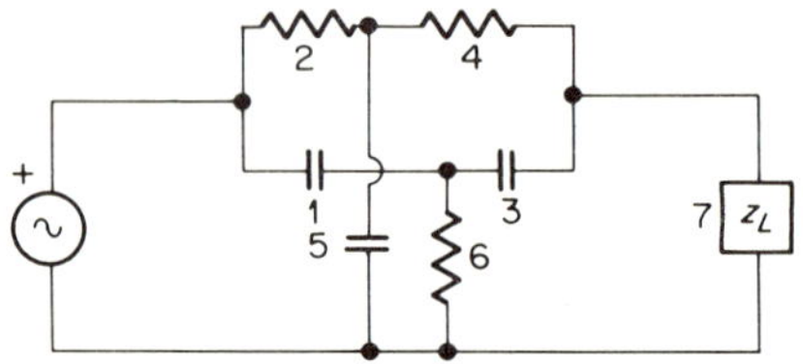

Fig. P4.22

4.23. Both N_1 and N_2 are grounded 4-terminal active networks. Represent each by a coupled graph. Draw the graph for the entire system, oriented so that all voltages to ground are positive. Choose a tree and find the fundamental-tie-set and cut-set matrices.

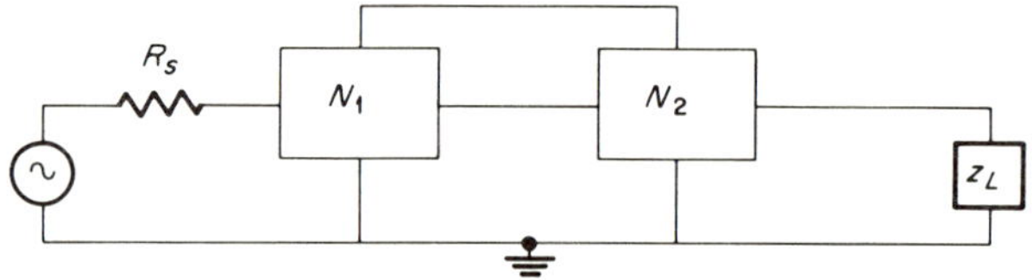

Fig. P4.23

5 NETWORK EQUILIBRIUM EQUATIONS

5.1 INTRODUCTION

In Chap. 4 a systematic procedure was developed for obtaining sets of independent loop currents and node-pair potentials in a given network. Tie-set and cut-set matrices were introduced as a systematic means of relating the branch currents and voltages to these independent current and voltage variables and to permit the KVL and KCL equations to be concisely written in matrix notation. In these representations, only the network geometry was considered, and no account was taken of the network elements or excitations.

The present chapter has a threefold purpose: (1) Network equilibrium equations are developed on a *branch* basis. These relate a given branch voltage and current to its branch impedance (admittance) and to the current and voltage sources feeding the branch. (2) Tie-set and cut-set matrices are used to develop network equilibrium equations on a *loop* and a *node-pair* basis. This results in L loop equations and N node-pair equations for a given network. (3) Generalized tie-set and cut-set matrices are introduced which permit generalized network equilibrium equations to be developed in matrix form on a *loop* and a *node-pair* basis. This yields the same number (B) of generalized loop and node-pair equations for a given network.

5.2 OHM'S LAW FOR A SINGLE BRANCH

The general representation of the kth branch of a network given in Fig. 2.1 is repeated in Fig. 5.1. Note that the branch has a voltage source in series with its impedance and a current source feeding its terminal nodes.

Neglecting coupling from other branches, the Ohm's law relations between the branch current and branch voltage are given by

$$v_k + e_{sk} = z_{kk}(j_k + i_{sk}) \tag{5.2.1}$$

and

$$j_k + i_{sk} = y_{kk}(v_k + e_{sk}) \tag{5.2.2}$$

Let us review the discussion in Sec. 2.1. First, branch-current and branch-voltage reference conditions are chosen so that a $+j_k$ produces a $+v_k$. Second, current and potential source reference directions are chosen so that all terms in Eqs. (5.2.1) and (5.2.2) are positive; in other words, the source current and voltage aid the branch current and voltage, respectively. Third, j_k is not the current in z_k, nor is v_k the voltage across z_k. Fourth, if in a given network all current and voltage sources do not appear across, or in series with, single branches, the network can be converted to this so-called standard form by the transformations of Figs. 2.6 and 2.7. Finally, Eqs. (5.2.1) and (5.2.2) can be modified to include coupling effects. This yields

$$v_k + e_{sk} = \sum_{l=1}^{B} z_{kl}(j_l + i_{sl}) \tag{5.2.3}$$

and

$$j_k + i_{sk} = \sum_{l=1}^{B} y_{kl}(v_l + e_{sl}) \tag{5.2.4}$$

When $z_{kl} = 0$ and $y_{kl} = 0$, $l \neq k$, these equations revert to the previous forms with no coupling.

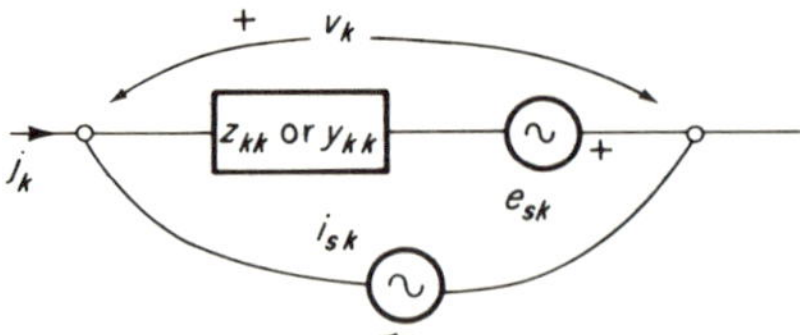

Fig. 5.1 The kth branch of a network.

5.3 BRANCH Z AND Y MATRICES

The relations presented in Sec. 5.2 apply to a single branch. We now want to express these relations in matrix form for a network containing B branches. To accomplish this, it is first necessary to formulate branch impedance and admittance matrices for the given network. This can be done in either of two ways.

Method 1 The branches of the network are chosen to contain one (occasionally more than one) element and are numbered from 1 through B in any convenient order (generally links are numbered 1 through L, tree branches $L+1$ through B). From this numbering, the $B \times B$ branch impedance and admittance matrices are written in the form

$$Z_B = [Z_{kl}] \qquad Y_B = [Y_{kl}] \tag{5.3.1}$$

These will be diagonal matrices if no coupling is present.

Method 2 The branches are chosen so that each branch contains only one element. Assume there are f inductive branches, g resistive branches, and h capacitive branches ($f + g + h = B$). Number the inductive branches 1 through f, the resistive branches $f + 1$ through $g + f$, and the capacitive branches $g + f + 1$ through B. Then Z_B will take the partitioned form

$$Z_B = \begin{bmatrix} [sL] & 0 & 0 \\ 0 & [R] & 0 \\ 0 & 0 & \left[\dfrac{1}{sC}\right] \end{bmatrix} \tag{5.3.2}$$

where sL is an $f \times f$ matrix, R is a $g \times g$ matrix, and $1/sC$ is an $h \times h$ matrix. The Z_B matrix is diagonal in this partitioned form. The inverse of a diagonal matrix is easily obtained from the inverses of the individual submatrices (see Prob. 1.13). Thus

$$Y_B = \begin{bmatrix} [sL]^{-1} & 0 & 0 \\ 0 & [R]^{-1} & 0 \\ 0 & 0 & \left[\dfrac{1}{sC}\right]^{-1} \end{bmatrix} \tag{5.3.3}$$

Example 5.1 We wish to formulate branch impedance and admittance matrices for the network in Fig. 5.2*a*.

Proceeding first according to method 1, we choose to let L_{11} be in branch 1, C in branch 2, and R and L_{22} in branch 3, as shown in Fig. 5.2*b*.

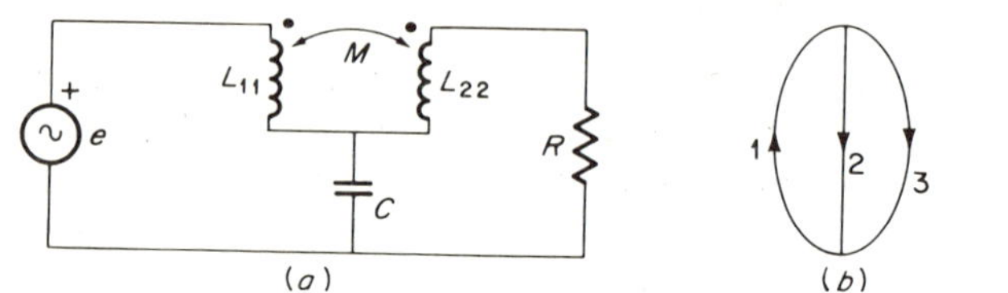

Fig. 5.2

Thus the branch impedance matrix is given by

$$Z_B = \begin{bmatrix} sL_{11} & 0 & -sM \\ 0 & \dfrac{1}{sC} & 0 \\ -sM & 0 & sL_{22} + R \end{bmatrix}$$

The branch admittance matrix can be found by taking the inverse of Z_B in partitioned form. Assuming the transformer to be perfect, the result is

$$Y_B = \left[\begin{array}{c|c|c} \dfrac{L_{22}}{L_{11}R} + \dfrac{1}{sL_{11}} & 0 & \dfrac{M}{L_{11}R} \\ \hline 0 & sC & 0 \\ \hline \dfrac{M}{L_{11}R} & 0 & \dfrac{1}{R} \end{array}\right]$$

If we proceed according to method 2, we must place L_{11} in branch 1, L_{22} in branch 2, R in branch 3, and C in branch 4, as shown in Fig. 5.2*c*. Thus $f = 2, g = 1$, and $h = 1$, and the Z_3 matrix is

$$Z_B = \begin{bmatrix} sL_{11} & -sM & 0 & 0 \\ -sM & sL_{22} & 0 & 0 \\ 0 & 0 & R & 0 \\ 0 & 0 & 0 & \dfrac{1}{sC} \end{bmatrix}$$

Its inverse by Eq. (5.3.3) gives

$$Y_B = \begin{bmatrix} \begin{bmatrix} sL_{11} & -sM \\ -sM & sL_{22} \end{bmatrix}^{-1} & \begin{matrix} 0 & 0 \\ 0 & 0 \end{matrix} \\ \begin{matrix} 0 & 0 \\ 0 & 0 \end{matrix} & \begin{matrix} \dfrac{1}{R} & 0 \\ 0 & sC \end{matrix} \end{bmatrix}$$

If the transformer is perfect, this inverse does not exist.

5.4 OHM'S LAW ON A BRANCH BASIS

Equation (5.2.3) can be combined with the Z matrix of Eq. (5.3.1) to express Ohm's law in matrix form for the entire network

$$\begin{bmatrix} v_1 \\ v_2 \\ \cdot \\ \cdot \\ \cdot \\ v_B \end{bmatrix} + \begin{bmatrix} e_{S1} \\ e_{S2} \\ \cdot \\ \cdot \\ \cdot \\ e_{SB} \end{bmatrix} = [z_{kl}] \left[\begin{bmatrix} j_1 \\ j_2 \\ \cdot \\ \cdot \\ \cdot \\ j_B \end{bmatrix} + \begin{bmatrix} i_{S1} \\ i_{S2} \\ \cdot \\ \cdot \\ \cdot \\ i_{SB} \end{bmatrix} \right] \tag{5.4.1}$$

In closed matrix form, Eq. (5.4.1) becomes

$$V_B + E_S = Z_B(J_B + I_S) \tag{5.4.2}$$

where V_B, E_S, J_B, and I_S are $B \times 1$ column matrices. The corresponding admittance equation, from Eq. (5.2.4) and Y_B in Eq. (5.3.1), is

$$J_B + I_S = Y_B(V_B + E_S) \tag{5.4.3}$$

5.5 OHM'S LAW ON A LOOP BASIS

Let Eq. (5.4.2) be premultiplied by β_F, where β_F is any fundamental-tie-set matrix of the given network†

$$\beta_F V_B + \beta_F E_S = \beta_F Z_B(J_B + I_S) \tag{5.5.1}$$

From Eq. (4.7.4), the $\beta_F V_B$ term is identically zero. Now Eq. (5.5.1) can be written in the form

$$\beta_F(E_S - Z_B I_S) = \beta_F Z_B J_B \tag{5.5.2}$$

But, from Eq. (4.7.2), $J_B = \beta_F{}^T I_L$. Incorporating this relation into Eq. (5.5.2) yields

$$\beta_F(E_S - Z_B I_S) = \beta_F Z_B \beta_F{}^T I_L \tag{5.5.3}$$

Equation (5.5.3) consists of L equations in L unknowns (the loop current variables). It is well to consider what each of the terms in this expression means:

$\beta_F E_S$ summation of the voltage sources around the loops defined by β_F in the direction of the loop currents

† If we number the branches by method 1 of Sec. 5.3 so that the first L are links, the first L columns of β_F form an $L \times L$ identity matrix. This choice generally simplifies calculations.

$Z_B I_S$ conversion of all current sources into equivalent voltage sources

$-\beta_F Z_B I_S$ summation of the converted current sources around the loops defined by β_F in the direction of the loop currents

$\beta_F Z_B J_B$ summation of element (not branch) voltages, in terms of element (not branch) currents, around the loops defined by β_F in the direction of the loop currents

$\beta_F Z_B \beta_F{}^T I_L$ summation of element (not branch) voltages, in terms of the loop currents, around the loops defined by β_F in the direction of the loop currents

With these interpretations in mind, Eq. (5.5.3) can be written in the compact matrix form

$$E_E = Z_E I_L \tag{5.5.4}$$

where the subscript E denotes equilibrium.

Example 5.2 Consider the simple network and its graph given in Fig. 5.3. We wish to write equilibrium equations on a loop basis.

Branch 3 has been chosen as the tree. The pertinent matrices are

$$E_S = \begin{bmatrix} e \\ 0 \\ 0 \end{bmatrix} \qquad I_S = \begin{bmatrix} 0 \\ 0 \\ i \end{bmatrix} \qquad I_L = \begin{bmatrix} i_1\,(j_1) \\ i_2\,(j_2) \end{bmatrix}$$

$$Z_B = \begin{bmatrix} 1 & 0 & 0 \\ 0 & 2 & 0 \\ 0 & 0 & 3 \end{bmatrix} \qquad \beta_F = \begin{bmatrix} 1 & 0 & 1 \\ 0 & 1 & -1 \end{bmatrix}$$

where the I_L notation should be interpreted to mean that loop current i_1 is defined as branch current j_1 and loop current i_2 as branch current j_2.

These matrices can be combined to give

$$\beta_F E_S = \begin{bmatrix} e \\ 0 \end{bmatrix} \qquad Z_B I_S = \begin{bmatrix} 0 \\ 0 \\ 3i \end{bmatrix} \qquad -\beta_F Z_B I_S = \begin{bmatrix} -3i \\ 3i \end{bmatrix}$$

$$\beta_F Z_B = \begin{bmatrix} 1 & 0 & 3 \\ 0 & 2 & -3 \end{bmatrix} \qquad \beta_F Z_B \beta_F{}^T = \begin{bmatrix} 4 & -3 \\ -3 & 5 \end{bmatrix}$$

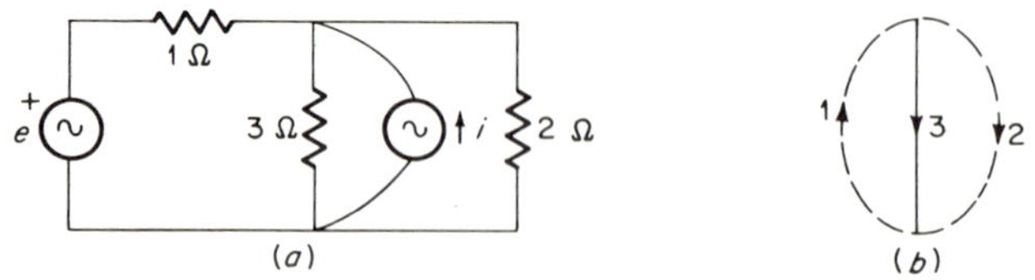

Fig. 5.3

Then the Ohm's law equilibrium equations on a loop basis are

$$\begin{bmatrix} e - 3i \\ 3i \end{bmatrix} = \begin{bmatrix} 4 & -3 \\ -3 & 5 \end{bmatrix} \begin{bmatrix} i_1 \\ i_2 \end{bmatrix}$$

5.6 OHM'S LAW ON A NODE-PAIR BASIS

A similar development can be carried out on a node-pair basis where, in this case, Eq. (5.4.3) is premultiplied by any fundamental cut-set matrix of the network† to yield

$$\alpha_F J_B - \alpha_F I_S = \alpha_F Y_B(V_B + E_S) \tag{5.6.1}$$

The $\alpha_F J_B$ term is zero and $V_B = \alpha_F{}^T E_N$ from Eqs. (4.8.4) and (4.8.2), respectively. With these changes, Eq. (5.6.1) becomes

$$\alpha_F(I_S - Y_B E_S) = \alpha_F Y_B \alpha_F{}^T E_N \tag{5.6.2}$$

Equation (5.6.2) consists of N equations in N unknowns (the node-pair voltage variables). The interpretations of the various terms are:

$\alpha_F I_S$ summation of the current sources entering positive reference node of each node pair defined by α_F

$Y_B E_S$ conversion of all voltage sources into equivalent current sources

$-\alpha_F Y_B E_S$ summation of converted voltage sources entering positive reference node of each node pair defined by α_F

$\alpha_F Y_B V_B$ summation of element (not branch) currents, in terms of element (not branch) voltages, leaving positive reference node of each node pair defined by α_F

$\alpha_F Y_B \alpha_F{}^T E_N$ summation of element (not branch) currents, in terms of node-pair potentials, leaving positive reference node of each node pair defined by α_F

With these interpretations, Eq. (5.6.2) can be put in the form

$$I_E = Y_E E_N \tag{5.6.3}$$

where again the subscript E denotes equilibrium.

† If the first L branches are links, the last N are tree branches and the last N columns of α_F form an identity matrix. As in the loop case, this choice generally simplifies calculations.

Example 5.3 Again consider the network of Fig. 5.3. To write Ohm's law equations on a node-pair basis for the tree chosen, the following matrices are needed:

$$E_S = \begin{bmatrix} e \\ 0 \\ 0 \end{bmatrix} \qquad I_S = \begin{bmatrix} 0 \\ 0 \\ i \end{bmatrix} \qquad E_N = e_1(v_3)$$

$$Y_B = \begin{bmatrix} 1 & 0 & 0 \\ 0 & 1/2 & 0 \\ 0 & 0 & 1/3 \end{bmatrix} \qquad \alpha_F = [-1 \quad 1 \quad 1]$$

Combining these matrices yields

$$\alpha_F I_S = i \qquad Y_B E_S = \begin{bmatrix} e \\ 0 \\ 0 \end{bmatrix} \qquad -\alpha_F Y_B E_S = e$$

$$\alpha_F Y_B = [-1 \quad 1/2 \quad 1/3] \qquad \alpha_F Y_B \alpha_F{}^T E_N = 11/6 e_1$$

Thus the Ohm's law equilibrium equation on a node-pair basis is

$$i + e = 11/6 e_1$$

5.7 ADDITIONAL POINTS CONCERNING THE EQUILIBRIUM EQUATIONS

Examine Eqs. (5.5.4) and (5.6.3). Are the Z_E and Y_E matrices always symmetric? To answer this question, return to Eq. (5.4.2). This equation can be premultiplied by any tie-set matrix of the graph being considered to yield

$$\beta_1 V_B + \beta_1 E_S = \beta_1 Z_B (J_B + I_S) \tag{5.7.1}$$

Now J_B is the $B \times 1$ branch-current matrix and can be expressed in terms of *any* independent set of current variables through the appropriate tie-set matrix (not necessarily β_1) according to Eq. (4.10.11). Thus let

$$J_B = \beta_2{}^T I \tag{5.7.2}$$

From Eq. (4.10.12), $\beta_1 V_B = 0$. Thus Eq. (5.7.1) becomes

$$\beta_1(E_S - Z_B I_S) = \beta_1 Z_B \beta_2{}^T I \tag{5.7.3}$$

Equation (5.7.3) is a more general form of Eq. (5.5.3). Similarly, a more general form of Eq. (5.6.2) is

$$\alpha_1(I_S - Y_B E_S) = \alpha_1 Y_B \alpha_2{}^T E \tag{5.7.4}$$

In Eq. (5.7.3), the loop-impedance matrix is given by

$$Z_E = \beta_1 Z_B \beta_2{}^T \tag{5.7.5}$$

To determine whether Z_E is symmetric, it is sufficient to show whether Z_E and Z_E^T are identical. Forming Z_E^T, we obtain

$$Z_E^T = \beta_2(\beta_1 Z_B)^T = \beta_2 Z_B^T \beta_1^T = \beta_2 Z_B \beta_1^T \tag{5.7.6}$$

where the last step is based on the assumption that Z_B is symmetric, which is true if we are considering a reciprocal network. By comparing the right-hand members of Eqs. (5.7.5) and (5.7.6) we see that, with Z_B symmetric, Z_E is symmetric if $\beta_1 = \beta_2$. Similarly, with Y_B symmetric, the node-pair admittance matrix

$$Y_E = \alpha_1 Y_B \alpha_2^T \tag{5.7.7}$$

is symmetric if $\alpha_1 = \alpha_2$.

The β_1 matrix in Eq. (5.7.5) defines the loops around which the equilibrium equations are to be written, and the β_2 matrix defines the current variables to be used in writing these equations. Similarly, the α_1 matrix in Eq. (5.7.7) defines the node pairs at which the equilibrium equations are to be written, and the α_2 matrix defines the voltage variables to be used in writing them.

We are now in a position to answer the question raised following Eq. (2.3.8). Recall that for one choice of nodes [Eq. (2.3.5)] we obtained an unsymmetric Y_E matrix whereas for our second choice of nodes [Eq. (2.3.7)] we obtained a symmetric Y_E matrix. In Eq. (2.3.5), the voltage variables are the voltages across branches 2, 3, and 5 in Fig. 2.2. This choice defines the α_2 matrix as the fundamental-cut-set matrix of tree 235. However, the KCL equations are written at nodes 1, 2, and 3, which implies the use of tree 135. This, in turn, defines the α_1 matrix as the fundamental cut-set matrix of tree 135. Thus α_1 and α_2 come from different trees, and the Y_E matrix in Eq. (2.3.6) is unsymmetric.

In Eq. (2.3.7), the α_2 matrix is again the fundamental-cut-set matrix of tree 235, since the voltage variables have not changed. However, the KCL equations are now written at nodes 1, 1-2, and 1-2-3, which also requires the use of tree 235. Hence α_1 and α_2 are the fundamental-cut-set matrices of the same tree, and the Y_B matrix in Eq. (2.3.8) is symmetric.

Example 5.4 Reconsider Example 5.2. Branch 3 was used as the tree for both β_1 and β_2. The result was a symmetric Z_E matrix. Now let branch 1 be the tree for β_1 and branch 3 the tree for β_2. The tie-set matrices are

$$\beta_1 = \begin{bmatrix} 1 & 0 & 1 \\ 1 & 1 & 0 \end{bmatrix} \qquad \beta_2 = \begin{bmatrix} 1 & 0 & 1 \\ 0 & 1 & -1 \end{bmatrix}$$

By proceeding as in Example 5.2, the Ohm's law equilibrium equations on a loop basis become

$$\begin{bmatrix} e - 3i \\ e \end{bmatrix} = \begin{bmatrix} 4 & -3 \\ 1 & 2 \end{bmatrix} \begin{bmatrix} i_1 \\ i_2 \end{bmatrix} \qquad \begin{aligned} i_1 &= j_1 \\ i_2 &= j_2 \end{aligned}$$

If branch 1 is used as the tree for both β_1 and β_2, the results are

$$\begin{bmatrix} e - 3i \\ e \end{bmatrix} = \begin{bmatrix} 4 & 1 \\ 1 & 3 \end{bmatrix} \begin{bmatrix} i_1' \\ i_2' \end{bmatrix} \qquad \begin{matrix} i_1' = j_3 \\ i_2' = j_2 \end{matrix}$$

Note that with different tie-set matrices, the Z_E matrix is unsymmetric whereas it returns to a symmetric matrix when only one tie-set matrix is used. However, this latter form is not identical with Z_E in Example 5.2 because different current variables were used. The results of Example 5.4 may be generalized by the following theorem.

Theorem I For any network there are infinite numbers of loop-impedance and node-pair admittance matrices, each depending on the definitions of the variables and the chosen loops or node pairs.

This theorem is of far-reaching importance in network synthesis and equivalent-network theory. In synthesis, the theorem implies that there are an infinite number of realizations of a given impedance or admittance matrix. In equivalent-network theory, the theorem implies that, given one network, there are an infinite number of equivalent networks. These implications are presented pictorially in Fig. 5.4.

5.8 EQUILIBRIUM EQUATIONS FOR NONRECIPROCAL NETWORKS

We have seen how to write equilibrium equations topologically for simple reciprocal networks. The reader should have little difficulty in extending these results to more complicated reciprocal structures. But what of nonreciprocal networks? Can these be handled topologically with equal ease? The answer is that they can, by representing each nonreciprocal device by a resistively coupled graph. In fact, topological analysis shows itself to best advantage when applied to nonreciprocal networks, as the following two examples will illustrate.

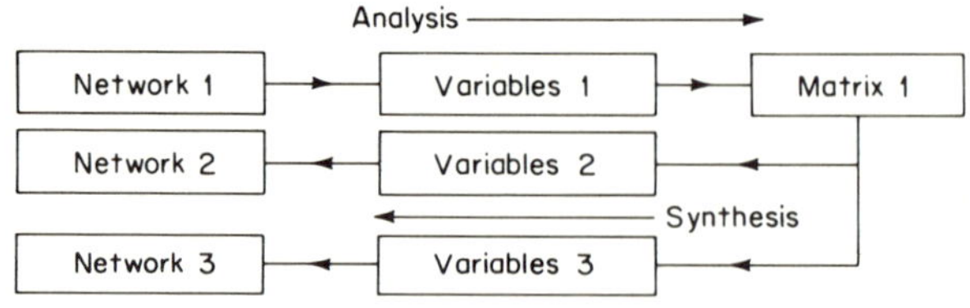

Fig. 5.4 Nonuniqueness of realizations.

Example 5.5 The boxes in Fig. 5.5 have the following 2-port representations:

$$N_1 \begin{bmatrix} v_1 \\ v_2 \end{bmatrix} = \begin{bmatrix} 0 & 1 \\ 2 & 0 \end{bmatrix} \begin{bmatrix} j_1 \\ j_2 \end{bmatrix} \qquad N_2 \begin{bmatrix} v_3 \\ v_4 \end{bmatrix} = \begin{bmatrix} 1 & 2 \\ 1 & 0 \end{bmatrix} \begin{bmatrix} j_3 \\ j_4 \end{bmatrix}$$

We wish to write equilibrium equations on a loop basis for the overall system.

First replace each box by a 2-branch resistively coupled network, as shown in Fig. 5.5*b*. Next draw an oriented graph and assign tree branches and links, as shown in Fig. 5.5*c*. In this diagram, branches 1 and 2 represent N_1, and branches 3 and 4 represent N_2. Voltage generators are arbitrarily assigned to the links (branches 1 and 2).

We now proceed exactly as we did in Example 5.2. The matrices needed are

$$E_S = \begin{bmatrix} -1 \\ -1 \\ 0 \\ 0 \\ 0 \end{bmatrix} \qquad I_L = \begin{bmatrix} i_1\,(j_1) \\ i_2\,(j_2) \end{bmatrix}$$

$$B_F = \begin{bmatrix} 1 & 0 & -1 & 0 & -1 \\ 0 & 1 & 0 & -1 & -1 \end{bmatrix} \qquad Z_B = \begin{bmatrix} 0 & 1 & 0 & 0 & 0 \\ 2 & 0 & 0 & 0 & 0 \\ 0 & 0 & 1 & 2 & 0 \\ 0 & 0 & 1 & 0 & 0 \\ 0 & 0 & 0 & 0 & 2 \end{bmatrix}$$

Note that the N_1 2-port matrix appears in the location in Z_B defined by rows 1 and 2 and columns 1 and 2 and that the N_2 matrix appears in the location defined by rows 3 and 4 and columns 3 and 4.

Now, through Eq. (5.5.3), we can write the desired equilibrium equations as

$$\begin{bmatrix} -1 \\ -1 \end{bmatrix} = \begin{bmatrix} 3 & 5 \\ 5 & 2 \end{bmatrix} \begin{bmatrix} i_1 \\ i_2 \end{bmatrix}$$

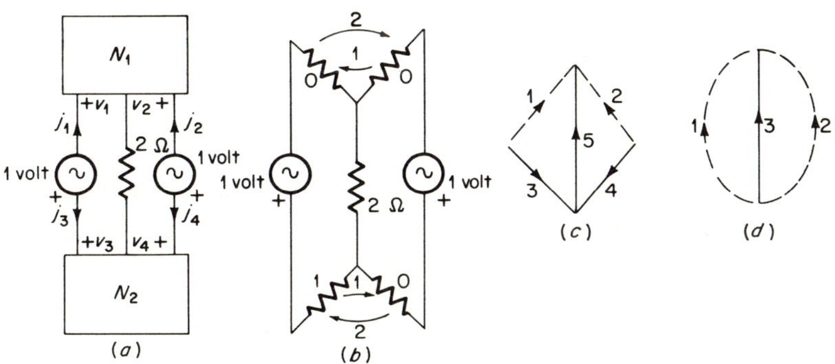

Fig. 5.5 A system with nonreciprocal elements.

The analysis can also be made using the 3-branch graph of Fig. 5.5*d*. The appropriate matrices are

$$E_S = \begin{bmatrix} -1 \\ -1 \\ 0 \end{bmatrix} \qquad I_L = \begin{bmatrix} i_1\ (j_1) \\ i_2\ (j_2) \end{bmatrix}$$

$$\beta_F = \begin{bmatrix} 1 & 0 & -1 \\ 0 & 1 & -1 \end{bmatrix} \qquad Z_B = \begin{bmatrix} 1 & 3 & 0 \\ 3 & 0 & 0 \\ 0 & 0 & 2 \end{bmatrix}$$

Again using Eq. (5.5.3), the equilibrium equations are

$$\begin{bmatrix} -1 \\ -1 \end{bmatrix} = \begin{bmatrix} 3 & 5 \\ 5 & 2 \end{bmatrix} \begin{bmatrix} i_1 \\ i_2 \end{bmatrix}$$

The 2×2 submatrix in the upper left of the 3×3 Z_B matrix requires further explanation. Note that this submatrix includes the effects of both N_1 and N_2. For example, row 1 of Z_B expresses the voltage across the elements in branch 1 (not the voltage across the branch itself) in terms of the link currents j_1 and j_2. Letting the element voltage of branch 1 be $\bar{v}_1$, we have

$$\begin{aligned} \bar{v}_1 = v_1 - v_3 &= j_2 - (j_3 - 2j_4) \\ &= j_2 - (-j_1 - 2j_2) \\ &= j_1 + 3j_2 \end{aligned}$$

from which the first row of Z_B can be written.

Example 5.5 illustrates a general procedure we can follow in the topological analysis of complicated networks. The steps are as follows:

1. Separate the network into parts consisting of the independent sources, the passive elements, and one or more grounded n-port active subnetworks.
2. Write voltage-current relations for each subnetwork in y-parameter or z-parameter matrix form.
3. Represent each subnetwork by a coupled $(n + 1)$-terminal n-branch graph (the coupling is resistive in Fig. 5.5*b*; in the general case inductive coupling would also be included). Assign each branch arrow in the direction of positive current flow in that branch.
4. Number the branches so that each subnetwork in step 3 has consecutively numbered branches.
5. Choose a tree and links. It is desirable to have links numbered 1 through L and tree branches $L + 1$ through N. It is also desirable to have all voltage sources in links and current sources across tree branches. We shall show in Sec. 5.9 that this can always be done.
6. Write equilibrium equations in the form of Eq. (5.5.4) or (5.6.3).

Example 5.6 Equilibrium equations on a node-pair basis are to be written for the two-stage transistor amplifier in Fig. 5.6*a* (dc excitations are omitted for simplicity). The

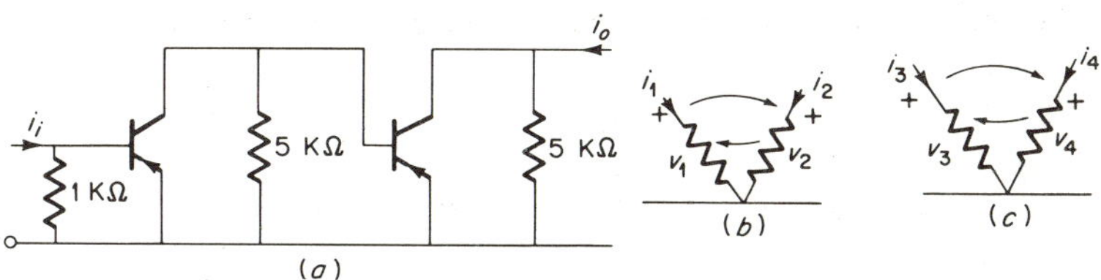

Fig. 5.6

transistors are identical and have the following common-emitter h parameters:

$$h_{11} = 2000 \text{ ohms} \qquad h_{12} = 6 \times 10^{-4}$$

$$h_{21} = 40 \qquad h_{22} = 30 \times 10^{-6} \text{ mho}$$

The choice of tree and links should be such that the input, output, and between-stage voltages are related to i_i and i_o in the final matrix equation.

The first step is to convert the transistor h parameters to y parameters, since we wish to write node-pair equations. Using Table 3.1, the results are

$$y_{11} = 5 \times 10^{-4} \text{ mho} \qquad y_{12} = -3 \times 10^{-7} \text{ mho}$$

$$y_{21} = 2 \times 10^{-2} \text{ mho} \qquad y_{22} = 1.8 \times 10^{-5} \text{ mho}$$

Next, represent each transistor as a conductively coupled grounded 2-port. The left-hand transistor is assigned branches 1 and 2 and appears in Fig. 5.6*b*; the right-hand transistor with branches 3 and 4 is shown in Fig. 5.6*c*. The overall network in Fig. 5.6*a* is now redrawn in Fig. 5.7*a*, where T_1 and T_2 denote the transistor boxes.

The next task is to draw an oriented graph, assign branch numberings, and choose a tree and links. From the statement of the problem, the three resistive branches are to be tree branches. We number them 5, 6, and 7. The graph is drawn in Fig. 5.7*b*. Note its petallike arrangement with the common-ground node at its "stem." This is typical of cascaded common-ground network graphs. Also note that all branch arrows have been chosen to point toward this common-ground node. This choice references all other node voltages positive with respect to ground. The currents i_i and i_o are treated as sources located across tree branches 5 and 7, respectively.

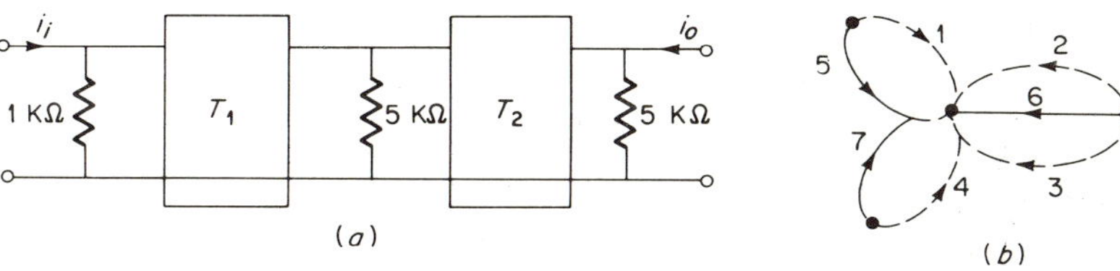

Fig. 5.7 Network in Fig. 5.6a redrawn.

To obtain the node-pair equilibrium equations, the pertinent matrices are

$$I_S = \begin{bmatrix} 0 \\ 0 \\ 0 \\ 0 \\ i_i \\ 0 \\ i_o \end{bmatrix} \qquad Y_B = \begin{bmatrix} 5\times10^{-4} & -3\times10^{-7} & 0 & 0 & 0 & 0 & 0 \\ 2\times10^{-2} & 1.8\times10^{-5} & 0 & 0 & 0 & 0 & 0 \\ 0 & 0 & 5\times10^{-4} & -3\times10^{-7} & 0 & 0 & 0 \\ 0 & 0 & 2\times10^{-2} & 1.8\times10^{-5} & 0 & 0 & 0 \\ 0 & 0 & 0 & 0 & 10^{-3} & 0 & 0 \\ 0 & 0 & 0 & 0 & 0 & 0.2\times10^{-3} & 0 \\ 0 & 0 & 0 & 0 & 0 & 0 & 0.2\times10^{-3} \end{bmatrix}$$

$$\alpha_F = \begin{bmatrix} 1 & 0 & 0 & 0 & 1 & 0 & 0 \\ 0 & 1 & 1 & 0 & 0 & 1 & 0 \\ 0 & 0 & 0 & 1 & 0 & 0 & 1 \end{bmatrix} \qquad E_N = \begin{bmatrix} e_1\,(v_5) \\ e_2\,(v_6) \\ e_3\,(v_7) \end{bmatrix}$$

The entries in Y_B are all in mhos. The first-transistor parameters are in the upper left, the second-transistor parameters are in the center, and the three resistors are in the lower right of Y_B.

The equilibrium equations, obtained from Eq. (5.6.2), are

$$\begin{bmatrix} i_i \\ 0 \\ i_0 \end{bmatrix} = \begin{bmatrix} 1.5\times10^{-3} & -3\times10^{-7} & 0 \\ 2\times10^{-3} & 7.2\times10^{-4} & -3\times10^{-7} \\ 0 & 2\times10^{-2} & 2.2\times10^{-4} \end{bmatrix} \begin{bmatrix} e_1 \\ e_2 \\ e_3 \end{bmatrix}$$

5.9 SOURCE TRANSFORMATIONS

In the following sections the equilibrium equations (5.5.4) and (5.6.3) are combined so that the current and voltage variables are expressed in terms of the source currents and voltages in a single matrix equation on both an impedance and admittance basis. To effect this combination, it is necessary to systematize the numbering of the branches so that the links are numbered from 1 through L and the tree branches from $L + 1$ through N. It is further necessary to transform the voltage and current sources in the network so that all voltage sources appear in the links and all current sources appear across the tree branches. The mechanisms for performing this transformation are now discussed.

In Fig. 5.8*a*, it is desired to have all current sources appear across the branches of the tree 245. The current source i_0 across branch 3 must be removed. This is accomplished by the insertion of additional i_0 current

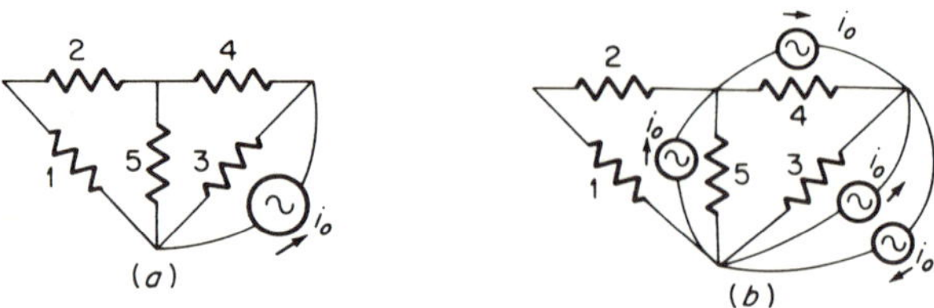

Fig. 5.8 Current source transformation.

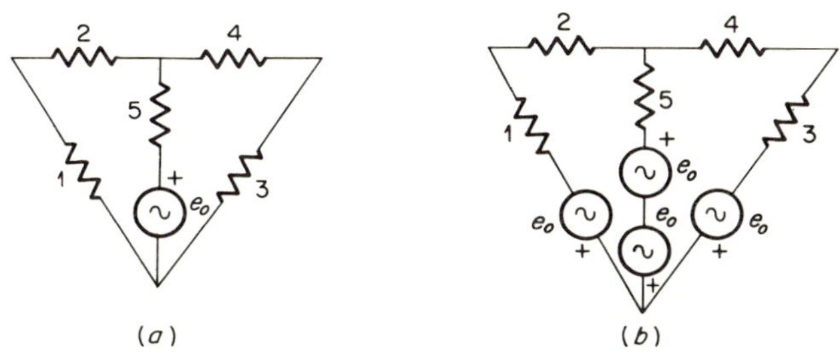

Fig. 5.9 Voltage source transformation

sources across branches 3, 4, and 5 to cancel the i_0 source across branch 3 without changing the current at any of the nodes of the network, as shown in Fig. 5.8*b*. Note that, in general, it is necessary to place an i_0 source across every branch included in the tie set defined by the link across which i_0 appears (tie set 345 in Fig. 5.8).

In Fig. 5.9*a*, it is desired to have all voltage sources appear in links 1 and 3. The voltage source e_0 in branch 5 must be removed. To accomplish this, it is necessary to insert additional e_0 voltage sources in branches 1, 3, and 5 to cancel the e_0 source in branch 5 without changing the voltages in any of the loops of the network. The result appears in Fig. 5.9*b*. Generally, it is necessary to place an e_0 source in every branch included in the cut set of the branch in which e_0 appears (cut set 135 in Fig. 5.9).

Special cases sometimes arise where there is no single branch directly across a given current source or no single branch directly in series with a given voltage source. These situations are handled in the manner shown in Figs. 2.6 and 2.7.

The foregoing results can be summarized in the following theorem.

Theorem 2 Any set of voltage sources in a network can be replaced by a set of voltage sources in the links of a tree without changing the voltage across any element in the network; similarly, any set of current sources in a network can be replaced by a set of current sources across the branches of a tree without changing the current through any element in the network.

5.10 GENERALIZED TIE-SET AND CUT-SET MATRICES †

In the following presentation, six equations from Chap. 4 will be used. They are repeated here for ease of reference.

$$J_B = \beta_F{}^T I_L \qquad V_B = \alpha_F{}^T E_N \tag{5.10.1}$$

$$I_L = \gamma_F J_B \qquad E_N = \delta_F V_B \tag{5.10.2}$$

$$\beta_F V_B = 0 \qquad \alpha_F J_B = 0 \tag{5.10.3}$$

† The presentations in this section and the next are adapted from Ref. 5.2.

The starting point in the development of generalized equilibrium equations for a given network is the selection of a tree. Next the links are numbered from 1 through L and the tree branches from $L + 1$ through N. Then all voltage sources are placed in the links and all current sources across the tree branches. Thus the voltage and current source $B \times 1$ column matrices take the form

$$E_S = \begin{bmatrix} e_{s1} \\ \cdot \\ \cdot \\ \cdot \\ e_{sL} \\ 0 \\ \cdot \\ \cdot \\ \cdot \\ 0 \end{bmatrix} = \begin{bmatrix} E_{sL} \\ \hline 0_N \end{bmatrix} \qquad I_S = \begin{bmatrix} 0 \\ \cdot \\ \cdot \\ \cdot \\ 0 \\ i_{s(L+1)} \\ \cdot \\ \cdot \\ \cdot \\ i_{sB} \end{bmatrix} = \begin{bmatrix} 0_L \\ \hline I_{sN} \end{bmatrix} \tag{5.10.4}$$

where 0_N and 0_L are $N \times 1$ and $L \times 1$ null matrices, respectively.

Furthermore, because of the numbering of the branches, the fundamental-tie-set and cut-set matrices take the form

$$\beta_F = [U_{LL} \mid \beta_{LN}] \qquad \alpha_F = [\alpha_{NL} \mid U_{NN}] \tag{5.10.5}$$

where U_{LL} and U_{NN} are $L \times L$ and $N \times N$ identity matrices, respectively. Also, because of this branch numbering, γ_F and δ_F matrices can be written as

$$\gamma_F = [U_{LL} \mid 0_{LN}] \qquad \delta_F = [0_{NL} \mid U_{NN}] \tag{5.10.6}$$

where 0_{LN} and 0_{NL} are $L \times N$ and $N \times L$ null matrices, respectively.

Now the second of Eqs. (5.10.2) can be combined with the first of Eqs. (5.10.3) to yield

$$\begin{bmatrix} \beta_F \\ \hline \delta_F \end{bmatrix} [V_B] = \begin{bmatrix} 0_L \\ \hline E_N \end{bmatrix} \tag{5.10.7}$$

and the first of Eqs. (5.10.2) can be combined with the second of Eqs. (5.10.3) to give

$$\begin{bmatrix} \gamma_F \\ \hline \alpha_F \end{bmatrix} [J_B] = \begin{bmatrix} I_L \\ \hline 0_N \end{bmatrix} \tag{5.10.8}$$

Equations (5.10.7) and (5.10.8) can be rewritten as

$$\beta_G V_B = \begin{bmatrix} 0_L \\ \hline E_N \end{bmatrix} \tag{5.10.9}$$

$$\alpha_G J_B = \begin{bmatrix} I_L \\ \hline 0_N \end{bmatrix} \tag{5.10.10}$$

where β_G and α_G are termed *generalized tie-set and cut-set matrices*, respectively. Note that β_G has the form

$$\beta_G = \left[\begin{array}{c|c} U_{LL} & \beta_{LN} \\ \hline 0_{NL} & U_{NN} \end{array}\right] \tag{5.10.11}$$

Thus β_G is an upper triangular matrix, in that all entries below the main diagonal are zero.

Similarly, α_G has the form

$$\alpha_G = \left[\begin{array}{c|c} U_{LL} & 0_{LN} \\ \hline \alpha_{NL} & U_{NN} \end{array}\right] \tag{5.10.12}$$

and is a lower triangular matrix.

Both β_G and α_G are square $B \times B$ matrices, unlike β_F and α_F, and have nonvanishing determinants. Therefore β_G^{-1} and α_G^{-1} exist, and it is possible to solve Eqs. (5.10.9) and (5.10.10) for V_B and J_B, respectively, giving

$$V_B = \beta_G^{-1} \begin{bmatrix} 0_L \\ \hline E_N \end{bmatrix} \tag{5.10.13}$$

$$J_B = \alpha_G^{-1} \begin{bmatrix} I_L \\ \hline 0_N \end{bmatrix} \tag{5.10.14}$$

It is also possible to express V_B in terms of α_g and J_B in terms of β_g and thus arrive at equations similar in form to Eqs. (5.10.1). These are

$$V_B = \alpha_G^T \begin{bmatrix} 0_L \\ \hline E_N \end{bmatrix} \tag{5.10.15}$$

$$J_B = \beta_G^T \begin{bmatrix} I_L \\ \hline 0_N \end{bmatrix} \tag{5.10.16}$$

Equations (5.10.15) and (5.10.16) can be justified by direct expansions of their right-hand sides in partitioned-matrix form. For example,

$$\beta_G{}^T \begin{bmatrix} I_L \\ \hline 0_N \end{bmatrix} = [\beta_F{}^T \mid \delta_F{}^T] \begin{bmatrix} I_L \\ \hline 0_N \end{bmatrix} = \beta_F{}^T I_L \tag{5.10.17}$$

Now Eqs. (5.10.13) and (5.10.15) or (5.10.14) and (5.10.16) yield identical results

$$\beta_G{}^{-1} = \alpha_G{}^T \qquad \alpha_G{}^{-1} = \beta_G{}^T \tag{5.10.18}$$

Four additional equations are needed. They are obtained from Eqs. (5.10.4), (5.10.11), and (5.10.12)

$$\beta_G E_S = \begin{bmatrix} E_{SL} \\ \hline 0_N \end{bmatrix} = E_S \qquad \beta_G{}^T I_S = \begin{bmatrix} 0_L \\ \hline I_{SN} \end{bmatrix} = I_S \tag{5.10.19}$$

$$\alpha_G I_S = \begin{bmatrix} 0_L \\ \hline I_{SN} \end{bmatrix} = I_S \qquad \alpha_G{}^T E_S = \begin{bmatrix} E_{SL} \\ \hline 0_N \end{bmatrix} = E_S \tag{5.10.20}$$

5.11 GENERALIZED EQUILIBRIUM EQUATIONS

All matrix relations necessary for the formulation of the generalized equilibrium equations have now been developed. To proceed with the derivation, reconsider Eq. (5.4.2), repeated here

$$V_B + E_S = Z_B(J_B + I_S) \tag{5.11.1}$$

Premultiply each term in Eq. (5.11.1) by β_G to obtain

$$\beta_G V_B + \beta_G E_S = \beta_G Z_B J_B + \beta_G Z_B I_S \tag{5.11.2}$$

Using Eq. (5.10.9), the first of Eqs. (5.10.19), Eq. (5.10.16), and the second of Eqs. (5.10.19), in that order, we can convert Eq. (5.11.2) to

$$\begin{bmatrix} 0_L \\ \hline E_N \end{bmatrix} + \begin{bmatrix} E_{SL} \\ \hline 0_N \end{bmatrix} = \beta_G Z_{BB} \beta_G{}^T \begin{bmatrix} I_L \\ \hline 0_N \end{bmatrix} + \beta_G Z_{BB} \beta_G{}^T \begin{bmatrix} 0_L \\ \hline I_{SN} \end{bmatrix} \tag{5.11.3}$$

which can be reduced to

$$\begin{bmatrix} E_{SL} \\ \hline E_N \end{bmatrix} = \beta_G Z_{BB} \beta_G{}^T \begin{bmatrix} I_L \\ \hline I_{SN} \end{bmatrix} \equiv Z_G \begin{bmatrix} I_L \\ \hline I_{SN} \end{bmatrix} \tag{5.11.4}$$

A similar development can be carried out starting with Eq. (5.4.3), repeated here

$$J_B + I_S = Y_{BB}(V_B + E_S) \tag{5.11.5}$$

Premultiplying by α_G gives

$$\alpha_G J_B + \alpha_G I_S = \alpha_G Y_{BB} V_B + \alpha_G Y_{BB} E_S \tag{5.11.6}$$

Using Eq. (5.10.10), the first of Eqs. (5.10.20), Eq. (5.10.15), and the second of Eqs. (5.10.20), in that order, we can put Eq. (5.11.6) in the form

$$\begin{bmatrix} I_L \\ \hline 0_N \end{bmatrix} + \begin{bmatrix} 0_L \\ \hline I_{SN} \end{bmatrix} = \alpha_G Y_{BB} \alpha_G{}^T \begin{bmatrix} 0_L \\ \hline E_N \end{bmatrix} + \alpha_G Y_{BB} \alpha_G{}^T \begin{bmatrix} E_{SL} \\ \hline 0_N \end{bmatrix} \tag{5.11.7}$$

which reduces to

$$\begin{bmatrix} I_L \\ \hline I_{SN} \end{bmatrix} = \alpha_G Y_{BB} \alpha_G{}^T \begin{bmatrix} E_{SL} \\ \hline E_N \end{bmatrix} \equiv Y_G \begin{bmatrix} E_{SL} \\ \hline E_N \end{bmatrix} \tag{5.11.8}$$

In Eqs. (5.11.4) and (5.11.8), Z_G and Y_G are square matrices whose determinants do not vanish. Hence their inverses may be taken. For Eq. (5.11.4),

$$\begin{bmatrix} I_L \\ \hline I_{SN} \end{bmatrix} = Z_G{}^{-1} \begin{bmatrix} E_{SL} \\ \hline E_N \end{bmatrix} \tag{5.11.9}$$

A comparison with Eq. (5.11.8) yields

$$Z_G{}^{-1} = Y_G \tag{5.11.10}$$

Thus Eqs. (5.11.4) and (5.11.8) are essentially the same equation, and all independent node-pair and loop equations are present in either set.

Note that an equation corresponding to Eq. (5.11.10) is not possible in the case of Z_E in Eq. (5.5.4) and Y_E in Eq. (5.6.3), since the variables in these two equations are not the same. Thus

$$Z_E{}^{-1} \neq Y_E \tag{5.11.11}$$

and it is not generally possible to go uniquely from loop equations to node-pair equations or vice versa.

Let us partition Z_g to obtain

$$Z_g = \left[\begin{array}{c|c} Z_{LL} & Z_{LN} \\ \hline Z_{NL} & Z_{NN} \end{array}\right] \tag{5.11.12}$$

where the subscripts denote the number of rows and columns in each submatrix. Inserting Eq. (5.11.12) into Eq. (5.11.4) and expanding gives

$$\begin{aligned} E_{SL} &= Z_{LL}I_L + Z_{LN}I_{SN} \\ E_N &= Z_{NL}I_L + Z_{NN}I_{SN} \end{aligned} \tag{5.11.13}$$

The first of Eqs. (5.11.13) is identical to Eq. (5.5.3) if we make the identifications

$$\begin{aligned} Z_{LL} &= \beta_F Z_B B_F{}^T \\ Z_{LN}I_{SN} &= \beta_F Z_B I_S \\ E_{SL} &= \beta_F E_S \end{aligned} \tag{5.11.14}$$

Thus Z_E in Eq. (5.5.4) is identical to Z_{LL} in the first of Eqs. (5.11.13). The second of Eqs. (5.11.13) relates the loop currents to the node-pair voltages.

Y_G can also be partitioned in the following way:

$$Y_G = \left[\begin{array}{c|c} Y_{LL} & Y_{LN} \\ \hline Y_{NL} & Y_{NN} \end{array}\right] \tag{5.11.15}$$

Substituting Eq. (5.11.15) into Eq. (5.11.8) and expanding, we obtain

$$\begin{aligned} I_L &= Y_{LL}E_{SL} + Y_{LN}E_N \\ I_{SN} &= Y_{NL}E_{SL} + Y_{NN}E_N \end{aligned} \tag{5.11.16}$$

The second of Eqs. (5.11.16) is identical to Eq. (5.6.2) if we note that

$$\begin{aligned} Y_{NN} &= \alpha_F Y_B \alpha_F{}^T \\ Y_{NL}E_{SL} &= \alpha_F Y_B E_S \\ I_{SL} &= \alpha_F I_S \end{aligned} \tag{5.11.17}$$

The first of these relations shows that Y_E in Eq. (5.6.3) and Y_{NN} in the second of Eqs. (5.11.16) are the same. The first of Eqs. (5.11.16) relates the loop currents and node-pair voltages.

The solution of Eqs. (5.11.13) for the variables E_N and I_L is carried out using the procedure of Sec. 2.9. Thus the first set of L equations is solved for the current variables. These can then be inserted in the second set of N equations to obtain the voltage variables. A similar procedure applies to Eqs. (5.11.16).

Example 5.7 Consider the network and tree of Fig. 5.3. Note that the network is already in standard form since all voltage sources appear in links and all current sources appear across tree branches. Also note that the branch numberings are correct, with links numbered from 1 through L and tree branches numbered from $L + 1$ through B.

To write generalized equilibrium equations in the form of Eq. (5.11.4), it is necessary to obtain β_G, which contains β_F and δ_F. From Fig. 5.3, these latter matrices are

$$\beta_F = \begin{bmatrix} 1 & 0 & 1 \\ 0 & 1 & -1 \end{bmatrix} \qquad \delta_F = [0 \quad 0 \quad 1]$$

and therefore B_G is given by

$$\beta_G = \left[\begin{array}{cc|c} 1 & 0 & 1 \\ 0 & 1 & -1 \\ \hline 0 & 0 & 1 \end{array}\right]$$

Note that its partitioned form agrees with that in Eq. (5.10.11). The source matrices are, from Eq. (5.10.4),

$$E_S = \left[\begin{array}{c} e \\ 0 \\ \hline 0 \end{array}\right] \qquad I_S = \left[\begin{array}{c} 0 \\ 0 \\ \hline i \end{array}\right]$$

and Z_B, as in Example 5.2, is

$$Z_B = \begin{bmatrix} 1 & 0 & 0 \\ 0 & 2 & 0 \\ 0 & 0 & 3 \end{bmatrix}$$

These matrices can now be assembled according to Eq. (5.11.4) in the form

$$\left[\begin{array}{c} e \\ 0 \\ \hline e_1 \end{array}\right] = \left[\begin{array}{cc|c} 4 & -3 & 3 \\ -3 & 5 & -3 \\ \hline 3 & -3 & 3 \end{array}\right] \left[\begin{array}{c} i_1 \\ i_2 \\ \hline i \end{array}\right] \tag{5.11.18}$$

Separating Eq. (5.11.18) according to Eq. (5.11.13) yields

$$\begin{bmatrix} e \\ 0 \end{bmatrix} = \begin{bmatrix} 4 & -3 \\ -3 & 5 \end{bmatrix} \begin{bmatrix} i_1 \\ i_2 \end{bmatrix} + \begin{bmatrix} 3 \\ -3 \end{bmatrix} i$$

and

$$e_1 = [3 \quad -3] \begin{bmatrix} i_1 \\ i_2 \end{bmatrix} + 3i$$

The first set of equations is identical to the equilibrium equations on a loop basis in Example 5.2, where

$$Z_{LL} = Z_E = \begin{bmatrix} 4 & -3 \\ -3 & 5 \end{bmatrix}$$

To write generalized equilibrium equations according to Eq. (5.11.8), α_G is required. This is composed of α_F and γ_F, as shown in Eq. (5.10.12). Referring again to Fig. 5.3, these latter matrices are

$$\alpha_F = [-1 \quad 1 \quad 1] \qquad \gamma_F = \begin{bmatrix} 1 & 0 & 0 \\ 0 & 1 & 0 \end{bmatrix}$$

and thus α_G is given by

$$\alpha_G = \left[\begin{array}{cc|c} 1 & 0 & 0 \\ 0 & 1 & 0 \\ \hline -1 & 1 & 1 \end{array}\right]$$

There is no change in the source matrices. The Y_B matrix, as given in Example 5.3, is

$$Y_B = \begin{bmatrix} 1 & 0 & 0 \\ 0 & \frac{1}{2} & 0 \\ 0 & 0 & \frac{1}{3} \end{bmatrix}$$

Combining these matrices in the form of Eq. (5.11.8) yields

$$\left[\begin{array}{c} i_1 \\ i_2 \\ \hline i \end{array}\right] = \left[\begin{array}{cc|c} 1 & 0 & -1 \\ 0 & \frac{1}{2} & \frac{1}{2} \\ \hline -1 & \frac{1}{2} & \frac{11}{6} \end{array}\right] \left[\begin{array}{c} e \\ 0 \\ \hline e_1 \end{array}\right] \tag{5.11.19}$$

Separating Eq. (5.11.19) according to Eq. (5.11.16), we obtain

$$\begin{bmatrix} i_1 \\ i_2 \end{bmatrix} = \begin{bmatrix} 1 & 0 \\ 0 & \frac{1}{2} \end{bmatrix} \begin{bmatrix} e \\ 0 \end{bmatrix} + \begin{bmatrix} -1 \\ \frac{1}{2} \end{bmatrix} e_1$$

$$i = [-1 \quad \tfrac{1}{2}] \begin{bmatrix} e \\ 0 \end{bmatrix} + \tfrac{11}{6} e_1$$

The second equation is recognized as the equilibrium equation on a node-pair basis developed in Example 5.3, where

$$Y_{NN} = Y_E = \tfrac{11}{6}$$

Equations (5.11.18) and (5.11.19) contain identical information. To illustrate this point, the inverse of the generalized impedance matrix in Eq. (5.11.18) is

$$\begin{bmatrix} 4 & -3 & 3 \\ -3 & 5 & -3 \\ 3 & -3 & 3 \end{bmatrix}^{-1} = \begin{bmatrix} 1 & 0 & -1 \\ 0 & \frac{1}{2} & \frac{1}{2} \\ -1 & \frac{1}{2} & \frac{11}{6} \end{bmatrix}$$

which is identical to the generalized admittance matrix in Eq. (5.11.19), in agreement with Eq. (5.11.10).

REFERENCES

5.1. Cote, A. J., and J. B. Oakes: "Linear Vacuum Tube and Transistor Circuits," McGraw-Hill Book Company, New York, 1961. In their last chapter, the authors develop a topological analysis procedure for transistor networks, using the equivalent circuit of the transistor and dependent sources rather than 2-branch coupled graphs.

5.2. Guillemin, E. A.: Some Generalizations of Linear Network Analysis That are Useful When Active and/or Non-bilateral Elements Are Involved, *MIT Electron. Res. Lab. Quart. Progr. Rept.* 57, 1954. The generalized equilibrium equations are derived and applied in the analysis of active and/or nonbilateral networks.

5.3. Reza, F. M., and S. Seely: "Modern Network Analysis," McGraw-Hill Book Company, New York, 1959. Chapter 5 presents a clear discussion of network equilibrium equations in a development paralleling that of Secs. 5.1 to 5.6.

5.4. Seshu, S., and M. B. Reed: "Linear Graphs and Electrical Networks," Addison-Wesley Publishing Company, Inc., Reading, Mass., 1961. Chapter 6 is devoted to the applications of linear graph theory to network analysis. Equilibrium equations, including initial conditions, are discussed.

5.5. Weinberg, L.: "Network Analysis and Synthesis," McGraw-Hill Book Company, New York, 1962. Equilibrium equations on a loop, node-pair, and node-to-datum basis are developed in Chapter 3.

PROBLEMS

All resistance values are in ohms unless otherwise stated.

Drill

5.1. Write equilibrium equations on both a loop and node-pair basis.

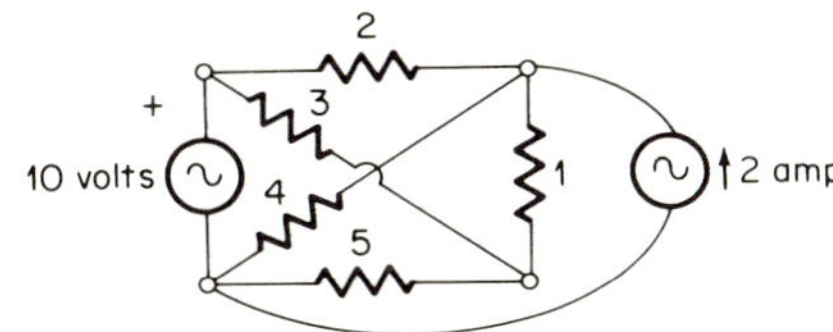

Fig. P5.1

5.2. Repeat Prob. 5.1 for the network in Fig. P5.2.

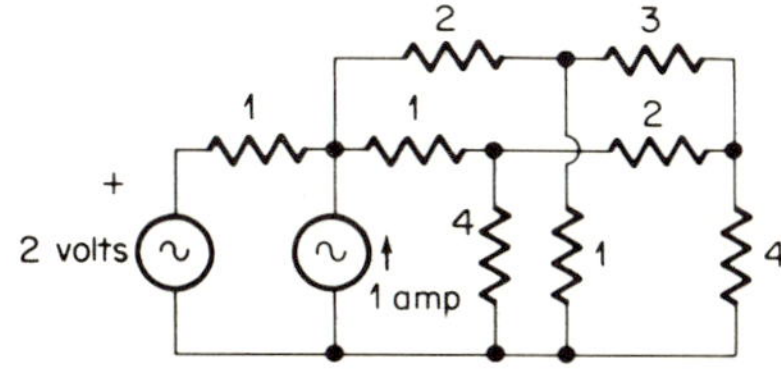

Fig. P5.2

5.3. One of the two Kuratowski nonplanar networks is shown. Write equilibrium equations on both a loop and a node-pair basis.

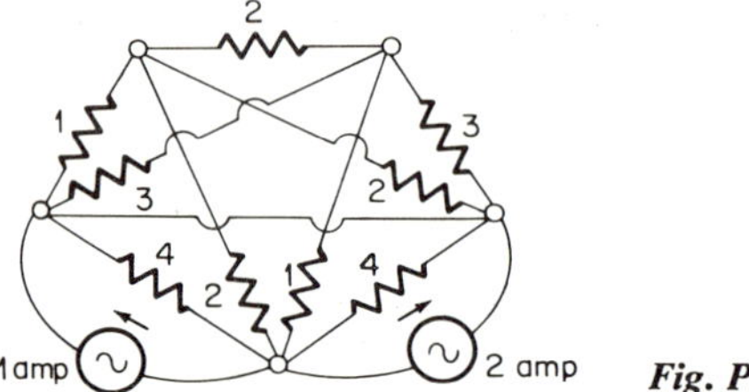

Fig. P5.3

5.4. The second Kuratowski nonplanar network is shown. Write equilibrium equations on both a loop and a node-pair basis.

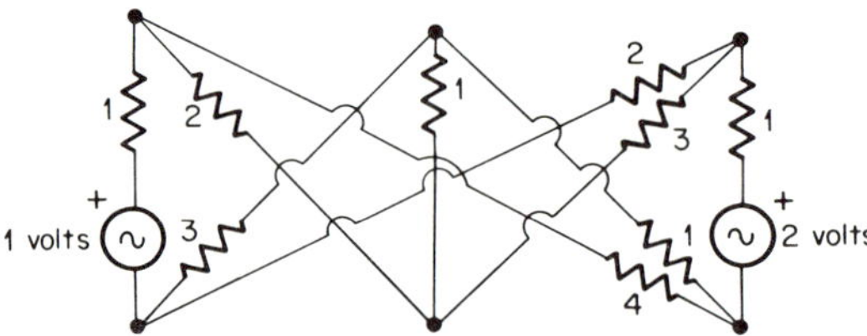

Fig. P5.4

5.5. For the given network write loop and node-pair equilibrium equations for the following conditions:

(*a*) The 4-, 5-, and 6-ohm resistors form the tree for β_1, α_1, β_2, and α_2.

(*b*) The 4-, 5-, and 6-ohm resistors form the tree for β_1 and α_1, and the 2-, 3-, and 6-ohm resistors form the tree for β_2 and α_2.

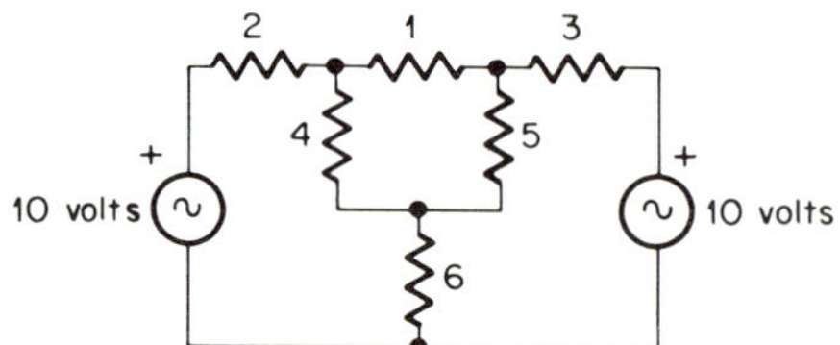

Fig. P5.5

5.6. Find Y_B in Example 5.1 if the transformer is not perfect.

5.7. For the network of Fig. P5.1, write generalized equilibrium equations on a loop basis.

5.8. For the network of Fig. P5.2, write generalized equilibrium equations on a node-pair basis.

5.9. For the given network, write generalized equilibrium equations on both a loop and a node-pair basis.

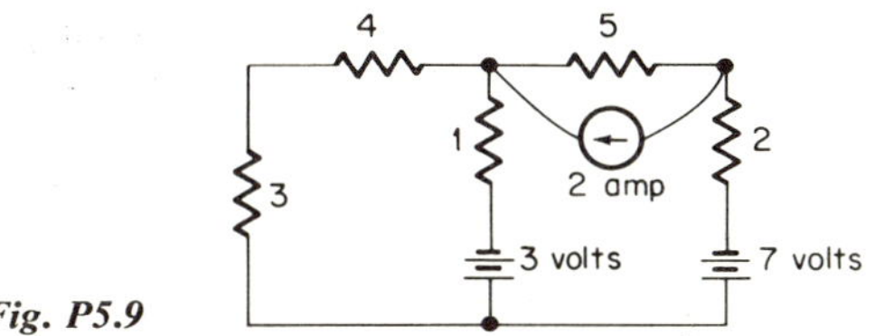

Fig. P5.9

5.10. Write equilibrium and generalized equilibrium equations on both a loop and a node-pair basis for the network shown.

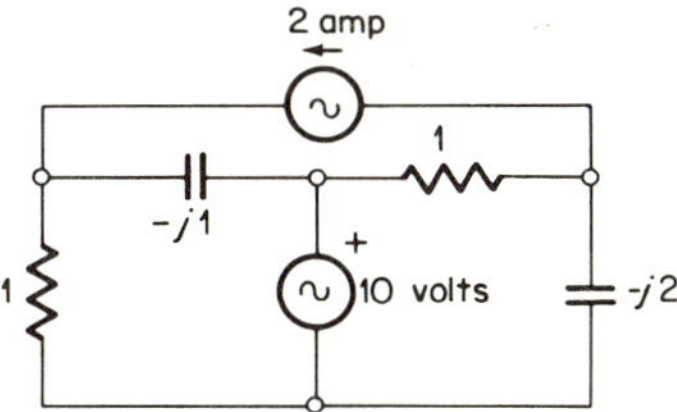

Fig. P5.10

Theory and proofs

5.11. A certain network has equilibrium equations on a loop and node-pair basis given by

(*a*) $E_E = Z_E I_L$

(*b*) $I_E = Y_E E_N$

where it is assumed that the same tree is used for each set of equations. Equation (*b*) can be solved for E_N to give

(*c*) $E_N = Y_E^{-1} I_E$

Equations (*a*) and (*c*) each define the network on an impedance-matrix basis, but they obviously do it in different ways. Discuss the differences between these two sets of equations with regard to (1) the point of application in the network of each source in the I_E and E_E matrices and (2) the physical meaning, in terms of the network, of each entry in the Z_E and Y_E^{-1} matrices.

5.12. For the network whose graph appears in Fig. P5.12, equilibrium equations on a loop basis and the branch impedance matrix are

$$\begin{bmatrix} 5 \\ -11 \end{bmatrix} = \begin{bmatrix} -19 & 17 \\ -2 & 13 \end{bmatrix} \begin{bmatrix} j_3 \\ j_5 \end{bmatrix} \qquad Z_B = \begin{bmatrix} 2 & & & & \\ & 4 & & & \\ & & 5 & & \mathbf{0} \\ \mathbf{0} & & & 3 & \\ & & & & 1 \end{bmatrix}$$

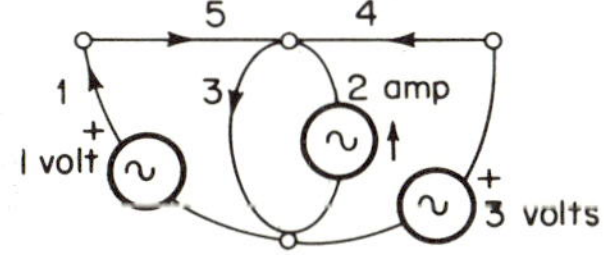

Fig. P5.12

Find the tie-set matrix (matrices) used in forming these equations. Is your answer unique? Is it possible, for any network, to find the tie-set matrix (matrices) from the information given?

5.13. The equilibrium equations on a loop basis for a certain network are

$$\begin{bmatrix} 1 \\ 6 \end{bmatrix} = \begin{bmatrix} 4 & -3 \\ -3 & 5 \end{bmatrix} \begin{bmatrix} i_1 \\ i_2 \end{bmatrix} \qquad \begin{array}{l} i_1 = j_1 \\ i_2 = j_2 \end{array}$$

The tree and links used in writing these equations, and the sources, appear in Fig. P5.13. Write equilibrium equations on a node-pair basis. State all assumptions necessary and whether or not your answer is unique. Is it generally possible to obtain node-pair equations from loop equations if the tree, links, and sources are known?

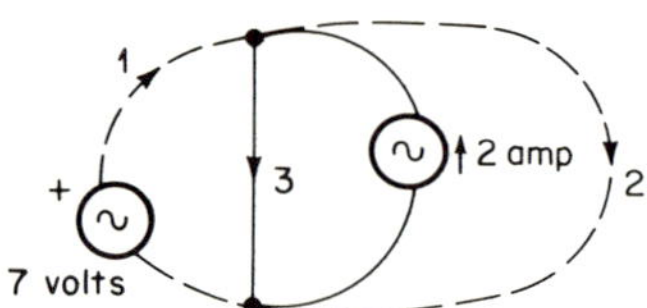

Fig. P5.13

5.14. For the network whose graph appears in Fig. P5.14, equilibrium equations on a loop basis are given. Derive generalized loop equations. State all assumptions made and whether or not your answer is unique. Is it always possible to derive generalized loop equations from loop equations if the graph and sources are known?

$$\begin{bmatrix} 2 & 1 \\ 1 & 1 \end{bmatrix} \begin{bmatrix} j_1 \\ j_2 \end{bmatrix} = \begin{bmatrix} 1 \\ 1 \end{bmatrix}$$

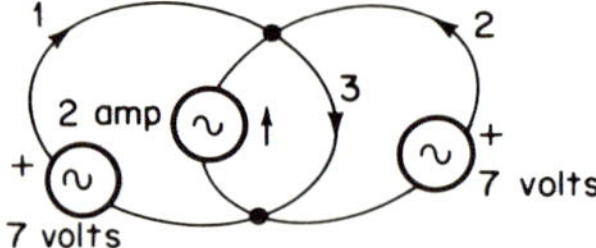

Fig. P5.14

Application

5.15. Using topological methods and appropriate equilibrium equations, find values for R and C such that the twin-T network has a voltage transmission zero at $f = 1000$ Hz.

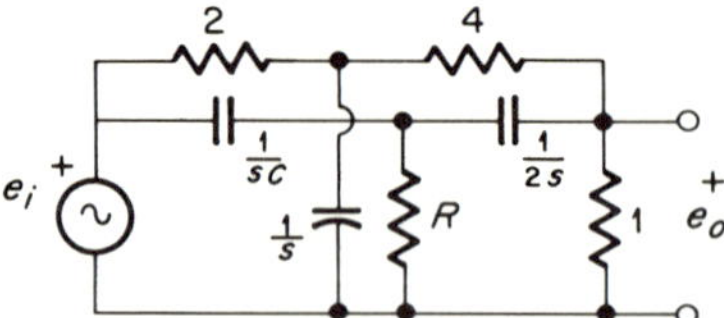

Fig. P5.15

5.16. For the system shown, write equilibrium equations on a node-pair basis.

$$\begin{bmatrix} j_1 \\ j_2 \\ j_3 \end{bmatrix} = \begin{bmatrix} 5 & 1 & 6 \\ 1 & 2 & 3 \\ 1 & 2 & 2 \end{bmatrix} \begin{bmatrix} v_1 \\ v_2 \\ v_3 \end{bmatrix}$$

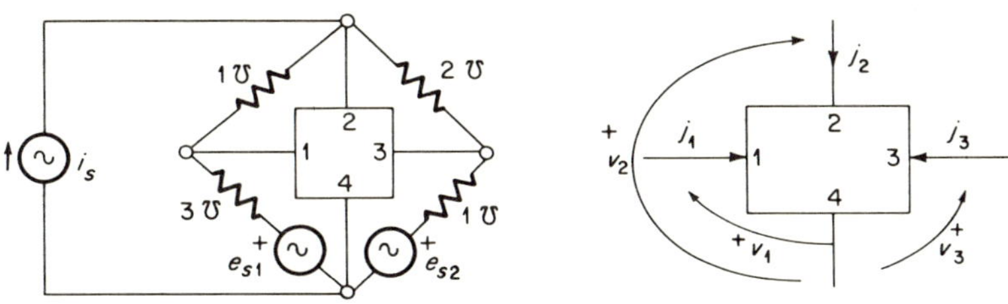

Fig. P5.16

5.17. Write equilibrium equations on both a loop and node-pair basis for the given transformer network. Assume the transformer is perfect. How are your answers affected if the transformer is imperfect?

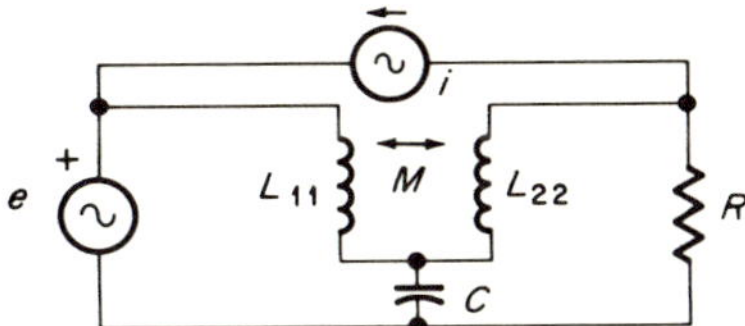

Fig. P5.17

5.18. For the networks in Fig. P5.18*a* and *b*, formulate equilibrium equations on a node-pair basis in the form

$$\begin{bmatrix} j_1 \\ j_2 \\ j_3 \end{bmatrix} = [Y_{E1}] \begin{bmatrix} v_1 \\ v_2 \\ v_3 \end{bmatrix} \quad \text{and} \quad \begin{bmatrix} j_4 \\ j_5 \\ j_6 \end{bmatrix} = [Y_{E2}] \begin{bmatrix} v_4 \\ v_5 \\ v_6 \end{bmatrix}$$

where Y_E is a 3×3 node-pair admittance matrix in each case. Now let the two networks be incorporated into the system in Fig. P5.18*c* in the manner indicated. Write equilibrium equations on a node-pair basis for the system. Your choice of tree should be such that the voltages to ground at the terminals of the N_1 and N_2 boxes appear in the final matrix equation.

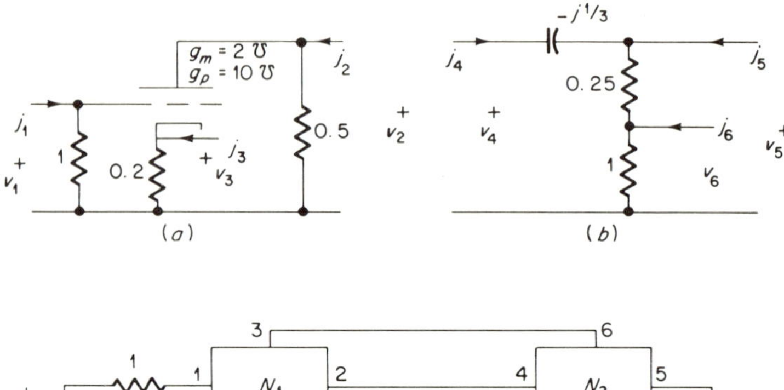

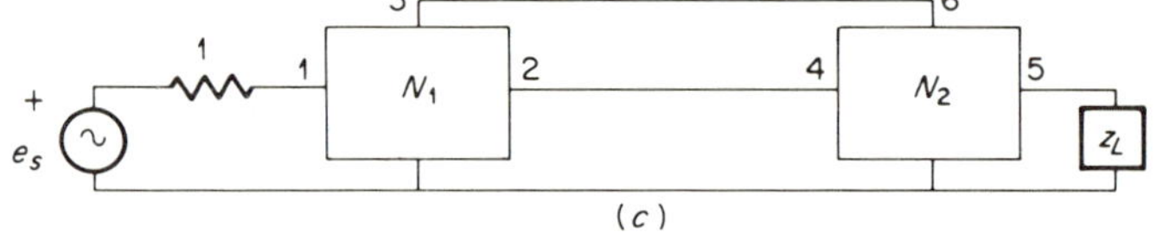

Fig. P5.18

5.19. The circuit is an instrument or oscilloscope preamplifier having characteristics of a high ac input impedance and excellent temperature stability. Perform a topological analysis of the circuit for small-signal ac conditions. Consider all capacitors as shorts to ac. It is desired that the voltages across the 100-, 500- and 10-kilohm resistors be related to i_i and i_o in the final set of matrix equations.

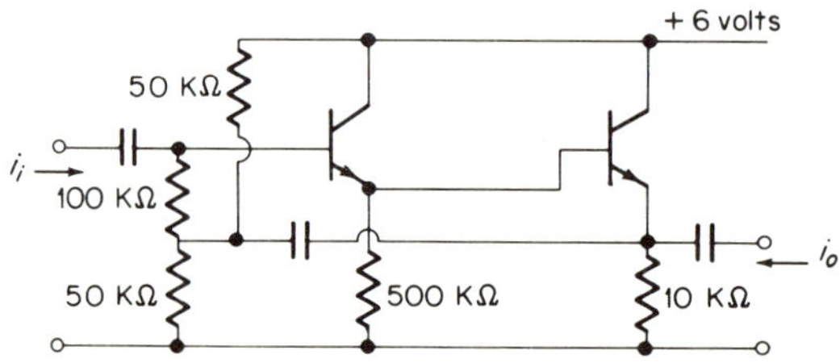

***Fig. P5.19** Common-emitter parameters are $h_{11} =$ 2000 ohms, $h_{12} = 6 \times 10^{-4}$, $h_{21} = 40$, and $h_{22} =$ 30×10^{-6} mho (identical transistors).*

5.20. Perform a small-signal ac topological analysis of the given transistor network.

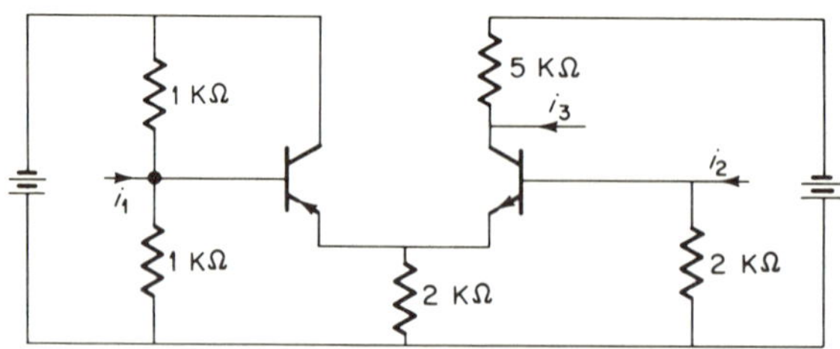

Fig. P5.20

Write equilibrium equations on a node-pair basis such that the voltages across the network resistors appear explicitly in the final matrix equations. Use the h parameters of Example 5.6.

5.21. Find the voltage gains e_x/e_i, e_o/e_x, and e_o/e_i for the triode amplifier in Fig. P3.12 by writing suitable node-pair equilibrium equations topologically.

5.22. For the network in Fig. P3.24, find the current gain i_o/i_i and the voltage gain e_o/e_i by writing suitable node-pair equilibrium equations topologically.

5.23. Write equilibrium equations on a node-pair basis for the triode amplifier network in Fig. 3.32. Choose your tree such that the voltages e_i and e_o are two tree-branch voltages in series and a single tree-branch voltage, respectively. Transform the 3×3 Y_E matrix into the 2×2 Y matrix given in Example 3.9.

5.24. Represent the feedback network within the dashed box and the transistor network within the solid box as 3-terminal networks (2-branch coupled graphs). Use the h parameters of Example 5.6. Form equilibrium equations on a node-pair basis using a 4-branch graph. Will the oscillator oscillate? If not, why not? If so, at what frequency?

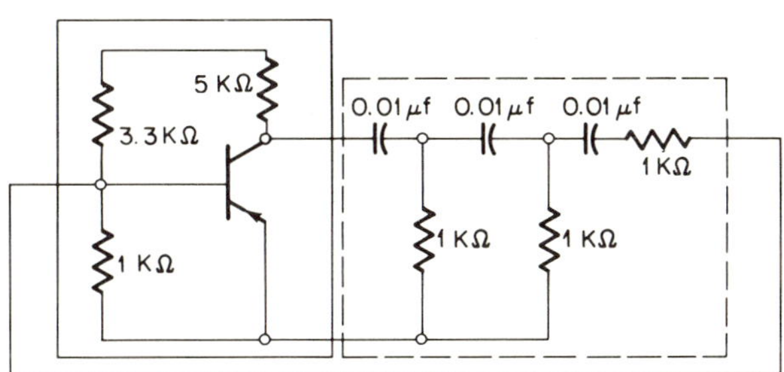

Fig. P5.24

5.25. The circuit shown is an Anderson bridge. Use topological methods to find the conditions for bridge balance at any frequency.

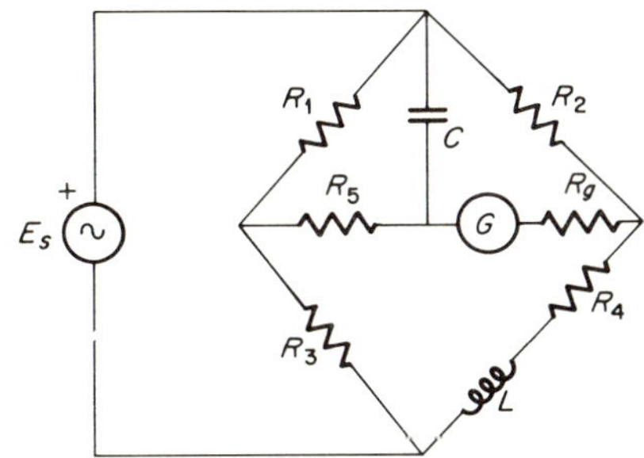

Fig. P5.25

5.26. The circuit of a transistorized voltage regulator is shown. Perform a topological analysis to obtain the ripple voltage across y_L as a function of e_r, the input ripple voltage. Calculate the ripple reduction factor for $y_L = 0.1$ mho.

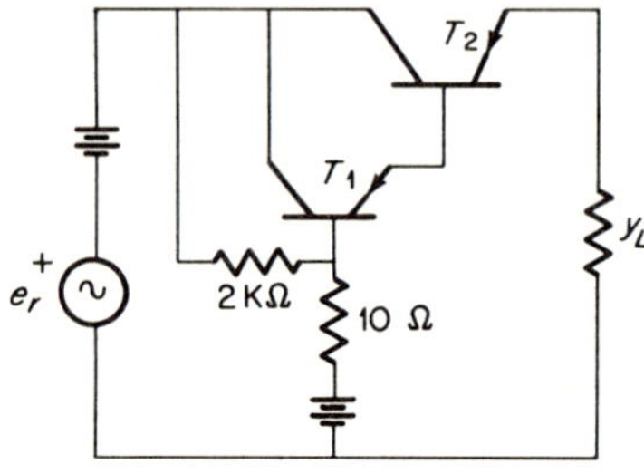

Fig. P5.26 Common-base h parameters for T_1 are $h_{11} = 30$ ohms, $h_{12} = 2 \times 10^{-4}$, $h_{21} = -0.98$, and $h_{22} = 10^{-5}$ mho; for T_2 they are $h_{11} = 1.15$ ohms, $h_{12} = 2 \times 10^{-3}$, $h_{21} = -0.97$, and $h_{22} = 5 \times 10^{-4}$ mho.

6 TOPOLOGICAL PROPERTIES OF NETWORKS

6.1 INTRODUCTION

In this chapter we use the results of the previous two chapters in the topological determination of properties of networks and network functions. First, expressions for the number of trees in a graph in terms of cut-set and tie-set matrices are derived. Next, topological formulas are developed that eliminate much of the labor from the evaluation of network determinants and cofactors. Finally, driving-point and transfer functions of networks are discussed, and topological formulas are developed for their determination.

6.2 UNIMODULAR CHARACTER OF α_F AND β_F†

In this section we show that any fundamental-cut-set or tie-set matrix of a linear graph is unimodular. We need to do this in order to derive simple expressions for the number of trees in a linear graph in the next section.

† Adapted from Ref. 6.6.

Recall from Sec. 1.14 that a unimodular matrix is a rectangular (or square) matrix of real elements each of whose subdeterminants of all orders is ± 1 or 0. Suppose we have a network $\mathcal{N}$ and we choose two sets of tree branches. From Eq. (4.8.2), we can write

$$V_B = \alpha_{F1}{}^T E_{N1} \tag{6.2.1}$$

and

$$V_B = \alpha_{F2}{}^T E_{N2} \tag{6.2.2}$$

where E_{N1} and E_{N2} are the two $N \times 1$ column matrices of chosen tree-branch voltages and α_{F1} and α_{F2} are the corresponding fundamental-cut-set matrices. Since each of the entries in E_{N1} and E_{N2} is a voltage across a single branch, each of these entries is contained in V_B. Thus we can rearrange the rows in Eqs. (6.2.1) and (6.2.2) and partition these equations to obtain

$$\begin{bmatrix} V_{L1} \\ E_{N2} \end{bmatrix} = \begin{bmatrix} \alpha_{L1}{}^T \\ \alpha_{N1}{}^T \end{bmatrix} [E_{N1}] \tag{6.2.3}$$

and

$$\begin{bmatrix} V_{L2} \\ E_{N1} \end{bmatrix} = \begin{bmatrix} \alpha_{L2}{}^T \\ \alpha_{N2}{}^T \end{bmatrix} [E_{N2}] \tag{6.2.4}$$

where V_{L1} is a column matrix containing the L branch voltages not contained in E_{N2}, and similarly for V_{L2} with respect to E_{N1}. From the lower portions of Eqs. (6.2.3) and (6.2.4) we can write

$$E_{N2} = \alpha_{N1}{}^T E_{N1} \tag{6.2.5}$$

and

$$E_{N1} = \alpha_{N2}{}^T E_{N2} \tag{6.2.6}$$

Combining Eqs. (6.2.5) and (6.2.6), we obtain

$$E_{N2} = \alpha_{N1}{}^T \alpha_{N2}{}^T E_{N2} \tag{6.2.7}$$

Thus

$$\alpha_{N1}{}^T \alpha_{N2}{}^T = U_{NN} \tag{6.2.8}$$

where U_{NN} is an $N \times N$ unit matrix. From Eqs. (6.2.8) and (1.11.2)

$$|\alpha_{N1}{}^T \alpha_{N2}{}^T| = |\alpha_{N1}{}^T|\,|\alpha_{N2}{}^T| = 1 \tag{6.2.9}$$

Recall now the composition of α_{F1} and α_{F2}. These are matrices all of whose entries are ± 1 or 0. As such, the determinant of any square submatrix of either one, if not zero, must be an integer. Thus in Eq. (6.2.9), $|\alpha_{N1}{}^T|$ and $|\alpha_{N2}{}^T|$ must be integers, and the only way their product can equal 1 is for

$$|\alpha_{N1}{}^T| = |\alpha_{N2}{}^T| = \pm 1 \tag{6.2.10}$$

The result in Eq. (6.2.10) proves that the determinant of any $N \times N$ submatrix of a fundamental-cut-set matrix is ± 1 if not zero. $A \pm 1$ is obtained if the N columns of α_F chosen define a tree (the case just considered); the result is 0 if they define a closed loop. We also know that each entry (subdeterminant of order 1) in a fundamental-cut-set matrix is ± 1 or 0. It remains to show that subdeterminants of all other orders are ± 1 or 0.

Suppose we short tree branch k of N. This eliminates row k from α_F and e_k from E_N. None of the other entries in α_F change; the matrix is merely reduced in order to $(N - 1) \times (B - 1)$. If we proceed through the steps of the previous derivation, we prove that all subdeterminants of order $(N - 1) \times (N - 1)$ of a fundamental-cut-set matrix must be ± 1 or 0. Now we can short another tree branch and continue the proof for subdeterminants of order $(N - 2) \times (N - 2)$, and so on. Our conclusion is that every fundamental-cut-set matrix of a linear graph is unimodular. By a similar proof (requested in the problems), we can show that any fundamental-tie-set matrix of a linear graph is also unimodular.

6.3 THE NUMBER OF TREES IN A GRAPH

In this section we derive simple expressions for the number of trees in a graph, first in terms of fundamental-cut-set and tie-set matrices and then in terms of nonfundamental ones.

To begin, let us define an $N \times N$ matrix T by

$$T = \alpha_F \alpha_F{}^T \tag{6.3.1}$$

The determinant of T can be found by using the Binet-Cauchy theorem, discussed in Sec. 1.11

$$|T| = |\alpha_F \alpha_F{}^T| = \sum_{\text{all majors}} |\alpha_{FNN}|\,|\alpha_{FNN}{}^T| \tag{6.3.2}$$

where $|\alpha_{FNN}|$ is any major ($N \times N$ submatrix determinant) of α_F and the summation of corresponding determinant products is taken over all majors.† The value of each $|\alpha_{FNN}|$ is $+1$, -1, or 0. Thus each product in Eq. (6.3.2) is $+1$ or 0, that is, $+1$ if the columns involved define a tree, 0 if they do not. If we add all such products, we obtain the total number of trees as $|T|$.

Does $|T| = |\alpha\alpha^T|$ yield the number of trees if the cut-set matrix in Eq. (6.3.2) is not fundamental? Our starting point in answering this question is Eq. (4.10.13), repeated here

$$V_B = \alpha^T E \tag{6.3.3}$$

† It is necessary in the following sections to distinguish submatrices from cofactors. We shall use uppercase double subscripts to denote the order of a submatrix and lowercase double subscripts to denote the row-column location of a cofactor.

where α is not a fundamental-cut-set matrix and E is not a set of tree-branch voltages. We partition Eq. (6.3.3) and obtain

$$\begin{bmatrix} V_L \\ V_N \end{bmatrix} = \begin{bmatrix} \alpha_{LN}{}^T \\ \alpha_{NN}{}^T \end{bmatrix} [E] \tag{6.3.4}$$

where V_B has been rearranged so that its last N entries define a set of tree branches, i.e., such that $|\alpha_{NN}{}^T|$ is nonvanishing. Solving for E gives

$$E = (\alpha_{NN}{}^T)^{-1} V_N \tag{6.3.5}$$

Substitution of Eq. (6.3.5) into Eq. (6.3.3) yields

$$V_B = \alpha^T (\alpha_{NN}{}^T)^{-1} V_N \equiv \alpha_F{}^T V_N \tag{6.3.6}$$

where the last equality follows because V_N is a set of tree-branch voltages. From Eq. (6.3.6),

$$\alpha_F = \alpha_{NN}{}^{-1} \alpha \tag{6.3.7}$$

Now inserting Eq. (6.3.7) into Eq. (6.3.2), we have for the number of trees

$$\begin{aligned} |T| &= |\alpha_{NN}{}^{-1}\alpha(\alpha_{NN}{}^{-1}\alpha)^T| \\ &= |\alpha_{NN}\alpha\alpha^T(\alpha_{NN}{}^{-1})^T| \\ &= |\alpha_{NN}{}^{-1}|\,|\alpha\alpha^T|\,|(\alpha_{NN}{}^{-1})^T| \end{aligned} \tag{6.3.8}$$

The second line of Eq. (6.3.8) follows from the first by the property of transposes. The third line is the customary representation of a determinant of a matrix product when the matrices are square and of the same order. Since E in Eq. (6.3.3) is not a set of tree-branch voltages, $|\alpha_{NN}{}^{-1}|$ is not in general ± 1. Thus $|T|$ is not given by $|\alpha\alpha^T|$.

Example 6.1 Determine the number of trees in the graph whose cut-set matrix is given by

$$\alpha = \begin{bmatrix} \tfrac{1}{2} & -1 & -\tfrac{1}{2} & -1 & \tfrac{1}{2} & 0 \\ \tfrac{1}{2} & -1 & \tfrac{1}{2} & 0 & \tfrac{1}{2} & -1 \\ \tfrac{1}{2} & 0 & \tfrac{1}{2} & 1 & -\tfrac{1}{2} & -1 \end{bmatrix}$$

The number of trees is given by Eq. (6.3.8). First find $|\alpha\alpha^T|$.

$$\alpha\alpha^T = \begin{bmatrix} \tfrac{11}{4} & \tfrac{5}{4} & -\tfrac{5}{4} \\ \tfrac{5}{4} & \tfrac{11}{4} & \tfrac{5}{4} \\ -\tfrac{5}{4} & \tfrac{5}{4} & \tfrac{11}{4} \end{bmatrix} \qquad |\alpha\alpha^T| = 4$$

Next $|\alpha_{NN}|$ is needed. Columns 1, 3, and 5 in α form one independent set of voltages.

Thus

$$\alpha_{NN} = \begin{bmatrix} \frac{1}{2} & -\frac{1}{2} & \frac{1}{2} \\ \frac{1}{2} & \frac{1}{2} & \frac{1}{2} \\ \frac{1}{2} & \frac{1}{2} & -\frac{1}{2} \end{bmatrix} \qquad \alpha_{NN}^{-1} = \begin{bmatrix} 1 & 0 & 1 \\ -1 & 1 & 0 \\ 0 & 1 & -1 \end{bmatrix} \qquad |\alpha_{NN}^{-1}| = -2$$

The number of trees is

$$|T| = -2 \times 4 \times -2 = 16$$

Thus far we have expressed the number of trees in a graph only in terms of cut-set matrices. A parallel development in terms of tie-set matrices is also possible. The results are

$$|T| = |\beta_F \beta_F^T| \tag{6.3.9}$$

for a fundamental-tie-set case and

$$|T| = |\beta_{LL}^{-1}|\,|\beta\beta^T|\,|(\beta_{LL}^{-1})^T| \tag{6.3.10}$$

for a nonfundamental-tie-set case, where β_{LL} in Eq. (6.3.10) is the counterpart of α_{NN} in Eq. (6.3.8). The derivations of Eqs. (6.3.9) and (6.3.10) are requested in the problems.

6.4 SOLUTIONS OF THE EQUILIBRIUM EQUATIONS BY CRAMER'S RULE

Network equilibrium equations on a loop and node-pair basis were derived in Chap. 5; we repeat Eqs. (5.5.4) and (5.6.3) here

$$E_E = Z_E I_L \tag{6.4.1}$$

and

$$I_E = Y_E E_N \tag{6.4.2}$$

Equation (6.4.1) is a square array consisting of L equations in L unknowns with a coefficient matrix Z_E of rank L. Thus Cramer's rule, in the form of Eq. (1.12.8), can be applied directly to obtain a solution for the loop currents as

$$i_l = \frac{Z_{Ekl} E_{Ek}}{|Z_E|} \tag{6.4.3}$$

where i_l is the lth loop current, E_{Ek} is the net source voltage in the kth loop, Z_{Ekl} is the klth cofactor of Z_E, and the repeated index indicates that a summation is to be made over all k (from 1 through L).

Similarly Eq. (6.4.2) is a square array of N equations in N unknowns with a coefficient matrix Y_E of rank N. Thus

$$e_l = \frac{Y_{Ekl} I_{Ek}}{|Y_E|} \tag{6.4.4}$$

where, in this case, k is summed from 1 through N. Here I_{Ek} is the net source current at the positive reference node of the kth tree branch.

6.5 TOPOLOGICAL SOLUTIONS OF THE EQUILIBRIUM EQUATIONS

Solutions of the equilibrium equations by Cramer's rule are tedious, especially if the order of Z_E or Y_E is high. This is true because the expansions of the numerators and denominators of Eqs. (6.4.3) and (6.4.4) contain a large number of terms, many of which cancel before the final forms are obtained. In this section we devise topological methods for obtaining $|Y_E|$, $|Z_E|$, Y_{Ekl}, and Z_{Ekl} for networks having no coupling (Y_B and Z_B diagonal). These topological methods are shown to represent a considerable saving in algebraic labor over the conventional determinant and cofactor expansions. In a later section we shall extend these results to the active- and/or coupled-network case.

DENOMINATOR EVALUATION

Consider first the topological evaluation of the denominators in Eqs. (6.4.3) and (6.4.4). Specifically, consider the determination of $|Y_E|$, given by

$$|Y_E| = |\alpha_F Y_B \alpha_F{}^T| \tag{6.5.1}$$

By the Binet-Cauchy theorem,

$$|\alpha_F Y_B \alpha_F{}^T| = \sum |(\alpha_F Y_B)_{NN}|\, |\alpha_{FNN}^T| \tag{6.5.2}$$

where the notation is similar to that in Eq. (6.3.2). Let us examine $|\alpha_{FNN}^T|$, or equivalently $|\alpha_{FNN}|$. We know from Sec. 6.2 that

$$|\alpha_{FNN}| = \pm 1 \text{ or } 0 \tag{6.5.3}$$

where the ± 1 holds if the columns of α_F included in α_{FNN} define a tree and the 0 holds if they define a closed loop.

Next consider $|(\alpha_F Y_B)_{NN}|$. Since Y_B is diagonal, $\alpha_F Y_B$ has the same structure as α_F; each column in $\alpha_F Y_B$ is obtained by multiplying the corresponding column of α_F by the branch admittance associated with that column. Thus column k in $\alpha_F Y_B$ is y_k times column k in α_F, where y_k is the admittance of branch k. It follows from property 4 of determinants that

$$\begin{aligned} |(\alpha_F Y_B)_{NN}| &= y_{i_1} y_{i_2} \cdots y_{iN} \, |\alpha_{FNN}| \\ &= \pm y_{i_1} y_{i_2} \cdots y_{iN} \text{ or } 0 \end{aligned} \tag{6.5.4}$$

where each set of y_i's is the product of the admittances of the branches of a tree. We call each of these products a *tree-admittance product* (abbreviated TYP). Now using Eqs. (6.5.2) to (6.5.4), Eq. (6.5.1) can be written in the form

$$|Y_E| = \sum \text{TYP of } \mathcal{N} \tag{6.5.5}$$

where $\mathcal{N}$ is the network being investigated.† Equation (6.5.5) is the sum of $|T|$ products (the number of trees) each of which contains N admittances (the number of branches in a tree).

By a similar development, we can show that

$$|Z_E| = \sum \text{LZP of } \mathcal{N} \tag{6.5.6}$$

where LZP is a *link-impedance product*, defined as the product of the impedances of the links of a tree. Equation (6.5.6) is also the sum of $|T|$ products, but in this case each product includes L impedances (the number of links in a tree).

Example 6.2 Find $|Y_E|$ for the graph in Fig. 6.1 (which is a repetition of Fig. 4.2) from a cofactor expansion of $|\alpha_F Y_B \alpha_F^T|$ and from Eq. (6.5.5). Repeat for $|Z_E|$.

(*a*) $|\alpha_F Y_B \alpha_F^T|$ Using the tree 235,

$$\alpha_F = \begin{bmatrix} -1 & 1 & 0 & 0 & 0 \\ 0 & 0 & 1 & 1 & 0 \\ -1 & 0 & 0 & -1 & 1 \end{bmatrix} \qquad Y_B = \begin{bmatrix} y_1 & & & & \mathbf{0} \\ & y_2 & & & \\ & & y_3 & & \\ \mathbf{0} & & & y_4 & \\ & & & & y_5 \end{bmatrix}$$

$$\alpha_F Y_B \alpha_F^T = \begin{bmatrix} y_1 + y_2 & 0 & y_1 \\ 0 & y_3 + y_4 & -y_4 \\ y_1 & -y_4 & y_1 + y_4 + y_5 \end{bmatrix}$$

$$\begin{aligned} |\alpha_F Y_B \alpha_F^T| = {} & y_1 y_2 y_3 + y_1 y_2 y_4 + y_1 y_3 y_4 + y_1 y_3 y_5 + y_1 y_4 y_5 \\ & + y_2 y_3 y_4 + y_2 y_3 y_5 + y_2 y_4 y_5 \end{aligned}$$

(*b*) $\sum$ *TYP of* $\mathcal{N}$ The trees are 123, 124, 134, 135, 145, 234, 235, and 245. Replacing each branch number by its branch admittance yields $|Y_E|$.

† We shall use a capital script $\mathcal{N}$ to denote network, reserving N, as before, for the number of tree branches.

Fig. 6.1

(*c*) $|\beta_F Z_B \beta_F^T|$ Again using the tree 235,

$$\beta_F = \begin{bmatrix} 1 & 1 & 0 & 0 & 1 \\ 0 & 0 & -1 & 1 & 1 \end{bmatrix} \qquad Z_B = \begin{bmatrix} z_1 & & & & \mathbf{0} \\ & z_2 & & & \\ & & z_3 & & \\ & & & z_4 & \\ \mathbf{0} & & & & z_5 \end{bmatrix}$$

$$\beta_F Z_B \beta_F^T = \begin{bmatrix} z_1 + z_2 + z_5 & z_5 \\ z_5 & z_3 + z_4 + z_5 \end{bmatrix}$$

$$|\beta_F Z_B \beta_F^T| = z_1 z_3 + z_1 z_4 + z_1 z_5 + z_2 z_3 + z_2 z_4 + z_2 z_5 + z_3 z_5 + z_4 z_5$$

(*d*) $\sum$ *LZP of* $\mathcal{N}$ The link sets are 45, 35, 25, 24, 23, 15, 14, 13. They can be obtained directly from the graph or as the complements of the trees in part (*b*). Replacing each branch number by its branch impedance gives $|Z_E|$.

Is there any simple relation between $|Y_E|$ in Eq. (6.5.5) and $|Z_E|$ in Eq. (6.5.6)? To answer this question, consider a particular tree-admittance product in Example 6.3, say $y_1 y_2 y_3$. If we multiply this product by the product of all the branch impedances, we obtain $y_1 y_2 y_3 z_1 z_2 z_3 z_4 z_5 = z_4 z_5$, which is the link-impedance product associated with that tree. But the product of the branch impedances is $|Z_B|$, since Z_B is diagonal. Thus $y_1 y_2 y_3 |Z_B| = z_4 z_5$. In general,

$$|Z_B|\,|Y_E| = |Z_B| \textstyle\sum \text{TYP of } \mathcal{N} = \sum \text{LZP of } \mathcal{N} = |Z_E| \tag{6.5.7}$$

Thus the network determinants on a loop and node-pair basis are related very simply through the determinant of the branch-impedance matrix. In a similar fashion

$$|Y_B|\,|Z_E| = |Y_E| \tag{6.5.8}$$

NUMERATOR EVALUATION

The numerators of Eqs. (6.4.3) and (6.4.4) will be considered next. Specifically, topological methods are needed for evaluating Y_{Ekl} and Z_{Ekl}, the *kl*th cofactors of Y_E and Z_E, respectively.

Consider first the determination of Y_{Ekl}. The magnitude of this cofactor is the determinant of the matrix that remains after deleting row *k* and column *l* from Y_E. Its sign is given by $(-1)^{k+l}$. But to delete row *k* and column *l* from Y_E, all we need do is delete row *k* from α_F and column *l* from α_F^T in Eq. (6.5.1). For deleting row *k* from α_F deletes row *k* from $\alpha_F Y_B$. Now postmultiplying this result by $\alpha_F{}^T$ with column l deleted and taking the determinant yields the desired result in the form

$$Y_{Ekl} = (-1)^{k+l} \, |\alpha_{F-k} Y_B \alpha^T_{F-l}| \tag{6.5.9}$$

where the $-k$ and $-l$ indicate the row and column deletions, respectively.

Example 6.3 Find Y_{E32} and Y_{E33} for the graph in Fig. 6.1, using the tree 235.

The fundamental cut-set matrix for the tree 235 is given in Example 6.3. Deleting row 3 and row 2, we obtain

$$\alpha_{F-3} = \begin{bmatrix} -1 & 1 & 0 & 0 & 0 \\ 0 & 0 & 1 & 1 & 0 \end{bmatrix} \qquad \alpha_{F-2} = \begin{bmatrix} -1 & 1 & 0 & 0 & 0 \\ -1 & 0 & 0 & -1 & 1 \end{bmatrix}$$

From Eq. (6.5.9),

$$Y_{E32} = (-1)^{3+2} \left| \begin{bmatrix} -y_1 & y_2 & 0 & 0 & 0 \\ 0 & 0 & y_3 & y_4 & 0 \end{bmatrix} \begin{bmatrix} -1 & -1 \\ 1 & 0 \\ 0 & 0 \\ 0 & -1 \\ 0 & 1 \end{bmatrix} \right|$$

$$= - \begin{vmatrix} y_1 + y_2 & y_1 \\ 0 & -y_4 \end{vmatrix} = y_1 y_4 + y_2 y_4$$

In a similar fashion,

$$Y_{E33} = (-1)^{3+3} \left| \begin{bmatrix} -y_1 & y_2 & 0 & 0 & 0 \\ 0 & 0 & y_3 & y_4 & 0 \end{bmatrix} \begin{bmatrix} -1 & 0 \\ 1 & 0 \\ 0 & 1 \\ 0 & 1 \\ 0 & 0 \end{bmatrix} \right|$$

$$= \begin{vmatrix} y_1 + y_2 & 0 \\ 0 & y_3 + y_4 \end{vmatrix} = y_1 y_3 + y_1 y_4 + y_2 y_3 + y_2 y_4$$

Equation (6.5.9) is not a topological formula for evaluating Y_{Ekl}, but it leads us to one. By the Binet-Cauchy theorem

$$|\alpha_{F-k} Y_B \alpha^T_{F-l}| = \sum |(\alpha_{F-k} Y_B)_{N-1,N-1}| \, |(\alpha^T_{F-l})_{N-1,N-1}| \tag{6.5.10}$$

where the notation is the same as in Eq. (6.5.2) except that now the majors are of order $N - 1$ rather than N. Consider α_{F-l}. Suppose we have the graph of a network $\mathcal{N}$ and we choose a tree and form α_F. Short the lth *tree* branch of $\mathcal{N}$ (*not usually the same as the lth branch*). The branch so shorted becomes a self-loop since the two nodes at its ends coalesce. Call the new network $\mathcal{N}_{sl}$, the subscripts denoting that *tree branch l is shorted.*

With the same tree as before (except for the shorted tree branch) the fundamental-cut-set matrix is now α_{F-l}, and from Eq. (6.5.3)

$$|(\alpha_{F-l})_{N-1,N-1}| = \pm 1 \text{ or } 0 \tag{6.5.11}$$

where the ± 1 holds if the particular columns included in a major of α_{F-l} define a tree of $\mathcal{N}_{sl}$ and the 0 holds if they define a closed loop.

Similarly, from the form of Eq. (6.5.4), we obtain

$$|(\alpha_{F-k} Y_B)_{N-1,N-1}| = \pm y_{i_1} y_{i_2} \cdots y_{i_{N-1}} \text{ or } 0 \tag{6.5.12}$$

where each set of y_i's is the product of tree-branch admittances of $\mathcal{N}_{sk}$, the network obtained by shorting tree branch k in $\mathcal{N}$.

When we form the products of corresponding majors in Eq. (6.5.10), a nonzero result is obtained if and only if a tree of $\mathcal{N}_{sl}$ is also a tree of $\mathcal{N}_{sk}$. This observation forms the basis of our topological method. We construct $\mathcal{N}_{sl}$ and $\mathcal{N}_{sk}$ and find the trees that are common to each. We denote the corresponding common tree-admittance products by CTYP. Then Eq. (6.5.10) can be written

$$|\alpha_{F-k} Y_B \alpha_{F-l}^T| = \sum \pm \text{CTYP of } \mathcal{N}_{sk} \text{ and } \mathcal{N}_{sl} \tag{6.5.13}$$

Incorporating Eq. (6.5.13) into Eq. (6.5.9) yields

$$Y_{E_{kl}} = (-1)^{k+l} \sum \pm \text{CTYP of } \mathcal{N}_{sk} \text{ and } \mathcal{N}_{sl} \tag{6.5.14}$$

Actually we can find the trees of $\mathcal{N}_{sk}$ and $\mathcal{N}_{sl}$ directly from $\mathcal{N}$ and thus avoid the necessity of constructing the graphs of the shorted networks. To this end we observe that any tree of $\mathcal{N}$ that includes tree branch k cannot possibly form a closed loop when this tree branch is shorted (except, of course, for the self-loop of tree branch k). Any tree of $\mathcal{N}$ that does not contain tree branch k will form a closed loop when this tree branch is shorted. Thus to find the trees of $\mathcal{N}_{sk}$, we need merely list the trees of $\mathcal{N}$ that contain tree branch k, with tree branch k deleted, and likewise for the trees of $\mathcal{N}_{sl}$.

There is no guarantee that a nonzero major of $\alpha_{F-k} Y_B$ in Eq. (6.5.12) will have the same sign as a corresponding major of α_{F-l}^T. Thus each product in Eq. (6.5.14) has a sign ambiguity, as indicated. There are two cases to consider. First, if $k = l$, Y_{Ekl} becomes a main-diagonal cofactor Y_{Ell}. In this case the sign is unambiguous and is plus because $(-1)^{l+l} = +1$ and because corresponding majors in Eqs. (6.5.11) and (6.5.12) are identical and thus have the same sign and a positive product. Thus Eq. (6.5.14) becomes, for main-diagonal cofactors,

$$Y_{Ell} = \sum \text{TYP of } \mathcal{N}_{sl} \tag{6.5.15}$$

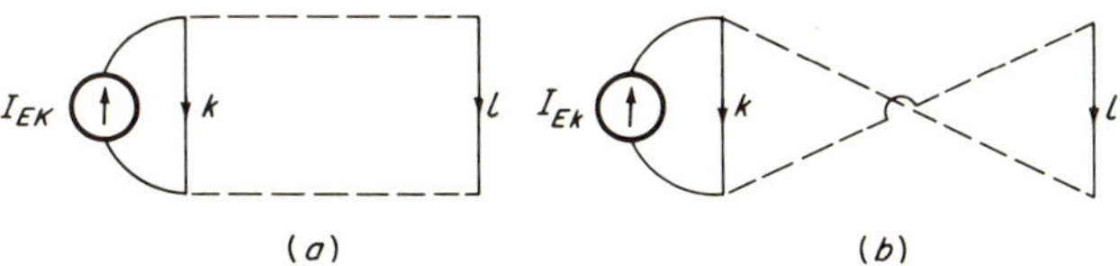

Fig. 6.2 Y_{Ekl} *sign determination.*

For off-diagonal cofactors, the proper sign can be obtained by a simple examination of the network graph.† Consider the diagrams in Fig. 6.2. These explicitly show the directed tree branches k and l of a network $\mathcal{N}$ and the two possible transmission paths between k and l. The voltage e_l across tree branch l is the lth node-pair voltage and is given by the summation in Eq. (6.4.4). Now assume that the only current source present is I_{Ek} across tree branch k. Then e_l is given by Eq. (6.4.4) *without* a summation on k, and we can treat $\mathcal{N}$ as a 2-port driven by a source current I_{Ek}, as indicated in Fig. 6.2. We note that, for $a + I_{Ek}$ and the indicated tree-branch direction arrows, e_l is positive in Fig. 6.2*a* and negative in Fig. 6.2*b*.

Now suppose we are seeking the overall sign (including the -1^{k+l}) of a certain CTYP of Y_{Ekl} in Eq. (6.5.14). If the branches included in that CTYP provide the transmission path of Fig. 6.2*a* between tree branches k and l, the overall sign of that CTYP is plus; if they provide the path of Fig. 6.2*b*, the sign is minus. Thus we can determine the overall sign of any CTYP by a simple examination of the transmission path that its branches provide between the specfied input and output tree branches. If the tree-branch direction arrows of these two tree branches are connected head to head and tail to tail by the CTYP, the sign is plus; if they are connected head to tail and tail to head, the sign is minus.

Example 6.4 Find Y_{E31}, Y_{E32}, and Y_{E33} topologically for the graph in Fig. 6.3*a* and the tree 235.

The required subgraphs are drawn in Fig. 6.3*b*, *c*, and *d*.

† This method of sign determination is due to W. S. Percival in The Solution of Passive Electrical Networks by Mathematical Trees, *J. IEE*, vol. 100, pt. III, 1953.

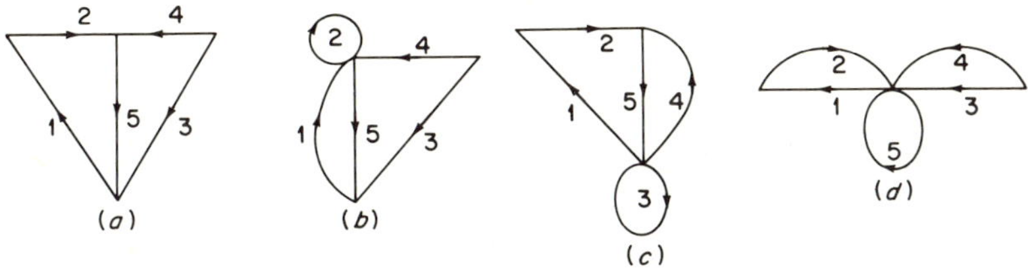

Fig. 6.3 η, η_{s1}, *and* η_{s3} *for Fig. 6.1.*

Note that branch 2 (the first tree branch) has become a self-loop in Fig. 6.3*b*, and likewise branch 3 (the second tree branch) in Fig. 6.3*c* and branch 5 (the third tree branch) in Fig. 6.3*d*.

The trees are

$\mathcal{N}_{s1}$ 13, 14, 34, 35, 45

$\mathcal{N}_{s2}$ 12, 14, 15, 24, 25

$\mathcal{N}_{s3}$ 13, 14, 23, 24

As previously explained, there is no need to draw $\mathcal{N}_{s1}$, $\mathcal{N}_{s2}$, and $\mathcal{N}_{s3}$ in Fig. 6.3 if we know the trees of $\mathcal{N}$. We have the latter from Example 6.3. The trees of $\mathcal{N}_{sk}$ are those trees of $\mathcal{N}$ which contain tree branch k, with tree branch k deleted. For example:

Trees of $\mathcal{N}$ 123, 124, 134, 135, 145, 234, 235, 245
Trees of $\mathcal{N}$ *containing tree-branch* 3 (*branch* 5) 135, 145, 235, 245
Trees of $\mathcal{N}_{s3}$ 13, 14, 23, 24

Now Y_{E33} is given by Eq. (6.5.15) as

$$Y_{E33} = y_1y_3 + y_1y_4 + y_2y_3 + y_2y_4$$

The common trees are 13 and 14 for Y_{E31} and 14 and 24 for Y_{E32}. Thus from Eq. (6.5.14),

$$Y_{E31} = (+1)(\pm y_1y_3 \pm y_2y_4)$$

$$Y_{E32} = (-1)(\pm y_1y_4 \pm y_2y_4)$$

We must now resolve the sign ambiguities. Consider the overall sign of the y_1y_4 term in Y_{E32}. The tree-branch arrows are connected head to head and tail to tail by this CTYP. Thus the sign of this product is plus. The same is true of y_2y_4 in Y_{E32}. In fact, the signs of the CTYPs for a given cofactor are always the same if one input and one output terminal are in common. Otherwise they may or may not be the same. The signs of both CTYPs in Y_{E31} are minus, since the tree-branch arrows in this case are connected head to tail and tail to head. Thus

$$Y_{E31} = -y_1y_3 - y_1y_4$$

and

$$Y_{E32} = \quad y_1y_4 + y_2y_4$$

The results agree with those of Example 6.3.

A completely similar development can be carried out on a loop basis to derive topological formulas for Z_{Ekk} and Z_{Ekl}. Thus

$$Z_{Ekl} = (-1)^{k+l}\,|\beta_{F-k}Z_B\beta^T_{F-l}| \tag{6.5.16}$$

By the Binet-Cauchy theorem,

$$|\beta_{F-k}Z_B\beta^T_{F-l}| = \sum |(\beta_{F-k}Z_B)_{L-1,L-1}|\,|(\beta^T_{F-l})_{L-1,L-1}| \tag{6.5.17}$$

where the majors are now of order $L - 1$. Choose a tree in $\mathcal{N}$ and open the lth link. This creates a dangling branch. Call the new network $\mathcal{N}_{ol}$

(network $\mathcal{N}$ with its lth link open). Its fundamental-tie-set matrix is β_{F-l}. Then

$$|(\beta_{F-l})_{L-1,L-1}| = \pm 1 \text{ or } 0 \tag{6.5.18}$$

± 1 for a link set, 0 for a tree. Replacing the lth link and opening the kth link, we obtain

$$|(\beta_{F-k}Z_B)_{L-1,L-1}| = \pm Z_{i_1}Z_{i_2}\cdots Z_{i_{L-1}} \text{ or } 0 \tag{6.5.19}$$

where each set of Z_i's is a link-impedance set of $\mathcal{N}_{ok}$. Thus

$$Z_{Ekl} = (-1)^{k+l} \sum \pm\text{CLZP of } \mathcal{N}_{ok} \text{ and } \mathcal{N}_{ol} \tag{6.5.20}$$

and

$$Z_{Ell} = \sum \text{LZP of } \mathcal{N}_{ol} \tag{6.5.21}$$

As it was unnecessary to draw the shorted graphs in the admittance case, so it is unnecessary to draw the open graphs in the impedance case. For any tree of a $\mathcal{N}$ that does not include link k will become a tree of $\mathcal{N}_{ok}$ if link k is added to the tree. Thus the trees of $\mathcal{N}_{ok}$ are the trees of $\mathcal{N}$ not containing link k with link k added, and the link sets of $\mathcal{N}_{ok}$ are the link sets of $\mathcal{N}$ that contain link k with link k deleted.

We can obtain the overall sign of any CLZP in a Z_{Ekl} cofactor by a slight modification of our method of overall sign determination for a CTYP of Y_{Ekl}. The lth link current i_l is given by Eq. (6.4.3) without a summation if the only source voltage is E_{Ek} in link k. Let us multiply that equation, top and bottom, by $|Y_B|$. We obtain

$$i_l = \frac{|Y_B|\, Z_{Ekl}E_{Ek}}{|Y_B|\, |Z_E|} = \frac{E_{Ek}(-1)^{k+l} \sum \pm\text{CTYP of } \mathcal{N}_{ok} \text{ and } \mathcal{N}_{ol}}{\sum \text{TYP of } \mathcal{N}} \tag{6.5.22}$$

The second form of the denominator follows from Eqs. (6.5.8) and (6.5.5). The second form of the numerator is justified by noting that Z_{Ekl} is given by Eq. (6.5.20); multiplying this result by $|Y_B| = y_1y_2\cdots y_B$ converts the CLZPs to CTYPs.

Now with the second form of Eq. (6.5.22) established, we can use the previous method of overall sign determination (Fig. 6.2). Thus the overall sign of any CLZP is found by deleting from $\mathcal{N}$ the branches included in the CLZP and observing whether the remaining transmission path connects the directed arrows of links k and l head to head and tail to tail or head to tail and tail to head. The overall sign of the CLZP is minus in the former case (currents bucking) and plus in the latter case (currents aiding).

Example 6.5 Find Z_{E21} and Z_{E22} for the graph of Fig. 6.1 and the tree 235.

Here $k = 2$ and $l = 1$, and thus the subgraphs needed are $\mathcal{N}_{o2}$ and $\mathcal{N}_{o1}$. They appear in Fig. 6.4. Branch 4 (link 2) is a dangling branch in Fig. 6.4*a* and branch 1 (link 1) in

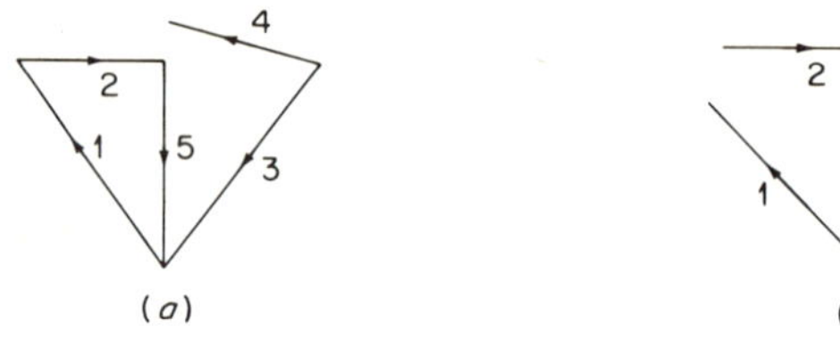

Fig. 6.4 η_{o2} *and* η_{o1} *derived from Fig. 6.1.*

Fig. 6.4*b*. The link sets are:

$\mathcal{N}_{o2}$ 1, 2, 5
$\mathcal{N}_{o1}$ 3, 4, 5

Note that the link sets of $\mathcal{N}_{ok}$ are the link sets of $\mathcal{N}$ containing link k with link k removed. Now

$$Z_{E22} = z_1 + z_2 + z_5$$

The only common link is 5. Thus

$$Z_{E21} = (-1)(\pm 1)z_5$$

The sign ambiguity is resolved by removing branch 5 from $\mathcal{N}$ and observing that the arrows of branches 1 and 4 (links 1 and 2) are connected head to head and tail to tail. Thus

$$Z_{E21} = -z_5$$

6.6 NETWORK ENTRY POINTS

In the discussion of driving-point and transfer functions of networks, the concepts of *pliers* and *solder* entries, as developed by Guillemin (Ref. 6.1), are useful. A pliers entry is the creation of a terminal pair by cutting into a branch of a network, as between the points x and x' in Fig. 6.5, whereas a solder entry is formed by attaching leads to two of the nodes of a network, as at points y and y' in Fig. 6.5.

Suppose we wish to excite the network from these entry points. There are four cases to consider:

1. Current source at solder entry
2. Voltage source at pliers entry
3. Current source at pliers entry
4. Voltage source at solder entry

In the first two cases the network topology is unaltered by the excitation. For if we connect a current source across a branch (or across several

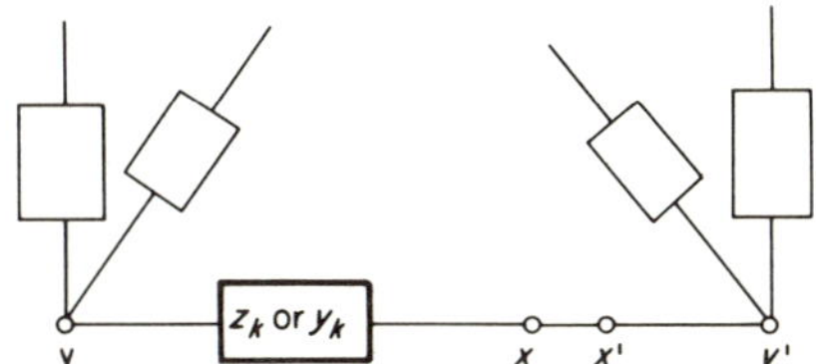

Fig. 6.5 Pliers and solder entries.

branches), the graph of the network does not change, and likewise for a voltage source in series with a branch (several branches). Excitations of the last two types do alter the network topology. A current source inserted in series with one or more branches increases the number of nodes by 1 and causes those branches to become dangling branches. A voltage source placed across one or more branches creates an additional closed loop in the network graph, and causes those branches to become self-loops.

6.7 TOPOLOGICAL DETERMINATION OF DRIVING-POINT FUNCTIONS

We now wish to reconsider Eqs. (6.4.3) and (6.4.4) from the point of view of determining driving-point impedances and admittances at pliers and solder entry points. Suppose we want the driving-point impedance at a pliers entry in a single branch of a network $\mathcal{N}$. We insert a voltage source there, choose a tree so that the branch in question is a link (the lth link), and deenergize all other sources in the network (replace voltage sources by their series impedances and current sources by their shunt impedances). As mentioned in Sec. 6.5, this causes the numerator in Eq. (6.4.3) to become a single product, rather than a summation, and the equation takes the form

$$i_l = \frac{Z_{Ell}E_{El}}{|Z_E|} \tag{6.7.1}$$

Solving Eq. (6.7.1) for the driving-point impedance yields

$$Z_{ll}{}^{P} = \frac{E_{El}}{i_l} = \frac{|Z_E|}{Z_{Ell}} \tag{6.7.2}$$

where the superscript on Z denotes a pliers entry. The numerator and denominator of Eq. (6.7.2) are expressed topologically by Eqs. (6.5.6) and (6.5.21), respectively. Thus

$$Z_{ll}{}^{P} = \frac{\sum \text{LZP of } \mathcal{N}}{\sum \text{LZP of } \mathcal{N}_{vl}} \tag{6.7.3}$$

In a similar fashion, we can derive a topological expression for the driving-point admittance at a solder entry across a single branch of $\mathcal{N}$. In this instance we choose a tree so that the branch in question is a tree branch (the lth tree branch). A current source is placed across this branch, and all other sources are deenergized. Then Eq. (6.4.4) takes the form

$$e_l = \frac{Y_{Ell} I_{El}}{|Y_E|} \tag{6.7.4}$$

with no summation on l intended. Solving for the driving-point admittance, we obtain

$$Y_{ll}{}^S = \frac{I_{El}}{e_l} = \frac{|Y_E|}{Y_{Ell}} \tag{6.7.5}$$

where the superscript on Y denotes a solder entry. Using Eqs. (6.5.5) and (6.5.15), Eq. (6.7.5) becomes

$$Y_{ll}{}^S = \frac{\sum \text{TYP of } \mathcal{N}}{\sum \text{TYP of } \mathcal{N}_{sl}} \tag{6.7.6}$$

The l subscripts denote the lth link in Eq. (6.7.3) and the lth tree branch in Eq. (6.7.6). Since these equations were derived from loop and node-pair equilibrium equations, it was inevitable that the results be expressed in terms of links and tree branches. But in using Eqs. (6.7.3) and (6.7.6), it is unnecessary to repeat the steps of the derivation; i.e., a formal definition of tree and links is not necessary each time we use these equations. Thus, in Eq. (6.7.3) we can consider $Z_{ll}{}^P$ as the driving-point impedance at a pliers entry in the lth *branch* and proceed directly to find the link sets of $\mathcal{N}$ and $\mathcal{N}_{ol}$. Similarly, in Eq. (6.7.6), $Y_{ll}{}^S$ is considered as the driving-point admittance at a solder entry across the lth *branch*, and we proceed directly to an enumeration of the trees of $\mathcal{N}$ and $\mathcal{N}_{sl}$.

Example 6.6 For the network whose graph appears in Fig. 6.1 find the driving-point impedance at a pliers entry in branch 1 and the driving-point admittance at a solder entry across branch 5.

For the driving-point impedance, Eq. (6.7.3) is used. The LZPs of $\mathcal{N}$ have been previously tabulated in Example 6.2 and those for $\mathcal{N}_{o1}$ in Example 6.5. Thus

$$Z_{11}{}^P = \frac{z_1z_3 + z_1z_4 + z_1z_5 + z_2z_3 + z_2z_4 + z_2z_5 + z_3z_5 + z_4z_5}{z_3 + z_4 + z_5}$$

To find $Y_{55}{}^S$, Eq. (6.7.6) is used. The tree-admittance products of $\mathcal{N}$ are given in Example 6.2 and those for $\mathcal{N}_{s3}$ in Example 6.4. The result is

$$Y_{55}{}^S = \frac{y_1y_2y_3 + y_1y_2y_4 + y_1y_3y_4 + y_1y_3y_5 + y_1y_4y_5 + y_2y_3y_4 + y_2y_3y_5 + y_2y_4y_5}{y_1y_3 + y_1y_4 + y_2y_3 + y_2y_4}$$

Several other driving-point cases should be considered. First is the situation of a driving-point impedance at a solder entry across a single branch. Of course we can solve this problem by exciting the network at the entry by a current source, using Eq. (6.7.6), and inverting the result. This will express the driving-point impedance in terms of branch admittances. If we want a result in terms of branch impedances, we can invert Eq. (6.7.6) prior to its application and multiply top and bottom by $|Z_B|$ to obtain [using Eq. (6.5.7)]

$$Z_{ll}{}^{S} = \frac{|Z_B| \sum \text{TYP of } \mathcal{N}_{sl}}{|Z_B| \sum \text{TYP of } \mathcal{N}} = \frac{\sum \text{LZP of } \mathcal{N}_{sl}}{\sum \text{LZP of } \mathcal{N}} \tag{6.7.7}$$

In a similar fashion we can derive from Eq. (6.7.3) a topological expression for the driving-point admittance at a pliers entry. The result is

$$Y_{ll}{}^{P} = \frac{\sum \text{TYP of } \mathcal{N}_{ol}}{\sum \text{TYP of } \mathcal{N}} \tag{6.7.8}$$

The numerators in Eqs. (6.7.7) and (6.7.8) can be obtained from the denominators. In Eq. (6.7.7), if we take the link-impedance products of $\mathcal{N}$ that do not contain z_l and multiply each by z_l, we have the link-impedance products of $\mathcal{N}_{sl}$. In Eq. (6.7.8), if we take the tree-admittance products of $\mathcal{N}$ that do not contain y_l and multiply each by y_l, we have the tree-admittance products of $\mathcal{N}_{ol}$.

Example 6.7 For the network whose graph appears in Fig. 6.1, find the driving-point impedance at a solder entry across branch 5 and the driving-point admittance at a pliers entry in branch 1.

For the driving-point impedance, Eq. (6.7.7) is used. The link sets of $\mathcal{N}$ are 13, 14, 15, 23, 24, 25, 35, 45. Those which do not contain 5 are 13, 14, 23, 24. Multiplying each of these by 5 gives the link sets of $\mathcal{N}_{S5}$ as 135, 145, 235, 245. Thus

$$Z_{55}{}^{S} = \frac{z_1z_3z_5 + z_1z_4z_5 + z_2z_3z_5 + z_2z_4z_5}{z_1z_3 + z_1z_4 + z_1z_5 + z_2z_3 + z_2z_4 + z_2z_5 + z_3z_5 + z_4z_5}$$

For the driving-point admittance, we use Eq. (6.7.8). The trees of $\mathcal{N}$ are 123, 124, 134, 135, 145, 234, 235, 245. Those which do not contain 1 are 234, 235, 245. Multiplying these by 1 gives the trees of $\mathcal{N}_{o1}$ as 1234, 1235, 1245. Then

$$Y_{11}{}^{P} = \frac{y_1y_2y_3y_4 + y_1y_2y_3y_5 + y_1y_2y_4y_5}{y_1y_2y_3 + y_1y_2y_4 + y_1y_3y_4 + y_1y_3y_5 + y_1y_4y_5 + y_2y_3y_4 + y_2y_3y_5 + y_2y_4y_5}$$

Finally we consider the two cases (Fig. 6.6) of a solder entry across more than one branch and a pliers entry in series with more than one branch. Are these cases different from the previous single-branch entry cases? The answer is "no"; Eqs. (6.7.6) and (6.7.3) still apply if $\mathcal{N}_{sl}$ and $\mathcal{N}_{ol}$ are considered as the networks obtained by shorting and opening, respectively, the *terminals* in question. The proof is requested in the problems.

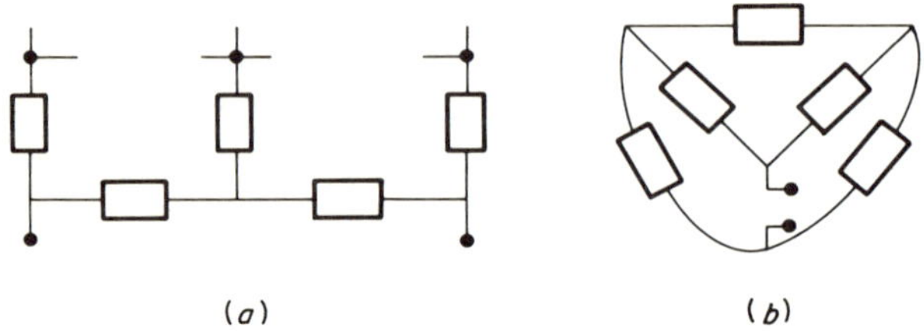

Fig. 6.6 Entries involving more than one branch.

6.8 TOPOLOGICAL DETERMINATION OF TRANSFER FUNCTIONS

Consider next the topological determination of transfer-impedance and transfer-admittance functions. Other transfer functions, such as voltage and current transfer ratios, can be obtained from these by the relationships in Sec. 3.5.

Expressions for transfer admittance and impedance functions are obtained from Eqs. (6.4.4) and (6.4.3), respectively,

$$Y_{kl} \equiv \frac{I_{Ek}}{e_l} = \frac{|Y_E|}{Y_{Ekl}} \tag{6.8.1}$$

and

$$Z_{kl} \equiv \frac{E_{Ek}}{i_l} = \frac{|Z_E|}{Z_{Ekl}} \tag{6.8.2}$$

In Eq. (6.8.1), e_l can be interpreted as a load voltage across an output branch and I_{Ek} as a source current across an input branch. Thus both "entries" are of the solder type. Likewise, in Eq. (6.8.2), i_l is a load current in an output branch, and E_{Ek} is a source voltage in series with an input branch. Here both entries are of the pliers type. Alternatively, we can think of the output entries as a voltmeter placed across the output branch or an ammeter placed in series with it.

There are other cases that could be considered. For example, the voltage across the lth tree branch as a function of the current in *series* with the kth tree branch or the current in the lth link as a function of the voltage *across* the kth link might be wanted. Topological equations for these ratios can be derived (and are requested in the problems). However, we must bear in mind that unless we associate voltage sources with pliers entries and current sources with solder entries, we are not dealing with the original graph but with one in which there is either a dangling branch (current source at a pliers entry) or a self-loop (voltage source at a solder entry).

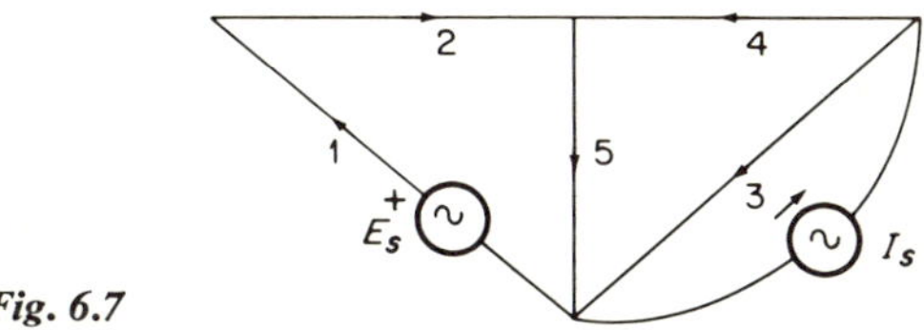

Fig. 6.7

It remains to express Eqs. (6.8.1) and (6.8.2) topologically. The numerator of Eq. (6.8.1) is given by Eq. (6.5.5); the denominator by Eq. (6.5.14). Thus

$$Y_{kl} = \frac{\sum \text{TYP of } \mathcal{N}}{(-1)^{k+l} \sum \pm \text{ CTYP of } \mathcal{N}_{sk} \text{ and } \mathcal{N}_{sl}} \tag{6.8.3}$$

Recall that the kl subscripts are tree-branch, not branch, numberings. To express Eq. (6.8.2) topologically, Eqs. (6.5.6) and (6.5.20) are used. The result is

$$Z_{kl} = \frac{\sum \text{LZP of } \mathcal{N}}{(-1)^{k+l} \sum \pm \text{ CLZP of } \mathcal{N}_{ok} \text{ and } \mathcal{N}_{ol}} \tag{6.8.4}$$

In this case the kl subscripts are link numberings.

Example 6.8 The graph of Fig. 6.1 is repeated in Fig. 6.7 with two sources, E_s and I_s, explicitly shown. Use topological formulas to find I_s/v_5 with $E_s = 0$ and E_s/j_4 with $I_s = 0$.

The first ratio requires Eq. (6.8.3). The numerator is given in Example 6.2. To evaluate the denominator a tree must be chosen such that 3 and 5 are tree branches. Let the tree be 235. Then $\mathcal{N}_{sk}$ and $\mathcal{N}_{sl}$ are $\mathcal{N}_{s2}$ and $\mathcal{N}_{s3}$, respectively (branches 3 and 5 are the second and third tree branches). The CTYPs were found in Example 6.4. Their signs are all plus. Thus

$$Y_{23} = \frac{y_1y_2y_3 + y_1y_2y_4 + y_1y_3y_4 + y_1y_3y_5 + y_1y_4y_5 + y_2y_3y_4 + y_2y_3y_5 + y_2y_4y_5}{y_1y_4 + y_2y_4}$$

For the second part of the problem, Eq. (6.8.4) is used. Again the numerator is given in Example 6.2. Tree 235 has links 1 and 4, which are the desired pair. The CLZP is given in Example 6.5. Its sign is minus. The result is

$$Z_{12} = \frac{z_1z_3 + z_1z_4 + z_1z_5 + z_2z_3 + z_2z_4 + z_2z_5 + z_3z_5 + z_4z_5}{-z_5}$$

6.9 EXTENSION TO ACTIVE AND/OR COUPLED NETWORKS†

The developments of Secs. 6.7 and 6.8 will now be extended on an admittance basis to include the cases of active and/or coupled networks. This will

† Based on the material in Ref. 6.2.

necessitate a revision of the topological formulas for the evaluation of the numerators and denominators in Eqs. (6.7.5) and (6.8.1).

Why do active and/or coupled networks require special treatment? The answer lies in the branch admittance matrix Y_B. This matrix is no longer diagonal when active and/or coupled elements are present (the network graph contains coupled branches). This is the only change, but it is a big one.

The starting point is the derivation of a new topological formula for $|Y_E|$. Separate Y_B into two $B \times B$ matrices and write

$$Y_B = Y_D + Y_0 \tag{6.9.1}$$

where Y_D is a $B \times B$ matrix containing only the main-diagonal terms in Y_B and Y_0 is a second $B \times B$ matrix containing only the off-diagonal terms in Y_B. Note that Y_B reduces to Y_D in the passive uncoupled case. Using Eq. (6.9.1), the node-pair admittance matrix can be expressed in the form

$$Y_E = \alpha_F Y_B \alpha_F{}^T = (\alpha_F Y_D + \alpha_F Y_0)\alpha_F{}^T \tag{6.9.2}$$

At this point it is convenient to convert to index notation (the reader may wish to review Chap. 1) and make the following identifications:

$$\begin{aligned} &(\alpha_F)_{ij} \equiv \alpha_{ij} \\ &(Y_D)_{ij} \equiv d_{ij} \qquad (Y_D)_{ij} = 0 \text{ for } i \neq j \\ &(Y_0)_{ij} \equiv y_{ij} \qquad (Y_0)_{ij} = 0 \text{ for } i = j \end{aligned} \tag{6.9.3}$$

Inserting these index forms into Eq. (6.9.2), we obtain

$$(Y_E)_{ij} = (\alpha_{ih} d_{hm} + \alpha_{ih} y_{hm})\alpha^T_{mj} \tag{6.9.4}$$

In Eq. (6.9.4), h and m are summed from 1 through B, the number of columns in α_F. Since Y_D is diagonal, d_{hm} is nonzero only when $h = m$. Let $d_{mm} \equiv y_m$, the admittance of branch m. The $\alpha_{ih} d_{hm}$ can be simplified to $\alpha_{im} y_m$, and Eq. (6.9.4) becomes

$$(Y_E)_{ij} = (\alpha_{im} y_m + \alpha_{ih} y_{hm})\alpha^T_{mj} \tag{6.9.5}$$

In Eq. (6.9.2), select any nonvanishing major ($N \times N$ subdeterminant) of the term in parentheses and the corresponding major of $\alpha_F{}^T$. From the Binet-Cauchy theorem, we know that the summation of the products of all such corresponding majors gives $|Y_E|$. Using the index-notation form of Eq. (6.9.5), we can write

$$|Y_E| = \sum |(\alpha_{im} y_m + \alpha_{ih} y_{hm})_{NN}|\, |\alpha^T_{mjNN}| \tag{6.9.6}$$

The expansion of the determinant of the sum of two square matrices of the same order was given in Eqs. (1.11.4) and (1.11.6). If, in Eq. (6.9.6), we think of the $N \times N$ submatrices of $\alpha_{im} y_m$ and $\alpha_{ih} y_{hm}$ as being A and B, respectively, we can use Eq. (1.11.4) to rewrite Eq. (6.9.6) in the form

$$|Y_E| = \sum (A_0 + A_1 + \cdots + A_N)_{NN} \, |\alpha^T_{mjNN}| \tag{6.9.7}$$

Consider a network $\mathcal{N}$ having three tree branches ($N = 3$), and consider a particular tree whose branch numbers are, say, 4, 6, and 7. Referring to Eq. (6.9.6), we are interested in columns 4, 6, and 7 of $\alpha_{im}y_m$ and $\alpha_{ih}y_{hm}$. These columns are obtained by letting m take on the values 4, 6, and 7. Call the product of the majors for this particular tree $|Y|$. Then, from Eq. (6.9.7),

$$|Y| = (A_0 + A_1 + A_2 + A_3)_{467} \, |\alpha^T_{mj}|_{467} \tag{6.9.8}$$

where the subscripts denote the tree involved. Consider the portion of $|Y|$ due to A_0 and call it $|Y|_0$. From Eq. (1.11.6), we note that A_0 is $|A|$. Thus $|Y|_0$ is simply the tree-admittance product of branches 4, 6, and 7. By summing over all majors, we find that the portion of $|Y_E|$ due to A_0 (call it $|Y_E|_0$) is the summation of all the TYPs of $\mathcal{N}$. This summation would be $|Y_E|$ if $\mathcal{N}$ were uncoupled.

We can write the expansion of $|Y|_0$ for tree 467 in terms of $\alpha_{im}y_m$ from the first row in Eq. (1.11.6). Thus

$$\begin{aligned} |Y|_0 &= e_{i_4 i_6 i_7}(\alpha_{i_4 4}y_4)(\alpha_{i_6 6}y_6)(\alpha_{i_7 7}y_7) \, |\alpha^T_{mj}|_{467} \\ &= y_4 y_6 y_7 \end{aligned} \tag{6.9.9}$$

Here i_4, i_6, and i_7 are summation indices, and each takes on the values 1, 2, and 3. These indices would be i_1, i_2, and i_3, as in Eq. (1.11.6), only if the branches of the tree under consideration were numbered 1, 2, and 3, respectively.

Consider next the contribution to $|Y|$ made by A_1 in Eq. (6.9.8), again for tree 467. From the second row in Eq. (1.11.6), we can write

$$\begin{aligned} |Y|_1 = e_{i_4 i_6 i_7}[&(\alpha_{i_4 h}y_{h4})(\alpha_{i_6 6}y_6)(\alpha_{i_7 7}y_7) \\ &+ (\alpha_{i_4 4}y_4)(\alpha_{i_6 h}y_{h6})(\alpha_{i_7 7}y_7) \\ &+ (\alpha_{i_4 4}y_4)(\alpha_{i_6 6}y_6)(\alpha_{i_7 h}y_{h7})] \, |\alpha^T_{mj}|_{467} \end{aligned} \tag{6.9.10}$$

In Eq. (6.9.10), the columns of Y_D associated with tree 467 are replaced, one at a time, by the corresponding columns of Y_0. Factoring out the admittances that multiply each column in the determinants gives

$$\begin{aligned} |Y|_1 = e_{i_4 i_6 i_7}[&y_{h4}y_6y_7(\alpha_{i_4 h}\alpha_{i_6 6}\alpha_{i_7 7}) + y_4y_{h6}y_7(\alpha_{i_4 4}\alpha_{i_6 h}\alpha_{i_7 7}) \\ &+ y_4y_6y_{h7}(\alpha_{i_4 4}\alpha_{i_6 6}\alpha_{i_7 h})] \, |\alpha^T_{mj}|_{467} \end{aligned} \tag{6.9.11}$$

To simplify the notation, let us represent subdeterminants of α_F by their column numbers (all these subdeterminants, being majors of α_F, have the same row numbers). Thus, for example, let

$$e_{i_4 i_6 i_7} \alpha_{i_4 h} \alpha_{i_6 6} \alpha_{i_7 7} \equiv |h67| \tag{6.9.12}$$

With these changes, Eq. (6.8.11) takes the form

$$|Y|_1 = (y_{h4} y_6 y_7 \, |h67| + y_4 y_{h6} y_7 \, |4h7| + y_4 y_6 y_{h7} \, |46h|) \, |467| \tag{6.9.13}$$

Equation (6.9.13) establishes a relationship between an arbitrary tree of a network and its contribution to $|Y_E|_1$, the portion of $|Y_E|$ due to each y_{hm} taken one at a time. To find $|Y_E|_1$, we would sum the $|Y|_1$ values for all the trees in the network.

We can use Eq. (6.9.13) to establish a topological method for obtaining $|Y_E|_1$. By inspection of that equation, we see that each product involves a subdeterminant with column m replaced by column h and y_m replaced by y_{hm}. For any one of these subdeterminants to be nonvanishing, branch h and the other branches defined by the column numbers must define a tree. Thus for each y_{hm} having $m = 4$, 6, or 7, we replace branch m in tree 467 by branch h and see whether the branches define a tree. For the network as a whole, *we find, for each* y_{hm}, *all the trees containing branch m that would remain trees if branch m were replaced by branch h. We then substitute* y_{hm} *for* y_m *in each TYP and sum the results. The sign of each term is the sign of the product of the subdeterminant of* α_F *for the particular tree and the subdeterminant of* α_F *with column m replaced by column h.*

Example 6.9 Given the transistor amplifier shown in Fig. 6.8, find the transfer admittance from terminals 1-1′ to 2-2′ and the driving-point admittance at terminals 1-1′.

We have not yet gone far enough in our topological development to work this problem in its entirety. We shall begin its solution here and complete it in steps as we develop the topological approach further.

The coupled graph of the network in Fig. 6.8 appears in Fig. 6.9, along with α_F for tree 2467 (we assume $y_{13} = 0$). The choice of tree is arbitrary, as long as branches 7 and 6, the input and output branches, are included. We wish to find $Y_{11'-22'}$, the transfer admittance between the fourth and third tree branches (branches 7 and 6), and $Y_{1-1'}$, the driving-point admittance at the terminals of the fourth tree branch. From Eq. (6.8.1), $|Y_E|$, Y_{E43}, and Y_{E44} are needed; $|Y_E|$ will be found first.

The trees of the network are:

1235, 1236, 1245, 1246, 1345, 1346, 1357, 1367, 1457, 1467, 2357, 2367, 2457, 2467, 3457, 3467

The summation of the TYPs gives $|Y_E|_0$, the first term in the expansion of $|Y_E|$. To find $|Y_E|_1$ we must find the contributions due to each of the y_{hm}'s, taken one at a time.

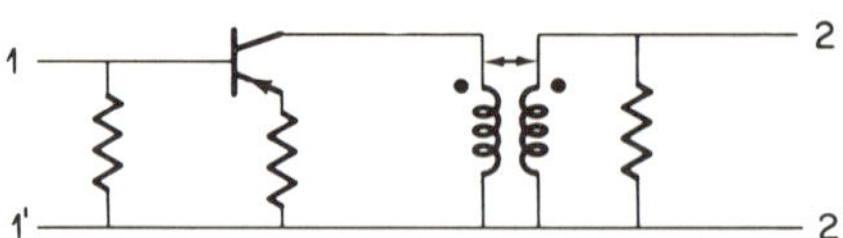

Fig. 6.8 Active-coupled network for Example 6.9.

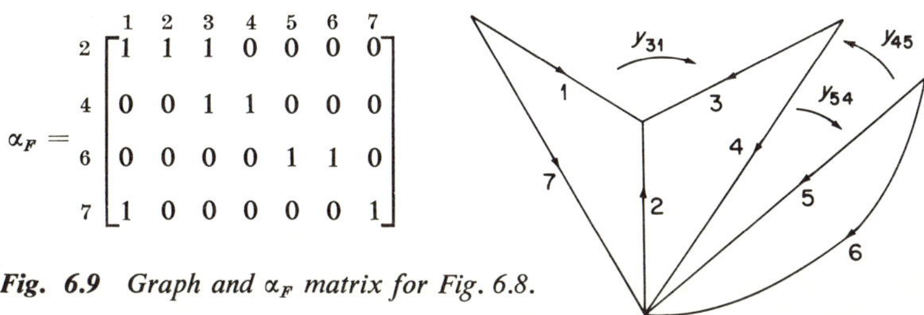

Fig. 6.9 Graph and α_F matrix for Fig. 6.8.

Consider y_{31}. We need the trees of the network containing branch 1 that would remain trees if branch 1 were replaced by branch 3. We then replace y_1 by y_{31} in each. Any tree containing branch 3 is immediately ruled out, since the new tree would contain two branch 3s.

Trees containing branch 1 1235, 1236, 1245, 1246, 1345, 1346, 1357, 1367, 1457, 1467
Those which would remain trees if branch 1 were replaced by branch 3 1457, 1467
Admittance products for these trees $y_{31}y_4y_5y_7$, $y_{31}y_4y_6y_7$
Signs of the products 1457: $|3457|\,|1457| = +1$; 1467: $|3467|\,|1467| = +1$

The terms due to y_{45} and y_{54} are similarly obtained. In this case, there are none.

Return now to the topological development and determine $|Y_E|_2$, the set of products resulting from a consideration of the y_{hm}'s taken two at a time. From the third row in Eq. (1.11.6), we can write an equation, similar to Eqs. (6.9.9) and (6.9.10), for the contribution to $|Y_E|_2$ from tree 467. Calling this contribution $|Y|_2$, there results

$$|Y|_2 = e_{i_4 i_6 i_7}[(\alpha_{i_4 h_1} y_{h_1 4})(\alpha_{i_6 h_2} y_{h_2 6})(\alpha_{i_7 7} y_7) + (\alpha_{i_4 h_1} y_{h_1 4})(\alpha_{i_6 6} y_6)(\alpha_{i_7 h_3} y_{h_3 7}) + (\alpha_{i_4 4} y_4)(\alpha_{i_6 h_2} y_{h_2 6})(\alpha_{i_7 h_3} y_{h_3 7})]\,|\alpha_{mj}^T|_{467} \tag{6.9.14}$$

where each h subscript has been indexed since now more than one h index appears in each product. Factoring out the admittance terms and using the notation of Eq. (6.9.12) yields

$$|Y|_2 = (y_{h_1 4} y_{h_2 6} y_7\,|h_1 h_2 7| + y_{h_1 4} y_6 y_{h_3 7}\,|h_1 6 h_3| + y_4 y_{h_2 6} y_{h_3 7}\,|4 h_2 h_3|)\,|467| \tag{6.9.15}$$

We now have an equation, similar to Eq. (6.9.13), that gives the contribution to $|Y_E|_2$ for one tree of the network. To find the entire $|Y_E|_2$, the portion of $|Y_E|$ due to the y_{hm}'s taken two at a time, *we find the set of trees containing both m branches that would remain trees if the m branches were replaced by the corresponding h branches. We then substitute each y_{hm} for its corresponding y_m in each TYP and sum. The sign of each term is the sign of the product of the subdeterminant of α_F for the particular tree and the subdeterminant of α_m with both m columns replaced by the corresponding h columns.*

The contribution to $|Y_E|$ from $|Y_E|_3$ is similarly found by considering the y_{hm}'s three at a time. The results can be summarized symbolically as

$$|Y_E| = \sum_{r=0}^{N} \epsilon_r \, |Y_E|_r \tag{6.9.16}$$

where $|Y_E|_r$ is the portion of $|Y_E|$ obtained by considering the y_{hm}'s r at a time and $\epsilon_r = \pm 1$ is the sign of *each* term in *each* $|Y_E|_r$. For $r = 0$, $|Y_E|_0 = \sum$ TYP of $\mathcal{N}$, and all ϵ_0's are $+1$.

Example 6.10 The determination of $|Y_E|$ begun in Example 6.9 will now be completed. We need to find $|Y_E|_2$, resulting from a consideration of the y_{hm}'s taken two at a time. For $y_{45}y_{54}$:

Trees containing 45 1245, 1345, 1457, 2457, 3457

All these trees remain trees if branches 5 and 4 are replaced by branches 4 and 5, respectively.

Admittance products for these trees $y_1y_2y_{54}y_{45}, y_1y_3y_{54}y_{45}, y_1y_{54}y_{45}y_7, y_2y_{54}y_{45}y_7, y_3y_{54}y_{45}y_7$

Each sign is minus since the product of two subdeterminants one of which differs from the other only in that two columns have been interchanged is minus.

There are no terms containing $y_{31}y_{45}$ or $y_{31}y_{54}$.

The last step is to determine $|Y_E|_3$:

Trees containing 145 1245, 1345, 1457
Trees that would remain trees if 3 replaced 1, 5 replaced 4, and 4 replaced 5 1457
Admittance product $y_{31}y_{54}y_{45}y_7$
Sign $|3547|\,|1457| = -1$

The final result is

$$\begin{aligned}|Y_E| = \Sigma\,\text{TYP} &+ y_{31}(y_4y_5y_7 + y_4y_6y_7) \\ &- y_{45}y_{54}(y_1y_2 + y_1y_3 + y_1y_7 + y_2y_7 + y_3y_7) - y_{31}y_{45}y_{54}y_7\end{aligned}$$

In determining the cofactor Y_{Ekl}, the procedure will be very similar to that just developed for finding $|Y_E|$. To form Y_{Ekl}, we delete row k and column l from Y_E, evaluate the determinant of the remaining submatrix, and affix the sign $(-1)^{k+l}$. The deletions are accomplished by deleting row k from α_F and column l from $\alpha_F{}^T$ in Eq. (6.9.2). Thus

$$Y_{Ekl} = (-1)^{k+l}\,|(\alpha_{F-k}Y_D + \alpha_{F-k}Y_0)\alpha^T_{F-l}| \tag{6.9.17}$$

Drawing from the form of Eq. (6.9.7), we can write

$$\begin{aligned}Y_{Ekl} = (-1)^{k+l} \sum (A_0 + A_1 + \cdots \\ + A_{N-1})^k_{N-1,N-1}\,|(\alpha^T_{mj})^l_{N-1,N-1}|\end{aligned} \tag{6.9.18}$$

where the notation indicates majors of order $N - 1$ with row k and row l, respectively, deleted from α_F, and the summation is over all such majors. Recall that deleting row k or row l from α_F corresponds to shorting tree branch k or tree branch l, respectively, in the network graph. Since we have previously shown that the contribution to Y_E due to A_0 in Eq. (6.9.7) is the summation of the TYPs of $\mathcal{N}$, we immediately see that the contribution to Y_{Ekl} due to A_0 in Eq. (6.9.18) is the summation, with proper signs, of the CTYPs of $\mathcal{N}_{sk}$ and $\mathcal{N}_{sl}$, the same summation previously given for the passive uncoupled case by Eq. (6.5.4).

Consider next the contribution of A_1 to Y_{Ekl} in Eq. (6.9.18). In the discussion following Eq. (6.9.13), a topological procedure was developed for finding the terms in Y_E due to each y_{hm} taken one at a time. In an analogous manner, to find the terms in Y_{Ekl} due to each y_{hm} taken one at a time, *we find the trees of the network with tree branch l shorted which contain branch m and which would also be trees of the network with tree branch k shorted if branch m were replaced by branch h. We replace y_m by y_{hm} in each resulting common tree admittance product. The sign of each product, apart from the $(-1)^{k+l}$, is the sign of the product of the subdeterminant of α_{F-l} and the subdeterminant of α_{F-k}, the latter with column m replaced by column h.*

The contribution of A_2 to Y_{Ekl} in Eq. (6.9.18) is found by a procedure analogous to the $|Y_E|$ case, as developed following Eq. (6.9.15). For each set of y_{hm}'s taken two at a time, *we find the trees of the network with tree branch l shorted which contain both m branches and which would also be trees of the network with tree branch k shorted if both m branches were replaced by the corresponding h branches. Upon substituting each corresponding y_{hm} for each y_m, we have the required products. The sign, again apart from the $(-1)^{k+l}$, is the sign of the product of the subdeterminant of α_{F-l} and the subdeterminant of α_{F-k}, the latter with both m columns replaced by the corresponding h columns.*

Contributions to Y_{Ekl} from A_3 through A_{N-1} are found similarly, and we represent the complete function by

$$Y_{Ekl} = (-1)^{k+l} \sum_{r=0}^{N-1} \epsilon_r'(Y_{Ekl})_r$$

where $(Y_{Ekl})_r$ is the portion of Y_{Ekl} obtained by considering the y_{hm}'s r at a time and $\epsilon_r' = \pm 1$ is the sign of each term in each $(Y_{Ekl})_r$. For $r = 0$, the signs, $(-1)^{k+l}\epsilon_0'$, are determined by inspection of the network graph, according to Fig. 6.2. For all other r values, the ϵ_r' are obtained from subdeterminant products and the $(-1)^{k+l}$ is then included.

If we are finding a main-diagonal cofactor, we can simplify this procedure considerably. For, if $k = l$, we observe that any product in $|Y_E|$ that includes tree branch l will, when tree branch l is deleted, also be a product in Y_{Ell}. Thus we can find Y_{Ell} by a direct examination of the products included in $|Y_E|$. Our rationale for this conclusion is the same as it was in the passive uncoupled case, discussed following Eq. (6.5.14).

Example 6.11 Let us now complete Example 6.9. We require the cofactors Y_{E43} associated with tree branches 4 and 3 (branches 7 and 6) and Y_{E44} associated with tree branch 4 (branch 7).

The cofactor Y_{E44} is obtained directly by examining our expression for $|Y_E|$. We find those products in $|Y_E|$ which contain y_7 and remove y_7 from them to obtain Y_{E44}.

Products in $|Y_E|$ containing y_7 1357, 1367, 1457, 1467, 2357, 2367, 2457, 2467, 3457, 3467, $y_{31}(457 + 467)$, $-y_{45}y_{54}(17 + 27 + 37)$, $-y_{31}y_{45}y_{54}7$

$$Y_{E44} = 135 + 136 + 145 + 146 + 235 + 236 + 245 + 246 + 345 + 346 + y_{31}(45 + 46) - y_{45}y_{54}(1 + 2 + 3) - y_{31}y_{45}y_{54}$$

Now the desired driving-point admittance is obtained from

$$Y_{1-1'} = \frac{|Y_E|}{Y_{E44}}$$

To find Y_{E43}, we recall that it will consist of the CTYPs of $\mathcal{N}_{s4}$ and $\mathcal{N}_{s3}$ plus terms resulting from a consideration of the y_{hm}'s taken one, two, and three at a time. To find the CTYPs we need the trees of $\mathcal{N}$ with tree branch 4 (branch 7) shorted that are also trees of $\mathcal{N}$ with tree branch 3 (branch 6) shorted.

Trees with branch 7 shorted 135, 136, 145, 146, 235, 236, 245, 246, 345, 346
Trees with branch 6 shorted 123, 124, 134, 137, 147, 237, 247, 347
Common trees None

Next we find the terms due to A_1 in Eq. (6.9.18).

	y_{hm}'s	*Trees for y_{hm}'s*	*Products*	ϵ_1'
$(Y_{E43})_1$	y_{31}	None		
	y_{45}	None		
	y_{54}	134	$y_1y_3y_{54}$	$\|134\|\,\|135\| = +1$

In determining the contribution of y_{54}, note that the trees with branch 6 shorted that contain branch 4 are 124, 134, 147, 247, 347. Replacing 4 by 5 in each gives 125, 135, 157, 257, 357. Only 135 is a tree with branch 7 shorted. Hence the only pertinent tree with 6 shorted is 134. We then replace y_4 by y_{54} in the product and obtain its sign, apart from $(-1)^7$, from the subdeterminant products shown. In $|135|$, row 3 of α_F (corresponding to branch 6) has been deleted, whereas in $|134|$, row 4 of α_F (corresponding to branch 7) has been deleted.

	y_{hm}'s	*Trees for y_{hm}'s*	*Products*	ϵ_2'
$(Y_{E43})_2$	$y_{31}y_{45}$	None		
	$y_{45}y_{54}$	None		
	$y_{31}y_{54}$	124	$y_{31}y_2y_{54}$	$\|325\|\,\|124\| = -1$
$(Y_{E43})_3$	$y_{31}y_{45}y_{54}$	None		

The final result is

$$Y_{E43} = (-1)^7(y_1y_3y_{54} - y_{31}y_2y_{54})$$
$$= y_{31}y_2y_{54} - y_1y_3y_{54}$$

Now

$$Y_{11'-22'} = \frac{|Y_E|}{Y_{E43}}$$

Example 6.12 Consider the active network shown in Fig. 6.10*a* and its coupled graph in Fig. 6.10*b*. We want the output admittance at terminals 2-2′ and the transfer admittance from terminals 1-1′ to 2-2′.

Assume that the y parameters of each transistor are identical and are given by the matrix

$$Y_T = \begin{bmatrix} 4.00 \times 10^{-4} & -8.00 \times 10^{-6} \\ 1.96 \times 10^{-2} & 1.08 \times 10^{-4} \end{bmatrix}$$

where all values are in mhos, and that R_S is 10 ohms. Thus in Fig. 6.10*b*,

$$y_1 = y_3 = 4.00 \times 10^{-4} \text{ mho}$$
$$y_2 = y_4 = 1.08 \times 10^{-4} \text{ mho}$$
$$y_5 = 0.1 \text{ mho}$$
$$y_{12} = y_{34} = -8.00 \times 10^{-6} \text{ mho}$$
$$y_{21} = y_{43} = 1.96 \times 10^{-2} \text{ mho}$$

Note that all the coupling terms are nonzero and that $y_{12} \neq y_{21}$ and $y_{34} \neq y_{43}$. The example is thus completely general.

Our first step is to find a suitable α_F matrix. Branches 4 and 5 must be included since they are the output and input branches, respectively. We choose tree 345 and write

$$\alpha_F = \begin{matrix} \\ 3 \\ 4 \\ 5 \end{matrix} \begin{matrix} 1 & 2 & 3 & 4 & 5 \\ \left[\begin{matrix} -1 & -1 & 1 & 0 & 0 \\ 0 & 1 & 0 & 1 & 0 \\ 1 & 0 & 0 & 0 & 1 \end{matrix}\right] \end{matrix}$$

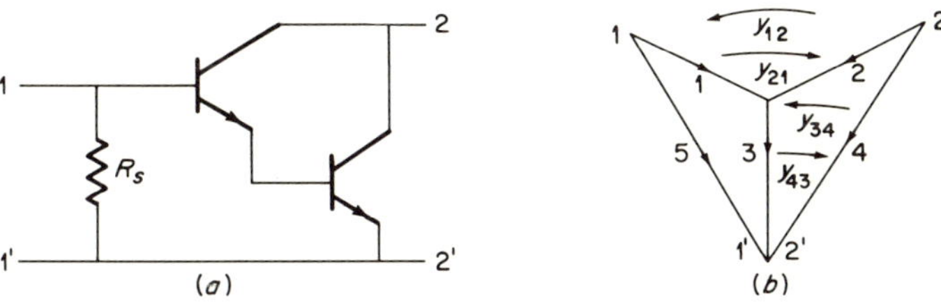

Fig. 6.10 Darlington circuit and its coupled graph.

Next an inspection of the graph yields the following trees: 123, 124, 125, 134, 145, 235, 245, 345. We now proceed to obtain $|Y_E|$. It will consist of the TYPs plus terms from $|Y_E|_1$, $|Y_E|_2$, and $|Y_E|_3$. Notice that, even though there are four coupling terms present, we do not include a term $|Y_E|_4$ (the y_{hm}'s taken four at a time) because $N = 3$.

	y_{hm}'s	*Trees for* y_{hm}'s	*Products*	*Signs*
$\lvert Y_E\rvert_1$	y_{12}	245	$y_{12}y_4y_5$	$+$
	y_{21}	145	$y_{21}y_4y_5$	$+$
	y_{34}	124, 245	$y_1y_2y_{34}$, $y_2y_{34}y_5$	$+, +$
	y_{43}	123, 235	$y_1y_2y_{43}$, $y_2y_{43}y_5$	$+, +$
$\lvert Y_E\rvert_2$	$y_{12}y_{21}$	123, 124, 125	$y_{21}y_{12}y_3$, $y_{21}y_{12}y_4$, $y_{21}y_{12}y_5$	$-, -, -$
	$y_{12}y_{34}$	None		
	$y_{12}y_{43}$	235	$y_{12}y_{43}y_5$	$+$
	$y_{21}y_{34}$	145	$y_{21}y_{34}y_5$	$+$
	$y_{21}y_{43}$	None		
	$y_{34}y_{43}$	None		
$\lvert Y_E\rvert_3$	$y_{12}y_{21}y_{34}$	124	$y_{21}y_{12}y_{34}$	$-$
	$y_{12}y_{21}y_{43}$	123	$y_{21}y_{12}y_{43}$	$-$
	$y_{12}y_{34}y_{43}$	None		
	$y_{21}y_{34}y_{43}$	None		

Now $|Y_E|$ can be written as

$$|Y_E| = \Sigma\,\text{TYP} + (y_{12} + y_{21})y_4y_5 + (y_{34} + y_{43})(y_1y_2 + y_2y_5)$$
$$- y_{12}y_{21}(y_3 + y_4 + y_5) + (y_{12}y_{43} + y_{21}y_{34})y_5 - y_{12}y_{21}(y_{34} + y_{43})$$

Substitution of numerical values yields

$$|Y_E| = 1.84 \times 10^{-8} + 0.215 \times 10^{-6} + 0.215 \times 10^{-6}$$
$$+ 1.57 \times 10^{-8} - 3.14 \times 10^{-8} + 0.310 \times 10^{-8}$$
$$= 0.424 \times 10^{-6}$$

The cofactor Y_{E22} for the output admittance is found next. We obtain this directly from $|Y_E|$ by summing the terms in $|Y_E|$ that contain y_4 with y_4 removed. Thus

$$Y_{E22} = y_1y_2 + y_1y_3 + y_1y_5 + y_2y_5 + y_3y_5 + (y_{12} + y_{21})y_5 - y_{12}y_{21}$$

Substituting numerical values.

$$Y_{E22} = 0.91 \times 10^{-4} + 0.196 \times 10^{-2} + 15.7 \times 10^{-8} = 0.205 \times 10^{-2}$$

The required cofactor for the transfer admittance is Y_{E32}. It consists of the CTYPs of $\mathcal{N}_{s3}$ and $\mathcal{N}_{s2}$ plus terms from $(Y_{E32})_1$ and $(Y_{E32})_2$.

CTYPs

Trees with branch 5 shorted 12, 14, 23, 24, 34
Trees with branch 4 shorted 12, 13, 15, 25, 35
Common trees 12
Sign $(-1)^5\epsilon_0' = +1$

	y_{hm}'s	*Trees for* y_{hm}'s	*Products*	ϵ_1'
$(Y_{E32})_1$	y_{12}	None		
	y_{21}	13	$y_{21}y_3$	+
	y_{34}	None		
	y_{43}	13	y_1y_{43}	+
$(Y_{E32})_2$	$y_{12}y_{21}$	12	$y_{21}y_{12}$	+
	$y_{12}y_{34}$	None		
	$y_{12}y_{43}$	None		
	$y_{21}y_{34}$	None		
	$y_{21}y_{43}$	13		
	$y_{34}y_{43}$	None	$y_{21}y_{43}$	+

The final result is

$$\begin{aligned} Y_{E32} &= y_1y_2 + (-1)^5(y_{21}y_3 + y_1y_{43} + y_{21}y_{12} + y_{21}y_{43}) \\ &= y_1(y_2 - y_{43}) - y_{21}(y_3 + y_{12} + y_{43}) \end{aligned}$$

Inserting numerical values gives

$$Y_{E32} = -7.80 \times 10^{-2} - 3.9 \times 10^{-4} = -7.84 \times 10^{-2}$$

The required parameters are

$$Y_{2-2'} = \frac{|Y_E|}{Y_{E22}} = 2.06 \times 10^{-4} \text{ mho}$$

$$Y_{11'-22'} = \frac{|Y_E|}{Y_{E32}} = -5.40 \times 10^{-6} \text{ mho}$$

REFERENCES

6.1. Guillemin, E. A.: "Synthesis of Passive Networks," John Wiley & Sons, Inc., New York, 1957. The discussion at the beginning of Chapter 1 of pliers and solder entries and the properties of network functions at these points is well worth reading.

6.2. Murdoch, J. B., and J. D. Rizzi: Topological Analysis of Active Networks, *Eighth Midwest Symposium on Circuit Theory, Fort Collins, Col.*, 1965. The material in Sec. 6.9 was adapted from this paper.

6.3. Mason, S. J., and H. J. Zimmerman: "Electronic Circuits, Signals, and Systems," John Wiley & Sons, Inc., New York, 1960. In Chapter 3, a topological method of analysis of active networks is developed using unistor and gyristor models. Chapters 4 and 5 give a detailed discussion of signal flow graphs, a topic not covered in this text.

6.4. Seshu, S., and N. Balabarian: "Linear Network Analysis," John Wiley & Sons, Inc., New York, 1959. Topological formulas for the analysis of passive networks with no coupling are presented in Chapter 9.

6.5. Seshu, S., and M. B. Reed: "Linear Graphs and Electrical Networks," Addison-Wesley Publishing Company, Inc., Reading, Mass., 1961. Chapter 7 presents a detailed and rigorous development of topological formulas for the analysis of passive and active networks. In the active-network case, the method presented is that of W. Mayeda.

6.6. Weinberg, L.: "Network Analysis and Synthesis," McGraw-Hill Book Company, New York, 1962. Chapter 3 discusses the unimodular character of fundamental-cut-set and tie-set matrices.

PROBLEMS

Drill

6.1 (*a*) Determine the number of trees in the graph whose cut-set matrix is

$$\alpha = \begin{bmatrix} 3 & -7 & 2 & -2 & 4 & -5 \\ 1 & -3 & 1 & -1 & 2 & -2 \\ 1 & -2 & 0 & -1 & 1 & -1 \end{bmatrix}$$

(*b*) Repeat part (*a*) for the graph whose tie-set matrix is

$$\beta = \begin{bmatrix} 1 & 1 & 2 & -2 & -1 \\ 2 & 2 & 1 & -1 & 1 \end{bmatrix}$$

6.2. For the given network use topological methods to find the following driving-point impedances: (*a*) at terminals 1-0 with 2-0 shorted and 3-0 open, (*b*) at terminals 2-0 with 1-0 shorted and 3-0 open, and (*c*) at terminals 3-0 with 1-0 and 2-0 shorted. Find these quantities by another method and compare the labor involved.

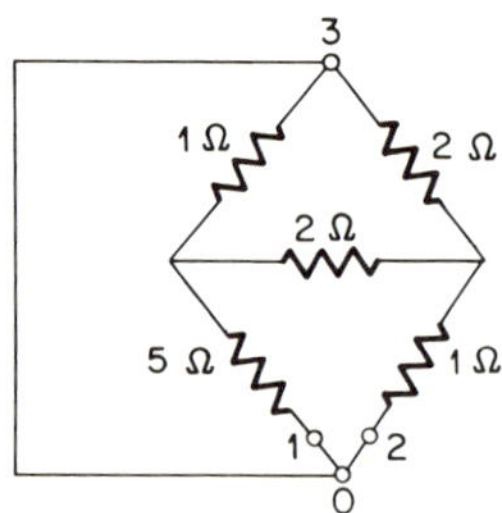

Fig. P6.2

6.3. For the tree 2357, find the cofactors Y_{B22} and Y_{B21} by (*a*) determinant expansions and (*b*) topological formulas. What is the effect on the cofactors of reversing the orientation of a branch?

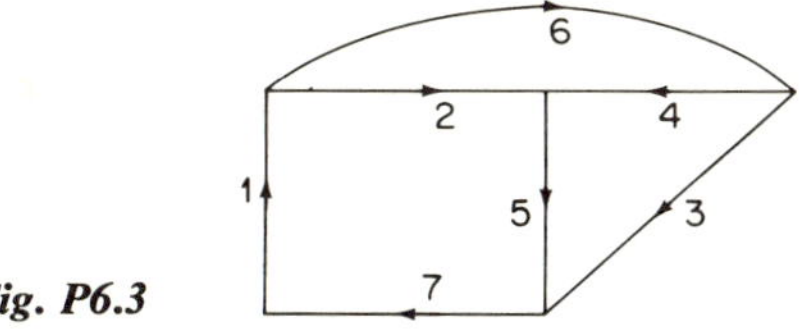

Fig. P6.3

6.4. Use topological formulas to find (*a*) Y_{10} with 2-0 shorted, (*b*) Z_{20} with 1-0 open, (*c*) j_4/i_s with 2-0 shorted and i_s applied at 1-0, and (*d*) v_4/e_s with 1-0 open and e_s applied at 2-0.

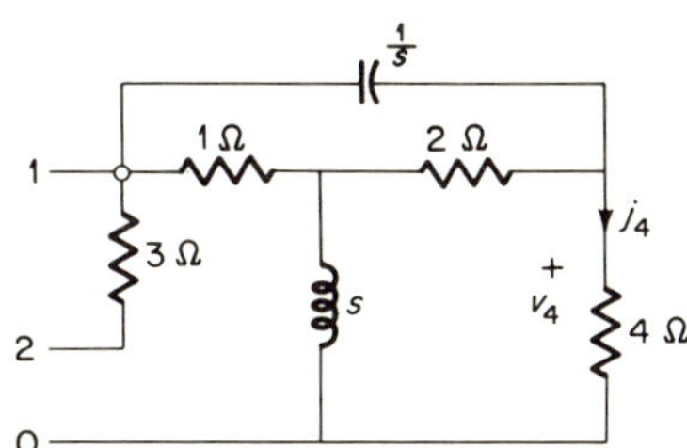

Fig. P6.4

6.5. Use topological methods to find the driving-point impedance e_i/i_i. Repeat for the transfer impedance between the input and R_3, that is, e_i/i_o.

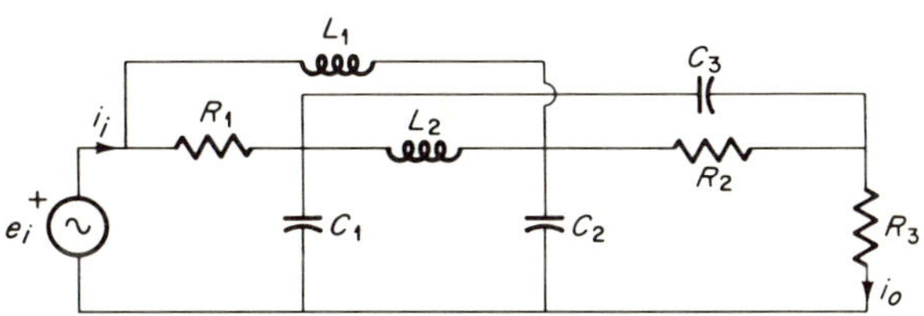

Fig. P6.5

6.6. For the network in Fig. P5.2, use topological methods to find the driving-point impedance seen by the voltage generator with the current generator opened. Repeat for the driving-point admittance seen by the current generator with the voltage generator shorted.

6.7. For the transistor amplifier in Example 5.6, find i_i/e_i and i_i/e_o by the topological procedure of Sec. 6.9, where e_i and e_o are the input and output voltages, respectively, and i_i is the input current.

6.8. Repeat Prob. 6.7 for the network in Fig. P3.27.

Theory and proofs

6.9. For what graph geometries is it always possible to write a fundamental-cut-set matrix containing only simple-node cut sets? Repeat for a fundamental-cut-set matrix each of whose columns has no more than two nonzero entries. Illustrate with examples.

6.10. In a network with no coupling, show that the network determinants on a loop and node-pair basis differ by a factor Ks^P, where K is a positive constant and P is a negative or positive integer or zero.

6.11. The vertex matrix A for the graph in Fig. 4.2 was given in Eq. (4.5.1). Delete any row, say row 3, from A to obtain a new matrix A_v. Prove that A_v is a cut-set matrix for the graph in Fig. 4.2. Prove that A_v is unimodular.

6.12. Derive Eqs. (6.3.9) and (6.3.10).

6.13. Prove the statements made in connection with Fig. 6.6 about entries involving more than one branch.

6.14. Assume that a current source is in series with a branch of a network and a voltage is measured across another branch. Assuming no coupled elements, develop a topological formula for the voltage-to-current ratio. Repeat for a voltage source across a branch and a current in a second branch. Find the current-to-voltage ratio in this case.

Application

6.15. Find the voltage transfer ratio e_o/e_s topologically.

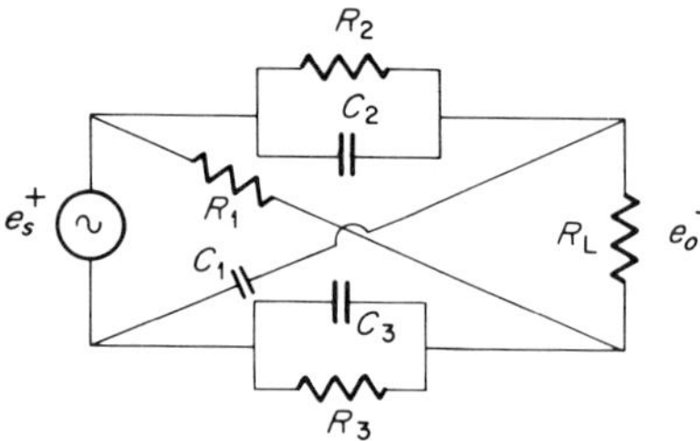

Fig. P6.15

6.16. Use topological formulas to find (*a*) the driving-point admittance at the i_s terminals, (*b*) the transfer admittance i_s/v_5, and (*c*) the y parameters of the 2-port within the box.

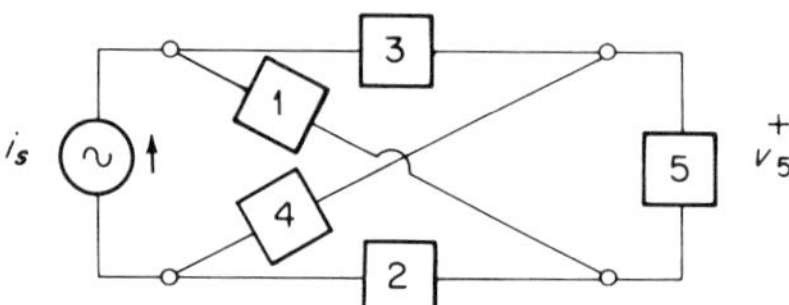

Fig. P6.16

6.17. The driving-point admittance at terminals 1-1′ can be represented in general form by

$$Y_{1-1'} = \frac{a_n s^n + a_{n-1}s^{n-1} + \cdots + a_1 s + a_0}{b_m s^m + b_{m-1}s^{m-1} + \cdots + b_1 s + b_0}$$

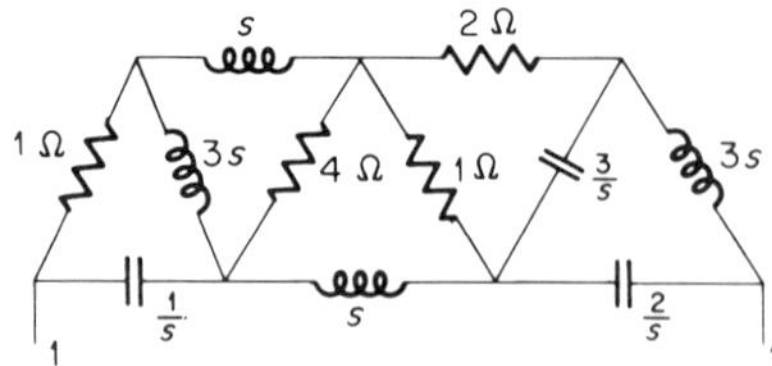

Fig. P6.17

Using topological methods, but without actually finding $Y_{1-1'}$, answer the following questions.

(*a*) What are the values of n and m? What is the ratio of a_n to b_m?

(*b*) Are a_0 and b_0 nonzero? What is their ratio?

(*c*) How many terms are there in the numerator before terms with like powers of s are combined? How many in the denominator?

(*d*) How many branches are included in each numerator term, again before combination of terms with like powers of s? How many in the denominator?

6.18. (*a*) Use topological formulas (without writing equilibrium equations) to find the frequency at which the network in Fig. 6.18*a* exhibits a phase shift of 180°.

(*b*) What is the e_2/e_1 ratio at the frequency of (*a*)?

(*c*) Find the driving-point impedance seen at terminals 1-1 of Fig. 6.18*a* at the frequency of (*a*), again by topological means.

(*d*) The networks in Fig. P6.18 are to be joined to form an *RC* oscillator. Find the minimum value of h_{21} to just sustain oscillations at the frequency of part (*a*). Assume that the amplifier load is a pure resistance equal in magnitude to the impedance of part (*c*). State any other assumptions made.

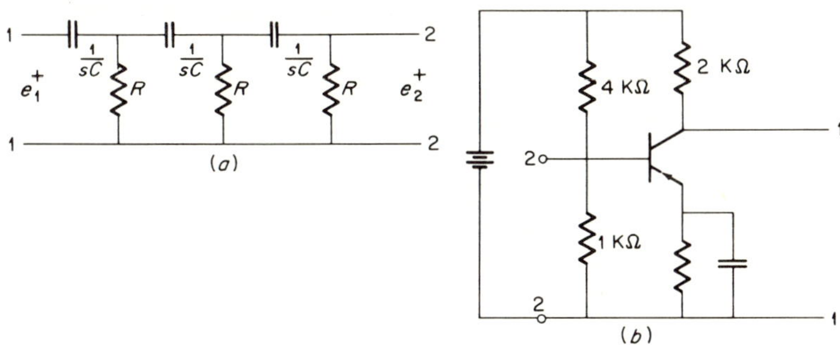

Fig. P6.18 Transistor $h_{11} \approx 2000$ ohms; $R = 1$ kilohm, $C = 0.01$ µf.

6.19. (*a*) Find the driving-point admittance at terminals 1-0 and the transfer admittance from terminals 1-0 to 2-0 topologically.

(*b*) Now assume $y_{12} \neq 0$. Find the additional terms that would be added to the numerators and denominators of the functions in part (*a*).

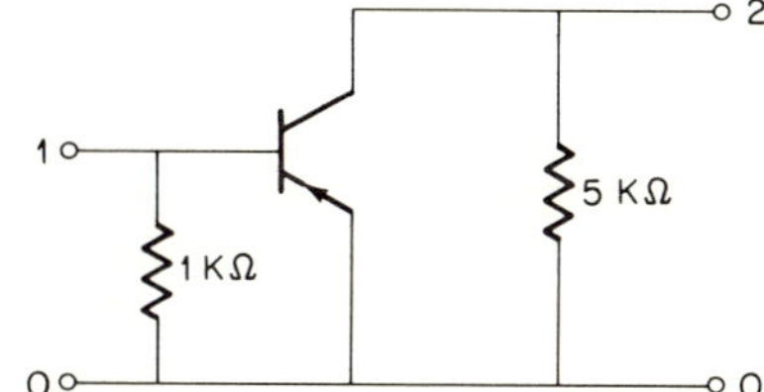

Fig. P6.19 Common-emitter y parameters are $y_{11} = 5 \times 10^{-4}$ mho, $y_{12} = 0$, $y_{21} = 2 \times 10^{-2}$ mho, and $y_{22} = 1.8 \times 10^{-5}$ mho.

6.20. (*a*) Compute i_1/e_2 for the network in Fig. P6.20*a* by any convenient method.

(*b*) y_1 and y_2 are replaced by a transistor in Fig. 6.20*b*. Let the common-emitter *y* parameters be given by

$$y_{11} = y_1 \qquad y_{12} = 0$$
$$y_{21} = K \qquad y_{22} = y_2$$

Use topological methods and the result of part (*a*) to find the new expression for i_1/e_2.

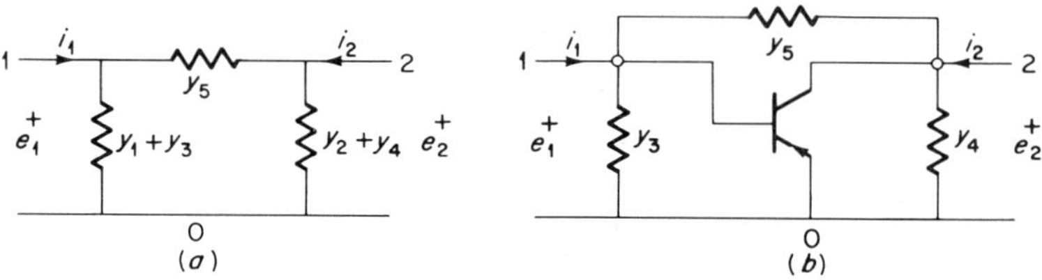

Fig. P6.20

6.21. Find the output impedance Z_0 and the voltage gain e_o/e_s by the methods of Sec. 6.9.

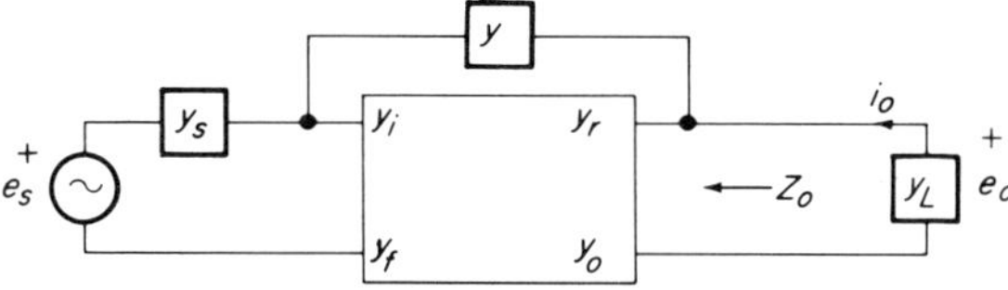

Fig. P6.21

7 NETWORK THEOREMS, DUALITY, AND SENSITIVITY

7.1 INTRODUCTION

In this chapter, which concludes the steady-state analysis of linear networks, we consider first four theorems which can often mean a considerable reduction in the amount of algebraic labor required in the solution of network problems. We then discuss the concept of duality and develop a procedure for finding the dual of a network. Finally we consider the important practical problem of the sensitivity of a network function and of its poles and zeros to changes in a network parameter. Throughout the chapter we treat these topics from both a conventional and a topological point of view.

A myriad of topics could have been included here. The ones chosen are important conceptually and at the same time are of practical significance and usefulness to the circuit analyst.

It is customary to use lowercase letters for time-domain quantities and capital letters for frequency-domain quantities. However, it is also common to use capital letters for matrices, cofactors, and determinants, and lowercase letters for elements of matrices (individual parameters). These two conventions conflict when the elements of a matrix are frequency-domain quantities. Thus far the second convention has been used, because of the predominance of matrices in the text. We shall continue this convention of assigning lowercase letters to individual s-domain quantities, but we shall assign capital letters to rms, maximum, and dc values, as well as matrices, cofactors, and determinants.

7.2 SUPERPOSITION THEOREM

A fundamental property of linear systems is given by the theorem of superposition. This theorem is treated in detail in Sec. 8.2 in connection with a discussion of the definition and properties of a linear system. For our purposes in this section, the theorem may be stated broadly as follows: The response of a linear system to n simultaneous excitations is the sum of n responses each caused by one of the excitations. Initial conditions, in the form of inductor currents and capacitor voltages, are treated as excitations in this context.

It is the superposition theorem that lets us determine the response of a linear system to a nonsinusoidal periodic excitation by expressing that excitation as a Fourier series of sinusoidal excitations of different frequencies in the time domain, transforming to the frequency domain, finding the response due to each excitation, transforming these responses back to the time domain, and summing. It is also the superposition theorem that permits us to find the response of a linear system to several excitations acting simultaneously at different points in the system by finding the response due to each excitation acting separately, with all other excitations deenergized, and summing the results. If the sources are sinusoidal and of the same frequency, the responses may be summed in the frequency domain. For excitations of different frequencies, the responses are added in the time domain. One common example of this use of the superposition principle is in electronic circuits, where we first find the response (the operating point) due to a dc excitation (bias) and then find the response (output signal) due to an ac excitation (input signal).

Let us try to formulate the principle of superposition quantitatively for a B-branch L-link N-tree-branch linear network excited by voltage and current sources, all of the same frequency. Our starting point will be Eq. (6.4.3), which we now repeat

$$i_l = \frac{Z_{Ekl}E_{Ek}}{|Z_E|} \tag{7.2.1}$$

Recall that i_l is the (s-domain) current in the lth link, E_{Ek} is the net source voltage (including transformed current sources) in the kth link, and a summation on k from 1 through L is required, as indicated by the repeated k

index. Performing this summation, we obtain

$$\begin{aligned} i_l &= \frac{Z_{E1l}E_{E1}}{|Z_E|} + \frac{Z_{E2l}E_{E2}}{|Z_E|} + \cdots + \frac{Z_{ELl}E_{EL}}{|Z_E|} \\ &= i_{l1} + i_{l2} + \cdots + i_{lL} \end{aligned} \tag{7.2.2}$$

Equation (7.2.2) shows that i_l may be considered as a summation of L currents, each caused by one net source voltage with the other net source voltages deenergized (shorted). Since any branch may be chosen as a link, we can just as well consider i_l as any branch current.

A completely similar development on a node-pair basis, starting with Eq. (6.4.4), yields

$$\begin{aligned} e_l &= \frac{Y_{E1l}I_{E1}}{|Y_E|} + \frac{Y_{E2l}I_{E2}}{|Y_E|} + \cdots + \frac{Y_{ENl}I_{EN}}{|Y_E|} \\ &= e_{l1} + e_{l2} + \cdots + e_{lN} \end{aligned} \tag{7.2.3}$$

where e_l, the lth tree-branch voltage (in the s domain), is composed of a summation of N voltages, each produced by one net source current with the other net source currents deenergized (opened). Since any branch may be a tree branch, e_l may be treated as a branch voltage.

The terms in the first forms of Eqs. (7.2.2) and (7.2.3) are easily obtained topologically. $|Z_E|$ and Z_{Ekl} are found from Eqs. (6.5.6) and (6.5.20), whereas to find $|Y_E|$ and Y_{Ekl}, we use Eqs. (6.5.5) and (6.5.14).

Example 7.1† Consider the full-wave rectifier with an LC filter shown in Fig. 7.1a. We want to find the ratio of the fourth-harmonic to second-harmonic voltage and the ratio of the second-harmonic voltage to the dc voltage, both at the load.

We shall apply the principle of superposition to the linear circuit to the right of the diodes in solving this problem. To begin, we assume that the diodes are ideal and that each diode conducts for a full half-cycle, i.e., that the inductor current is continuous. With these assumptions, the voltage e_i will be the rectified sine wave shown in Fig. 7.1b, where E_{sm} is the maximum value of e_s, the voltage across one-half of the transformer

† This problem is adapted from an excellent discussion of rectifiers and filters in T. W. Gray, "Applied Electronics," John Wiley & Sons, Inc., New York, 1954.

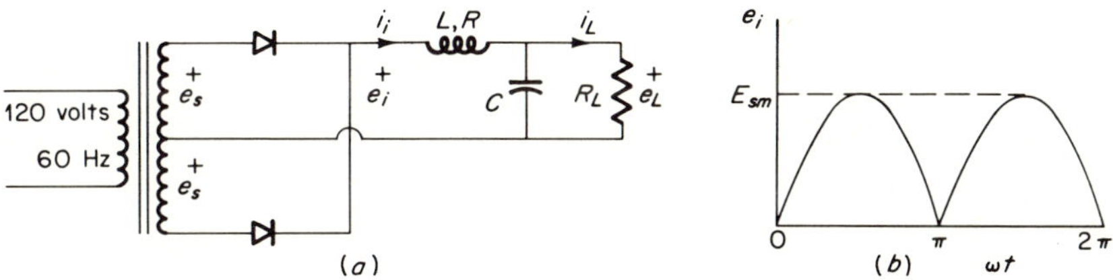

Fig. 7.1 Rectifier-filter network to illustrate superposition.

secondary. The Fourier series representation of e_i is

$$e_i = E_{sm}\left(\frac{2}{\pi} - \frac{4}{3\pi}\cos 2\,\omega t - \frac{4}{15\pi}\cos 4\,\omega t - \cdots\right)$$

showing that e_i is composed of a dc term plus terms containing even harmonics of ω ($\omega = 2\pi \times 60$ in this problem).

We now find the voltage at the load due to each of the three explicitly shown components of e_i acting separately

$$E_{Ldc} = \frac{2E_{sm}}{\pi}\frac{R_L}{R + R_L}$$

where the resistance ratio is simply the voltage transfer function of the filter under dc conditions. Under ac conditions, the voltage transfer function is

$$\frac{e_L(s)}{e_i(s)} = \frac{R_L/(1 + sCR_L)}{R + sL + R_L/(1 + sCR_L)}$$

which, for $s = jn\omega$ ($n = 2, 4, \ldots$), becomes

$$\frac{e_L(n)}{e_i(n)} = \frac{R_L/(1 + jn\omega CR_L)}{R + jn\omega L + R_L/(1 + jn\omega CR_L)}$$

Most practical LC filters are designed such that

$$\frac{1}{n\omega C} \ll R_L$$

$$\frac{1}{n\omega C} \ll n\omega L$$

$$R \ll n\omega L \qquad \text{coil } Q \text{ high}$$

which causes most of the ripple voltage to appear across the inductor. Incorporating these inequalities into the transfer function yields the simple relationship

$$\frac{e_{L(n)}}{e_{i(n)}} = -\frac{1}{n^2\omega^2 LC}$$

Thus the rms harmonic load voltages are

$$E_{L2} = \frac{4E_{sm}}{3\pi\sqrt{2}}\frac{1}{4\omega^2 LC}$$

and

$$E_{L4} = \frac{4E_{sm}}{15\pi\sqrt{2}}\frac{1}{16\omega^2 LC}$$

and the desired ratios are

$$\frac{E_{L4}}{E_{L2}} = \frac{1}{20}$$

and

$$\frac{E_{L2}}{E_{Ldc}} = \frac{R_L + R}{6\sqrt{2}\, R_L \omega^2 LC} \approx \frac{0.83 \times 10^{-6}}{LC}$$

Example 7.2 As a second example, superposition will be used to specify the rms current I_s in Fig. 7.2*a* required to make the ammeter read zero if $E_{s1} = 2$ mv and $E_{s2} = 1$ mv, both rms. The problem will be worked as nearly as possible by inspection, with a few diagrams used along the way to illustrate the procedure.

First consider the response due to the voltage sources. From Fig. 7.2*b* we obtain by inspection

$$I_1 = \frac{E_{s1} - E_{s2}}{3 + (5 \times 7)/12}\,\frac{7}{12} = \frac{7}{71}(E_{s1} - E_{s2}) = 7/71 \text{ ma rms}$$

Next I_s is split into two sources, as indicated by the dashed generators in Fig. 7.2*a*. Then, by inspection of Fig. 7.2*c* and *d*, we obtain

$$I_2 = I_s \frac{1}{3 + (5 \times 7)/12}\,\frac{7}{12} = \frac{7}{71} I_s$$

$$I_3 = -I_s \frac{3}{7 + (5 \times 3)/8}\,\frac{3}{8} = -\frac{9}{71} I_s$$

$$I_2 + I_3 = -2/71\, I_s$$

For a zero-current ammeter reading,

$$-2/71\, I_s + 7/71 = 0$$

$$I_s = 7/2 \text{ ma rms}$$

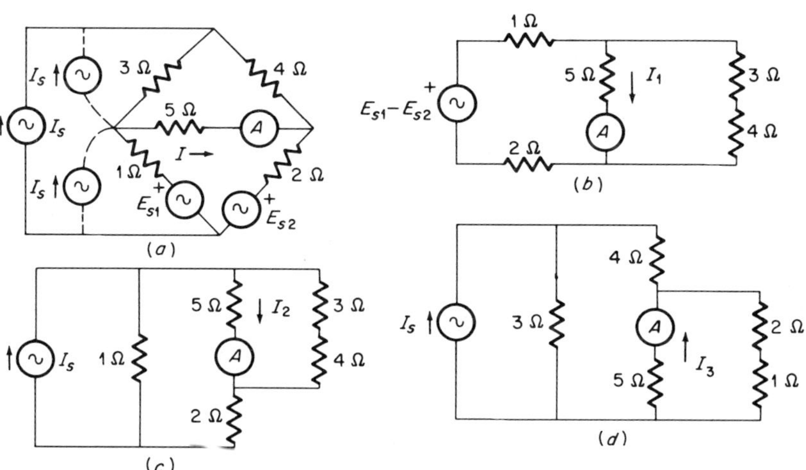

Fig. 7.2 Bridge circuit illustrating superposition.

7.3 RECIPROCITY THEOREM

In our discussion of 2-port parameters in Sec. 3.4, we noted that if the 2-port is reciprocal, certain relations exist between the parameters. These were summarized in Eqs. (3.4.11).

Later, in Sec. 5.7, we noted that the loop-impedance and node-pair admittance matrices of a linear reciprocal network are symmetric if $\beta_1 = \beta_2$ and $\alpha_1 = \alpha_2$ in Eqs. (5.7.5) and (5.7.7), respectively.

A useful theorem stemming from these results is the reciprocity theorem. Consider Fig. 7.3*a*, which depicts the *m*th and *n*th links of a linear reciprocal network. The only source in the network is e_{sm} in link *m*. The orientation of this source is, for convenience, such that it aids the link voltage v_m and produces a current i_n in link *n*. Now let us move e_{sm} to link *n*, orienting it so that it aids the link voltage v'_n and produces a current i'_m in link *m*. The currents i_n and i'_m are given by Eq. (7.2.1) as

$$i_n = \frac{Z_{Emn}e_{sm}}{|Z_E|} \qquad i'_m = \frac{Z_{Enm}e_{sm}}{|Z_E|} \tag{7.3.1}$$

where no summation on *m* is intended. Since the network is reciprocal, $Z_{Emn} = Z_{Enm}$ and the currents i_n and i'_m are identical. Thus the following conclusion is reached: In a linear reciprocal network, the current in link *m* (voltage across tree branch *m*) due to a voltage source in link *n* (current source across tree branch *n*) is equal to the current in link *n* (voltage across tree branch *n*) due to the same voltage source acting in the same reference sense in link *m* (same current source acting in the same reference sense across tree branch *m*). This is the reciprocity theorem. Let us make several observations about it.

1. We have anticipated, by the statements in parentheses in the definition, that the theorem is valid on a node pair as well as on a loop basis, since in a linear reciprocal network $Y_{Emn} = Y_{Enm}$.
2. It is essential in applying the theorem that the network topology remain unchanged when the voltage (current) source is moved. If the topology were to change, $|Z_E|$ would change and Eqs. (7.3.1) would not yield the

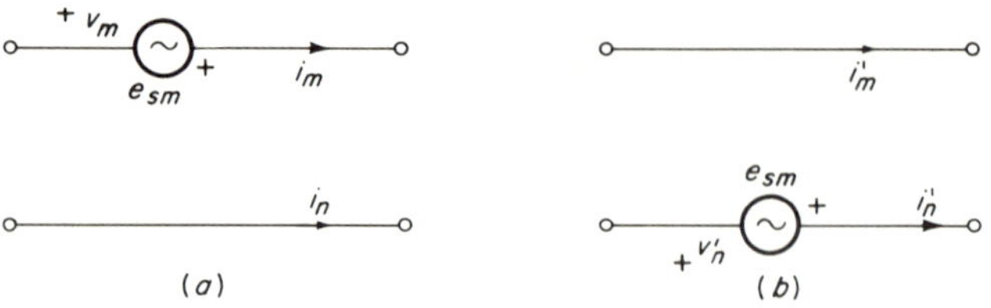

Fig. 7.3 Diagrammatic representation of the reciprocity theorem.

same current. For example, it is not permissible to move the voltage source from its position in *series* with a link to a new position in *shunt* with another link, since the network graph would change.

3. When the voltage (current) source is moved, the currents (voltages) in other parts of the network will change.
4. Only one source is present in the network (if additional sources are present, the superposition theorem can be used in conjunction with the reciprocity theorem to find the total response). Thus all initial conditions must be zero (unless superposition is used to handle these as separate sources).
5. The reciprocity theorem relates a current (voltage) to an applied voltage (current). It cannot be used directly to find, say, the voltage (current) response due to a voltage (current) excitation.

We now work two examples to illustrate the use of the theorem and the points we have made concerning it.

Example 7.3 In the *RC* ladder network of Fig. 7.4*a*, we wish to find the ammeter reading with applied rms voltages E_{s1} and E_{s2}. We could solve the problem directly by loop or node-pair analysis, but considerable algebraic labor would be required. Let us try instead to use the superposition and reciprocity theorems.

We interchange the voltage source E_{s1} and ammeter, as shown in Fig. 7.4*b*, and replace E_{s2} by a short circuit. By the reciprocity theorem, the currents I_1 and I_1' are identical (the subscript 1 denoting currents due to E_{s1}), and it is easier to find I_1' than it is to find I_1. In fact, we can write the expression for I_1' by inspection

$$I_1' = \frac{E_{s1}}{1/SC + (2R/sC)/(R + 1/sC)} \frac{R}{R + 1/sC} = \frac{E_{s1}RC^2s^2}{3RCs + 1}$$

Next we interchange the voltage source E_{s2} and the ammeter and replace E_{s1} by a short circuit, as in Fig. 7.4*c*. Once again the current I_2' is easier to compute than I_2. We have

$$I_2' = \frac{E_{s2}}{1/sC + (2R/sC)/(R + 1/sC)} \frac{1/sC}{R + 1/sC} = \frac{E_{s2}Cs}{3RCs + 1}$$

By superposition, the ammeter reading is

$$I_1 - I_2 = \frac{Cs(E_{s1}RCs - E_{s2})}{3RCs + 1}$$

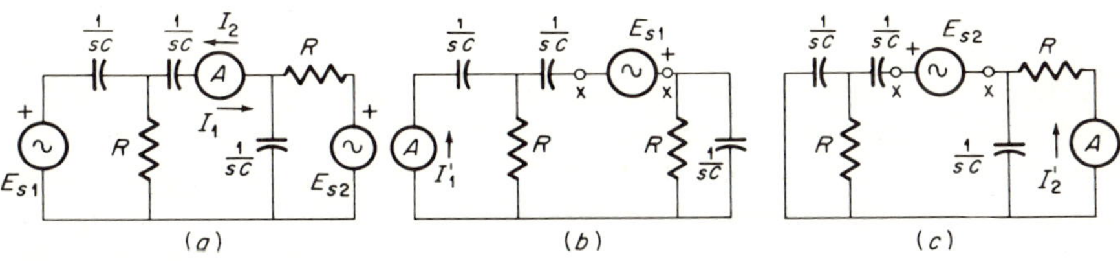

Fig. 7.4 Using reciprocity to find a current response.

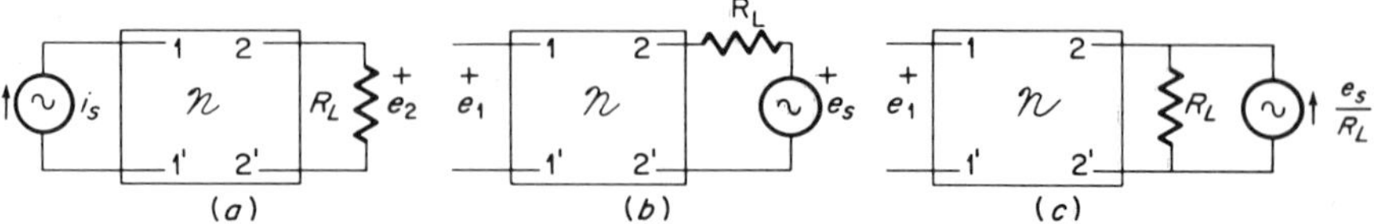

Fig. 7.5 Using reciprocity to compare transfer functions.

In general, whenever the impedance (admittance) seen looking into the network from the point at which the response is to be obtained is easy to calculate, it will pay to use the reciprocity theorem in finding that response.

Example 7.4 The problem is to show that the transfer function $Z_{21} = e_2/i_s$ of the configuration in Fig. 7.5*a* is equal, within a constant, to the transfer function $E_{12}{}^0 = e_1/e_s$ of the configuration in Fig. 7.5*b* if $\mathcal{N}$ is a linear reciprocal network.

From an examination of the reciprocity theorem, we know that the voltage e_2 in Fig. 7.5*a* can be found by placing i_s across R_L and measuring the open-circuit voltage at terminals 1-1′. If we can transform Fig. 7.5*b* to a current-source drive at terminals 2-2′, we shall be able to compare the two responses. This we have done in Fig. 7.5*c*. Now we can relate the transfer functions by writing

$$Z_{21} = \frac{e_2}{i_s} = \frac{e_1}{e_s/R_L} = E_{12}{}^0 R_L$$

Thus the transfer functions are equal within a constant, and that constant is the resistor R_L.

7.4 HELMHOLTZ EQUIVALENT-SOURCE THEOREMS

In dealing with complicated networks, it often happens that we are interested in the voltages and currents in only a portion of the network. In such cases, the Helmholtz equivalent-source theorems, more popularly known as *Thévenin's* and *Norton's theorems*, are useful in reducing the complexity of the portion of the network in which we are *not* interested. Thévenin's theorem states that, subject to certain constraints (which we shall presently enumerate), a linear 2-terminal network containing any number of elements and sources can be replaced at its terminals by an ideal voltage source and series impedance. Norton's theorem states that, subject to the same constraints, the same 2-terminal network can be replaced at its terminals by an ideal current source and shunt admittance.

Before developing a general procedure for obtaining the Thévenin and Norton equivalent circuits, let us illustrate their use in a simple network. Consider the resistive circuit in Fig. 7.6*a*. It is desired to represent the circuit at its *X*-*X* terminals by an ideal voltage source and series resistance and by an ideal current source and shunt admittance. In Fig. 7.6*b*, the voltage generator e_s in series with R_1 is converted to a current source e_s/R_1 in parallel with

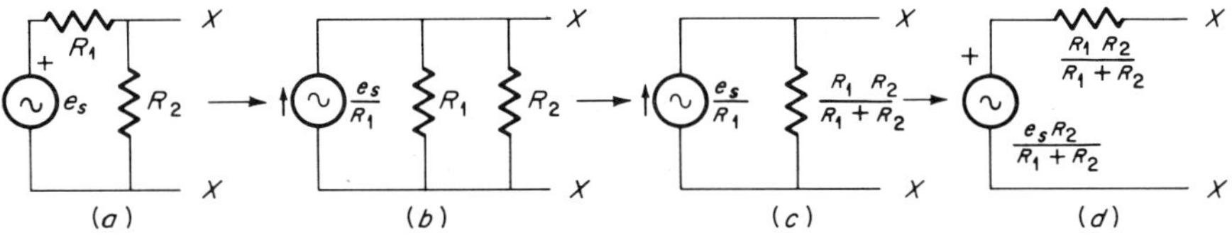

Fig. 7.6 Thévenin and Norton equivalent circuits from source transformations.

R_1. In Fig. 7.6*c*, R_1 and R_2 are combined to obtain the Norton equivalent circuit. Then in Fig. 7.6*d*, a reconversion to a voltage source and series resistance yields the Thévenin equivalent circuit.

The reader should not interpret Fig. 7.6 as being a general procedure for obtaining the Thévenin and Norton equivalent circuits. Nor should he interpret the result as a general proof of the existence of these circuits. Actually the method of Fig. 7.6 *is* completely general for linear *reciprocal* networks (even for coupled circuits, as will be shown in Example 7.5), but it can be applied only with extreme difficulty, if at all, in the nonreciprocal case. The reason is that it is always possible to replace a reciprocal coupled 2-port by an equivalent three-element T or π without the use of dependent generators. It is the presence of dependent generators in the nonreciprocal case that makes the source transformation procedure difficult, if not impossible, to apply. Even in the reciprocal case, it would be most laborious and time consuming if we had to obtain the Thévenin and Norton equivalent circuits for complex networks by the means we have just described.

A more direct and general method is needed. Consider a B-branch $(N + 1)$-node L-link network $\mathcal{N}$ and separate it into two parts, $\mathcal{N}_1$ and $\mathcal{N}_2$, as shown in Fig. 7.7*a*. We assume the following:

1. The two leads shown are the *only* connections between $\mathcal{N}_1$ and $\mathcal{N}_2$.
2. No element in $\mathcal{N}_1$ is coupled (inductively or resistively) to an element in $\mathcal{N}_2$.
3. $\mathcal{N}_1$ contains any number of linear reciprocal and/or nonreciprocal elements and linear independent sources.
4. All initial conditions in $\mathcal{N}_1$ are zero. This constraint may be waived if we are willing to use superposition to handle these initial conditions as

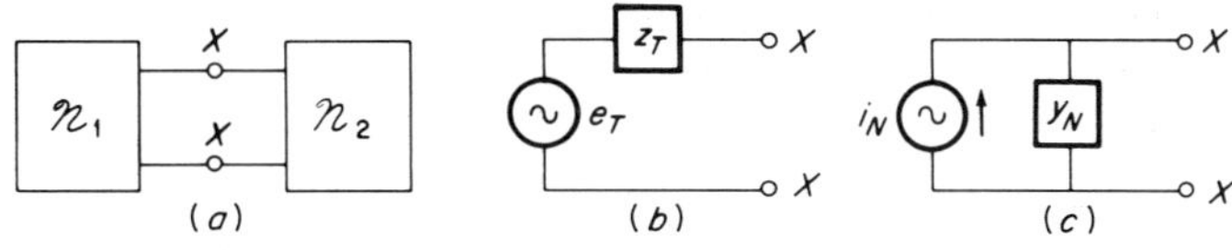

Fig. 7.7 A simpler procedure for finding the Thévenin and Norton equivalent circuits.

dc sources. Then each equivalent circuit will contain a dc as well as an ac source.

5. All sources inside $\mathcal{N}_1$ are of the same frequency. Again, this constraint may be waived if we use superposition and more than one equivalent source.

Assuming that these constraints are satisfied, the plan is to represent the $\mathcal{N}_1$ portion of $\mathcal{N}$ by the Thévenin and Norton equivalent circuits shown in Fig. 7.7*b* and *c* such that the same currents and voltages are produced in $\mathcal{N}_2$ as would be produced there by the original $\mathcal{N}_1$. The procedure for obtaining these circuits is:

1. Measure or calculate the open-circuit voltage of $\mathcal{N}_1$ across the *X-X* terminals in Fig. 7.7*a*. A glance at Fig. 7.7*b* shows that this voltage is e_T.
2. Measure or calculate the short-circuit current of $\mathcal{N}_1$ between the *X-X* terminals in Fig. 7.7*a*. This is i_N of Fig. 7.7*c*.
3. The quantities z_T and y_N can be calculated from e_T and i_N of steps 1 and 2. With terminals *X-X* shorted in Fig. 7.7*b*,

$$z_T = \frac{e_T}{i_N} \tag{7.4.1}$$

and with terminals *X-X* open in Fig. 7.7*c*,

$$y_N = \frac{i_N}{e_T} \tag{7.4.2}$$

Alternatively, z_T can be measured directly by applying a voltage at terminals *X-X* of $\mathcal{N}_1$, with all independent sources inside $\mathcal{N}_1$ replaced by their internal impedances and noting the current that flows. To measure y_N directly, a current is applied at terminals *X-X* of $\mathcal{N}_1$, again with all independent sources inside $\mathcal{N}_1$ replaced by their internal impedances, and the voltage across the *X-X* terminals noted.

Is the procedure just developed a general one? Actually, we have not proved that it is. What we have shown is that when the input impedance of $\mathcal{N}_2$ is either zero or infinite, the circuits of Fig. 7.7*b* and *c* produce the same currents and voltages in $\mathcal{N}_2$ as are produced by $\mathcal{N}_1$ in Fig. 7.7*a*. We have not proved this result for other values of input impedance of $\mathcal{N}_2$. Let us do so now. Consider Fig. 7.8*a*. Here we specify that $\mathcal{N}_2$ of Fig. 7.7*a* is the *l*th link of $\mathcal{N}$. This element has been removed from $\mathcal{N}$ and the remaining network labeled $\mathcal{N}_{-l}$. The five assumptions previously made concerning $\mathcal{N}_1$ (now $\mathcal{N}_{-l}$) still apply.

The current i_l in Fig. 7.8*a* is given by Eq. (7.2.1). We wish to separate the denominator of this equation into two parts so as to display z_l explicitly.

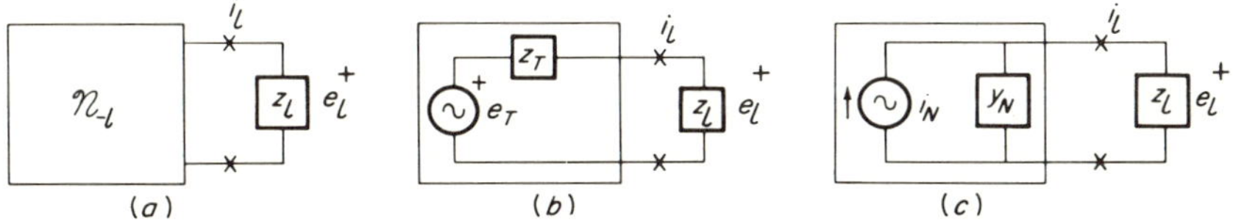

Fig. 7.8 Diagrams used in deriving general proofs of Thévenin's and Norton's theorems.

Performing a cofactor expansion of $|Z_E|$ down column l yields

$$|Z_E| = z_{1l}Z_{E1l} + \cdots + (z'_{ll} + z_l)Z_{Ell} + \cdots + z_{Ll}Z_{ELl} \tag{7.4.3}$$

which can be written as

$$|Z_E| = |Z'_E| + z_l Z_{Ell} \tag{7.4.4}$$

Here we have separated the self-impedance of loop l (z_{ll}) into two parts (z'_{ll} and z_l) and recombined all terms not containing z_l into a single term $|Z'_E|$. Inserting the second form of Eq. (7.4.3) into Eq. (7.2.1) gives

$$i_l = \frac{Z_{Ekl}E_{Ek}}{|Z'_E| + z_l Z_{Ell}} \tag{7.4.5}$$

The voltage across z_l in Fig. 7.8*a* is $i_l z_l$. Calling this voltage e_l and solving for it in Eq. (7.4.5), we obtain

$$e_l = \frac{Z_{Ekl}}{Z_{Ell}}E_{Ek} - \frac{|Z'_E|}{Z_{Ell}}i_l \tag{7.4.6}$$

We now note that the first term on the right side of Eq. (7.4.6) is a summation of voltages (Z_{Ekl} and Z_{Ell} are of the same rank), whereas the coefficient of i_l has the dimensions of impedance ($|Z'_E|$ is one rank higher than Z_{Ell}). Thus we can rewrite Eq. (7.4.6) in the abbreviated form

$$e_l = e_T - z_T i_l \tag{7.4.7}$$

and arrive at the Thévenin equivalent circuit of Fig. 7.8*b*. On the other hand, we can solve Eq. (7.4.6) for i_l and obtain

$$i_l = \frac{Z_{Ekl}}{|Z'_E|}E_{Ek} - \frac{Z_{Ell}}{|Z'_E|}e_l \tag{7.4.8}$$

This time the summation term on the right side has the dimensions of current and the e_l coefficient represents an admittance. Thus we can rewrite Eq. (7.4.8) as

$$i_l = i_N - y_N e_l \tag{7.4.9}$$

and obtain the Norton equivalent circuit of Fig. 7.8*c*. This completes the general proof.

An interesting and useful application of Thévenin's and Norton's theorems is in the analysis of active 2-ports. Assume an active network is given and we derive its 2-port equations (by conventional loop or node analysis, by matrix analysis, or by topological means). The final result, in y-parameter notation, can be written in the matrix form

$$\begin{bmatrix} i_i \\ i_o \end{bmatrix} = \begin{bmatrix} y_i & y_r \\ y_f & y_o \end{bmatrix} \begin{bmatrix} e_i \\ e_o \end{bmatrix} \tag{7.4.10}$$

The Thévenin and Norton equivalent circuits can be obtained in terms of these y parameters. The Thévenin voltage source at the output is e_o with $i_o =$ 0 and is given by

$$e_T = \frac{y_f}{y_o} e_i = -\frac{y_f}{|y|} i_i \tag{7.4.11}$$

where

$$|y| = y_i y_o - y_f y_r \tag{7.4.12}$$

The Norton current source is $i_N = -i_o$ with $e_o = 0$. In terms of the y parameters,

$$i_N = -y_f e_i = -\frac{y_f}{y_i} i_i \tag{7.4.13}$$

The two alternative forms in each of Eqs. (7.4.11) and (7.4.13) permit us to consider either a voltage (e_i) or current (i_i) source excitation at the input terminals of the 2-port. The Thévenin impedance and Norton admittance are found directly from Eqs. (7.4.7) and (7.4.9). For an e_i excitation,

$$z_T = \frac{1}{y_N} = \frac{1}{y_o} \tag{7.4.14}$$

whereas, for an i_i excitation,

$$z_T = \frac{1}{y_N} = \frac{y_i}{|y|} \tag{7.4.15}$$

It is often desirable to include the impedance (or admittance) seen by the signal source in the equivalent circuits so that the loading effect of the 2-port on a previous stage can be easily accounted for. Computing y_{in} for a load admittance y_L from Eq. (7.4.10) yields

$$y_{\text{in}} = \frac{y_i y_L + |y|}{y_o + y_L} \tag{7.4.16}$$

Assembling all these quantities yields the two Thévenin and two Norton equivalent circuits of Fig. 7.9.

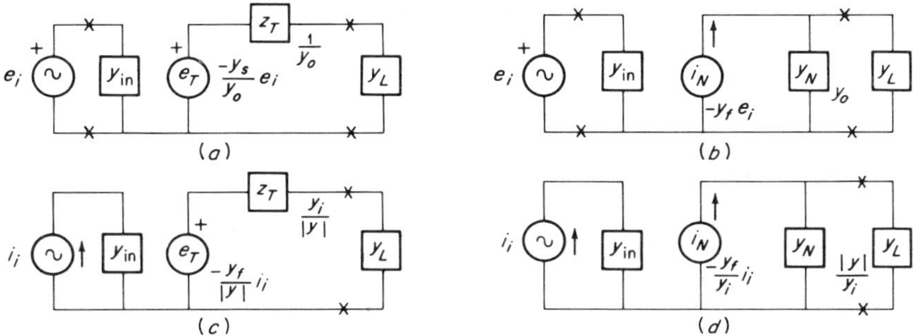

Fig. 7.9 Thévenin and Norton equivalent circuits for active 2-port networks.

This concludes the discussion of Thévenin's and Norton's theorems. Their use will be illustrated through three examples.

Example 7.5 The Thévenin equivalent circuit looking to the left from terminals *X-X* in Fig. 7.10*a* is desired. It will be found in two ways, first by the procedure relating to Fig. 7.7 and second by the source transformation method of Fig. 7.6. The 2-port equations for the coupled circuit are

$$e_s = \left(sL_1 + \frac{1}{sC}\right)i_1 + \left(sM + \frac{1}{sC}\right)i_2$$

$$e_l = \left(sM + \frac{1}{sC}\right)i_1 + \left(sL_2 + \frac{1}{sC}\right)i_2$$

The open-circuit voltage e_T is obtained by setting $i_2 = 0$ and is

$$e_T = e_s \frac{MCs^2 + 1}{L_1Cs^2 + 1}$$

To find z_T, we set $e_s = 0$ and solve for e_2/i_2 to obtain

$$z_T = \frac{C(L_1L_2 - M^2)s^3 + (L_1 + L_2 - 2M)s}{L_1Cs^2 + 1}$$

Alternatively, to find the Thévenin equivalent circuit through source transformations, we first convert the transformer to its *T* equivalent, as shown in Fig. 7.10*b*, to get rid of

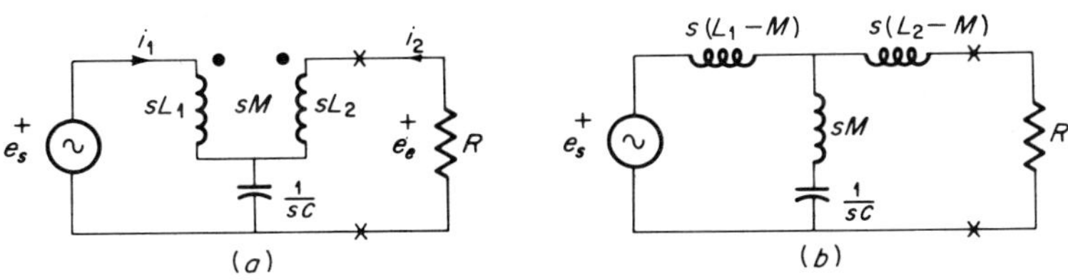

Fig. 7.10 Applying Thévenin's theorem to a coupled circuit.

the mutual coupling. Then we proceed through the steps indicated in Fig. 7.6 to obtain e_T and z_T.

Let the parameters in Fig. 7.10 be

$$L_1 = 1 \text{ henry} \qquad L_2 = 4 \text{ henrys} \qquad M = 1.5 \text{ henrys} \qquad C = 1 \text{ farad}$$

Then e_T and z_T become

$$e_T = e_s \frac{1.5s^2 + 1}{s^2 + 1}$$

$$z_T = \frac{1.75s^3 + 2s}{s^2 + 1} = 1.75s + \frac{0.25s}{s^2 + 1}$$

If we increase M to 2 henrys, the transformer becomes perfect and

$$e_T = e_s \frac{2s^2 + 1}{s^2 + 1}$$

$$z_T = \frac{s}{s^2 + 1}$$

The Thévenin equivalents for these two numerical cases appear in Fig. 7.11.

Example 7.6 The problem is to find the Thévenin equivalent circuit looking to the left from the X-X terminals in Fig. 7.12a. Note that both a 1000-Hz and a dc source are included in the network. Thus we must find two Thévenin voltage generators and two series impedances. The capacitor blocks dc but has negligible impedance at 1000 Hz. Direct calculations yield

$$e_{T\text{dc}} = 25 \times \frac{6}{12} = \frac{25}{2} \text{ volts}$$

$$z_{T\text{dc}} = \frac{6 \times 6}{12} = 3 \text{ kilohms}$$

$$e_{T\text{ac}} = \frac{5}{11/3} \frac{2}{12} \times 6 = \frac{15}{11} \text{ volts}$$

$$z_{T\text{ac}} = \frac{6 \times 5}{11} = \frac{30}{11} \text{ kilohms}$$

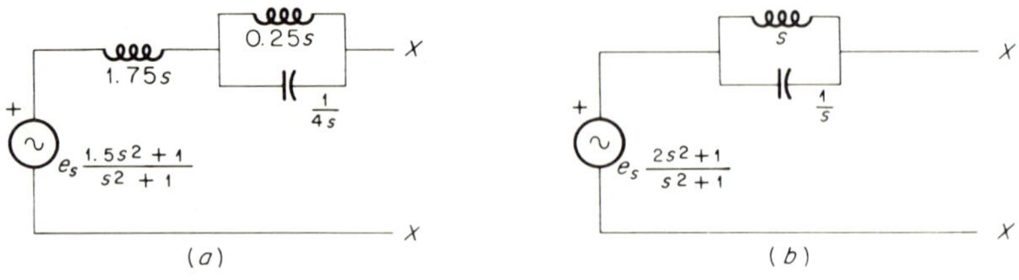

Fig. 7.11 Thévenin equivalent circuits for Example 7.5.

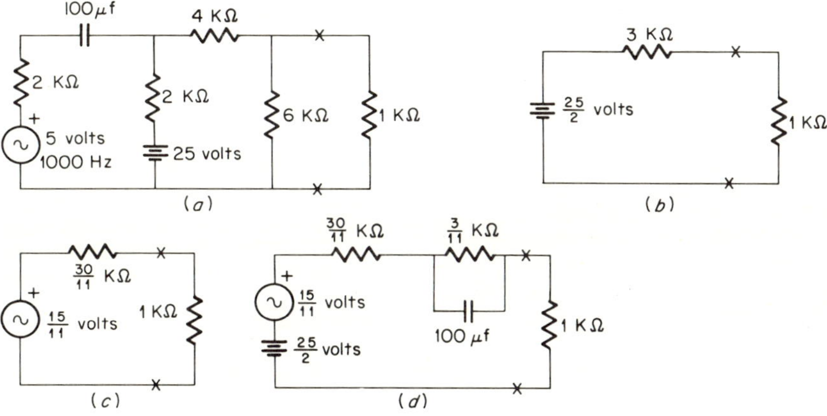

Fig. 7.12 Thévenin's theorem when sources of different frequency are included in the network.

The first two quantities create the dc Thévenin equivalent circuit in Fig. 7.12*b*, whereas the last two are used to form the ac circuit in Fig. 7.12*c*. We can combine these two circuits into one and obtain the overall Thévenin equivalent circuit of Fig. 7.12*d*. Note that the effect of the capacitor is to make the dc and 1000-Hz impedances differ, as required.

Example 7.7 The Thévenin and Norton equivalent circuits are desired for the active 2-port of Fig. 7.13*a*. These circuits will have the form of those in Fig. 7.9*c* and *d*, since the excitation is a current source.

A matrix analysis of the network in Fig. 7.13*a* was carried out in Example 3.7. From the results of that example (assuming that the same transistor *y* parameters apply)

$$\begin{bmatrix} i_i \\ i_o \end{bmatrix} = 10^{-5} \begin{bmatrix} 122 & -0.071 \\ 93 & 50 \end{bmatrix} \begin{bmatrix} e_i \\ e_o \end{bmatrix}$$

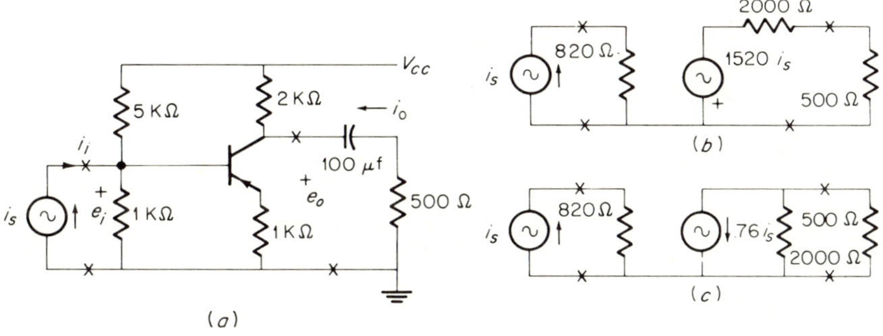

Fig. 7.13 Applying Thévenin's and Norton's theorems to an active network.

Now it is a simple matter to use Eqs. (7.4.11) to (7.4.16) to find

$$|y| = 10^{-10}(122 \times 50 + 0.071 \times 93) = 0.611 \times 10^{-6}$$

$$e_T = -\frac{93 \times 10^{-5}}{0.611 \times 10^{-6}} i_s = -1520 i_s$$

$$i_N = -{}^{93}\!/\!_{122} i_s = -0.76 i_s$$

$$z_T = \frac{1}{y_N} = \frac{122 \times 10^{-5}}{0.611 \times 10^{-6}} = 2000 \text{ ohms}$$

$$z_{\text{in}} = \frac{1}{y_{\text{in}}} = \frac{50 \times 10^{-5} + 2 \times 10^{-3}}{122 \times 10^{-5} \times 2 \times 10^{-3} + 0.611 \times 10^{-6}} = 820 \text{ ohms}$$

The resulting equivalent circuits appear in Fig. 7.13*b* and *c*.

7.5 DUALITY

Although we have not mentioned it by name, the concept of *duality* has been present throughout our discussion of linear network analysis. Whenever we have said that a similar development can be carried out on a loop (or node-pair) basis, we have meant that the node-pair analysis of a network is equivalent to the loop analysis of its dual network. It is the purpose of this section to formalize the concept of duality and develop a procedure for finding the dual of a given network.

Let us begin by stating what is meant by a dual network. Consider the graph in Fig. 7.14*a* of a linear passive uncoupled network $\mathcal{N}$. We observe that $B = 7$, $N = 4$, and $L = 3$, that a current and a voltage source are included, and that the three simple loops and the circuit exterior are numbered. The KVL equations are

$$\begin{aligned} &① \quad -v_2 + v_4 + v_6 = 0 \\ &② \quad v_1 + v_2 + v_5 + v_7 = 0 \\ &③ \quad v_3 - v_4 - v_5 = 0 \end{aligned} \tag{7.5.1}$$

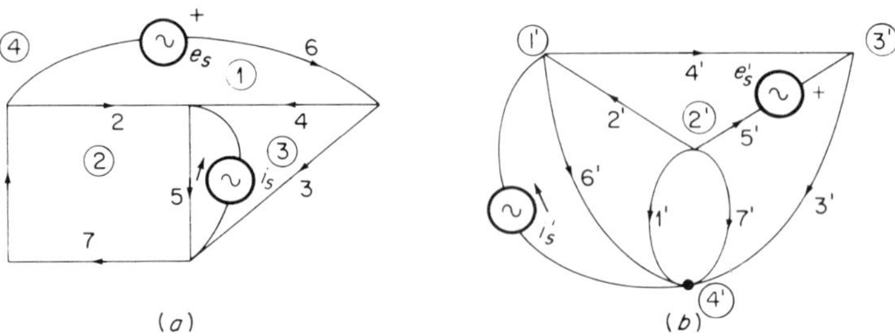

Fig. 7.14 Graphs of dual networks.

Using the Ohm's law equation for each branch, these can be written as

$$\begin{aligned} &① \quad -j_2z_2 + j_4z_4 + j_6z_6 - e_s = 0 \\ &② \quad j_1z_1 + j_2z_2 + j_5z_5 + i_sz_5 + j_7z_7 = 0 \\ &③ \quad j_3z_3 - j_4z_4 - j_5z_5 - i_sz_5 = 0 \end{aligned} \tag{7.5.2}$$

Now in Eqs. (7.5.1) and (7.5.2) replace each v by a j', each j by a v', each z by a y', i_s by e'_s, and e_s by i'_s. Thus each voltage (current) becomes a current (voltage), and each impedance becomes an admittance. The results are

$$\begin{aligned} &①' \quad -j'_2 + j'_4 + j'_6 = 0 \\ &②' \quad j'_1 + j'_2 + j'_5 + j'_7 = 0 \\ &③' \quad j'_3 - j'_4 - j'_5 = 0 \end{aligned} \tag{7.5.3}$$

$$\begin{aligned} &①' \quad -v'_2y'_2 + v'_4y'_4 + v'_6y'_6 - i'_s = 0 \\ &②' \quad v'_1y'_1 + v'_2y'_2 + v'_5y'_5 + e'_sy'_5 + v'_7y'_7 = 0 \\ &③' \quad v'_3y'_3 - v'_4y'_4 - v'_5y'_5 - e'_sy'_5 = 0 \end{aligned} \tag{7.5.4}$$

Now examine the graph in Fig. 7.14*b* of a network that will be called $\mathcal{N}'$. Note that Eqs. (7.5.3) and (7.5.4) are its KCL and Ohm's law equations (the encircled numbers now denote nodes instead of loops) and that $B' = 7$, $L' = 4$, and $N' = 3$. This second network is the dual of the first. We make the following observations:

1. A simple (super) loop in $\mathcal{N}$ becomes a simple (super) node in $\mathcal{N}'$. The exterior of $\mathcal{N}$ becomes the datum node of $\mathcal{N}'$.
2. A branch common to simple loops k and l of $\mathcal{N}$ is common to simple nodes k' and l' of $\mathcal{N}'$.
3. An impedance in $\mathcal{N}$ becomes an admittance in $\mathcal{N}'$. Thus resistances become conductances, inductors become capacitors, and capacitors become inductors.
4. Series branches in $\mathcal{N}$ bcome parallel branches in $\mathcal{N}'$. Thus a series *RLC* circuit in $\mathcal{N}$ becomes a parallel *GCL* circuit in $\mathcal{N}'$.
5. Clockwise loop sense in $\mathcal{N}$ is diverging node sense in $\mathcal{N}'$. If a branch current has clockwise direction in a loop, its dual is a branch voltage (drop) positive at the dual node. Similarly, if a branch voltage (drop) is positive in a clockwise direction in a loop, its dual is a branch current diverging from the dual node.
6. If a voltage source aids a branch voltage in $\mathcal{N}$, its dual is a current source aiding the dual branch current in $\mathcal{N}'$. Similarly, if a current source aids a branch current in $\mathcal{N}$, its dual is a voltage source aiding the dual branch voltage in $\mathcal{N}'$.
7. The number of branches in a tree of $\mathcal{N}$ is the number of branches in a link sct of $\mathcal{N}'$, and vice versa.

With these observations as a foundation, we can develop a general procedure for finding the dual of a network. This procedure will be valid for all planar graphs, i.e., those that can be drawn on a sphere without branches crossing each other. The Kuratowski graphs in Probs. 5.3 and 5.4 are the two basic nonplanar graphs. These two graphs have no duals, and any network containing either or both of them will have no dual. The reason is that at least one branch in a nonplanar graph must be part of more than two simple loops and therefore its dual is a branch incident at more than two simple nodes, which has no physical meaning.† The procedure for creating a dual network is the following:

1. Place a simple node inside each simple loop of $\mathcal{N}$ and an additional simple node outside $\mathcal{N}$ (do not place a simple node inside a loop formed by a current source and its branch admittance).
2. Connect adjacent simple nodes by a line through each branch of $\mathcal{N}$. Give each line the number primed of the branch it cuts.
3. Adopt a duality of branch polarity such that a branch arrow oriented in a clockwise direction in a loop becomes, in the dual branch, a branch arrow diverging from the dual node.
4. A voltage (current) source aiding a branch voltage (current) in $\mathcal{N}$ becomes a current (voltage) source aiding the dual branch current (voltage) in $\mathcal{N}'$.

Example 7.8 To illustrate the procedure just outlined, find the dual of the network in Fig. 7.15*a*. At first glance, it appears that this network is not planar because of the crossing branches 6, 7, 8, and 9. However its graph can be drawn with no crossing branches, as shown in Fig. 7.15*b* (solid-line graph). Actually, if a branch were added between points X and Y in Fig. 7.15*b*, one of the Kuratowski nonplanar graphs would result and it would be impossible to draw it without crossing branches.

To find the dual, we place a simple node inside each simple loop in Fig. 7.15*b* and one simple node outside the graph. Adjacent nodes are joined by dashed lines. The dashed graph is extracted and repeated in Fig. 7.15*c*, with the required dual sources and primed branch directions added. For example, the arrow for branch 9 is clockwise around loop 947 in the original graph. Therefore the arrow for branch 9′ diverges from the dual node (the node enclosed by loop 947) in the dual graph. Also the voltage source bucks the voltage of branch 1 in the original network. Thus its dual current source bucks the current of branch 1′ in the dual network. In Fig. 7.15*d*, we convert the graph of Fig. 7.15*c* to an orderly circuit configuration.

Before closing this section, it is appropriate to assemble in one package several of the properties of dual networks and graphs (see Table 7.1). Most of these properties have been discussed, but a few require further comment.

† The incompleteness of our knowledge of duality is brought out from the point of view of network synthesis in The Realization of N-port Networks without Transformers: A Panel Discussion, *IRE Profess. Group Circuit Theory*, vol. CT-9, Sept., 1962.

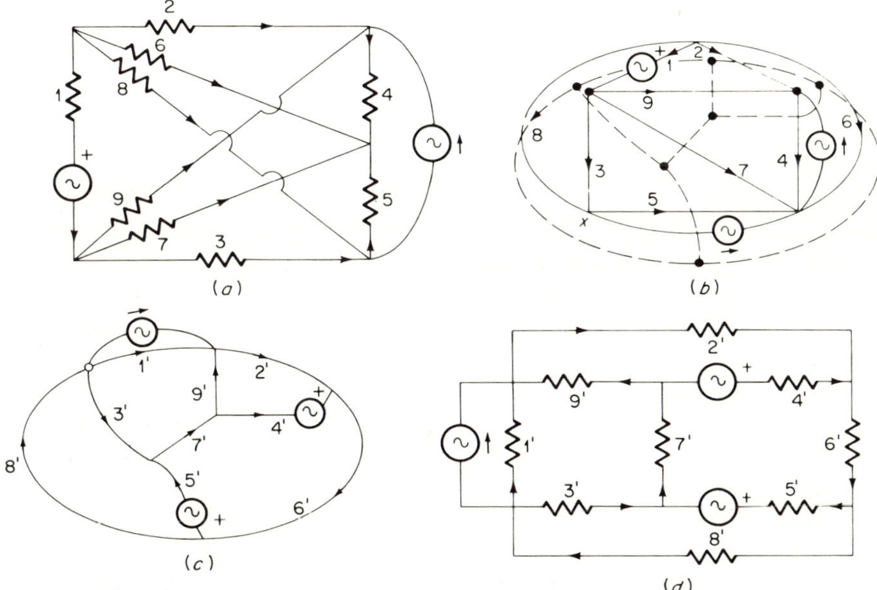

Fig. 7.15 Systematic formulation of a dual network.

TABLE 7.1 Dual properties of networks and graphs

1	Impedance	Admittance
2	Resistance	Conductance
3	Inductance	Capacitance
4	Voltage	Current
5	Series	Parallel
6	Simple loop	Simple node
7	Loop current	Node-pair voltage
8	Circuit exterior	Datum node
9	Circuit-exterior current	Node-to-datum voltage
10	Link	Tree branch
11	Link set	Tree
12	L (number of links)	N (number of tree branches)
13	Voltage source	Current source
14	$v = jz$	$j = vy$
15	KVL	KCL
16	Open circuit	Short circuit
17	Dangling branch	Self-loop
18	Tie set (β_F)	Cut set (α_F)
19	$\beta_F V_B = 0$	$\alpha_F J_B = 0$
20	$J_B = \beta_F{}^T I_L$	$V_B = \alpha_F{}^T E_N$
21	$E_E = Z_E I_L$	$I_E - Y_E E_N$

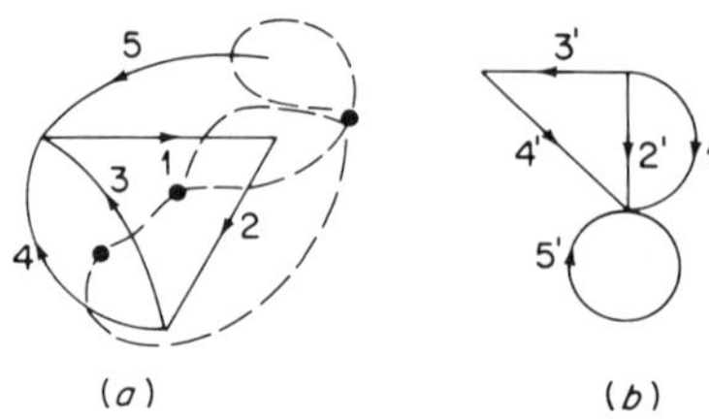

Fig. 7.16 Dual of a dangling branch.

For example, properties 8 and 9 relate the datum node to the circuit exterior and voltages to the datum node to currents in the circuit exterior. In Fig. 7.14*b*, 4′ is the datum node. The dual of any voltage from another node to 4′ in Fig. 7.14*b* is a current in the external loop (1637) in Fig. 7.14*a*. Next, property 10 tells us that a link and a tree branch are duals; property 11 that a link set and a tree are also duals. Choose any tree, say 2457, in the graph of Fig. 7.14*a*. Then 2′, 4′, 5′, 7′ form a link set in the graph of Fig. 7.14*b*. Properties 18 to 21 follow immediately from properties 10 and 11. Finally, properties 16 and 17 require justification. Consider the graph in Fig. 7.16*a*. Branch 5 is a dangling branch. To form the dual of the graph, we proceed as before except that we draw an extra closed dashed line through branch 5 and to the external node. This yields a self-loop as the dual of the dangling branch, as shown in the dual graph in Fig. 7.16*b*. Also we observe that the open circuit (infinite impedance) in series with branch 5 has as its dual a short circuit (infinite admittance) in parallel with branch 5′.

7.6 SENSITIVITY†

The sensitivity of a network function, and of its poles and zeros, to deviations of elements from their prescribed values is a subject of considerable interest to the circuit designer. He wants to know, for example, how the performance of his network will be affected if resistors with 10 percent tolerance are used. Or he may wish to compare two networks he has designed with respect to their sensitivity to environmental factors, such as temperature changes.

In this section we present quantitative definitions of classical and pole-zero sensitivity, show the relationship between them, and develop a topological procedure for computing the classical sensitivity of a linear uncoupled network, using the formulas of Secs. 6.7 and 6.8.

The first of the two sensitivities to be discussed, classical, or function, sensitivity, was first defined by Bode.‡ It relates a per-unit deviation in a

† The material in this discussion is adapted from Refs. 7.2 and 7.3.

‡ H. W. Bode, "Network Analysis and Feedback Amplifier Design," D. Van Nostrand Company, Inc., Princeton, N.J., 1945.

network parameter to a per-unit change in a network function. Mason† inverted Bode's definition so that an ideal network would have zero, rather than infinite, sensitivity. It is Mason's definition that we shall use here

$$S_x^T = \frac{dT/T}{dx/x} = \frac{x}{T}\frac{dT}{dx} \tag{7.6.1}$$

where S_x^T is defined as the sensitivity of the network function T to changes in the network parameter x.

The second of the sensitivities is pole-zero, or root, sensitivity. It was defined by Truxal‡ and is a measure of the change that occurs in the s-plane location of a pole or zero of the network function due to a per-unit change in a network parameter. Formally, we write

$$S_x^{s_k} = \frac{ds_k}{dx/x} = x\frac{ds_k}{dx} \tag{7.6.2}$$

where $S_x^{s_k}$ is defined as the sensitivity of the root location $s = s_k$ to changes in the network parameter x.

Before working an example to illustrate these definitions by finding the classical and root sensitivities of a particular network, let us simplify Eq. (7.6.1) for computational purposes. The function T is a ratio of polynomials in s and also varies with x. We can note this by writing

$$T(x,s) = \frac{P(x,s)}{Q(x,s)} \tag{7.6.3}$$

Inserting Eq. (7.6.3) into Eq. (7.6.1), we obtain

$$S_x^T = \frac{xQ}{P}\frac{d(P/Q)}{dx} = \frac{xQ}{P}\frac{Q\,dP/dx - P\,dQ/dx}{Q^2}$$

$$= \frac{x}{P}\frac{dP}{dx} - \frac{x}{Q}\frac{dQ}{dx} = S_x^P - S_x^Q \tag{7.6.4}$$

The last form of Eq. (7.6.4) will be used for calculating classical sensitivity. In it we describe the classical sensitivity of the function $T(x,s)$ as the difference in the classical sensitivities of its numerator P and its denominator Q.

Example 7.9 For the network in Fig. 7.17, find the classical sensitivity of the driving-point impedance Z and the root sensitivity of the pole at $s_1 = -(C_1 + C_2)/R_2C_1C_2$, both with respect to changes in the parameter R_2.

To compute the classical sensitivity, we use Eq. (7.6.4) and write

$$S_{R_2}^Z = S_{R_2}^P - S_{R_2}^Q = \frac{R_2}{P}\frac{dP}{dR_2} - \frac{R_2}{Q}\frac{dQ}{R_2}$$

† S. J. Mason and H. J. Zimmerman, "Electric Circuits, Signals and Systems," p. 579, John Wiley & Sons, Inc., New York, 1960.

‡ J. G. Truxal and I. M. Horowitz, Sensitivity Considerations in Active Network Synthesis, *Second Midwest Symposium on Circuit Theory, East Lansing, Mich.*, 1956.

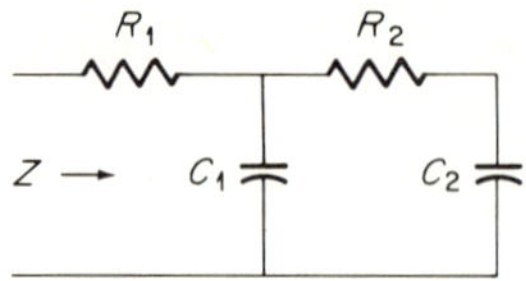

Fig. 7.17 Network for which classical and root sensitivities are to be obtained.

To simplify calculations, we insert the values of R_1, C_1, and C_2 into the expression for Z (R_2 must be retained explicitly until after we differentiate) and obtain

$$Z = \frac{12R_2s^2 + (7 + 4R_2)s + 1}{12R_2s^2 + 7s}$$

Now we proceed to calculate

$$S_{R_2}{}^Z = \frac{2}{24s^2 + 15s + 1}(12s^2 + 4s) - \frac{2}{24s^2 + 7s}12s^2$$

$$= \frac{32s}{(24s + 7)(24s^2 + 15s + 1)}$$

The root sensitivity can be calculated directly from Eq. (7.6.2).

$$S_{R_2}{}^{s_1} = R_2\left(-\frac{C_1 + C_2}{C_1C_2}\right)\left(-\frac{1}{R_2{}^2}\right)$$

$$= \frac{C_1 + C_2}{R_2C_1C_2} = \frac{7}{24}$$

Note that the root sensitivity is a number (real in this case) independent of s whereas the classical sensitivity is a ratio of polynomials in s. To investigate the latter sensitivity further, let us compute its value at $s = j1$. We find

$$S_{R_2}{}^Z|_{s=j1} = \frac{32j}{(7 + j24)(-23 + j15)} = 0.0465\underline{/-130.6^\circ}$$

This complex number, being obtained from a derivative operation, is a slope. Thus the result is valid only for *small* changes in the parameter R_2. To illustrate this point quantitatively, let us compute Z at $s = j1$ for $R_2 = 2$ ohms and 3 ohms, formulate $\Delta Z/\Delta R_2$, compute the classical sensitivity from Eq. (7.6.1), and compare the result with that just obtained for $S_{R_2}{}^Z$.

$$Z\Big|_{\substack{R_2=2\text{ ohms}\\ s=j1}} = \frac{-23 + j15}{-24 + j7} = 1.10\underline{/-16.9^\circ}$$

$$Z\Big|_{\substack{R=3\text{ ohms}\\ s=j1}} = \frac{-35 + j19}{-36 + j7}$$

$$\Delta Z = \frac{-35 + j19}{-36 + j7} - \frac{-23 + j15}{-24 + j7}$$

$$= \frac{-16}{815 - j420} = 0.0176\underline{/-152.6^\circ}$$

$$S_{R_2}{}^Z = \frac{2}{1.10\underline{/-16.9^\circ}}\,\frac{0.0176\underline{/-152.6^\circ}}{1}$$

$$= 0.0321\underline{/-135.7^\circ}$$

This result does not compare favorably with the actual sensitivity. There is a 31 percent error in the magnitude and a 4 percent error in the angle. Let us repeat the calculation for a smaller ΔR_2 by using $R_2 = 2.1$ ohms.

$$Z|_{\substack{R_2=2.1 \text{ ohms} \\ s=j1}} = \frac{-24.2 + j15.4}{-25.2 + j7}$$

$$\Delta Z = \frac{-24.2 + j15.4}{-25.2 + j7} - \frac{-23 + j15}{-24 + j7}$$

$$= \frac{-1.6}{555.8 - j344.4} = 0.00245\underline{/-148.2^\circ}$$

$$S_{R_2}{}^Z = \frac{2}{1.10\underline{/-16.9^\circ}} \frac{0.00245\underline{/-148.2^\circ}}{0.10}$$

$$= 0.0445\underline{/-131.3^\circ}$$

This result is much better. The magnitude and angle errors are only 4.3 and 0.5 percent, respectively. In general, we note that the classical sensitivity is only a first-order approximation to the true deviation of the function and is thus valid only for small values of Δx.

In Eqs. (7.6.1) and (7.6.2) definitions of the two types of sensitivity have been given. We now relate these sensitivities to each other analytically and begin by rewriting $T(x,s)$ in Eq. (7.6.3) in the form

$$T(x,s) = \frac{\sum_{i=0}^{n} (a_i + xb_i)s^i}{\sum_{j=0}^{m} (c_j + xd_j)s^j} \tag{7.6.5}$$

which assumes that each coefficient of $T(x,s)$ can vary no more than linearly with x. This is by far the most common case. Observe, for example, that the coefficients of Z in Fig. 7.17 have no more than a linear dependence on any of the four parameters in the network. The most common exception to this assumption occurs when a mutual coupling M is present. Then M^2 will appear as part of a coefficient of one or more powers of s.† This case will not be considered here.

Let us next substitute Eq. (7.6.5) into the expression for classical sensitivity in Eq. (7.6.4). The result is

$$S_x{}^T = \frac{x \sum_{i=0}^{n} b_i s^i}{\sum_{i=0}^{n} (a_i + xb_i)s^i} - \frac{x \sum_{j=0}^{m} d_j s^j}{\sum_{j=0}^{m} (c_j + xd_j)s^j} \tag{7.6.6}$$

† This situation is considered by Mikulski (Ref. 7.3). He generalizes each coefficient of $T(x,s)$ to the form $[a_k + b_k f(x)]s^k$, where $f(x)$ is a function of x.

If we perform a partial-fraction expansion of Eq. (7.6.6), each term in that expansion will contain one of the zeros or one of the poles of $T(x,s)$. This may permit us to express the classical sensitivity as some sort of a summation of root sensitivities. Carrying out this expansion, we have

$$S_x{}^T = \sum_h \frac{k_h}{s - z_h} - \sum_l \frac{k_l}{s - p_l} \tag{7.6.7}$$

where the k_h's and the k_l's are functions of x and the z_h's and p_l's denote the zeros and poles, respectively, of $T(x,s)$. Note that we have assumed that all these poles and zeros are simple (of first degree). We have also assumed that $b_n = d_m = 0$ with $a_n \neq 0$ and $c_m \neq 0$, so that no constant term (independent of s but a function of x) appears in either partial-fraction expansion. The consequences of the presence of multiple roots and/or constant terms will be discussed later.

The problem now is to try to relate the k_h's and the k_l's to the root sensitivities. Consider the sensitivity of a particular pole of $T(x,s)$ at $s = p_k$. The poles of $T(x,s)$ in Eq. (7.6.5) are the roots of

$$\sum_{j=0}^{m} (c_j + xd_j)s^j = 0 \tag{7.6.8}$$

When the parameter x changes to $x + \Delta x$, the poles become the roots of

$$\sum_{j=0}^{m} [c_j + (x + \Delta x)d_j]s^j = 0 \tag{7.6.9}$$

which can be written as

$$1 + \frac{\Delta x}{x} \frac{\sum_{j=0}^{m} xd_js^j}{\sum_{j=0}^{m} (c_j + xd_j)s^j} = 0 \tag{7.6.10}$$

We perform a partial-fraction expansion of the second quantity in Eq. (7.6.10) and obtain [using Eq. (7.6.6)]

$$1 + \frac{\Delta x}{x} \sum_l \frac{k_l}{s - p_l} = 0 \tag{7.6.11}$$

Here again we assume that all roots are simple and that no constant term is present. Note that the k_l's are the same constants as in Eq. (7.6.7). Near the pole $s = p_k$ the term $k_k/(s - p_k)$ dominates the summation in Eq. (7.6.11). Hence, near that pole, at a point $s = p_k'$, we can approximate Eq. (7.6.11) by

$$1 + \frac{\Delta x}{x} \frac{k_k}{p_k' - p_k} = 0 \tag{7.6.12}$$

But $p_k' - p_k$ is just the shift in the pole location. Calling it Δp_k and rearranging Eq. (7.6.12), we have

$$x \frac{\Delta p_k}{\Delta x} = -k_k = S_x^{p_k} \tag{7.6.13}$$

The result in Eq. (7.6.13) shows that the sensitivity of a pole at $s = p_k$ is the negative of the coefficient k_k obtained in the partial-fraction expansions of Eqs. (7.6.11) and (7.6.7).

A similar result is obtained for the sensitivity of a zero at $s = z_k$. We merely replace Eq. (7.6.8) by the numerator of Eq. (7.6.5) and repeat the derivation.

Using the result of Eq. (7.6.13), we can now rewrite Eq. (7.6.7) in the final form

$$S_x^T = \sum_l \frac{S_x^{p_l}}{s - p_l} - \sum_h \frac{S_x^{z_h}}{s - z_h} \tag{7.6.14}$$

Thus the classical sensitivity is the summation of the pole sensitivities, weighted according to the pole locations, minus the summation of the zero sensitivities, weighted according to the zero locations.

Two special cases remain to be discussed, namely, the presence of multiple roots and constant terms. When one multiple pole of degree r at $s = p_k$ is present, the partial-fraction expansion in Eq. (7.6.11) becomes

$$1 + \frac{\Delta x}{x} \sum_l \frac{k}{s - p_l} + \frac{\Delta x}{x}\left[\frac{k_r}{(s - p_k)^r} + \cdots + \frac{k_1}{s - p_k}\right] = 0 \tag{7.6.15}$$

Near the pole at $s = p_k$, all terms under the summation sign are negligible, and we can write

$$1 + \frac{\Delta x}{x}\left[\frac{k_r}{(\Delta p)^r} + \cdots + \frac{k_1}{\Delta p}\right] = 0 \tag{7.6.16}$$

This yields an rth-order equation in Δp. Its solution gives the changes in the r roots at $s = p_k$ as x is changed. Each of the r root sensitivities is then found by multiplying each Δp by $x/\Delta x$. Note that none of these sensitivities can be expressed in terms of a single partial-fraction-expansion coefficient, as in the simple-root case.

For one multiple pole of degree r at $s = p_k$, Eq. (7.6.7) becomes

$$S_x^T = \sum_h \frac{k_h}{s - z_h} - \sum_l \frac{k_l}{s - p_l} - \left[\frac{k_r}{(s - p_k)^r} + \cdots + \frac{k_1}{s - p_k}\right] \tag{7.6.17}$$

Then Eq. (7.6.14) becomes

$$S_x^T = \sum_l \frac{S_x^{p_l}}{s - p_l} - \sum_h \frac{S_x^{z_h}}{s - z_h} - \left[\frac{k_r}{(s - p_k)^r} + \cdots + \frac{k_1}{s - p_k}\right] \tag{7.6.18}$$

Equation (7.6.18) does relate the classical and root sensitivities, but not as nicely as in the simple-root case.

If the partial-fraction expansions in Eqs. (7.6.7) and (7.6.11) each contain a constant term, these equations take the form

$$S_x{}^T = \sum_h \frac{k_h}{s - z_h} + k_z - \sum_l \frac{k_l}{s - p_l} - k_p$$
$$1 + \frac{\Delta x}{x}\left(\sum_l \frac{k_l}{s - p_l} + k_p\right) = 0 \tag{7.6.19}$$

where k_z and k_p are generally functions of x. The arguments following Eq. (7.6.11) still apply, and Eqs. (7.6.12) and (7.6.13) remain valid. Then Eq. (7.6.14) becomes

$$S_x{}^T = \sum_l \frac{S_x{}^{p_l}}{s - p_l} - k_p - \sum_h \frac{S_x{}^{z_h}}{s - z_h} + k_z \tag{7.6.20}$$

Example 7.10 For the network in Fig. 7.17 compute the classical sensitivity $S_{R_2}{}^Z$ for the given parameter values from the root sensitivities.

The driving-point impedance Z in Fig. 7.17 has two finite zeros and one finite pole. The pole at the origin does not contribute to the classical sensitivity. The pole sensitivity was computed in Example 7.9. To find the two zero sensitivities, we must first find the zeros of Z. These are

$$s_2, s_3 = \frac{-(7 + R_2) \pm \sqrt{49 + 8R_2 + 16R_2{}^2}}{24R_2}$$

The zero sensitivities are computed from Eq. (7.6.2)

$$S_{R_2}^{s_{2,3}} = R_2 \frac{ds_{2,3}}{dR_2} = \frac{7}{24R_2} \pm \frac{-49 - 4R_2}{24R_2\sqrt{49 + 8R_2 + 16R_2{}^2}}$$

For $R_2 = 2$ ohms, the zeros are

$$s_2, s_3 = \frac{-15 \pm \sqrt{129}}{48}$$

and the zero sensitivities are

$$S_2^{s_{2,3}} = \frac{7}{48} = \frac{57}{48\sqrt{129}}$$

An examination of Z shows that Eq. (7.6.20), rather than Eq. (7.6.14), is the correct one to use here. To find k_p and k_z, we note from Eq. (7.6.5) that $n = m = 2$. Also $a_n = c_m = 0$, but $b_n \neq 0$ and $d_m \neq 0$. Thus $k_p = k_z = 1$, and these coefficients cancel in Eq. (7.6.20). Inserting the calculated root sensitivities into Eq. (7.6.20) gives

$$\begin{aligned} S_{R_2}{}^Z &= \frac{7/24}{s + 7/24} - \frac{7/48 - 57/(48\sqrt{129})}{s + (15 - \sqrt{129})/48} - \frac{7/48 + 57/(48\sqrt{129})}{s + (15 + \sqrt{129})/48} \\ &= \frac{7}{24s + 7} - \frac{7s + 1}{24s^2 + 15s + 1} \\ &= \frac{32s}{(24s + 7)(24s^2 + 15s + 1)} \end{aligned}$$

The resulting classical sensitivity is the same as that found in Example 7.9.

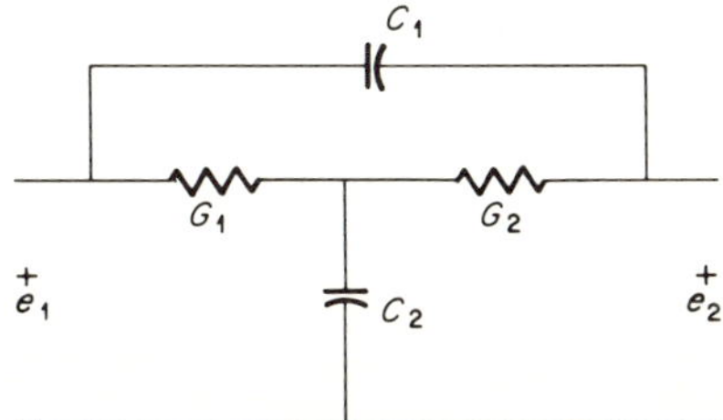

Fig. 7.18

Example 7.11 Find the open-circuit voltage transfer ratio e_2/e_1 for the bridged T network in Fig. 7.18. For $G_1 = G_2 = 1$ mho and $C_2 = 1$ farad, find the sensitivity of the transmission (numerator) zeros of e_2/e_1 to changes in C_1. For $C_1 = C_2 = 1$ farad and $G_2 = 1$ mho, find this same sensitivity to changes in G_1. From the results, compare the sensitivity of the transmission zeros to changes in the capacitors and resistors in the network.

The desired voltage transfer ratio can be found by conventional or topological circuit analysis to be

$$\frac{e_2}{e_1} = \frac{C_1C_2s^2 + C_1(G_1 + G_2)s + G_1G_2}{C_1C_2s^2 + C_1[G_1 + G_2 + (C_2/C_1)G_2]s + G_1G_2}$$

For $G_1 = G_2 = 1$ mho and $C_2 = 1$ farad, the ratio becomes

$$\frac{e_2}{e_1} = \frac{C_1s^2 + 2C_1s + 1}{C_1s^2 + C_1(2 + 1/C_1)s + 1}$$

and for $C_1 = C_2 = 1$ farad and $G_2 = 1$ mho, the ratio is

$$\frac{e_2}{e_1} = \frac{s^2 + (1 + G_1)s + G_1}{s^2 + (2 + G_1)s + G_1}$$

The numerator roots are $-1 \pm \sqrt{1 - 1/C_1}$ for the C_1 transfer ratio and -1, $-G_1$ for the G_1 transfer ratio.

The desired sensitivities are obtained from Eq. (7.6.2) as

$$S_{C_1}{}^s = \pm \frac{1}{2C_1\sqrt{1 - 1/C_1}}$$

and

$$S_{G_1}{}^s = 0 \qquad \text{and} \qquad -G_1$$

The transmission zeros are sensitive nonlinearly to C_1 and linearly and not at all to G_1. A plot of the sensitivities appears in Fig. 7.19. The sensitivity to changes in C_1 is less than that due to G_1 for G_1 and C_1 greater than 1.2 mhos and farads, respectively. The sensitivity to C_1 is worst (infinite) at $C_1 = 1$ farad, corresponding to a second-order transmission zero at $s = -1$.

To conclude this section, a topological procedure will be developed for computing classical sensitivity in a linear network with no coupling. Suppose we want the sensitivity of a driving-point admittance with respect to a branch admittance y_k. The driving-point admittance at a solder entry across

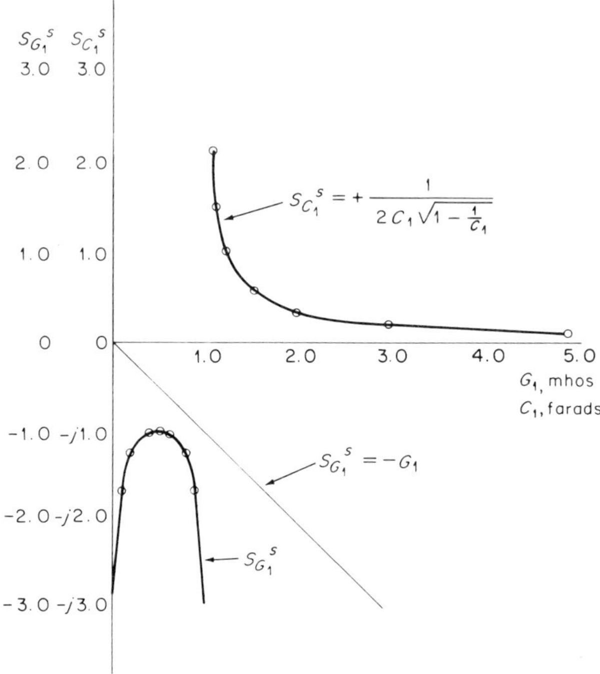

Fig. 7.19 Sensitivity plot for Fig. 7.18.

branch l is given topologically by Eq. (6.7.6), which we repeat here

$$Y_{ll}{}^s = \frac{\sum \text{TYP of } \mathcal{N}}{\sum \text{TYP of } \mathcal{N}_{sl}} \tag{7.6.21}$$

Recall that $\mathcal{N}$ is the original network and $\mathcal{N}_{sl}$ is the network obtained by shorting the terminals in question. Insert Eq. (7.6.21) into the expression for classical sensitivity in Eq. (7.6.4) to obtain

$$S_{y_k}^{Y_{ll}{}^s} = \frac{y_k}{\sum \text{TYP of } \mathcal{N}} \frac{d}{dy_k} \left(\sum \text{TYP of } \mathcal{N}\right) - \frac{y_k}{\sum \text{TYP of } \mathcal{N}_{sl}} \frac{d}{dy_k} \left(\sum \text{TYP of } \mathcal{N}_{sl}\right) \tag{7.6.22}$$

We now need a method of evaluating the derivatives in Eq. (7.6.22) topologically. Consider the first of the derivatives. Given $\sum$ TYP of $\mathcal{N}$, we find those terms which contain y_k and delete y_k from them to obtain this derivative. From the previous discussion in Sec. 6.7, we see that the result is the $\sum$ TYP of $\mathcal{N}_{sk}$, the summation of tree-admittance products of the network formed by shorting branch k. Consider next the second of the derivatives. By a similar argument, this derivative is the $\sum$ TYP of $\mathcal{N}_{sk,l}$, the summation of tree-admittance products of the network formed by shorting both the input terminals and the branch k. Assembling these results in Eq. (7.6.22) gives

$$S_{y_k}^{Y_{ll}^{s}} = y_k\left(\frac{\sum \text{TYP of } \mathcal{N}_{sk}}{\sum \text{TYP of } \mathcal{N}} - \frac{\sum \text{TYP of } \mathcal{N}_{sk,l}}{\sum \text{TYP of } \mathcal{N}_{sl}}\right) \tag{7.6.23}$$

There are other sensitivity functions that can be computed topologically, e.g., the sensitivity of a transfer impedance or admittance to changes in a branch impedance or admittance. These will be left to the problems.

The result in Eq. (7.6.23) gives the sensitivity of a driving-point admittance to changes in a branch admittance. What if we are interested in an impedance sensitivity? Given $S_{y_k}^{\ Y}$, can we find $S_{y_k}^{\ Z}$, $S_{z_k}^{\ Y}$, and $S_{z_k}^{\ Z}$ directly? Given Eq. (7.6.4), to find $S_{y_k}^{\ Z}$, we merely exchange the identities of P and Q. Thus

$$S_{y_k}^{\ Z} = -S_{y_k}^{\ Y} \tag{7.6.24}$$

To obtain $S_{z_k}^{\ Y}$, we note from Eq. (7.6.1) that

$$S_x^{\ T} = \frac{x}{T}\frac{dT}{dx} = \frac{x}{T}\frac{dT}{d(1/x)}\frac{d(1/x)}{dx} = -\frac{1/x}{T}\frac{dT}{d(1/x)} = -S_{1/x}^{T} \tag{7.6.25}$$

Thus

$$S_{z_k}^{\ Y} = -S_{y_k}^{\ Y} \tag{7.6.26}$$

and

$$S_{z_k}^{\ Z} = S_{y_k}^{\ Y} \tag{7.6.27}$$

It also often happens that the item of interest is the sensitivity to changes in a network parameter rather than to changes in the impedance or admittance of that parameter. For example, suppose we want the sensitivity with respect to changes in a capacitor, given the sensitivity with respect to changes in the admittance of that capacitor. Note that

$$\begin{aligned} S_{sC}^{\ Y} &= \frac{sC}{Y}\frac{dY}{d(sC)} \\ S_C^{\ Y} &= \frac{C}{Y}\frac{dY}{dC} = \frac{C}{Y}\frac{dY}{d(sC)}\frac{d(sC)}{dC} = \frac{sC}{Y}\frac{dY}{d(sC)} \end{aligned} \tag{7.6.28}$$

Thus

$$S_{sC}^{\ Y} = S_C^{\ Y} \tag{7.6.29}$$

If the network element is an inductor, rather than a capacitor,

$$S_{1/sL}^{Y} = \frac{1}{sLY}\frac{dY}{d(1/sL)}$$
$$S_L{}^Y = \frac{L}{Y}\frac{dY}{dL} = \frac{L}{Y}\frac{dY}{d(1/sL)}\frac{d(1/sL)}{dL} = -\frac{1}{sLY}\frac{dY}{d(1/sL)} \qquad (7.6.30)$$

Thus

$$S_{1/sL}^{Y} = -S_L{}^Y \qquad (7.6.31)$$

Similar relations can be derived on an impedance basis. The point is that Eq. (7.6.23) can always be used for finding the sensitivity of a driving-point function because of the simple relations in Eqs. (7.6.24) to (7.6.31).

Example 7.12 Find topologically the sensitivity of Z in Example 7.9 to changes in the parameter R_2.

The graphs of $\mathcal{N}$, $\mathcal{N}_{sl}$, $\mathcal{N}_{sk}$, and $\mathcal{N}_{sk,l}$ appear in Fig. 7.20. From Eq. (7.6.27), $S_{R_2}{}^Z = S_{1/R_2}^{Y}$. Using Eq. (7.6.23) directly, the summations needed are (numbers substituted for branch admittances)

$$\sum \text{TYP of } \mathcal{N} = 123 + 124 + 134$$
$$\sum \text{TYP of } \mathcal{N}_{sl} = 12 + 14 + 23 + 24 + 34$$
$$\sum \text{TYP of } \mathcal{N}_{sk} = 13 + 14$$
$$\sum \text{TYP of } \mathcal{N}_{sk,l} = 1 + 3 + 4$$
$$S_{R_2}{}^Z = y_2\left(\frac{13 + 14}{123 + 124 + 134} - \frac{1 + 3 + 4}{12 + 14 + 23 + 24 + 34}\right)$$

Noting that $y_1 = 1$, $y_2 = \frac{1}{2}$, $y_3 = 3s$, and $y_4 = 4s$, we have

$$S_{R_2}{}^Z = \frac{1}{2}\left(\frac{7s}{12s^2 + \frac{7}{2}s} - \frac{7s + 1}{12s^2 + \frac{15}{2}s + \frac{1}{2}}\right)$$
$$= \frac{32s}{(24s + 7)(24s^2 + 15s + 1)}$$

which checks the previous classical sensitivity calculation.

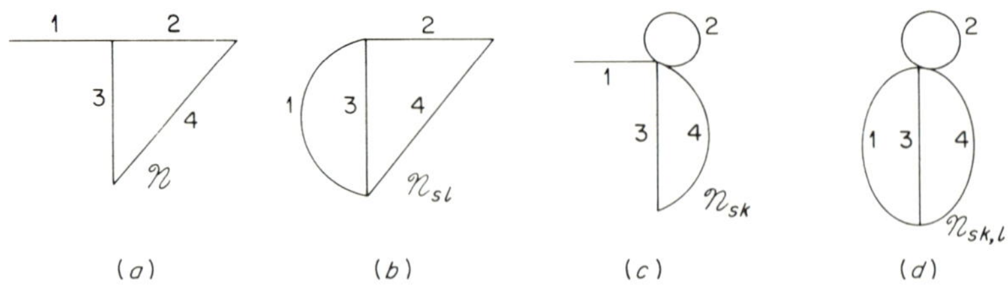

Fig. 7.20 Graphs for Example 7.12.

REFERENCES

7.1. de Pian, L.: "Linear Active Network Theory," Prentice-Hall, Inc., Englewood Cliffs, N.J., 1962. Chapter 4 contains an excellent and detailed discussion of reciprocity.

7.2. Melvin, D. W.: Sensitivity and Stability in Active Network Synthesis, *Syracuse Univ. Res. Inst. Rept.* EE 975-6401SR2, 1964. Includes an excellent section on the use of topological formulas in the determination of classical sensitivity of driving-point and transfer functions.

7.3. Mikulski, J. J.: A Correlation between Classical and Pole-zero Sensitivity, *Circuit Theory Group, Univ. Ill., ASTIA Rept.* 209847, 1959. A well-written discussion of classical and pole-zero sensitivities.

7.4. Millman, J.: "Vacuum-tube and Semiconductor Electronics," McGraw-Hill Book Company, New York, 1958. Recommended for the reader who wishes to review the use of Thévenin's and Norton's theorems in the solution of active-network problems.

7.5. Reza, F. M., and S. Seely: "Modern Network Analysis," McGraw-Hill Book Company, New York, 1959. Network theorems are discussed in Chapter 5. Chapter 4 contains an interesting discussion of duality, with special emphasis on the two basic Kuratowski nonplanar graphs.

7.6. Seshu, S., and N. Balabanian: "Linear Network Analysis," John Wiley & Sons, Inc., New York, 1959. Network theorems are discussed in Chapter 5.

7.7. Truxal, J. G.: "Automatic Feedback Control System Synthesis," McGraw-Hill Book Company, New York, 1955. Chapter 2 contains a well-written discussion of the sensitivity of the overall gain of a control system to parameter changes.

PROBLEMS

Drill

7.1. Find i without writing loop or node equations.

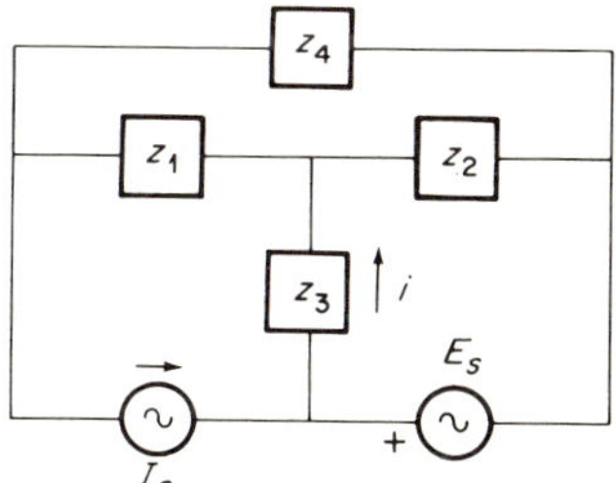

Fig. P7.1

7.2. Find the Thévenin and Norton equivalent circuits looking to the left from terminals X-X. All sources are of the same frequency.

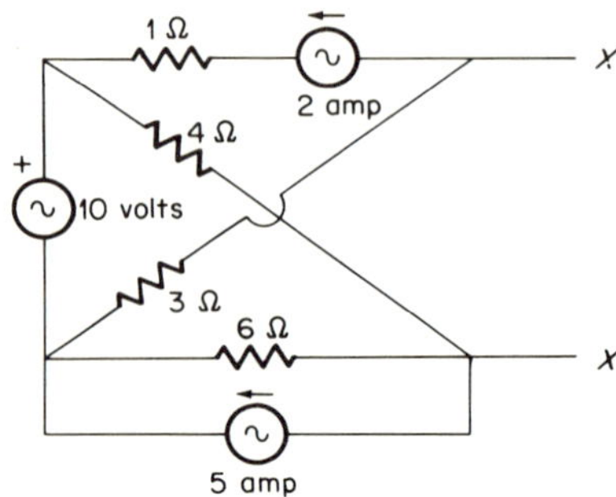

Fig. P7.2

7.3. Find the dual of the network in Fig. P6.5.

7.4. Find the sensitivity of Z_{20} with 1-0 open in Fig. P6.4, and of its poles, to changes in the value of the inductor. Repeat for Y_{10} with 2-0 shorted.

7.5. Perform successive voltage and current source transformations to convert the given network to a single source and resistance that will produce the correct current through R_L for any value of R_L. Assume all sources are of the same frequency.

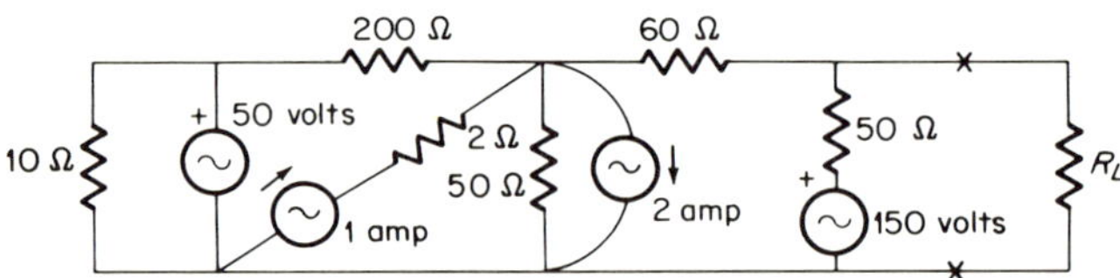

Fig. P7.5

7.6. Find the dual of the network in Fig. P7.5.

7.7. Find the current in the 1-ohm resistor without writing loop or node equations. All sources are 1000 Hz.

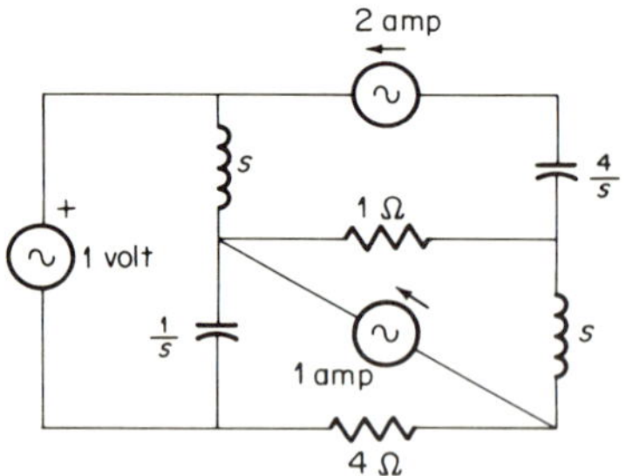

Fig. P7.7

7.8. Find the dual of the given network.

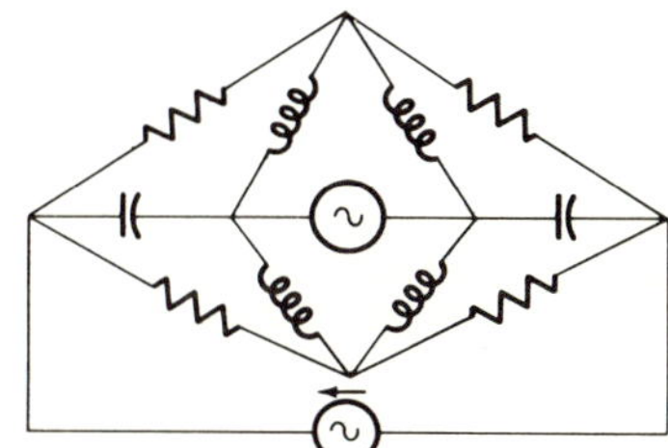

Fig. P7.8

7.9. Find the sensitivity of the driving-point impedance Z_1 to changes in the value of C_2. Work the problem in two ways, directly from Eq. (7.6.4) and topologically from Eq. (7.6.23). $R_1 = R_2 = 1$ ohm, $L_1 = L_2 = L_3 = 1$ henry, $C_1 = C_2 = 1$ farad.

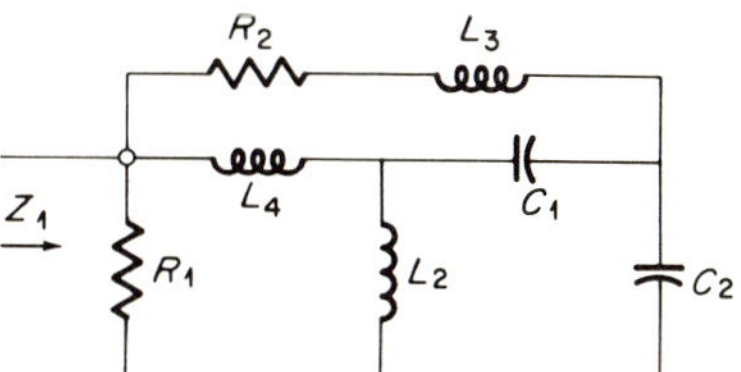

Fig. P7.9

7.10. Find the Thévenin and Norton equivalent circuits as seen from the output terminals of the amplifier.

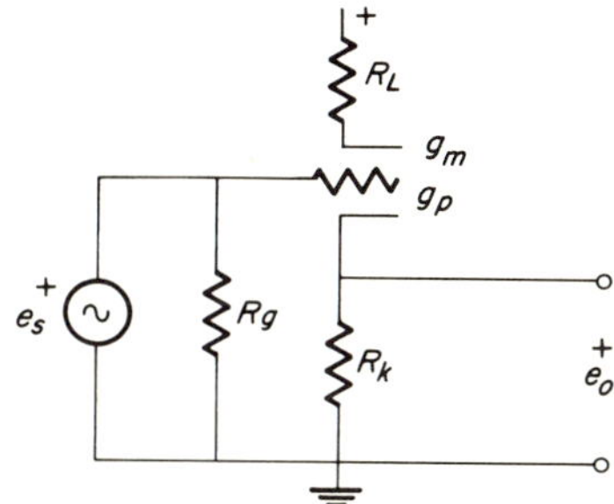

Fig. P7.10

Theory and proofs

7.11. Starting with Eq. (6.8.3) for the topological determination of the transfer admittance Y_{kl} of a linear uncoupled network, derive a topological expression, similar to Eq. (7.6.23), for the sensitivity function $S_{y_k}^{Y_{kl}}$.

7.12. Develop topological expressions for the coefficients of E_{Ek} and i_l in Eq. (7.4.6). Use these results to find the Thévenin equivalent circuit looking to the left from the terminals of R in the given network.

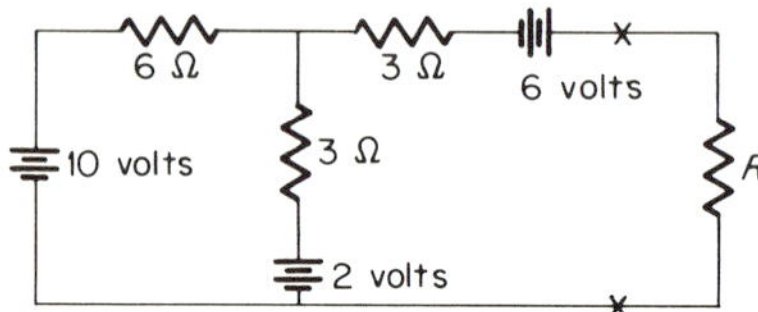

Fig. P7.12

7.13. Given a network whose driving-point impedance at terminals 1-1′ is Z_1, use duality concepts to prove that it is always possible to construct a second network whose driving-point impedance is k/Z_1, where k is a positive real constant. Find such a second network for the network in Fig. P7.9.

Application

7.14. For the transistor amplifier shown, find the parameters of the Norton equivalent circuit in Fig. 7.9*d*. Assume small-signal ac operation and negligible reactance of the capacitor at the signal frequency. Compute the current gain for the 50-ohm load.

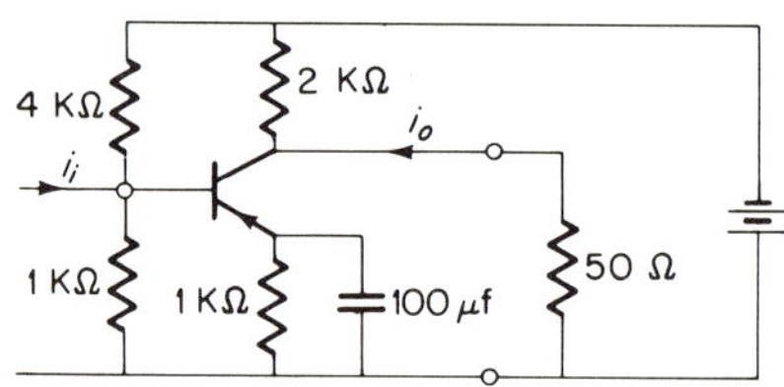

Fig. P7.14 Transistor common-emitter y parameters are $y_i = 5 \times 10^{-4}$ mho, $y_r = -3 \times 10^{-7}$ mho, $y_f = 2 \times 10^{-2}$ mho, and $y_o = 1.8 \times 10^{-5}$ mho.

7.15. Find the Thévenin equivalent circuit for the feedback amplifier shown. Calculate the voltage gain at 1000 Hz.

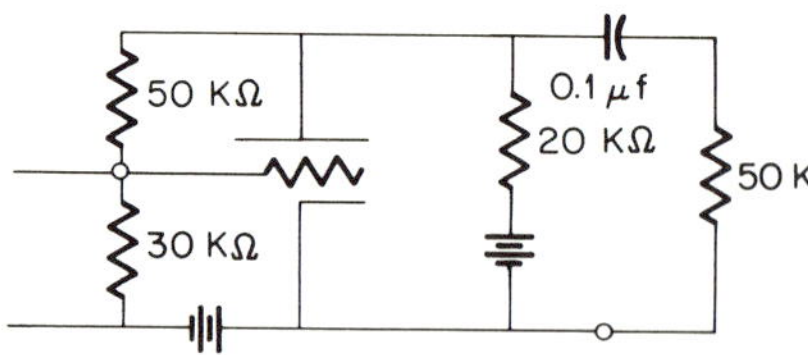

Fig. P7.15 $g_m = 2000$ μmhos, and $g_p = 200$ μmhos.

7.16. The waveform of the voltage generator e_s in Fig. P7.16*a* is shown in Fig. P7.16*b*. It represents the output of a half-wave thyratron rectifier. The output voltage e_o will contain a dc term and several ac terms. Choose L so that the largest amplitude of an ac term is less than 5 percent of the dc term. Repeat the problem if e_s is a 60-Hz square wave of maximum amplitude 50 volts and minimum amplitude 0 volts. (This problem assumes a prior knowledge of Fourier series.)

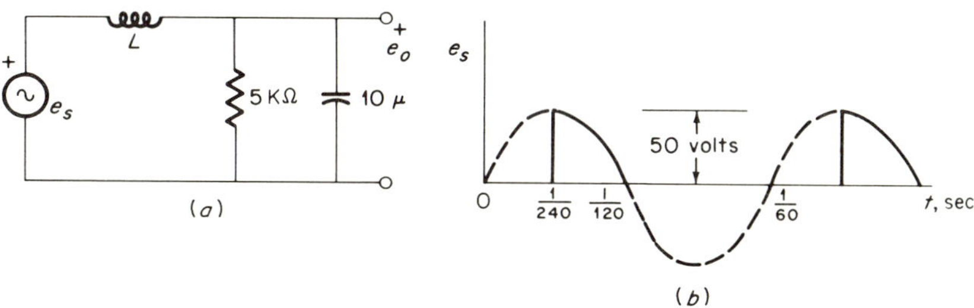

Fig. P7.16

7.17. (*a*) With terminal 3 and the resistances connected to it removed in each network, find values for r_0, r_1, and r_2 such that the two networks are equivalent. Is this π-to-T conversion always possible? Explain.

(*b*) With all terminals and resistances present, find values for r_0, r_1, r_2, and r_3 such that the two grounded 3-port networks in Fig. P7.17 are equivalent. Is this mesh-to-star transformation always possible? Explain.

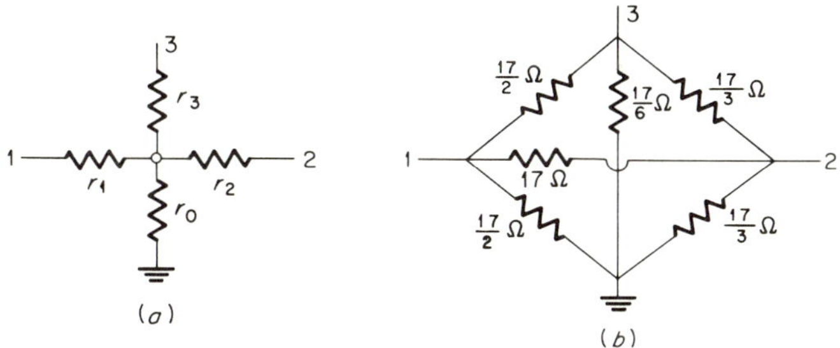

Fig. P7.17

7.18. Find the sensitivity of Y_{10} in Prob. 6.4 to changes in the 4-ohm resistor. What value of that resistor will yield a minimum sensitivity magnitude at $f = 1000$ Hz?

7.19. In the network in Fig. 7.17, replace the resistor R_2 by two resistors of value $2R_2$ in parallel. Compute the sensitivity of Z to one of these resistors. Compare your result with that obtained in Example 7.9.

7.20. Each of the networks has the driving-point impedance $Z = (s^2 + 4s + 2)/(s^2 + 6s + 12)$. Compare the sensitivity of Z in each network to changes in the $\frac{1}{6}$-ohm resistor.

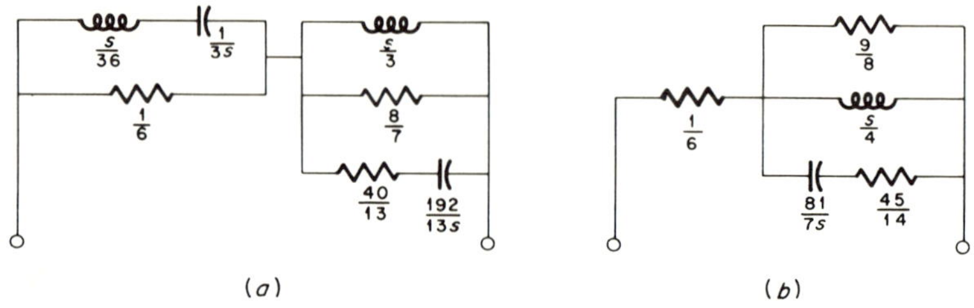

Fig. P7.20

7.21. For the given network, find 2-generator Thévenin and Norton equivalent circuits.

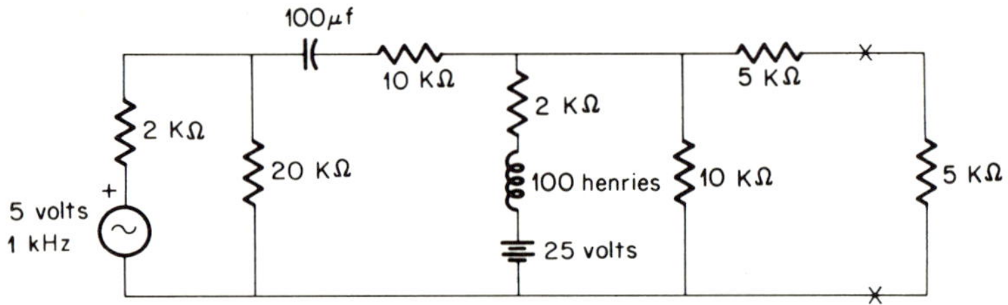

Fig. P7.21

8 NATURAL FREQUENCIES AND FREE RESPONSE

8.1 INTRODUCTION

Chapter 7 completed the treatment of the steady-state analysis of linear networks. The second major topic of the text, namely, the free and forced responses of linear networks, begins in this chapter. Actually it is impossible to separate these subjects completely, and thus Chap. 8 also serves as a bridging chapter between the two.

We start by defining a linear system and outlining its properties. Next the free response of an nth-order linear system is derived, first from the nth-order differential equation and then from n first-order differential equations derived from the nth-order equation. The latter formulation leads to a study of eigenvalues and related topics. We then discuss natural frequencies of linear networks, which are considered from the points of view of the poles and zeros of network functions and the concepts of dynamic and geometric independence and degrees of freedom. Finally, procedures for finding the number of natural frequencies and dynamically independent variables in a network are developed.

8.2 DEFINITION AND PROPERTIES OF A LINEAR SYSTEM

One way of defining a linear system is to state that it is one that can be characterized by a linear differential equation, i.e., by a differential equation whose dependent variable and its derivatives are to the first degree only, with no products of derivatives present, and whose coefficients are not functions of the dependent variable. The philosophy behind this definition is important. It implies that instead of dealing with the physical system directly, we shall discuss it in terms of its mathematical representation.

The most general linear differential equation to describe a linear system has the form

$$a_n \frac{d^n y}{dt^n} + a_{n-1} \frac{d^{n-1} y}{dt^{n-1}} + \cdots + a_1 \frac{dy}{dt} + a_0 y = b_m \frac{d^m x}{dt^m} + b_{m-1} \frac{d^{m-1} x}{dt^{m-1}} + \cdots + b_1 \frac{dx}{dt} + b_0 x \qquad (8.2.1)$$

In Eq. (8.2.1), x (the excitation) and y (the response) are both functions of t (time).† Throughout an interval I, defined by $t_0 \le t \le t_f$, if the a and b coefficients and x are single-valued continuous functions of t or constants and $a_n \neq 0$, then corresponding to each point $t = t_k$ in I and each prescribed set of n values of

$$y, \frac{dy}{dt}, \frac{d^2 y}{dt^2}, \ldots, \frac{d^{n-1} y}{dt^{n-1}} \qquad \text{at } t = t_0$$

there is a unique solution y of the differential equation. This conclusion is both an existence theorem and a uniqueness theorem; existence in the sense that if certain conditions are met, there is a solution; uniqueness in the sense that there is only one solution satisfying the prescribed constraints.‡

Example 8.1 Formulate the differential equation for the simple network in Fig. 8.1 to verify the general form of Eq. (8.2.1) [for example, the reader may wonder why there are no integral terms in Eq. (8.2.1) and why the right-hand member contains derivatives of the excitation].

† The notation x and y, instead of $x(t)$ and $y(t)$, will be used except where confusion about time dependence might arise.

‡ Existence and uniqueness theorems are discussed in most books on differential equations. Two examples are H. Wayland, "Differential Equations Applied in Science and Engineering," D. Van Nostrand Company, Inc., Princeton, N.J., 1957, and L. M. Kells, "Elementary Differential Equations," 6th ed., McGraw-Hill Book Company, New York, 1965. It is pointed out in the latter reference that the conditions we have given for the existence of a unique solution are sufficient but not always necessary.

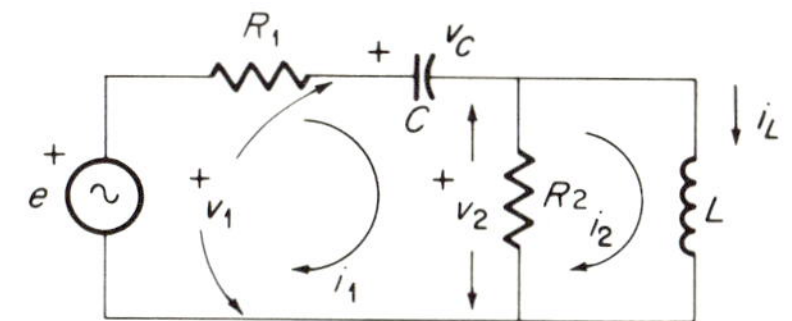

Fig. 8.1

The two loop equations in Fig. 8.1 are

$$e = i_1(R_1 + R_2) + \frac{1}{C}\int_0^t i_1\,dt + v_0 - i_1R_2$$

$$0 = -i_1R_2 + i_2R_2 + L\frac{di_2}{dt}$$

where v_0 is the initial capacitor voltage. We want a single differential equation relating i_1 (a response) to e (the excitation). Solving the first equation for i_2 gives

$$i_2 = -\frac{e}{R_2} + \frac{R_1 + R_2}{R_2}i_1 + \frac{1}{R_2C}\int_0^t i_1\,dt + \frac{v_0}{R_2}$$

Substituting this result into the second equation yields

$$i_1R_1 + \frac{1}{C}\int_0^t i_1\,dt + v_0 + L\frac{R_1 + R_2}{R_2}\frac{di_1}{dt} + \frac{L}{R_2C}i_1 = e + \frac{L}{R_2}\frac{de}{dt}$$

Differentiating once and collecting terms gives the final form

$$LC(R_1 + R_2)\frac{d^2i_1}{dt^2} + (R_1R_2C + L)\frac{di_1}{dt} + R_2i_1 = LC\frac{d^2e}{dt^2} + R_2C\frac{de}{dt}$$

Note that the right-hand member does contain derivatives of e and that there are no integral terms on either side of the equation; i.e., integral terms have been differentiated away.

The statements following Eq. (8.2.1) are mathematically sound but give little insight into the physical behavior of a system. This is largely because, in a system problem, the initial values of n parameters are generally given, rather than the initial values of a single parameter and its first $n - 1$ derivatives. Consider the series *RLC* circuit in Fig. 8.2. It has two voltage excitations and two initial conditions, the latter being the values of the

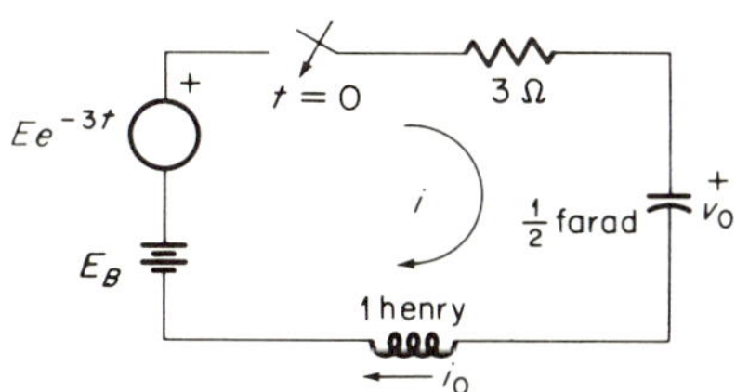

Fig. 8.2

capacitor voltage and inductor current just prior to the closing of the switch at $t = 0$. The response of interest is the current i for $t > 0$.

If we know the initial *states* of the system, i.e., the two initial conditions at $t = 0$, and if we know the total excitation for $t > 0$, then the response for $t > 0$ is completely determined. This statement is completely equivalent to the statement following Eq. (8.2.1) and serves equally well to define the existence and uniqueness of a solution. Thus in Fig. 8.2, knowledge of v_0 and i_0 and of the two sources for $t > 0$ tells us exactly what the response i will be for $t > 0$.

Let us inquire into the makeup of the response. It is easy to show by conventional transient-circuit analysis that if each excitation and initial condition acts separately, the current responses will be

$$i = v_0(-e^{-t} + e^{-2t}) \qquad \text{due to } v_0 \text{ alone} \qquad (8.2.2)$$

$$i = i_0(-e^{-t} + 2e^{-2t}) \qquad \text{due to } i_0 \text{ alone} \qquad (8.2.3)$$

$$i = E_B(e^{-t} - 2e^{-2t}) \qquad \text{due to } E_B \text{ alone} \qquad (8.2.4)$$

$$i = E(-\tfrac{1}{2}e^{-t} + 2e^{-2t} - \tfrac{3}{2}e^{-3t}) \qquad \text{due to } Ee^{-3t} \text{ alone} \qquad (8.2.5)$$

If the initial conditions act simultaneously, with each excitation zero,

$$i = -(v_0 + i_0)e^{-t} + (v_0 + 2i_0)e^{-2t} \qquad (8.2.6)$$

and finally if the two excitations act simultaneously, with each initial condition zero,

$$i = (E_B - \tfrac{1}{2}E)e^{-t} - (E_B - 2E)e^{-2t} - \tfrac{3}{2}Ee^{-3t} \qquad (8.2.7)$$

The response in Eq. (8.2.6) is called the *zero-input response* or *free response*. It is the response produced by the initial conditions in the circuit with no excitations present. The response in Eq. (8.2.7) is termed the *zero-state response* or *forced response*. It is the response resulting from the application of the excitations with the circuit initially at rest, i.e., with its initial states (its initial conditions) zero.

We have noted that the linearity or nonlinearity of a system can be determined from an examination of its nth-order differential equation. An alternative and physically more meaningful way of defining linearity is by ascertaining whether or not the system's zero-input and zero-state responses satisfy the conditions of *superposition* and *homogeneity*. These two characteristics of a linear system may be stated as follows.

Superposition condition If, in a linear system, an excitation or initial state x_1 yields a response y_1 and a second excitation or initial state x_2 yields a response y_2, then an excitation or initial state $x_1 + x_2$ must yield a response $y_1 + y_2$.

Homogeneity condition If, in a linear system, an excitation or initial state x yields a response y, then an excitation or initial state kx must yield a response ky.

If the zero-input response of a system satisfies these two conditions, the system is zero-input linear. If the zero-state response of the system satisfies the two conditions, the system is zero-state linear. A system is linear if and only if it is both zero-input linear and zero-state linear.

Let us examine the circuit in Fig. 8.2 for linearity in terms of these conditions. The zero-input response due to v_0 is given in Eq. (8.2.2), that due to i_0 in Eq. (8.2.3), and that due to v_0 and i_0 in Eq. (8.2.6). The response in Eq. (8.2.6) is seen to be the sum of those in Eqs. (8.2.2) and (8.2.3), and thus the superposition condition is satisfied. Also if v_0 and i_0 are each multiplied by k, the current in Eq. (8.2.6) is multiplied by k. Hence the homogeneity condition is satisfied. Thus the system in Fig. 8.2 is zero-input linear. Completely identical reasoning can be used with Eqs. (8.2.4), (8.2.5), and (8.2.7) to show that the circuit is also zero-state linear. It follows that the circuit in Fig. 8.2 is linear.

Do the principles of superposition and homogeneity hold in the linear time-varying case? What causes them to fail in the nonlinear case?

Example 8.2 Let two voltage excitations e_1 and e_2 be applied separately to the simple network in Fig. 8.3, where R is a function of time. We can write

$$L\frac{di_1}{dt} + R(t)i_1 = e_1$$

$$L\frac{di_2}{dt} + R(t)i_2 = e_2$$

To find the response due to an excitation $e_1 + e_2$, we add these two equations

$$L\frac{di_1}{dt} + L\frac{di_2}{dt} + R(t)i_1 + R(t)i_2 = e_1 + e_2$$

Because the coefficients are independent of i_1 and i_2, we can combine terms to obtain

$$L\frac{d(i_1 + i_2)}{dt} + R(t)(i_1 + i_2) = e_1 + e_2$$

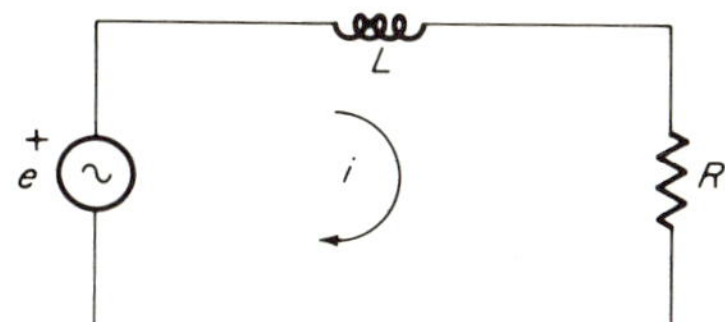

Fig. 8.3

which shows that $e_1 + e_2$ yields $i_1 + i_2$, thus proving that the superposition principle does hold for the linear time-varying case. By a similar argument, using e_1 and ke_1, the homogeneity principle can be shown to hold.

Now let R be a function of i, so that the system is nonlinear. We obtain upon adding the equations

$$L\frac{d(i_1 + i_2)}{dt} + R(i_1)i_1 + R(i_2)i_2 = e_1 + e_2$$

The principle of superposition fails because $R(i_1) \neq R(i_2)$, and thus we cannot combine these two terms as we did previously. The same is true of the homogeneity principle.

A third characteristic of a linear system should be mentioned.

Related-sources condition If, in a linear system with constant coefficients, an excitation x yields a response y, then an excitation dx/dt will yield a response dy/dt and likewise an excitation $\int x\,dt$ will yield a response $\int y\,dt + k$, where k is an integration constant.

Let us raise questions concerning the principle of related sources similar to those we raised concerning the superposition and homogeneity principles: Does it hold in the linear time-varying case, and why does it fail in the nonlinear case?

Example 8.3 Consider again the network in Fig. 8.3 with R a function of time. For an excitation e, we can write

$$L\frac{di}{dt} + R(t)i = e$$

Taking the derivative of both sides yields the correct equation for obtaining the response due to an excitation de/dt

$$L\frac{d(di/dt)}{dt} + R(t)\frac{di}{dt} + i\frac{dR(t)}{dt} = \frac{de}{dt}$$

If we merely replace e by de/dt and i by di/dt in the first equation, we obtain

$$L\frac{d(di/dt)}{dt} + R(t)\frac{di}{dt} = \frac{de}{dt}$$

These two results do not check. The second one is correct only if R is independent of time. The principle of related sources does not hold in the linear time-varying case.

In the nonlinear case R is a function of i, which in turn is a function of t, and the principle again fails.

One interesting linear time-varying situation is the so-called *switched linear mode of operation.* In the system depicted in Fig. 8.4*a* the wiper on the rotary switch travels at a rate that is a function of time (the independent

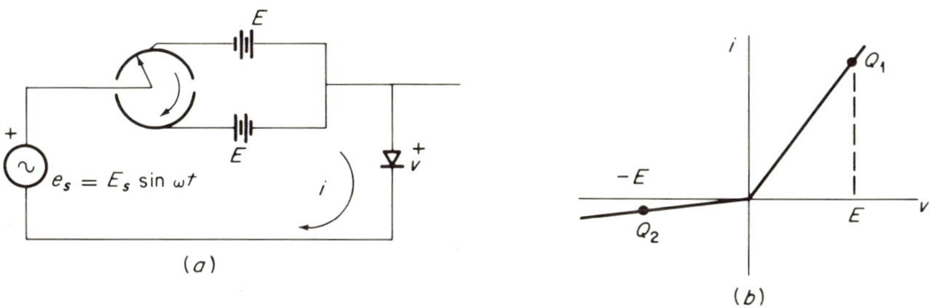

Fig. 8.4 *Switched linear operation.*

variable) but not a function of current (the dependent variable). With the wiper on the upper half of the switch, the operating point is Q_1; on the lower half, Q_2. If $E_s \leq E$, the operation is linear about either operating point. The output voltage $e = Ri$, where R is the diode resistance. But R is a function of the switch position and hence is a function of time. Thus the system is linear time-varying.

To summarize results, we note that the principles of superposition and homogeneity hold for both the linear time-varying and non-time-varying cases but that the principle of related sources holds only for the linear non-time-varying case. From this point on, we shall restrict our attention to the non-time-varying (the constant-coefficient) case, and thus all three principles will be valid.

8.3 FREE RESPONSE FROM *n*th-ORDER EQUATION

The overall purpose of this chapter and Chaps. 9 and 10 is to explore ways of analyzing and utilizing the total solution of the general nth-order linear differential equation with constant coefficients given in Eq. (8.2.1). This total solution has two parts and is, from the superposition principle, their sum. The first part is the solution of the *homogeneous* equation, the equation obtained by setting the right-hand member of Eq. (8.2.1) equal to zero. This solution is the zero-input, or free, response and is generated by the initial states in the system. The second part is the solution for a particular x. This solution is the zero-state, or forced, response, the response generated by applying an excitation to a system initially at rest.†

† In texts on differential equations, the names complementary function and particular integral are given to the two solutions. Here it seems preferable to use the more physically descriptive terms we have chosen.

The homogeneous equation corresponding to Eq. (8.2.1) is given by

$$a_n \frac{d^n y}{dt^n} + a_{n-1} \frac{d^{n-1} y}{dt^{n-1}} + \cdots + a_1 \frac{dy}{dt} + a_0 y = 0 \tag{8.3.1}$$

where the a's are now assumed to be constants. We seek a function y that, when substituted in Eq. (8.3.1), will make the left-hand member vanish. Intuition tells us that an exponential function will work, since its derivatives are the function itself multiplied by constants. Accordingly we choose

$$y_h = ce^{st} \tag{8.3.2}$$

where c and s are complex constants and the h subscript denotes that y_h is the solution of the homogeneous equation. Inserting Eq. (8.3.2) into Eq. (8.3.1) yields

$$(a_n s^n + a_{n-1} s^{n-1} + \cdots + a_1 s + a_0) ce^{st} = 0 \tag{8.3.3}$$

Since e^{st} cannot be zero for all t, we must require

$$a_n s^n + a_{n-1} s^{n-1} + \cdots + a_1 s + a_0 = 0 \tag{8.3.4}$$

Thus the choice of an exponential solution has reduced the problem from the solution of an nth-order differential equation to the solution of an nth-degree algebraic one. Equation (8.3.4) is called the *characteristic* equation of the system. The fundamental law of algebra states that a polynomial of highest degree n will, when equated to zero, have n roots, assuming that any multiple root of order k is counted as k roots. Thus we can rewrite Eq. (8.3.4) in the factored form

$$a_n (s - s_1)(s - s_2) \cdots (s - s_n) = 0 \tag{8.3.5}$$

where we assume, for the moment, no multiple roots. The roots s_1 through s_n are called the *natural frequencies* of the system described by Eq. (8.2.1).

The coefficients in Eqs. (8.2.1) and (8.3.1) are real numbers since they are derived from a physical system. It follows that the natural frequencies in Eq. (8.3.5) must be real, imaginary conjugates, or complex conjugates. For if $s_k = \sigma_k + j\omega_k$ is a complex root, it is necessary to have a second root $s_l = s_k^* = \sigma_k - j\omega_k$ if the product $(s - s_k)(s - s_l)$ is to yield real-number coefficients in Eq. (8.3.4).

After finding the natural frequencies, we can use Eq. (8.3.2) and the principle of superposition to write the free response of the system [the solution of Eq. (8.3.1)] as

$$y_h = c_1 e^{s_1 t} + c_2 e^{s_2 t} + \cdots + c_n e^{s_n t} \tag{8.3.6}$$

Thus each natural frequency of Eq. (8.3.5) provides one solution to the homogeneous equation.

We now show that each of the solutions in Eq. (8.3.6) is *linearly independent*, i.e., that there is no choice of the n constants, other than $c_1 = c_2 = \cdots = c_n = 0$, that will make the right-hand side of Eq. (8.3.6) vanish (the reader may wish to review the discussion of linear independence given in Sec. 2.5 for the case of m algebraic equations in n unknowns). To do this we write, from Eq. (8.3.6), the following set of equations:

$$\begin{aligned} 0 &= c_1 e^{s_1 t} + c_2 e^{s_2 t} + \cdots + c_n e^{s_n t} \\ 0 &= s_1 c_1 e^{s_1 t} + s_2 c_2 e^{s_2 t} + \cdots + s_n c_n e^{s_n} \\ &\cdots\cdots\cdots\cdots\cdots\cdots\cdots\cdots \\ 0 &= s_1^{n-1} c_1 e^{s_1 t} + s_2^{n-1} c_2 e^{s_2 t} + \cdots + s_n^{n-1} c_n e^{s_n t} \end{aligned} \tag{8.3.7}$$

The first of Eqs. (8.3.7) is the equation for *linear dependence.* The other $n - 1$ equations are obtained by repeated differentiation of this first equation. The result is a set of n homogeneous equations in n unknown c's. A nonzero solution for the c's will exist if and only if the coefficient determinant of the c's vanishes. If it does, the solutions in Eq. (8.3.6) are linearly dependent. Thus the requirement for the *linear independence* is

$$\begin{vmatrix} 1 & 1 & \cdots & 1 \\ s_1 & s_2 & \cdots & s_n \\ \cdots & \cdots & \cdots & \cdots \\ s_1^{n-1} & s_2^{n-1} & \cdots & s_n^{n-1} \end{vmatrix} \neq 0 \tag{8.3.8}$$

where use has been made of property 4 of determinants in Sec. 1.11 and the fact that the product $e^{s_1 t} e^{s_2 t} \cdots e^{s_n t}$ cannot vanish for all t. The determinant in Eq. (8.3.8) is known as the *wronskian* of the set of functions in Eq. (8.3.6).

The nonvanishing of the wronskian is the requirement for linear independence. Is this requirement always met for the wronskian in Eq. (8.3.8)? Consider the third-order case, for which we have

$$\begin{aligned} \begin{vmatrix} 1 & 1 & 1 \\ s_1 & s_2 & s_3 \\ s_1^2 & s_2^2 & s_3^2 \end{vmatrix} &= \begin{vmatrix} 1 & 0 & 0 \\ s_1 & s_2 - s_1 & s_3 - s_1 \\ s_1^2 & s_2^2 - s_1^2 & s_3^2 - s_1^2 \end{vmatrix} = \begin{vmatrix} s_2 - s_1 & s_3 - s_1 \\ s_2^2 - s_1^2 & s_3^2 - s_1^2 \end{vmatrix} \\ &= (s_2 - s_1)(s_3 - s_1) \begin{vmatrix} 1 & 1 \\ s_2 + s_1 & s_3 + s_1 \end{vmatrix} \\ &= (s_2 - s_1)(s_3 - s_1)(s_3 - s_2) \neq 0 \end{aligned} \tag{8.3.9}$$

Use has been made of properties 4 and 7 of determinants in this expansion. Assuming no repeated roots, the wronskian does not vanish.† This result

† The multiple-root case will be considered shortly.

can be extended to the nth order case, permitting the generalization that the solutions in Eq. (8.3.6) are linearly independent.

It will be shown in Sec. 8.6 that the values of c_1 through c_n are determined by the system's initial conditions and excitations. In the meantime note that just as the n roots in Eq. (8.3.5) were constrained to be real, imaginary conjugates, or complex conjugates, so also are the n constants in Eq. (8.3.6), but for a different reason, namely, that y_n must be a real function of time. For example, let y_h for a third-order system be given by

$$\begin{aligned} y_h &= c_1e^{s_1t} + c_2e^{s_2t} + c_3^{s_3t} \\ &= c_1e^{\sigma_1t} + c_2e^{(\sigma_2+j\omega_2)t} + c_3e^{(\sigma_2-j\omega_2)t} \end{aligned} \tag{8.3.10}$$

The constant c_1 must be real for the first term to be a real function of time. Replacing $e^{\pm j\omega_2t}$ by $\cos\omega_2t \pm j\sin\omega_2t$ and combining the second and third terms, gives

$$c_2e^{s_2t} + c_3e^{s_3t} = e^{\sigma_2t}[(c_2+c_3)\cos\omega_2t + j(c_2-c_3)\sin\omega_2t] \tag{8.3.11}$$

For these terms to yield a real function of time, $c_2 + c_3$ must be real and $c_2 - c_3$ must be imaginary. This can be true only if

$$c_3 = c_2^* \tag{8.3.12}$$

Then y_h in Eq. (8.3.10) takes the form

$$y_h = c_1e^{\sigma_1t} + 2e^{\sigma_2t}(\text{Re } c_2\cos\omega_2t - \text{Im } c_2\sin\omega_2t) \tag{8.3.13}$$

where Re and Im denote real and imaginary, respectively. The magnitude and angle of c_2 are given by

$$|c_2| = \sqrt{(\text{Re } c_2)^2 + (\text{Im } c_2)^2} \qquad \text{and} \qquad \theta_2 = \tan^{-1}\frac{\text{Im } c_2}{\text{Re } c_2} \tag{8.3.14}$$

Inserting these in Eq. (8.3.13) yields the useful form

$$\begin{aligned} y_h &= c_1e^{\sigma_1t} + 2e^{\sigma_2t}\,|c_2|\left(\frac{\text{Re } c_2}{|c_2|}\cos\omega_2t - \frac{\text{Im } c_2}{|c_2|}\sin\omega_2t\right) \\ &= c_1e^{\sigma_1t} + 2e^{\sigma_2t}\,|c_2|\cos(\omega_2t+\theta_2) \end{aligned} \tag{8.3.15}$$

Assume now that Eq. (8.3.5) contains one multiple root of order 2 (call it s_p) and $n - 2$ simple roots. Rather than assuming Eq. (8.3.2) as the solution, try

$$y_h = c_1e^{st} + c_2te^{st} \tag{8.3.16}$$

The first term in Eq. (8.3.16) is the previous solution. The second term can be justified if we can show that it satisfies Eq. (8.3.1). Inserting this term into that equation yields

$$\begin{aligned} &(a_ns^n + a_{n-1}s^{n-1} + \cdots + a_1s + a_0)c_2te^{st} \\ &\qquad + (na_ns^{n-1} + \cdots + 2a_2s + a_1)c_2e^{st} = 0 \end{aligned} \tag{8.3.17}$$

The first term in parentheses is the left side of Eq. (8.3.4). It contains $(s - s_p)^2$ and thus vanishes at $s = s_p$. The second term in parentheses can be rewritten in the form

$$\frac{d}{ds}(a_n s^n + a_{n-1}s^{n-1} + \cdots + a_1 s + a_0) \tag{8.3.18}$$

Now $(s - s_p)^2$ is a factor of the term in parentheses in Eq. (8.3.18). It follows that $s - s_p$ is a factor of the second term in parentheses in Eq. (8.3.17). Thus the entire left-hand side of Eq. (8.3.17) vanishes at $s = s_p$, and Eq. (8.3.16) *is the solution for that root.* In general if Eq. (8.3.5) contains a multiple root of order p and $n - p$ simple roots, a free response of the form

$$y_h = (c_1 + c_2 t + \cdots + c_p t^{p-1})e^{s_p t} + c_{p+1}e^{s_{p+1}t} + \cdots + c_n e^{s_n t} \tag{8.3.19}$$

will result. Then linear independence can be proved by a development similar to that for the simple-root case.

We know that for the linear systems being discussed to be stable they must have bounded time responses to bounded inputs. This means that all terms in y_h must decay with time or at least reach some (periodically) steady amplitude. Consider this stability constraint in terms of the roots of Eq. (8.3.5). Recall that, in general, these roots may be real, imaginary conjugates, or complex conjugates and that they may be simple or multiple. In Table 8.1, we have listed these various cases and given a stability criterion for each based on not having y_h increase indefinitely with time. From this table, we can compose the following general rules. For stability, all simple roots must have nonpositive real parts and all multiple roots must have negative real parts. In terms of the s plane, all roots must lie in the left half-plane or on the $j\omega$ axis, and those on the $j\omega$ axis must be simple.

TABLE 8.1 Time response for various root types

Roots	*Response*	*Stability criteria*
Simple real	$ce^{\sigma t}$	$\sigma \leq 0$
Simple imaginary conjugates	$c\cos(\omega t + \theta)$	Always stable
Simple complex conjugates	$ce^{\sigma t}\cos(\omega t + \theta)$	$\sigma \leq 0$
Multiple real (second order)	$(c_1 + c_2 t)e^{\sigma t}$	$\sigma < 0$
Multiple imaginary conjugates (second order)	$(c_1 + c_2 t)\cos(\omega t + \theta)$	Always unstable
Multiple complex conjugates (second order)	$(c_1 + c_2 t)e^{\sigma t}\cos(\omega t + \theta)$	$\sigma < 0$

Example 8.4 Find the solution of the following homogeneous equation and check it for stability.

$$\frac{d^5y}{dt^5} + 6\frac{d^4y}{dt^4} + 17\frac{d^3y}{dt^3} + 28\frac{d^2y}{dt^2} + 24\frac{dy}{dt} + 8y = 0$$

The characteristic equation is

$$s^5 + 6s^4 + 17s^3 + 28s^2 + 24s + 8 = 0$$

which factors into

$$(s + 1)^2(s + 2)(s + 1 + j\sqrt{3})(s + 1 - j\sqrt{3}) = 0$$

Each of the roots satisfies the stability criteria. The homogeneous solution is

$$y_h = (c_1 + c_2t)e^{-t} + c_3e^{-2t} + 2\,|c_4|\,e^{-t}\cos(\sqrt{3}t + \theta_4)$$

where c_1, c_2, and c_3 are real and c_4 is complex.

8.4 FREE RESPONSE FROM n FIRST-ORDER EQUATIONS

In the analysis of a physical network or system, the starting point is *not*, in general, an nth-order differential equation. Rather we begin by writing several loop and/or node differential equations of lesser order and then combine them to obtain a single nth-order equation. This procedure was illustrated in Example 8.1, where the first step was the writing of two first-order differential equations around the loops of the network, from which a single second-order equation was derived. It would be convenient if we could obtain the free response of a system, and ultimately its total response, from these lesser-order equations rather than always having to derive the nth-order equation.

It is the purpose of this section to develop a procedure for obtaining the natural frequencies and free response of a linear system from a set of n first-order differential equations. The starting point is the nth-order homogeneous differential equation (8.3.1). In that equation, let

$$\begin{aligned} y_1 &= y \\ y_2 &= dy/dt \\ y_3 &= d^2y/dt^2 \\ &\cdots\cdots\cdots \\ y_n &= d^{n-1}y/dt^{n-1} \end{aligned} \tag{8.4.1}$$

From Eqs. (8.3.1) and (8.4.1), we obtain

$$\begin{aligned}
\dot{y}_1 &= y_2 \\
\dot{y}_2 &= y_3 \\
&\cdots\cdots \\
\dot{y}_{n-1} &= y_n \\
\dot{y}_n &= -\frac{a_0}{a_n}y_1 - \frac{a_1}{a_n}y_2 - \cdots - \frac{a_{n-1}}{a_n}y_n
\end{aligned} \tag{8.4.2}$$

where the dots over the y's denote differentiation with respect to time. Equations (8.4.2) can be written in matrix form as

$$\begin{bmatrix} \dot{y}_1 \\ \dot{y}_2 \\ \cdot \\ \cdot \\ \cdot \\ \dot{y}_{n-1} \\ \dot{y}_n \end{bmatrix} = \begin{bmatrix} 0 & 1 & 0 & \cdots & 0 \\ 0 & 0 & 1 & \cdots & 0 \\ \cdots & \cdots & \cdots & \cdots & \cdots \\ 0 & 0 & 0 & \cdots & 1 \\ -\frac{a_0}{a_n} & -\frac{a_1}{a_n} & -\frac{a_2}{a_n} & \cdots & -\frac{a_{n-1}}{a_n} \end{bmatrix} \begin{bmatrix} y_1 \\ y_2 \\ \cdot \\ \cdot \\ \cdot \\ y_{n-1} \\ y_n \end{bmatrix} \tag{8.4.3}$$

which can be abbreviated to

$$\dot{Y} = AY \tag{8.4.4}$$

where $\dot{Y}$ and Y are $n \times 1$ column matrices and A is an $n \times n$ matrix of coefficients.

Equation (8.4.4) is the general matrix form for a set of n homogeneous first-order differential equations describing an nth-order linear system. These equations are called a set of homogeneous *state equations* of the system, and the variables contained therein (the y's) are called a set of *state variables.*

The concept of state was introduced in Sec. 8.2 in the discussion of the free and forced responses of a linear system. The state of a system may be considered as the least amount of information that must be known about the system to determine its subsequent performance completely. For example, in Fig. 8.2, if we know the two initial conditions at $t = 0$ and the excitations for $t > 0$, the total response for $t > 0$ is determined. In general, if we know a set of n independent initial conditions for a system and we know its excitation, a unique response is determined.

What quantities are suitable as state variables for an nth-order system? Certainly, from our discussion in Secs. 8.2 and 8.3, a single variable and the first $n - 1$ of its derivatives are an acceptable set. Also acceptable is an independent set of capacitor voltages and inductor currents, since they completely describe the energy stored in the system at any given time.

Are sets of node-voltage and loop-current variables also acceptable? The answer is "yes" if the proposed set of voltage or current variables can be written as a nonsingular linear transformation of the independent set of capacitor voltages and inductor currents. Thus, if V and I are $n \times 1$ column matrices of proposed sets of node-voltage and loop-current variables, respectively, and Y represents the $n \times 1$ column matrix of independent capacitor voltages and inductor currents, then for V and I to be acceptable sets of state variables we must be able to write

$$Y = PV \qquad \text{and} \qquad Y = QI \tag{8.4.5}$$

where P and Q are $n \times n$ square matrices whose determinants do not vanish.

The use of state variables, especially capacitor-voltage and inductor-current state variables, in the analysis of linear systems has become extremely popular in the past few years. We shall devote Chap. 11 entirely to a discussion of state, writing state equations, and obtaining the total response of a system from these equations by time- and frequency-domain methods. We conclude this section by deriving the free response of a linear system from Eqs. (8.4.4).

As in Sec. 8.3, we assume a solution of the form

$$Y = Ce^{st} \tag{8.4.6}$$

where C is an $n \times 1$ column matrix of (complex) constants. Differentiating Eq. (8.4.6) yields

$$\dot{Y} = sCe^{st} = sY \tag{8.4.7}$$

Then, from Eqs. (8.4.4) and (8.4.7), we have

$$sY = AY$$

or

$$(A - sU)Y = 0 \tag{8.4.8}$$

where U is an $n \times n$ identity matrix. Equation (8.4.8) is a set of n homogeneous equations in n unknowns. For a nonzero solution for Y to exist, the determinant of the Y coefficient matrix must vanish. Forming this determinant, we obtain

$$|A - sU| = \begin{vmatrix} -s & 1 & 0 & \cdots & 0 \\ 0 & -s & 1 & \cdots & 0 \\ 0 & 0 & -s & \cdots & 0 \\ \cdots & \cdots & \cdots & \cdots & \cdots \\ -\dfrac{a_0}{a_n} & -\dfrac{a_1}{a_n} & -\dfrac{a_2}{a_n} & \cdots & -\dfrac{a_{n-1}}{a_n} - s \end{vmatrix} = 0 \tag{8.4.9}$$

which, upon expansion, yields

$$(-1)^n\left(s^n + \frac{a_{n-1}}{a_n}s^{n-1} + \cdots + \frac{a_1}{a_n}s + \frac{a_0}{a_n}\right) = 0 \tag{8.4.10}$$

Comparing Eq. (8.4.10) with Eq. (8.3.4), we see that the former is the characteristic equation of the system, and thus our present development *does* provide an alternative procedure for obtaining the system's natural frequencies.

When a system is analyzed in the way we have just shown, we call each root of Eq. (8.4.10) an *eigenvalue* of the matrix A (from the German word meaning characteristic). One solution of Eq. (8.4.8) is generated by each eigenvalue. Each of these solutions is called an *eigenvector* of A and has the form

$$Y_i = C_i e^{s_i t} \tag{8.4.11}$$

where Y_i and C_i are $n \times 1$ column matrices and s_i is the ith eigenvalue of A. The total solution of the n homogeneous equations is then the summation of n solutions of the form of Eq. (8.4.11)

$$Y = \sum_1^n C_i e^{s_i t} = \begin{bmatrix} c_{11} & c_{12} & \cdots & c_{1n} \\ c_{21} & c_{22} & \cdots & c_{2n} \\ \cdots & \cdots & \cdots & \cdots \\ c_{n1} & c_{n2} & \cdots & c_{nn} \end{bmatrix} \begin{bmatrix} e^{s_1 t} \\ e^{s_2 t} \\ \cdot \\ \cdot \\ \cdot \\ e^{s_n t} \end{bmatrix} \tag{8.4.12}$$

There are not n^2 independent constants in Eq. (8.4.12). The components of each eigenvector (entries in each column of C) are not independent. Since from Eq. (8.4.2) $y_i = dy_{i-1}/dt$, then from Eq. (8.4.12) $c_{ij} = s_j c_{i-1,j}$. This relation permits Eq. (8.4.12) to be written in the form

$$Y = \begin{bmatrix} c_1 & c_2 & \cdots & c_n \\ s_1 c_1 & s_2 c_2 & \cdots & s_n c_n \\ \cdots & \cdots & \cdots & \cdots \\ s_1^{n-1} c_1 & s_2^{n-1} c_2 & \cdots & s_n^{n-1} c_n \end{bmatrix} \begin{bmatrix} e^{s_1 t} \\ e^{s_2 t} \\ \cdot \\ \cdot \\ \cdot \\ e^{s_n t} \end{bmatrix} \tag{8.4.13}$$

where we have replaced c_{ij} by c_j in each column of C. Now it is clear that there are n independent constants.

Example 8.5 Find the eigenvalues and eigenvectors of the following linear homogeneous differential equation:

$$\frac{d^3y}{dt^3} + 5\frac{d^2y}{dt^2} + 12\frac{dy}{dt} + 8y = 0$$

Let

$$y_1 = y \qquad y_2 = \frac{dy}{dt} \qquad y_3 = \frac{d^2y}{dt^2}$$

from which we can write, using Eqs. (8.4.4) and (8.4.7),

$$\begin{bmatrix} \dot{y}_1 \\ \dot{y}_2 \\ \dot{y}_3 \end{bmatrix} = \begin{bmatrix} 0 & 1 & 0 \\ 0 & 0 & 1 \\ -8 & -12 & -5 \end{bmatrix} \begin{bmatrix} y_1 \\ y_2 \\ y_3 \end{bmatrix} = s \begin{bmatrix} y_1 \\ y_2 \\ y_3 \end{bmatrix}$$

For a nonzero solution,

$$\begin{vmatrix} -s & 1 & 0 \\ 0 & -s & 1 \\ -8 & -12 & -5-s \end{vmatrix} = 0$$

which yields

$$-(s^3 + 5s^2 + 12s + 8) = 0$$

or, in factored form,

$$-(s+1)(s+2-j2)(s+2+j2) = 0$$

Thus the eigenvalues are

$$s_1 = -1 \qquad s_2 = -2 + j2 \qquad s_3 = -2 - j2$$

and these yield the solutions, according to Eq. (8.4.13),

$$\begin{bmatrix} y_1 \\ y_2 \\ y_3 \end{bmatrix} = \begin{bmatrix} c_1 & c_2 & c_2^* \\ -c_1 & (-2+j2)c_2 & (-2-j2)c_2^* \\ c_1 & -j8c_2 & j8c_2^* \end{bmatrix} \begin{bmatrix} e^{-t} \\ e^{(-2+j2)t} \\ e^{(-2-j2)t} \end{bmatrix}$$

Note that in column 3 of C we have used the relation $c_3 = c_2^*$ and that in column 1, c_1 must be real since s_1 is real.

If we had not previously derived Eq. (8.4.13), we could still find the eigenvectors quite easily once the eigenvalues were known. For example, for the eigenvalue s_2, Eq. (8.4.8) gives

$$\begin{bmatrix} 0 & 1 & 0 \\ 0 & 0 & 1 \\ -9 & -12 & -5 \end{bmatrix} \begin{bmatrix} y_1 \\ y_2 \\ y_3 \end{bmatrix} = (-2+j2) \begin{bmatrix} y_1 \\ y_2 \\ y_3 \end{bmatrix}$$

which yields

$$\begin{bmatrix} 2-j2 & 0 & 0 \\ 0 & 2-j2 & 1 \\ -8 & -12 & -3-j2 \end{bmatrix} \begin{bmatrix} y_1 \\ y_2 \\ y_3 \end{bmatrix} = \begin{bmatrix} 0 \\ 0 \\ 0 \end{bmatrix}$$

The rank of this set of equations is 2. Thus the value of one variable may be arbitrarily chosen and the others expressed in terms of it. Choosing $y_1 = c_2e^{s_2t}$, for example, gives

$$y_2 = (-2 + j2)y_1 = (-2 + j2)c_2e^{s_2t}$$

and

$$y_3 = (-2 + j2)y_2 = -j8c_2e^{s_2t}$$

Thus the eigenvector is

$$\begin{bmatrix} c_2 \\ (-2 + j2)c_2 \\ -j8c_2 \end{bmatrix}$$

which checks the previous result using Eq. (8.4.13).

8.5 PROPERTIES OF NETWORK FUNCTIONS

In this section several s-domain properties of network functions are derived from an analysis of a network's natural frequencies. These properties may be considered as a set of necessary realizability conditions that any ratio of polynomials in s must satisfy if it is to be the driving-point or transfer function of a linear stable network.

Consider a linear stable network $\mathcal{N}$ excited at a selected terminal pair 1-1′ by an ideal current source i_s, as shown in Fig. 8.5*a*. The voltage v_1 across the terminals is given by i_sZ_1, where Z_1 is the driving-point impedance. Let i_s become zero (the terminals become open). Then v_1 becomes zero unless Z_1 is infinite; i.e., for a voltage to exist at the open terminals with no excitation, it is necessary that the driving-point impedance there be infinite. Thus the poles of Z_1 (zeros of its denominator) are the natural frequencies of $\mathcal{N}$ with terminals 1-1′ open.

Consider next the same network excited by an ideal voltage source e_s, as shown in Fig. 8.5*b*. The current i_1 is e_s/Z_1. Now let e_s become zero (the terminals become shorted). Then i_1 becomes zero unless Z_1 is zero; i.e., for a current to exist at the shorted terminals with no excitation present, it is necessary that the driving-point impedance there be zero. We conclude that

Fig. 8.5 Open- and short-circuit natural frequencies.

the zeros of Z_1 (zeros of its numerator) are the natural frequencies of $\mathscr{N}$ with terminals 1-1′ shorted.

From the discussion of the free response of linear systems in the previous sections of this chapter, we know that with i_s zero in Fig. 8.5*a*, the voltage v_1 is given by

$$v_1 = c_1 e^{s_1 t} + c_2 e^{s_2 t} + \cdots \tag{8.5.1}$$

where $s_1, s_2, \ldots$ are the natural frequencies (roots of the characteristic equation) of $\mathscr{N}$ with terminals 1-1′ open. Likewise we know that with e_s zero in Fig. 8.5*b* the current i_1 has the form

$$i_1 = c_a e^{s_a t} + c_b e^{s_b t} + \cdots \tag{8.5.2}$$

where $s_a, s_b, \ldots$ are the natural frequencies (roots of the characteristic equation) of $\mathscr{N}$ with terminals 1-1′ shorted. Since the network is assumed stable, these natural frequencies must lie in the left half of the s plane or on the $j\omega$ axis (have nonpositive real parts). If roots lie on the $j\omega$ axis, they must be simple. Our conclusion is that *the zeros and poles of* Z_1 *must lie in the left half of the s plane or on the* $j\omega$ *axis, with the latter required to be simple.* We also conclude that *the highest powers of s in the numerator and denominator of* Z_1 *can differ by no more than* 1, *since a pole or zero at infinity is on the* $j\omega$ *axis and thus must be simple. The same is true of the lowest powers of s*, since a pole *or zero* at the origin is also on the $j\omega$ axis.

To sum up, a linear network $\mathscr{N}$ with one terminal pair has two natural modes of operation, one with the terminals open and the other with them shorted. For the network to be stable in both modes, i.e., open-circuit stable and short-circuit stable, it is necessary and sufficient that the poles and zeros of its driving-point impedance have nonpositive real parts and that the $j\omega$ axis poles and zeros be simple.

The reader should not confuse stability with passivity. A passive network is stable, but the reverse is not necessarily true. Passivity of a linear 2-terminal network requires that, for any excitation, the total energy delivered to the network be nonnegative. It is shown in texts on network synthesis† that this, in turn, requires that the network's driving-point impedance be positive real, i.e., that:

1. $Z(s)$ is real for s real.
2. $\operatorname{Re} Z(s) \geq 0$ for $\operatorname{Re} s = 0$.
3. $\operatorname{Re} Z(s) > 0$ for $\operatorname{Re} s > 0$.

† See, for example, N. Balabanian, "Network Synthesis," Prentice-Hall, Inc., Englewood Cliffs, N.J., 1958, or D. Hazony, "Elements of Network Synthesis," Reinhold Publishing Corporation, 1963.

Further it can be shown that these three conditions can be replaced by the two equivalent ones:

1. $Z(s)$ is open- and short-circuit stable.
2. $\operatorname{Re} Z(j\omega) \geq 0$.

Condition 1 is the condition for stability. Condition 2 is the additional condition that must be met for passivity.

Two examples will illustrate the concepts of open- and short-circuit natural frequencies.

Example 8.6 Find the driving-point impedance Z_1 at terminals 1-1′ in Fig. 8.6 from a consideration of the open- and short-circuit natural frequencies of the network.

The poles of Z_1 are the natural frequencies of the network with terminals 1-1′ open. The latter are the roots of

$$sL + R_1 + R_2 + \frac{1}{sC} = 0 \qquad \text{or} \qquad LCs^2 + (R_1 + R_2)Cs + 1 = 0$$

because a current must flow around the network loop to obtain a voltage at terminals 1-1′ with no excitation. The zeros of Z_1 are the natural frequencies of the network with terminals 1-1′ shorted. The latter are given by

$$Ls + R_1 = 0 \qquad \text{and} \qquad R_2 + \frac{1}{sC} = 0$$

because there must be a zero-impedance path through the network for a current to exist at the terminals 1-1′ with no excitation. We can compose Z_1, within a constant, from a knowledge of these poles and zeros. Thus

$$Z_1 = \frac{K(Ls + R_1)(CR_2s + 1)}{LCs^2 + (R_1 + R_2)Cs + 1}$$

Note that all roots lie in the left half of the s plane and that the highest powers of s in the numerator and denominator differ by no more than 1, and likewise for the lowest powers.

K is determined by finding the value of Z_1 at one frequency. For example, at $s = 0$, $Z_1 = R_1$ in the network in Fig. 8.6. Thus K must be 1 in the equation for Z_1.

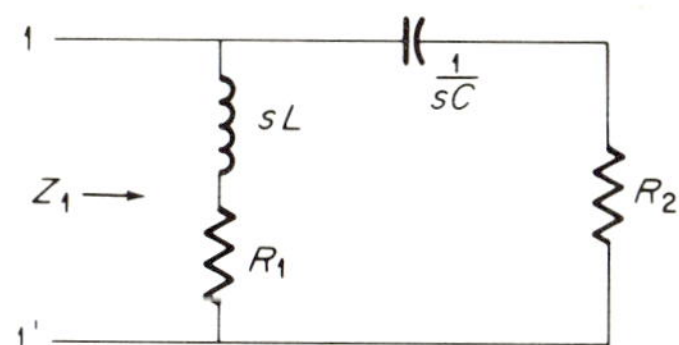

Fig. 8.6

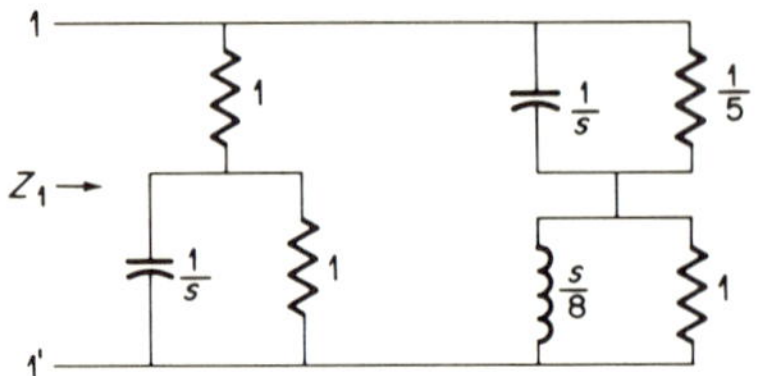

Fig. 8.7

Example 8.7 Repeat Example 8.6 for the network in Fig. 8.7.

The zeros of Z_1 are given by

$$1 + \frac{1}{s+1} = 0 \qquad \text{and} \qquad \frac{1}{s+5} + \frac{s}{s+8} = 0$$

For values of s satisfying either of these equations, a current can exist in the short between terminals 1 and 1′ with no excitation present.

The poles of Z_1 are given by

$$1 + \frac{1}{s+1} + \frac{1}{s+5} + \frac{s}{s+8} = 0$$

Values of s satisfying this equation permit a voltage to exist across the opened terminals 1 and 1′ in the absence of an excitation.

The first two equations yield

$$s + 2 = 0 \qquad \text{and} \qquad s^2 + 6s + 8 = 0$$

The third equation gives

$$2s^3 + 22s^2 + 80s + 88 = 0$$

Thus Z_1 is given by

$$Z_1 = \frac{K(s+2)(s^2+6s+8)}{2s^3 + 22s^2 + 80s + 88}$$

At $s = \infty$, $Z_1 = ½$ in Fig. 8.7. Thus, from the equation for Z_1, $K = 1$. Or at $s = 0$, $Z_1 = 1/(5 + ½) = 2/11$ in Fig. 8.7. This again yields $K = 1$.

We find that Z_1 contains the common factor $s + 2$ in its numerator and denominator. Eliminating this factor gives

$$Z_1 = \frac{s^2 + 6s + 8}{2s^2 + 18s + 44}$$

Again, no roots lie in the right half of the s plane, the highest powers of s differ by no more than 1, and the lowest powers of s differ by no more than 1.

The common $s + 2$ polynomial in Example 8.7 is known as a *surplus factor*. Such factors, which are often quite difficult to detect in system analysis, are very often used in system synthesis procedures.† When surplus

† See, for example, Hazony, *op. cit.*

factors are present, not all the natural frequencies of the network are poles and zeros of Z_1. Any natural frequency with terminals 1-1′ opened that is also a natural frequency with terminals 1-1′ shorted will cancel in the final expression for Z_1.

In Chap. 6, several formulas were developed for finding driving-point functions at pliers and solder entries topologically. These results can be used in the present discussion. Consider again Fig. 8.5 and assume that the entry at terminals 1-1′ is of the pliers variety. Then Z_1 is given topologically by Eq. (6.7.3) in the form

$$Z_1{}^p = \frac{\sum LZP \text{ of } \mathcal{N}}{\sum LZP \text{ of } \mathcal{N}_{o1}} \tag{8.5.3}$$

where $\mathcal{N}_{o1}$ is the network $\mathcal{N}$ with terminals 1-1′ open.

Thus the natural frequencies of $\mathcal{N}$ are the roots of $\sum LZP$ of $\mathcal{N} = 0$ and the natural frequencies of $\mathcal{N}_{o1}$ are the roots of $\sum LZP$ of $\mathcal{N}_{o1} = 0$.

If terminals 1 and 1′ in Fig. 8.5 are a solder entry, Eq. (6.7.7) applies and

$$Z_1{}^s = \frac{\sum LZP \text{ of } \mathcal{N}_{s1}}{\sum LZP \text{ of } \mathcal{N}} \tag{8.5.4}$$

where $\mathcal{N}_{s1}$ is the network with terminals 1-1′ shorted and the natural frequencies of $\mathcal{N}_{s1}$ are the roots of $\sum LZP$ of $\mathcal{N}_{s1} = 0$.

Example 8.8 For the network in Fig. 8.8, find topologically the natural frequencies with terminals 1-1′ opened and with them shorted.

The entry is of the solder variety, and so Eq. (8.5.4) applies. The graphs of $\mathcal{N}$ and $\mathcal{N}_{s1}$ are shown in Fig. 8.8*b* and *c*, respectively. We tabulate the following:

Link sets of $\mathcal{N}$ 1235, 1236, 1245, 1246, 1345, 1346, 1356, 1456, 2345, 2346, 2356, 2456
Link sets of $\mathcal{N}_{s1}$ 12345, 12346, 12356, 12456

LZP's of $\mathcal{N}$ $\frac{1}{s}, 1, \frac{1}{s^3}, \frac{1}{s^2}, \frac{1}{2s^2}, \frac{1}{2s}, \frac{1}{2s}, \frac{1}{2s^3}, \frac{2}{s}, 2, 2, \frac{2}{s^2}$

LZP's of $\mathcal{N}_{s1}$ $\frac{1}{s^2}, \frac{1}{s}, \frac{1}{s}, \frac{1}{s^3}$

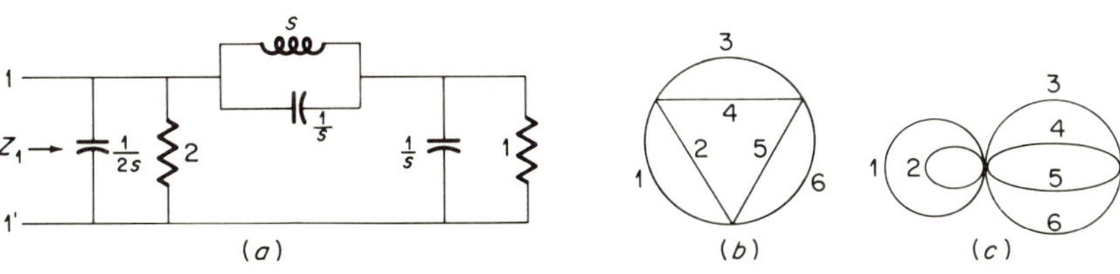

Fig. 8.8

$$\sum LZP\ of\ \mathcal{N} \quad \frac{5s^3 + 4s^2 + 7/2 s + 3/2}{s^3}$$

$$\sum LZP\ of\ \mathcal{N}_{s1} \quad \frac{2s^2 + s + 1}{s^3}$$

The roots of these two equations give

$$\text{Natural frequencies of } \mathcal{N} = -0.537, -0.131 \pm j0.736$$

$$\text{Natural frequencies of } \mathcal{N}_{s1} = \frac{-1 \pm j\sqrt{7}}{2}$$

The driving-point impedance is

$$Z_1 = \frac{2s^2 + s + 1}{5s^3 + 4s^2 + 7/2 s + 3/2}$$

To extend the discussion of the poles and zeros of driving-point functions a bit further, suppose in a network $\mathcal{N}$ we create two solder entries. Call the terminals so created 1-1′ and 2-2′ and the corresponding driving-point impedances Z_1 and Z_2. The poles of Z_1 are obtained with both sets of terminals open and thus are the natural frequencies of $\mathcal{N}$. By the same reasoning, the poles of Z_2 are also the natural frequencies of $\mathcal{N}$. The zeros of Z_1 are the natural frequencies of $\mathcal{N}_{s1}$, the network with terminals 1-1′ shorted. The zeros of Z_2 are the natural frequencies of $\mathcal{N}_{s2}$, the network with terminals 2-2′ shorted, and in general differ from those of Z_1. We conclude that *the poles of driving-point impedances at solder entries are the same, but their zeros, in general, differ.*

Suppose next that the two entries are of the pliers type. The zeros of Z_1 are obtained with both sets of terminals shorted and are the natural frequencies of $\mathcal{N}$. The same is true of the zeros of Z_2. The poles of Z_1 are the natural frequencies of $\mathcal{N}_{o1}$, those of Z_2 are the natural frequencies of $\mathcal{N}_{o2}$. These, in general, differ. We conclude that *the zeros of driving-point impedances at pliers entries are the same but their poles, in general, differ.* We also conclude that *the poles of driving-point impedances at solder entries are the same as the zeros of driving-point impedances at pliers entries and are the natural frequencies of $\mathcal{N}$. The zeros at solder entries and the poles at pliers entries, in general, differ and are not the natural frequencies of $\mathcal{N}$.*

Example 8.9 In the network of Fig. 8.9, terminals 1-1′ and 2-2′ are solder entries (terminals normally open), but 3-3′ are a pliers entry (terminals normally shorted). Find the driving-point impedance at each set of terminals with the other sets of terminals in their normal condition.

The impedances desired are Z_1 with 2-2′ open and 3-3′ shorted, Z_2 with 1-1′ open and 3-3′ shorted, and Z_3 with 1-1′ and 2-2′ open. The zeros of Z_1 are the natural frequencies of $\mathcal{N}_{s1}$, the network with 1-1′ shorted. These are given by $sRC + 1 = 0$ and $sL = 0$.

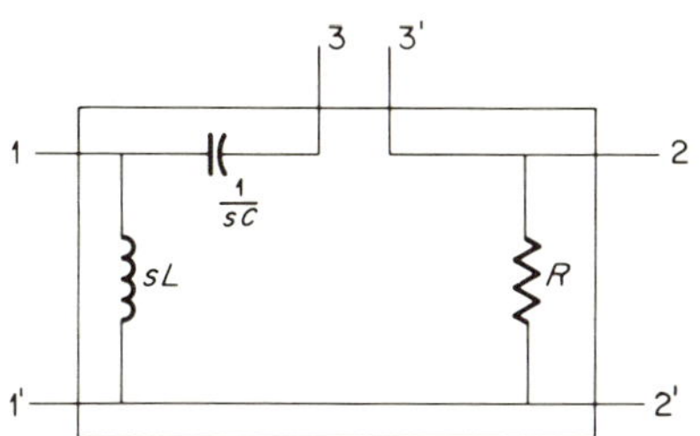

Fig. 8.9

The poles of Z_1 are the natural frequencies of $\mathcal{N}$. These are obtained from $s^2LC + sRC + 1 = 0$. Thus

$$Z_1 = \frac{sL(sRC + 1)}{s^2LC + sRC + 1}$$

where the scaling constant is 1.

Z_2 is a driving-point impedance at a solder entry. Thus its poles are the same as those of Z_1. Its zeros are the natural frequencies of $\mathcal{N}_{s2}$, and are given by $s^2LC + 1 = 0$. Thus

$$Z_2 = \frac{R(s^2LC + 1)}{s^2LC + sRC + 1}$$

where the scaling constant is R.

Z_3 is the driving-point impedance at a pliers entry. Thus its zeros are the poles of Z_1 and Z_2. Its poles are the natural frequencies of $\mathcal{N}_{03}$. These are given by $sC = 0$ and $sL = \infty$. Then

$$Z_3 = \frac{s^2LC + sRC + 1}{sC}$$

We conclude this section with a brief discussion of the zeros and poles of the z parameters of a 2-port, thus bringing transfer impedances into the picture.

Consider the linear passive 2-port network $\mathcal{N}$ in Fig. 8.10. Its z-parameter equations are

$$e_1 = z_{11}i_1 + z_{12}i_2$$

$$e_2 = z_{12}i_1 + z_{22}i_2 \tag{8.5.5}$$

The driving-point impedances z_{11} and z_{22} are obtained with the opposite terminal pair open. From our previous discussion, their poles are the natural frequencies of $\mathcal{N}$ with 1-1′ and 2-2′ open and are identical. The

Fig. 8.10 Natural frequencies of a 2-port.

parameter z_{12} is e_2/i_1 with $i_2 = 0$. To measure it, we apply a current i_1 at 1-1′ and obtain the voltage e_2 at 2-2′ with the latter terminals open. As i_1 tends to zero, e_2 tends to zero unless z_{12} becomes infinite. *Thus the poles of z_{12} are the natural frequencies of $\mathcal{N}$ with 1-1′ and 2-2′ open and are identical with the poles of z_{11} and z_{22}. Since the network $\mathcal{N}$ is passive, these poles must lie in the left half of the s plane or on the $j\omega$ axis and be simple.*

The zeros of z_{11} are the natural frequencies of $\mathcal{N}_{s1}$; the zeros of z_{22} are the natural frequencies of $\mathcal{N}_{s2}$. *These zeros, in general, differ but are the natural frequencies of linear passive networks and therefore must lie in the left half of the s plane or on the $j\omega$ axis and be simple.* The zeros of z_{12} are a different matter. If a voltage e_2 is applied at terminals 2-2′ and allowed to approach zero in order to obtain the zeros of z_{12}, the current i_2 is no longer zero and the definition of z_{12} is violated. *Thus the zeros of z_{12} are in no way related to the natural frequencies of $\mathcal{N}$ and can lie anywhere in the s plane.*

To illustrate these conclusions, note that with terminals 3-3′ shorted in Fig. 8.9, the network becomes a 2-port. Then the Z_1 and Z_2 of Example 8.9 are the z_{11} and z_{22} parameters of the 2-port. Note that their poles are the same but their zeros differ. The transfer impedance is given by $z_{12} = s^2LCR/(s^2LC + sRC + 1)$. Its poles are those of z_{11} and z_{22}. There is a second-degree zero at $s = 0$ that is unrelated to the natural frequencies of the network.

8.6 FORCED RESPONSE TO e^{s_0t} EXCITATION

Although our major concern in this chapter is the free response of a linear system, we find it desirable at this point to obtain the forced response of such a system to one general type of excitation. This will enable us to answer some of the questions raised in previous sections about initial conditions and the n arbitrary constants in the free response.

The excitation of a system can take many forms. It can be discrete or continuous, specified or random, periodic or nonperiodic, analytical or graphical. Here we consider one continuous specified nonperiodic analytical type, the general exponential excitation of the form

$$x = x_0e^{s_0t} \tag{8.6.1}$$

where x_0 and s_0 are constants that may be real or complex and x is presumed not to exist prior to $t = 0$.

The excitation in Eq. (8.6.1) is much more general than it may seem. If s_0 is negative real (then x_0 must be real), it represents a decaying exponential. If s_0 is imaginary, and we include a second excitation of the form $x_0^*e^{s_0^*t}$, then, by superposition, we have a sinusoidal excitation [see Eqs. (8.3.11) to (8.3.15)]. If s_0 is complex, the product of a sinusoidal variation and a decaying exponential results. If $s_0 = 0$ (x_0 again real), the excitation is a step function. Then, through the principle of related sources, the responses due to ramp, parabolic, and impulse-function excitations can be obtained directly from the step response.

Let the excitation of Eq. (8.6.1) be applied to the system described by Eq. (8.2.1), assumed initially at rest. We already know that a portion of the total solution, the free response, is given by Eq. (8.3.6). Our task is to find the forced response. Following our previous inclination, let us assume a solution of the form

$$y_p = y_0 e^{s_0 t} \tag{8.6.2}$$

where y_0 is a real or complex constant and it is assumed that s_0 is not a natural frequency of the system. The p subscript denotes the solution for the particular excitation. Inserting Eqs. (8.6.1) and (8.6.2) into Eq. (8.2.1) yields

$$y_0 e^{s_0 t}(a_n s_0{}^n + \cdots + a_0) = x_0 e^{s_0 t}(b_m s_0{}^m + \cdots + b_0) \tag{8.6.3}$$

Define a quantity $T(s_0)$, called the system *transfer function*, by

$$T(s_0) = \frac{y_p}{x} = \frac{b_m s_0{}^m + \cdots + b_0}{a_n s_0{}^n + \cdots + a_0} \tag{8.6.4}$$

Then y_0 can be expressed as

$$y_0 = \frac{b_m s_0{}^m + \cdots + b_0}{a_n s_0{}^n + \cdots + a_0} x_0 = T(s_0) x_0 \tag{8.6.5}$$

and inserting this result into Eq. (8.6.2) yields the forced response as

$$y_p = T(s_0) x_0 e^{s_0 t} \tag{8.6.6}$$

The transfer function depends on s_0 but not on t. Its poles are the natural frequencies of the system and, for the system to be stable, must conform to the stability criteria of Table 8.1. Its zeros may, in general, lie anywhere in the s plane.

$T(s_0)$ completely describes the forced response of an initially relaxed linear system to a single exponential forcing function whose frequency is not a natural frequency of the system. Does it also completely describe the free response of the system? What if the system is not initially relaxed? Suppose the forcing-function frequency is also a natural frequency of the system? What if the system has more than one exponential excitation? These questions are best answered by an example, but first we digress to discuss the role of initial conditions.

8.7 INITIAL CONDITIONS: I

Assume a relaxed series *RLC* circuit to which is applied a voltage excitation at $t = 0$ given by

$$e = e_0 e^{s_0 t} \tag{8.7.1}$$

The system equation is

$$L\frac{d^2i}{dt^2} + R\frac{di}{dt} + \frac{1}{C}i = \frac{de}{dt} \tag{8.7.2}$$

The total solution for $t > 0$ is the sum of the free and forced responses [(Eqs. (8.3.6) and (8.6.6)] and has the form

$$i = c_1 e^{s_1 t} + c_2 e^{s_2 t} + T(s_0)e_0 e^{s_0 t} \tag{8.7.3}$$

where c_1 and c_2 are arbitrary constants.

To find c_1 and c_2 requires two equations. The first is obtained by evaluating Eq. (8.7.3) just after $t = 0$. This gives

$$i_+ = c_1 + c_2 + T(s_0)e_0 \tag{8.7.4}$$

where the subscript plus denotes evaluation at $t = 0_+$, just after $t = 0$. The second equation is obtained by differentiating Eq. (8.7.3) to obtain

$$\frac{di}{dt} = c_1 s_1 e^{s_1 t} + c_2 s_2 e^{s_2 t} + T(s_0)e_0 s_0 e^{s_0 t} \tag{8.7.5}$$

and then evaluating at $t = 0_+$. The result is

$$\left.\frac{di}{dt}\right|_+ = c_1 s_1 + c_2 s_2 + T(s_0)e_0 s_0 \tag{8.7.6}$$

If the current and its derivative are known at $t = 0_+$, the only unknowns in Eqs. (8.7.4) and (8.7.5) are c_1 and c_2, and they can be solved for. Our problem, then, is to find i_+ and $di/dt\,|_+$ in terms of their values at $t = 0_-$, just prior to $t = 0$.

We integrate both sides of Eq. (8.7.2) from $t = 0_-$ to $t = 0_+$ and obtain

$$\int_-^+ L\frac{d^2i}{dt^2}\,dt + \int_-^+ R\frac{di}{dt}\,dt + \int_-^+ \frac{1}{C}\,i\,dt = \int_-^+ \frac{de}{dt}\,dt \tag{8.7.7}$$

Integrating Eq. (8.7.2) a second time gives

$$\int_-^+\!\!\int L\frac{d^2i}{dt^2}\,dt^2 + \int_-^+\!\!\int R\frac{di}{dt}\,dt^2 + \int_-^+\!\!\int \frac{1}{C}\,i\,dt^2 = \int_-^+\!\!\int \frac{de}{dt}\,dt^2 \tag{8.7.8}$$

which can be rewritten as

$$\int_-^+ L\frac{di}{dt}\,dt + \int_-^+ Ri\,dt + \int_-^+\!\!\int \frac{1}{C}\,i\,dt^2 = \int_-^+ e\,dt \tag{8.7.9}$$

Because of the infinitesimal time from $t = 0_-$ to $t = 0_+$, any integral between these limits must certainly vanish unless the integrand is infinite there (contains an impulse at the origin).† This amounts to saying that the area under a curve of infinitesimally small base is zero unless the curve has infinite height. Thus the third integral on the left side of Eq. (8.7.7) vanishes, as do the second and third integrals on the left side and the integral on the right side of Eq. (8.7.9).

In Eq. (8.7.9) we are left with

$$Li_+ - Li_- = 0 \qquad \text{or} \qquad i_+ = i_- \tag{8.7.10}$$

and, inserting this result in Eq. (8.7.7), we obtain

$$L\frac{di}{dt}\bigg|_+ - L\frac{di}{dt}\bigg|_- = e_+ - e_- \qquad \text{or} \qquad \frac{di}{dt}\bigg|_+ = \frac{di}{dt}\bigg|_- + \frac{e_0}{L} \tag{8.7.11}$$

Note that although the current is zero at $t = 0_+$ if the series RLC circuit is initially relaxed, its first derivative is not.

In the general case of an nth-order differential equation, the nth integration yields the value of the dependent variable at $t = 0_+$, and so on, until the first integration yields the value of the $(n - 1)$st derivative at $t = 0_+$. This gives n independent initial conditions at $t = 0_+$ from which to evaluate the n independent constants in the free response.

We are now ready to work an example to illustrate the evaluation of the initial conditions and the arbitrary constants in the free response of a linear system and to answer the questions raised at the close of Sec. 8.6.

Example 8.10 Find the total solution of

$$\frac{d^2y}{dt^2} + 3\frac{dy}{dt} + 2y = \frac{dx}{dt} + 3x$$

using the following conditions:

(*a*) $x = 2e^{-4t} \qquad y_- = \dfrac{dy}{dt}\bigg|_- = 0$

(*b*) $x = 2e^{-4t} \qquad y_- = -1 \qquad \dfrac{dy}{dt}\bigg|_- = 0$

(*c*) $x = 2e^{-t} \qquad y_- = \dfrac{dy}{dt}\bigg|_- = 0$

(*d*) $x = 2(e^{-t} + e^{-4t}) \qquad y_- = \dfrac{dy}{dt}\bigg|_- = 0$

(*e*) $x = 2e^{-3t} \qquad y_- = \dfrac{dy}{dt}\bigg|_- = 0$

† The case of impulses at the origin is discussed in detail in Sec. 10.2.

(*a*) Employing the techniques of this section, we find

$$y_+ = 0 \qquad \left.\frac{dy}{dt}\right|_+ = 2$$

The characteristic equation is

$$(s + 1)(s + 2) = 0$$

from which the free response is

$$y_h = c_1 e^{-t} + c_2 e^{-2t}$$

The transfer function is

$$T(s) = \frac{s + 3}{s^2 + 3s + 2}$$

giving a forced response, from Eq. (8.6.6), of

$$y_p = T(-4)2e^{-4t} = -\tfrac{1}{3}e^{-4t}$$

Thus the total solution has the form

$$y = c_1 e^{-t} + c_2 e^{-2t} - \tfrac{1}{3}e^{-4t}$$

To evaluate the constants, we note that at $t = 0_+$ the response becomes

$$0 = c_1 + c_2 - \tfrac{1}{3}$$

Differentiating the total solution once and evaluating at $t = 0_+$ yields a second equation

$$2 = -c_1 - 2c_2 + \tfrac{4}{3}$$

Simultaneous solution gives $c_1 = \tfrac{4}{3}$ and $c_2 = -1$. Thus the total response is

$$y = \tfrac{4}{3}e^{-t} - e^{-2t} - \tfrac{1}{3}e^{-4t}$$

Note that there are two arbitrary constants, requiring the specification of two independent initial conditions.

(*b*) The only change here is in the initial conditions and thus in the simultaneous equations used to evaluate the arbitrary constants in the free response. We find that $y_+ = -1$, $dy/dt\,|_+ = 2$. The resulting equations are

$$-1 = c_1 + c_2 - \tfrac{1}{3} \qquad \text{and} \qquad 2 = -c_1 - 2c_2 + \tfrac{4}{3}$$

yielding $c_1 = -\tfrac{2}{3}$ and $c_2 = 0$. Thus

$$y = -\tfrac{2}{3}e^{-t} - \tfrac{1}{3}e^{-4t}$$

The only effect of changing the initial conditions is a change in the amplitudes of the terms in the free response.

(*c*) Here the system is initially relaxed so that the initial conditions at $t = 0_+$ are those of part (*a*). But now the driving frequency is a natural frequency of the system, a pole of the transfer function.

The free response has the same form as that in solution (*a*). To obtain the forced response, we use the same technique that was used in finding the free response when multiple roots were present in the homogeneous equation. Thus we assume

$$y_p = y_0te^{s_0t} = y_0te^{-t}$$

Inserting this quantity and the given excitation into the differential equation, defining the numerator and denominator of $T(s)$ as $P(s)$ and $Q(s)$, respectively, and referring to Eqs. (8.3.17) and (8.3.18), we obtain

$$y_0 = x_0 \frac{P(s)}{Q'(s)}\bigg|_{s=s_0} = 2\frac{s+3}{2s+3}\bigg|_{-1} = 4$$

where $Q'(s)$ is the derivative of $Q(s)$ with respect to s. Note that y_0 is not given by $T(s_0)x_0 = 2T(-1)$, the latter being infinite in this case since $s + 1$ is a factor of $Q(s)$.

The total solution now takes the form

$$y = c_1e^{-t} + c_2e^{-2t} + 4te^{-t}$$

Evaluating c_1 and c_2 in the same manner as before, we find

$$c_1 = -2 \qquad c_2 = 2$$

and the response is

$$y = -2e^{-t} + 2e^{-2t} + 4te^{-t}$$

(*d*) Here the excitation is the sum of excitations in parts (*a*) and (*c*). Since the system has the same initial conditions in parts (*a*) and (*c*), we can use the superposition principle and write the response as the sum of the responses of those two parts

$$y = -\tfrac{2}{3}e^{-t} + e^{-2t} - \tfrac{1}{3}e^{-4t} + 4te^{-t}$$

(*e*) In this case, the driving frequency is a zero of the transfer function. Thus y_0 in Eq. (8.6.5) is zero, and the forced response is zero. The total solution is given by

$$y = c_1e^{-t} + c_2e^{-2t}$$

The initial conditions at $t = 0_+$ are the same as in part (*a*). Solving for c_1 and c_2 yields

$$c_1 = -2 \qquad \text{and} \qquad c_2 = 2$$

Thus

$$y = -2e^{-t} + 2e^{-2t}$$

We posed some questions at the close of Sec. 8.6 that we can now answer. Recall that the transfer function completely describes the forced response of a relaxed linear system to a single exponential forcing function whose frequency is not a natural frequency of the system. We see from solution (*a*) that it also describes the free response of the system if the system is initially relaxed, in that it determines the amplitudes of the arbitrary constants of the free response. If the system is *not* initially relaxed, the transfer function, by itself, does *not* determine these constants. The initial conditions also enter into their determination as is shown by solution (*b*). When the driving frequency is a natural frequency of the system, neither the forced nor the free response is directly related to the transfer function, as solution (*c*) illustrates. When more than one exponential excitation is present, we can use the transfer function to find the free and forced responses due to each excitation separately and then use superposition as we did in solution (*d*). But we cannot conveniently use the transfer function if we consider the sum of the several excitations as a single excitation. Finally, in solution (*e*), where the frequency of the driving function is a zero of the transfer function, the transfer function does not influence the values of the arbitrary constants of the free response.

Our purpose in briefly considering the forced response at this point was to interrelate initial conditions, the driving function, and the arbitrary constants in the free response. This we have now done. A second, unstated purpose was to show that the present method of finding the total response is not particularly versatile. There are too many special cases to remember, as shown by Example 8.10. The present method can be generalized to the nth-order case,† but the generalization is cumbersome. In the next several chapters we shall proceed to the more straightforward Laplace transform technique for solving a linear differential equation. Initial conditions, multiple roots, multiple exponential excitations, and many nonexponential excitations are handled with relative ease by Laplace methods.

There is an alternative method for finding the initial conditions. Recall that up to now we have dealt directly with the nth-order differential equation. Specifically, we have described how to obtain from this equation the values of the dependent variable and the first $n - 1$ of its derivatives at $t = 0_+$ in terms of their values at $t = 0_-$ and the value of the excitation at $t = 0_+$. We turn now to a procedure for finding the initial conditions directly from the network itself. Such a procedure logically stems from the values of voltages across capacitors and currents through inductors, for it is in these elements that the energy of the system resides.

Consider the inductor first. Its voltage is given by $v = L\,di/dt$. Solving for the current yields

$$i(t) = \frac{1}{L}\int_{t_0}^{t} v\,dt + i(t_0) \tag{8.7.12}$$

In this equation, all events prior to time t_0 are summed up in the value of the current at time t_0. The integral then gives the contribution to $i(t)$ from t_0 onward. For our purposes in evaluating initial conditions we let $t_0 = 0_-$ and $t = 0_+$ and rewrite Eq. (8.7.12) to obtain

$$i_+ = \frac{1}{L}\int_{-}^{+} v\,dt + i_- \tag{8.7.13}$$

Using exactly the same argument as was used following Eq. (8.7.9), the

† See, for example, Ref. 8.8.

integral in Eq. (8.7.13) will vanish unless the voltage applied to the inductor at $t = 0$ is an impulse. If the voltage is a unit impulse,

$$i_+ = \frac{1}{L} + i_- \tag{8.7.14}$$

Otherwise

$$i_+ = i_- \tag{8.7.15}$$

Equation (8.7.15) is the familiar concept of conservation of flux linkage, which is valid in the absence of a voltage impulse. It means that at the instant of connecting an inductor to an excitation, the inductor appears as an open circuit if $i_- = 0$ and as a current generator of value i_- otherwise.

The case of the capacitor is very similar. Its current is given by $i = C\,dv/dt$; its voltage by

$$v(t) = \frac{1}{C}\int_{t_0}^{t} i\,dt + v(t_0) \tag{8.7.16}$$

Letting $t_0 = 0_-$ and $t = 0_+$, we have

$$v_+ = \frac{1}{C}\int_{-}^{+} i\,dt + v_- \tag{8.7.17}$$

With a unit current impulse at $t = 0$,

$$v_+ = \frac{1}{C} + v_- \tag{8.7.18}$$

Otherwise

$$v_+ = v_- \tag{8.7.19}$$

Equation (8.7.19) is the familiar principle of conservation of charge. This result, that in the absence of a current impulse the voltage across a capacitor does not change instantaneously, means that at the instant of application of an excitation the capacitor appears as a short circuit if $v_- = 0$ and as a voltage generator of value v_- otherwise.

In finding initial conditions directly from a circuit, the procedure is to redraw the circuit so that all inductors are replaced by open circuits (or current generators) and all capacitors by short circuits (or voltage generators). From the redrawn circuit, the values of currents and voltages at $t = 0_+$ can be determined in terms of their values at $t = 0_-$. To find the values of the derivatives of these quantities at $t = 0_+$, appropriate loop and node equations are written and evaluated at $t = 0_+$. The proper equations to write depend on the circuits, and no general procedure can be formulated. A good deal of intuition is required. A specific example will illustrate.

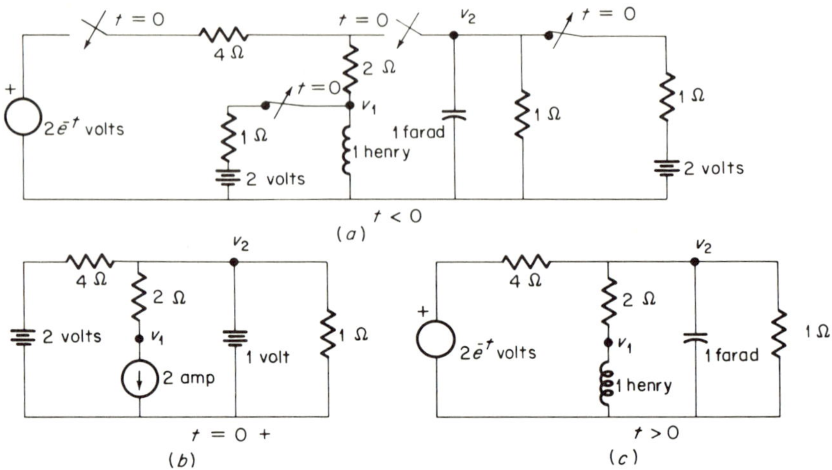

Fig. 8.11

Example 8.11 In the circuit shown in Fig. 8.11*a*, two switches shut and the other two switches open at $t = 0$. Find the values at $t = 0_+$ of v_1, v_2, dv_1/dt, and dv_2/dt.

From Fig. 8.11*a*, the values of the capacitor voltage and inductor current at $t = 0_-$ are 1 volt and 2 amp. From the conservation of flux linkage and charge, these values will also apply at $t = 0_+$. To find the values of v_{1+} and v_{2+}, the circuit is redrawn as in Fig. 8.11*b*. Note that the inductor has been replaced by a current generator and the capacitor by a battery. From Fig. 8.11*b*, it is easy to see that $v_{1+} = -3$ volts and $v_{2+} = 1$ volt.

To find the derivatives at $t = 0_+$, appropriate circuit equations in v_1 and v_2 must be written using the redrawn circuit in Fig. 8.11*c*. At the v_2 node

$$v_2 + \frac{dv_2}{dt} + \frac{1}{2}(v_2 - v_1) + \tfrac{1}{4}(v_2 - 2e^{-t}) = 0 \qquad t > 0$$

Evaluating at $t = 0_+$ yields

$$1 + \left.\frac{dv_2}{dt}\right|_+ + \frac{1}{2}(1 + 3) + \tfrac{1}{4}(1 - 2) = 0 \qquad t = 0_+$$

$$\left.\frac{dv_2}{dt}\right|_+ = -\,{}^{11}\!/_{4} \text{ volts/sec}$$

Turning next to the v_1 node, we have

$$\tfrac{1}{2}(v_1 - v_2) + \int_0^t v_1\,dt + 2 = 0 \qquad t > 0$$

Differentiating yields

$$\frac{1}{2}\frac{dv_1}{dt} - \frac{1}{2}\frac{dv_2}{dt} + v_1 = 0 \qquad t > 0$$

and evaluating at $t = 0_+$ gives

$$\frac{1}{2}\frac{dv_1}{dt}\bigg|_+ - \tfrac{1}{2}(-\tfrac{11}{4}) - 3 = 0 \qquad t = 0_+$$

$$\frac{dv_1}{dt}\bigg|_+ = \tfrac{13}{4} \text{ volts/sec}$$

8.8 THE MANY FACETS OF n

We saw in Sec. 8.3 that an nth-order system is described by an nth-order differential equation and has an algebraic characteristic equation of degree n whose roots are the n natural frequencies of the system. From this we found that the system's free response contains n linearly independent terms, each of which displays one of the n natural frequencies and contains an arbitrary multiplicative constant. In Sec. 8.4, we showed that the nth-order system can be alternatively described by n first-order differential equations through a matrix A whose rank is n.

In Sec. 8.7 we established that the determination of the n arbitrary constants requires the specification of n independent initial conditions. Also we indicated that the latter may be the values of a single variable and $n - 1$ of its derivatives or the values of n independent variables.

In Sec. 8.4 we suggested that a set of n independent variables that has considerable physical significance in an RLC network is a set of independent inductive currents and capacitive voltages, because the values of these at any time are a set of n independent initial conditions that give the total stored energy in the network at that time. We called these variables a set of state variables, specifically a set of dynamically independent variables the values of which at any time are the least amount of information necessary to determine the future response of the network to any excitation. Furthermore, each state variable represents one degree of freedom in the network. An nth-order network has n degrees of freedom.

Let us summarize the observations of the last three paragraphs by noting that in an nth-order linear system the number attached to each of the following quantities is the same and is equal to n:

Order of single differential equation
Number of first-order differential equations
Degree of characteristic equation
Number of independent initial conditions
Number of independent points of energy storage
Number of natural frequencies
Number of state variables

Number of arbitrary constants in free response
Number of degrees of freedom
Rank of the A matrix

We now want to develop a procedure for finding n in a given network quickly and easily. Of course we can always derive the characteristic equation and note its degree, but we want a shorter, almost-by-inspection way. We are tempted to count the number of energy-storage elements in the network and call this number n. But this is incorrect, for what we really want is the number of *independent* energy-storage elements. Certain topological arrangements of inductances and capacitances rob a network of some of its degrees of freedom. After we find what these topological constraints are and how to spot them, we can find n by subtracting their number from the total number of energy-storage elements.

8.9 DYNAMIC CONSTRAINTS

Suppose there is a tie set (loop) of capacitors embedded in a network, as shown in Fig. 8.12*a*. Writing the KVL equation around the capacitive loop gives

$$V_0 = \frac{1}{C_1}\int_0^t (i_1 - i_4)\,dt + \frac{1}{C_2}\int_0^t (i_2 - i_4)\,dt + \frac{1}{C_3}\int_0^t (i_3 - i_4)\,dt \quad (8.9.1)$$

where V_0 is the net initial-condition voltage around the loop. Differentiation of Eq. (8.9.1) yields

$$i_4\left(\frac{1}{C_1} + \frac{1}{C_2} + \frac{1}{C_3}\right) = \frac{1}{C_1} i_1 + \frac{1}{C_2} i_2 + \frac{1}{C_3} i_3 \quad (8.9.2)$$

If the network in Fig. 8.12*a* has just the four loops indicated, we conclude that there are four geometrically independent loop current variables. But

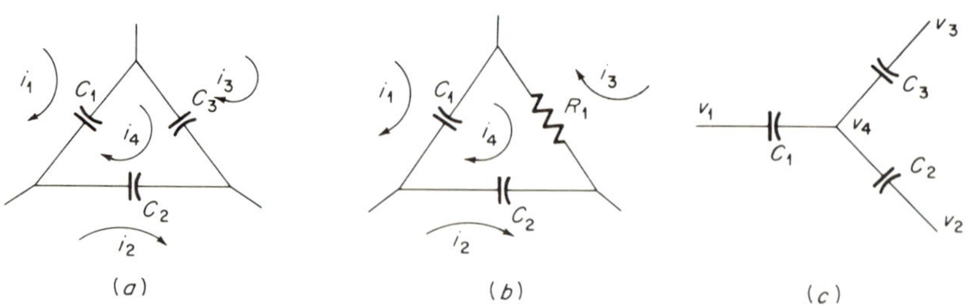

Fig. 8.12 Dynamic constraints.

not all these are independent in a dynamic sense. From Eq. (8.9.2), at least one of these currents is dynamically dependent on the others. We say "at least" because there may be additional dynamic constraints due to the other three loops that have not been considered. Thus there are at most three dynamically independent current variables in the network of Fig. 8.12*a*.

Constraint relations of the sort described by Eq. (8.9.2) arise whenever there exists in a network a closed path containing only one kind of element. Thus all-inductive and all-resistive tie sets also create dynamic constraints. For in such situations, when we write the KVL equation around the closed path in question, and perhaps differentiate or integrate it, we obtain an algebraic, rather than a differential, equation. To clarify this point, consider Fig. 8.12*b*, where we have replaced one capacitor by a resistor so that we no longer have a loop containing just one kind of element. The KVL equation is

$$V_0 = \frac{1}{C_1}\int_0^t (i_1 - i_4)\,dt + \frac{1}{C_2}\int_0^t (i_2 - i_4)\,dt + R_1(i_3 - i_4) \tag{8.9.3}$$

Differentiating gives

$$R_1\frac{di_4}{dt} + \left(\frac{1}{C_1} + \frac{1}{C_2}\right)i_4 = R_1\frac{di_3}{dt} + \frac{1}{C_1}i_1 + \frac{1}{C_2}i_2 \tag{8.9.4}$$

In Eq. (8.9.2), fixing the values of i_1, i_2, and i_3 determined i_4. This is not true in Eq. (8.9.4). Equation (8.9.2) is a dynamic constraint equation; Eq. (8.9.4) is not.

How else can constraint relations arise? Consider the cut set of capacitors in Fig. 8.12*c*, where the voltages are all with respect to some reference node, not shown. The KCL equation at the internal node is

$$0 = C_1\frac{d(v_4 - v_1)}{dt} + C_2\frac{d(v_4 - v_2)}{dt} + C_3\frac{d(v_4 - v_3)}{dt} \tag{8.9.5}$$

Integrating from 0 to t yields the constraint relation

$$(C_1 + C_2 + C_3)v_4 = C_1v_1 + C_2v_2 + C_3v_3 + Q_0 \tag{8.9.6}$$

where Q_0 is the net initial-condition charge. If we replace one of the capacitors by another element, no constraint relation results. As in the tie-set case, we can also have inductive and resistive cut-set constraint relations.

8.10 DETERMINATION OF n

Returning now to the problem of finding n as the number of independent energy-storage elements in a network, we consider an RLC network having one separate part and B single-element branches. Let B_{LC} be the total number of capacitive and inductive branches. Also let

$$L_L = \text{number of independent inductive tie sets}$$
$$L_C = \text{number of independent capacitive tie sets}$$
$$N_L = \text{number of independent inductive cut sets}$$
$$N_C = \text{number of independent capacitive cut sets}$$

Then

$$n = B_{LC} - (L_L + L_C + N_L + N_C) \tag{8.10.1}$$

Equation (8.10.1) states that the number of independent energy-storage elements in a network is the difference between the total number of energy-storage elements and the total number of independent constraints. [We, of course, do not include resistive tie-set and cut-set constraints in Eq. (8.10.1) because resistors are not energy-storage elements and hence do not enter into the determination of n.]

Our problem now is to find the four numbers L_L, L_C, N_L, and N_C. For uncomplicated networks they can be found by inspection. Consider, for example, the network in Fig. 8.13, which contains eight energy-storage elements. It is quite easy to see that there is one each of the four possible constraints, namely, one L loop, one C loop, one L node, and one C node. These are labeled by dashed lines. With $B_{LC} = 8$ and with four constraints, the value of n is $8 - 4 = 4$.

With more complicated networks, a systematic procedure for evaluating L_L, L_C, N_L, and N_C is desirable. Consider the first, the number of independent inductive tie sets. Each such tie set is a closed path of inductive branches in the network graph. If all the resistive and capacitive branches are removed from this graph, these inductive closed paths will remain and be clearly visible, and no new inductive closed paths will be created. Thus to find the number of independent inductive tie sets, we simply draw an inductive subgraph with all resistive and capacitive branches absent, called the L^∞ *subgraph* (because all resistances and capacitive reactances are infinite), and count the number of independent loops. This number is L_L. Similarly,

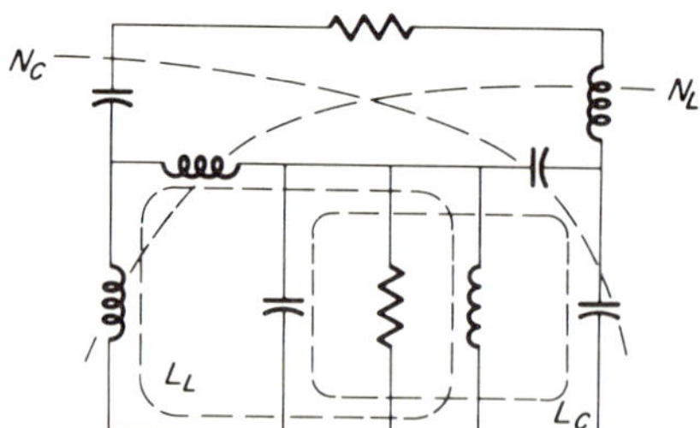

Fig. 8.13

to find the number of independent capacitive tie sets, we form the C^∞ subgraph, with all resistive and inductive branches absent, and count the number of independent capacitive loops. This number is L_C.

Turning now to cut-set constraints, we note that each inductive cut set is a set of inductive branches the removal of which severs the network into two separate parts. If we shrink each resistive and capacitive branch to nothing and let each pair of terminal nodes coalesce into one, all inductive cut sets present will remain and be easy to identify. No new inductive cut sets are created in this process. Thus to find N_L, we form an inductive subgraph with all resistive and capacitive branches shrunk to nothing, called the L^0 *subgraph* (because all resistances and capacitive reactances are zero) and count the number of independent inductive cut sets. Similarly, to find N_C, we form a capacitive subgraph with all resistive and inductive branches shrunk to nothing, called the C^0 *subgraph*, and count the number of independent capacitive cut sets. Now Eq. (8.10.1) can be used to find n.

Example 8.12 Find n for the network in Fig. 8.14*a*.

The network has seven inductances and five capacitances. Thus $B_{LC} = 12$, and this would be the value of n if there were no dynamic constraints. To find the tie-set constraints, we require the open-circuit subgraphs. These are obtained from the graph in Fig. 8.14*b* and are shown in Fig. 8.14*c* and *d* (isolated nodes are included). There are two independent tie sets in the L^∞ subgraph and one in the C^∞ subgraph. Thus $L_L = 2$ and $L_C = 1$.

To obtain the cut-set constraints, the short-circuit subgraphs are required. These are not as easy to draw as the open-circuit graphs. However, if we shrink each R and C branch to nothing, the L^0 subgraph of Fig. 8.14*e* results, where all inductive branches have become self-loops. Thus there are no independent inductive cut sets, and $N_L = 0$. Shrinking each R and L branch to nothing yields the C^0 subgraph of Fig. 8.14*f*. Here there is one self-loop, but the remaining four capacitive branches are not shorted and form one independent capacitive cut set. Thus $N_C = 1$. Finally

$$n = 12 - (2 + 1 + 0 + 1) = 8$$

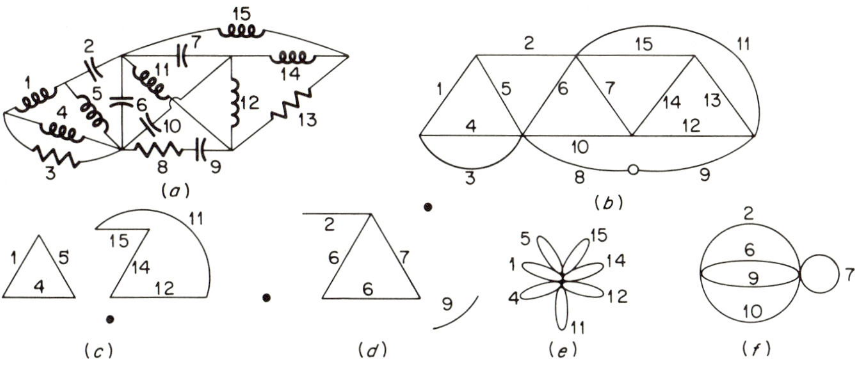

Fig. 8.14

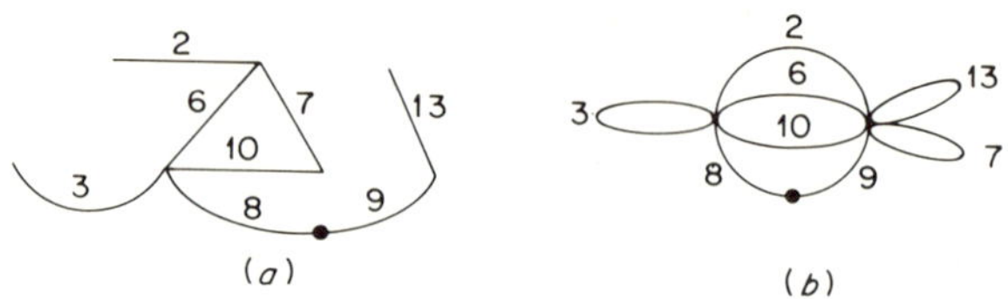

Fig. 8.15 *Changes resulting from replacing R's by C's in Fig. 8.14a.*

Let each capacitance in Fig. 8.14*a* become a resistance. Again find n.

The circuit is now an *RL* circuit. The two inductive subgraphs do not change; the two capacitive subgraphs are nonexistent. Thus there are two dynamic constraints and, with $B_{LC} = 7$,

$$n = 7 - (2 + 0) = 5$$

Let each of the resistors in Fig. 8.14*a* become a capacitor and again find n.

The circuit is now a lossless *LC* circuit. The L^{∞} subgraph does not change and therefore $L_L = 2$. The C^{∞} subgraph becomes that of Fig. 8.15*a*, and $L_C = 1$.

The L^0 subgraph does not change, and so $N_L = 0$. The new C^0 subgraph appears in Fig. 8.15*b*. There are two independent cut sets; thus $N_C = 2$. With B_{LC} now 15, we have

$$n = 15 - (2 + 1 + 0 + 2) = 10$$

8.11 DYNAMICALLY INDEPENDENT VARIABLES

In this section, the relationship between the number of geometrically independent and dynamically independent variables in a given network is considered. We know that the number of geometrically independent variables depends on the topology of the network and which variables we chose (loop currents and node voltages) and does not depend on the kinds of elements present in the various branches of the network. The same cannot be said for the number of dynamically independent variables. This number is n, and it depends very much on the kinds of elements and their locations in the network, as we have just seen in Sec. 8.10.

An alternative to the procedure of Sec. 8.10 for finding n now suggests itself. If we know the number of geometrically independent variables on some basis and we subtract the number of independent dynamic constraints, we should obtain the number of dynamically independent variables on that same basis. For example, on a loop basis, we would subtract from L (the number of geometrically independent loop currents) the number of independent inductive, capacitive, and resistive tie sets and get the number of dynamically independent loop-current variables, which we shall call L_D. On a node basis, the number of inductive, capacitive, and resistive cut sets would be subtracted from N (the number of geometrically independent node voltages) to obtain the number of dynamically independent node-voltage variables. We shall label this N_D. Now we are inclined to say that L_D and N_D are the same number and that this number is n.

Let us test this last premise with the simple circuit of Fig. 8.16*a*. Our task is to find L_D, N_D, and n.

We note that $L = 1$ and $N = 2$ for this network. There are no dynamic constraints. Hence $L_D = 1$ and $N_D = 2$. These numbers are not the same, thus refuting the premise. There are two independent energy-storage elements. Thus $n = 2$ and $n = N_D > L_D$.

Why does L_D not equal N_D? Why does one equal n and the other not? The answers lie in the loop and node-pair equations themselves. Writing these for the choice of tree, link, and orientation in Fig. 8.16*b* and assuming zero initial conditions, we have

$$L\frac{di_1}{dt} + Ri_1 + \frac{1}{C}\int_0^t i_1\,dt = 0 \tag{8.11.1}$$

and

$$\frac{1}{R}(e_2 + e_3) + C\frac{de_3}{dt} = 0$$

$$\frac{1}{R}(e_2 + e_3) + \frac{1}{L}\int_0^t e_2\,dt = 0 \tag{8.11.2}$$

In the node-pair equations, only the derivative of e_3 appears with e_3, and only the integral of e_2 appears with e_2. In the loop equations, the variable i_1, its derivative, *and* its integral appear. Apparently we need to weight (as well as count) our dynamically independent loop-current and node-voltage variables if we are to use them in determining the value of n. For example, on a loop-current basis, we would need to follow each loop current around the network to see what kinds of elements it encounters. If it encounters only R's and C's or only R's and L's, its weight is 1. If it encounters L's and C's or R's, L's, and C's, its weight is 2. If it encounters just R's, just L's, or just C's, a dynamic constraint results and a weight of 0 is assigned. Similar conclusions are reached on a node basis. Thus in Fig. 8.16*b* each node-pair voltage has a weight of 1, whereas the single loop current gets a weight of 2. The dynamically independent node-pair voltages e_2 and e_3 are a

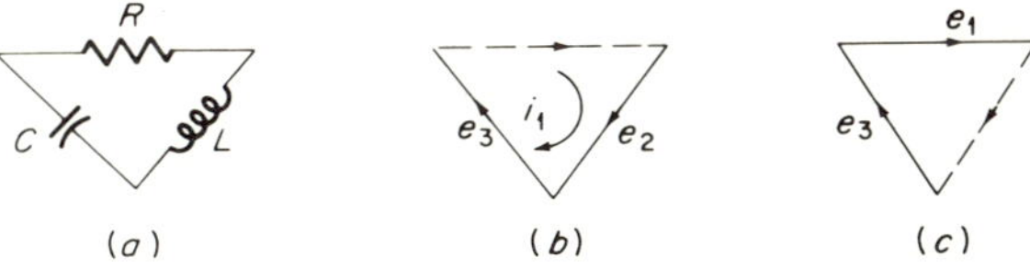

Fig. 8.16 Weighted variables.

full set of dynamically independent variables; the single dynamically independent loop current is not. This loop current and its first derivative are a full set of dynamically independent variables.

One further problem can arise. Suppose we had chosen the tree and link of Fig. 8.16*c*. Now the node-pair equations are

$$\frac{1}{R}e_1 + \frac{1}{L}\int_0^t (e_1 + e_3)\,dt = 0$$

$$\frac{1}{L}\int_0^t (e_1 + e_3)\,dt + C\frac{de_3}{dt} = 0 \qquad (8.11.3)$$

The variable e_1 has a weight of 1; e_3 has a weight of 2. This suggests that n is 3, which we know is incorrect. We conclude that to find n, not only must we weight each variable, but also we must choose our variables so that the least total number results when the weighted variables are added. This may not be easy.

This method of finding n from L_D and N_D is obviously inferior to the method of Sec. 8.10. It has been presented for three other reasons. First, it permitted us to introduce L_D and N_D and to show that these two quantities are not in general equal to each other or to n. Thus a set of dynamically independent loop currents or node voltages is not in general a full set of dynamically independent variables for a network. We often need derivatives (or integrals) of some of these variables to complete the full set.

Second, it enabled us to establish the concept of weighted variables. We found that weights of 2, 1, and 0 were possible, the latter representing a dynamic constraint. We can see now one of the benefits of writing sets of first-order differential state equations in the form $\dot{Y} = AY$ of Eq. (8.4.4). Note that no integral terms are present in this equation. Thus each of the variables contained therein has a weight of 1, and we can be sure that a set of state variables is indeed a full set of dynamically independent variables for a network.

Third, we can use our conclusions about weighting of variables to derive some simple relations between L_D, N_D, and n for two-element-kind networks. In *RL* and *RC* networks, all dynamically independent variables must have a weight of 1, because only derivatives or integrals, but not both, of each variable are present. It follows that $L_D = N_D = n$ for these networks. In the *LC* case, each dynamically independent variable must have a weight of 2, and thus $L_D = N_D = \frac{1}{2}n$.

Example 8.13 Find n, N_D, L_D, and one set each of voltage and current variables having a total weighted value of n for the network whose graph and elements appear in Fig. 8.17*a*.

Let us first use the method of Sec. 8.10 to find n. There are seven energy-storage elements in the network. The only dynamic constraint among them is a single capacitive

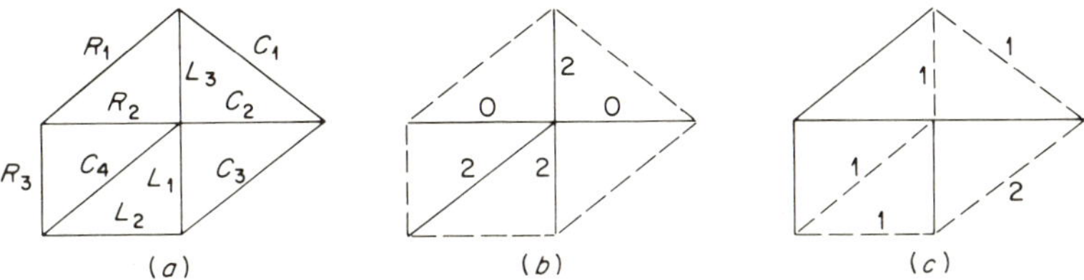

Fig. 8.17

cut set. Thus

$$n = 7 - 1 = 6$$

To find N_D, we subtract from N the total number of cut-set constraints. In this case, N is 5, and there are two cut-set constraints, one resistive and one capacitive. Thus

$$N_D = 5 - 2 = 3$$

Now to find L_D, we note that $L = 5$. There are no tie-set constraints, so that

$$L_D = 5 - 0 = 5$$

To find n independently using first N_D, we must discover what weighting of voltage variables is necessary. In Fig. 8.17*b*, we show a set of tree-branch voltage variables that gives a minimum total weight of 6. On an L_D basis, a set of link-current variables provides this same minimum total weight, as is shown in Fig. 8.17*c*.

Example 8.14 For the network in Fig. 8.18*a*, find L_D, N_D, and n. Repeat if each resistance is replaced by a capacitance to obtain the network of Fig. 8.18*b*.

The network in Fig. 8.18*a* is an *RL* network. Hence we know that $L_D = N_D = n$. We note that $B_{LC} = 4$ and that there is one inductive tie set and one inductive cut set. Hence $n = 4 - 2 = 2$. (The absence of capacitors does not invalidate the method of Sec. 8.10 for finding n.) To check the result by finding L_D directly, we note that $L = 3$ and that there is one (inductive) tie set. Thus $L_D = 3 - 1 = 2 = n$. To find N_D directly, observe that $N = 4$ and that there are two (one resistive and one inductive) cut sets, yielding $N_D = 4 - 2 = 2 = n$. Here all variables have a weight of 1.

The network in Fig. 8.18*b* is an *LC* network. It follows at once that $L_D = N_D = \frac{1}{2}n$. We note that $B_{LC} = 7$ and that there is one inductive tie set, one inductive cut set,

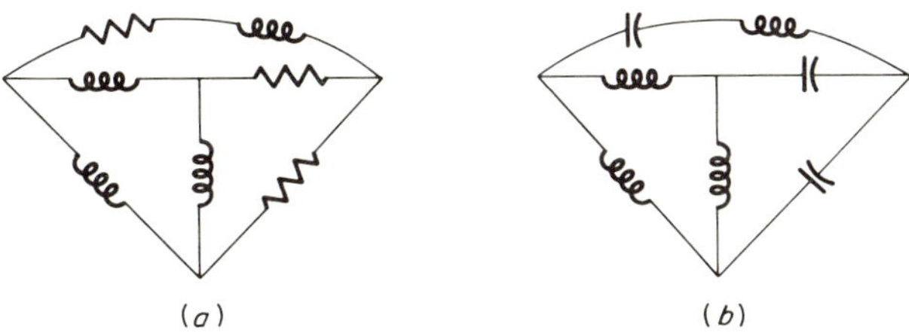

Fig. 8.18

and one capacitive cut set. Hence $n = 7 - 3 = 4$. L_D and N_D can be found directly in exactly the same way as for the RL case, and each is 2. In this case, all variables have a weight of 2.

REFERENCES

8.1. Bers, A.: The Degrees of Freedom in *RC* Networks, *IRE Trans. Circuit Theory*, vol. CT-6, pp. 91–94, March, 1959. A topological method is presented for determining *n* that involves counting the increases in the number of separate parts in the capacitive and inductive subgraphs to find the cut-set constraints. The method of determining tie-set constraints is that of Sec. 8.10.

8.2. de Pian, L.: "Linear Active Network Theory," Prentice-Hall, Inc., Englewood Cliffs, N.J., 1962. Three excellent chapters are devoted to reciprocity, stability, and passivity.

8.3. Del Toro, V., and S. R. Parker: "Principles of Control System Engineering," McGraw-Hill Book Company, New York, 1960. Chapter 2 provides an excellent review of the classical solution of linear differential equations with constant coefficients, including a clear discussion of initial-condition evaluation.

8.4. Desoer, C. A.: Modes in Linear Circuits, *IRE Trans. Circuit Theory*, vol. CT-7, pp. 211–223, Sept. 1960. This comprehensive theoretical paper reviews in depth the classical theory of modes of oscillation in a linear system in terms of quadratic forms, eigenvalues, eigenvectors, and linear vector spaces.

8.5. Guillemin, E. A.: "Synthesis of Passive Networks," John Wiley & Sons, Inc., New York, 1957, and "Theory of Linear Physical Systems," John Wiley & Sons, Inc., New York, 1963. Poles and zeros of network functions are discussed in Chapter 1 of the first reference. Chapter 5 contains a procedure for determining *n* in a two-element-kind network and for making pole-zero plots of lossless driving-point functions. The second reference considers the topics of normal coordinates and natural frequencies in detail.

8.6. Hildebrand, F. B.: "Methods of Applied Mathematics," Prentice-Hall, Inc., Englewood Cliffs, N.J., 1952. Chapter 1 presents a thorough and readable discussion of eigenvalues, eigenvectors, and linear vector spaces.

8.7. Kuh, E. S., and R. A. Rohrer: The State-variable Approach to Network Analysis, *Proc. IEEE*, July, 1965, pp. 672–686. This comprehensive paper describes the state approach, stressing the concepts of zero-input response and zero-state response.

8.8. Reza, F. M., and S. Seely: "Modern Network Analysis," McGraw-Hill Book Company, New York, 1959. The material of Secs. 8.3 and 8.6 is discussed in Chapter 9, and a derivation of the Heaviside expansion theorem, which extends the development of Sec. 8.6 to the *n*th-order case, is presented. Natural frequencies are considered in Chapter 12.

PROBLEMS

Drill

8.1. Find eigenvalues and eigenvectors for the following set of linear algebraic equations:

$$\begin{aligned} -x - 2y - 2z &= \lambda x \\ -x + y + 4z &= \lambda y \\ -23x - 17y + 8z &= \lambda z \end{aligned}$$

8.2. Find the solutions of the given homogeneous differential equations by the method of Sec. 8.3. Repeat by the method of Sec. 8.4. Assume all initial conditions zero.

(a) $$\frac{d^3y}{dt^3} + 7\frac{d^2y}{dt^2} + 14\frac{dy}{dt} + 8y = 0$$

(b) $$\frac{d^4y}{dt^4} + 4\frac{d^3y}{dt^3} + 7\frac{d^2y}{dt^2} + 6\frac{dy}{dt} + 2y = 0$$

8.3. Find the driving-point impedance of the given network using the methods of Sec. 8.5.

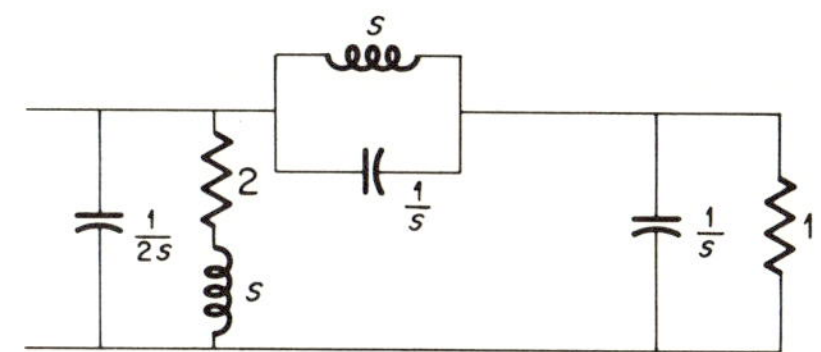

Fig. P8.3

8.4. Repeat Prob. 8.3 for the network shown.

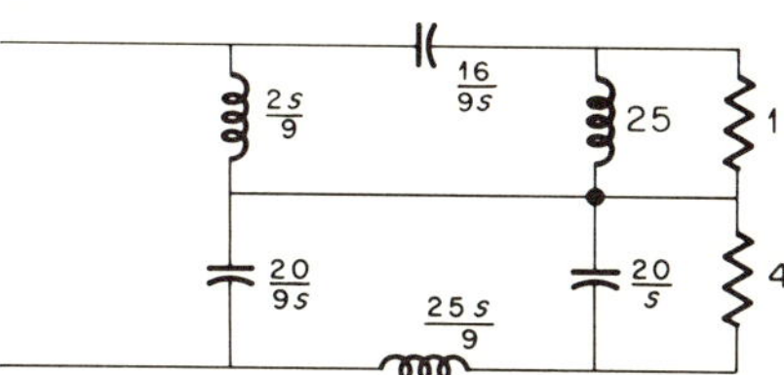

Fig. P8.4

8.5. For the given differential equation and excitation, find i_+, $di/dt\,|_+$, and $d^2i/dt^2\,|_+$, assuming all these quantities are zero at $t = 0_-$.

$$CRL\frac{d^2i}{dt^2} + L\frac{di}{dt} + Ri = CL\frac{d^2v}{dt^2} + CR\frac{dv}{dt}$$

$$v = \begin{cases} 0 & t < 0 \\ v_0 e^{s_0 t} & t > 0 \end{cases}$$

8.6. A system is described by

$$\frac{d^2y}{dt^2} + 2\frac{dy}{dt} + 2y = x$$

and is initially at rest. Using the method of Sec. 8.6, find the total response to the following excitations:

(a) $$x = \begin{cases} e^{jt} & t > 0 \\ 0 & t < 0 \end{cases}$$

(b) $$x = \begin{cases} \cos t & t > 0 \\ 0 & t < 0 \end{cases}$$

8.7. A system is described by

$$\frac{d^2y}{dt^2} + 3\frac{dy}{dt} + 2y = \frac{dx}{dt} + 3x$$

and is initially at rest. Its excitation is

$$x = \begin{cases} 2e^t & t > 0 \\ 0 & t < 0 \end{cases}$$

(*a*) Solve for y for a right-hand member of just x; then use the principle of related sources to find the total response.

(*b*) Solve for y directly using the entire right-hand member at the outset.

(*c*) Insert the given x into the right-hand member at the outset and then solve for y.

8.8. Find, independently, n, L_D, and N_D for the given lossless network. Repeat if each capacitor is replaced by a resistor.

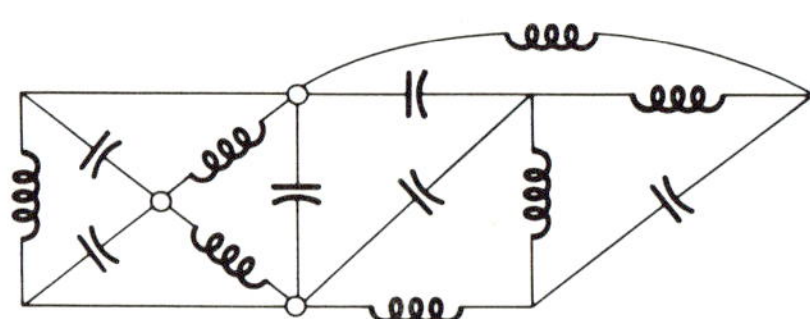

Fig. P8.8

8.9. Find n, L_D, N_D, and one set each of voltage and current variables having a total weight of n for the network whose graph is given.

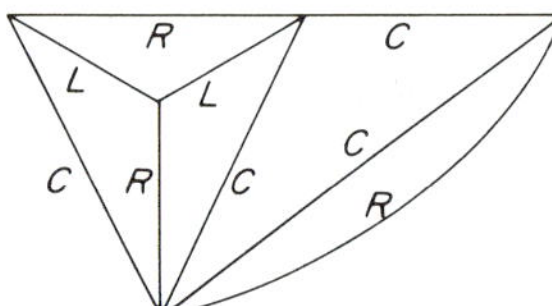

Fig. P8.9

8.10. Find, independently, n, L_D, and N_D for the given network. If both series resistors are shorted and the shunt resistor is opened, what do these values become?

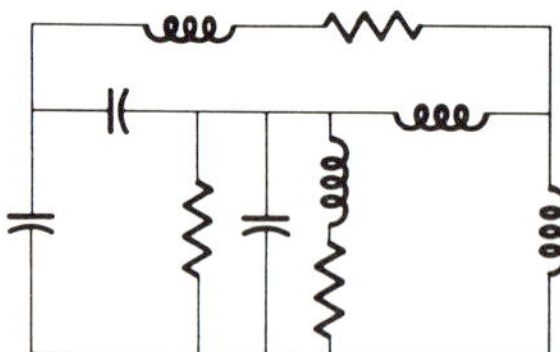

Fig. P8.10

8.11. For the network shown, find n, L_D, N_D, and one set each of voltage and current variables whose total weight is n.

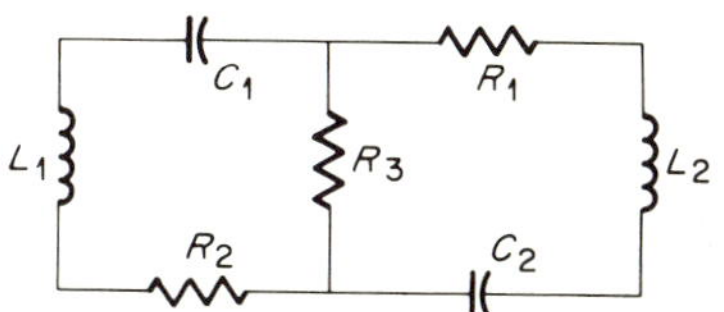

Fig. P8.11

8.12. For the network in Fig. 8.1, with $L = 1$ henry, $C = \frac{1}{4}$ farad, $R_1 = 2$ ohms, and $R_2 = 6$ ohms, write equations in the form $\dot{Y} = AY$ using the variables (*a*) i_1 and $\dot{i}_1$, (*b*) v_1 and $\dot{v}_1$, (*c*) i_1 and i_2, (*d*) v_1 and v_2, (*e*) v_c and i_L.

8.13. For the circuit whose graph and elements are shown, the value of n is 4. Find N_D and define a set of node-pair voltage variables that has a total weight of 4. Find L_D and define a set of loop current variables with a total weight of 4.

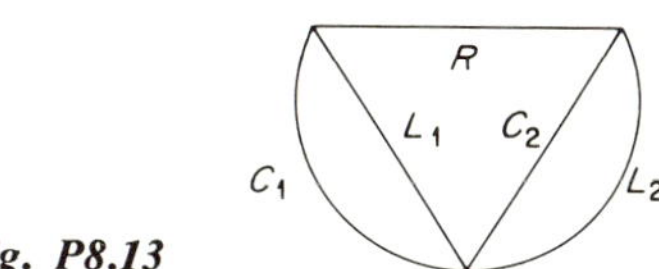

Fig. P8.13

Theory and proofs

8.14. (*a*) Prove that the eigenvalues of a real symmetric matrix are real.

(*b*) Prove that the eigenvectors of two different eigenvalues of a real symmetric matrix are orthogonal; i.e., if A and B are the two eigenvectors, then $\sum a_i b_i = 0$, where the summation runs from 1 through the order of the symmetric matrix.

8.15. Carry out a development similar to that following Eq. (8.5.5) to relate the zeros and poles of the y parameters of a 2-port to its natural frequencies.

8.16. A linear system has an excitation $v = V_0 e^{s_0 t}$ for $t > 0$, where s_0 is one of the natural frequencies of the system. Let the system transfer function be given by $T(s) = P(s)/Q(s)$. Verify the expression in Example 8.10*c* for finding the coefficient of the forced-response term. Use this result to find the total solution of

$$\frac{d^2 i}{dt^2} + 3\frac{di}{dt} + 2i = 3e^{-t}$$

assuming the system is initially at rest.

8.17. For the general third-order system defined by

$$\frac{d^3 i}{dt^3} + a_2\frac{d^2 i}{dt^2} + a_1\frac{di}{dt} + a_0 i = \frac{d^2 v}{dt^2} + b_1\frac{dv}{dt} + b_0 v$$

with an excitation given by

$$v = V_0 s^{s_0 t} \qquad t > 0$$

and with all initial conditions zero, prove that

$$c_k = \left.\frac{V_0 T(s)(s - s_k)}{s - s_0}\right|_{s=s_k,\, k=0,\ldots,3}$$

where c_k is the coefficient of the kth term in the system's free response and c_0 is the coefficient of the forced-response term. What change would occur in the expression for the c_k's if i at $t = 0_-$ had the value I_i, all other initial conditions at $t = 0_-$ remaining zero?

8.18. You are given an LC network.

(*a*) Show that the inductive and capacitive cut-set constraints can be obtained respectively from the capacitive and inductive open-circuit subgraphs.

(*b*) Derive a formula based on (*a*) for determining N_D for a lossless network directly from the open-circuit subgraphs.

(*c*) Using the formula derived in (*b*) and without considering the weighting of variables, prove that $N_D = L_D$ for a lossless network.

(*d*) Extend these results to the RC and RL cases.

8.19. Show that i_1 and i_2 in Prob. 8.12 are related to v_c and i_L through a nonsingular linear transformation, as indicated in Eq. (8.4.5). Repeat for v_1 and v_2. Find the P and Q matrices.

8.20. Given $\dot{Y} = AY$, define a linear transformation $Y = QZ$, where Q is an $n \times n$ square matrix and Z is an $n \times 1$ column matrix equation of the form $\dot{Z} = BZ$, B an $n \times n$ square matrix, and show that the eigenvalues of B are identical to those of A.

Application

8.21. (*a*) In Fig. 8.1, let $R_1 = R_2 = 1$ ohm, $C = 1$ farad, and $L = 1$ henry. Find the current i_1 for an excitation of $e = e^{-t}$ applied at $t = 0$, assuming $i_1 = 1$ amp, $di/dt = 0$ amp/sec just prior to that time.

(*b*) Insert a resistor R in series with e. Choose R so that the combined network has only real-axis natural frequencies. For your choice of R and the excitation and initial conditions of (*a*), find i_1.

8.22. The passive network shown is initially at rest. It is excited at $t = 0$ by a 1-volt battery having a variable internal resistance of R ohms. When $R = 1$ ohm, the current is $i = e^{-t} - 2e^{-3t} + e^{-4t}$. Choose a new setting for R such that i has an e^{-2t} term. What other natural frequencies result for this setting of R?

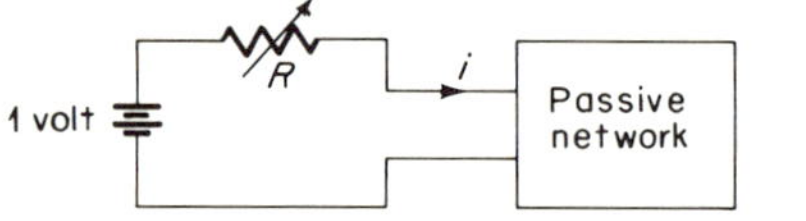

Fig. P8.22

8.23. Using natural-frequency concepts, find the circuits within the boxes in the network. The driving-point impedance and RC products of the network are

$$Z = \frac{s^2 + 13s + 32}{3s^2 + 27s + 44} \qquad R_1C_1 = 1 \qquad R_2C_2 = \tfrac{1}{4}$$

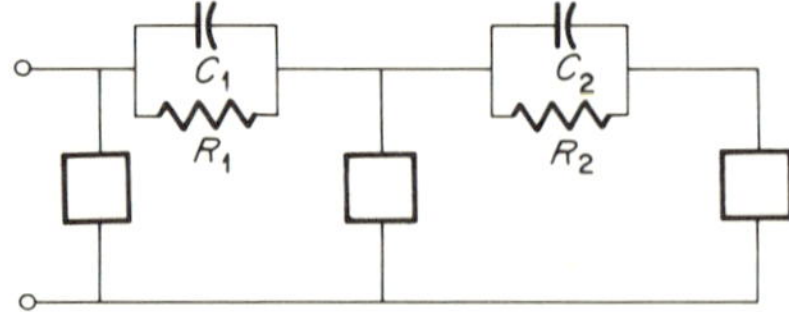

Fig. P8.23

8.24. The driving-point impedance of a passive network is

$$Z = \frac{6s^3 + 13s^2 + 12s + 4}{5s^3 + 14s^2 + 14s + 4}$$

A realization in the form shown is desired, in which each Z box can contain interconnected R, L, and C elements but with no coupling. Using natural-frequency concepts, find R_1 and R_2 and networks for Z_a and Z_b. *Note:* For a driving-point impedance of the form

$$Z = \frac{as^2 + bs + c}{ds^2 + es + f}$$

to be realizable, it is necessary that all coefficients be positive and that $(as^2 + c)(ds^2 + f) - bes^2 \geq 0$ for all $s = j\omega$.

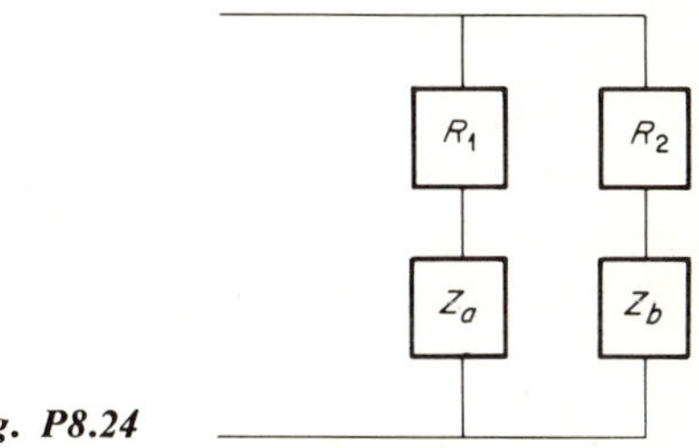

Fig. P8.24

8.25. Entries have been created at points 1-1′ and 2-2′ in the given network. Consider these as the input and output terminals, respectively, of a 2-port.

(*a*) Accepting that the poles and zeros of a driving-point function of a lossless 2-port must alternate, draw plots of the pole-zero locations of the Z parameters, to the extent possible.

(*b*) Determine the number of natural frequencies that the network has with the top inductor shorted in two ways: (1) the method of Sec. 8.10 and (2) the relations in Sec. 8.5, after having found n for the network with the inductor unshorted.

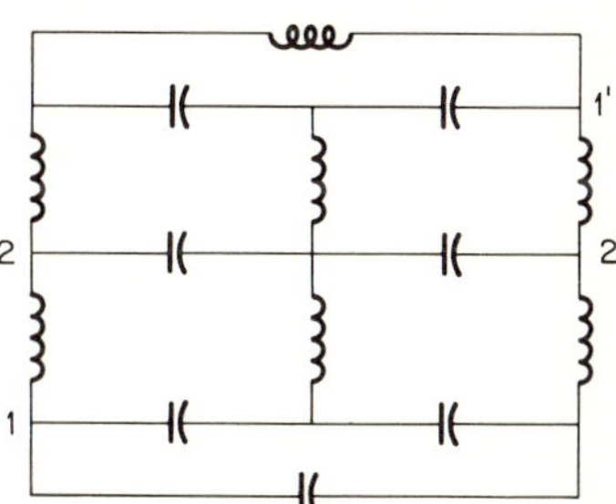

Fig. P8.25

8.26. The network within the box is lossless. The unattached leads inside the box connect with each other through additional capacitors and inductors, but there are no completely inductive or completely capacitive paths between terminals, except as shown. Given that the driving-point impedance at terminals 1-1′ with 2-2′ open and 3-3′ shorted has three finite, nonzero poles and knowing that the poles and zeros of a lossless network must

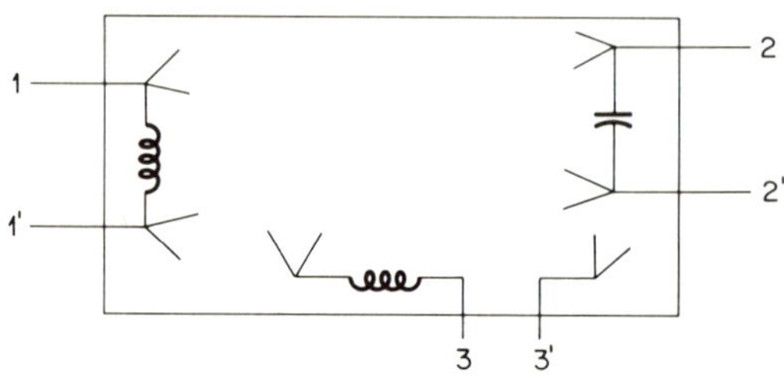

Fig. P8.26

alternate, draw the following:

(*a*) The pole-zero pattern of $Z_{1\text{-}1'}$ with 2-2′ open and 3-3′ shorted
(*b*) The pole-zero pattern of $Z_{2\text{-}2'}$ with 1-1′ open and 3-3′ shorted
(*c*) The pole-zero pattern of $Z_{3\text{-}3'}$ with 1-1′ and 2-2′ open
(*d*) The pole-zero pattern of $Z_{2\text{-}2'}$ with 1-1′ and 3-3′ shorted

8.27. (*a*) Find n for the network in Fig. P8.25. Now let the top L and the bottom C be lossy, i.e., replaced by a series RL and a shunt RC circuit, respectively. Again find n.

(*b*) Replace each L with a C and each C with an L in the network in Fig. P8.25. Find n. Now let the top C and the bottom L be lossy and again find n.

(*c*) From the results of parts (*a*) and (*b*) what conclusions can you reach concerning the effect of loss on the number of natural frequencies that a given lossless network has?

9 TRANSFORM METHODS

9.1 INTRODUCTION

In this chapter we review and deepen the reader's understanding of the mathematical tools used in finding the total response (free plus forced) of a linear system to a prescribed excitation. We begin with a derivation of the open (sine and cosine) form of the Fourier series, stressing the fact that the coefficients of the series are chosen so as to minimize the mean square error between the actual waveform and its approximant. The closed (exponential) form of the Fourier series is then developed, and the equivalence of the two forms is demonstrated. Fourier spectra are considered, and from them certain properties of the Fourier series of pulse trains are derived in terms of pulse width and pulse period. Two limiting cases are considered, namely, the impulse train and the single nonrecurring pulse.

It is shown that the Fourier series is not useful in describing a single nonrecurring pulse. This deficiency leads us to the Fourier transform pairs, which are derived from the Fourier series pairs. We then point out that the Fourier transform pairs, though very useful, have the limitation of being incapable of easily handling functions that do not vanish at $t = \pm\infty$. This deficiency leads to the Laplace transform pairs, which are derived from the Fourier transform pairs by introducing two converging factors into the latter.

At this point in the chapter we digress briefly to consider the important family of singularity functions—impulse, step, ramp, parabola—and to develop their properties. A table of Laplace transforms of these and other functions is then composed, not by using the forward Laplace transform integral directly but by developing a set of Laplace transform theorems (extremely useful in their own right) and using them.

The chapter concludes with a discussion of the convolution theorem, presented as an alternative to the Laplace transform method of finding the response of a system, such that the user is able to make his analysis completely in the time domain, instead of transforming to the s domain and then back. Particular attention is paid to graphical convolution, primarily because of the insight it provides into the convolution process.

9.2 FOURIER SERIES PAIRS

Assume that a function $f(x)$ satisfies a set of conditions, called the *Dirichlet conditions*, within the interval $-\pi < x < \pi$. These conditions are:

1. $f(x)$ must be finite and single-valued at every point within the interval.
2. $f(x)$ must be continuous within the interval except for a finite number of finite discontinuities. At a discontinuity inside the interval at $x = a$, the value of $f(x)$ is defined to be

$$f(a) = \lim_{\Delta x \to 0} \tfrac{1}{2}[f(a + \Delta x) + f(a - \Delta x)] \tag{9.2.1}$$

These conditions may require explanation. The adjectives finite and single-valued rule out the waveforms in Fig. 9.1a and b, respectively. The waveform in Fig. 9.1c is permitted: it has a finite number of discontinuities, and

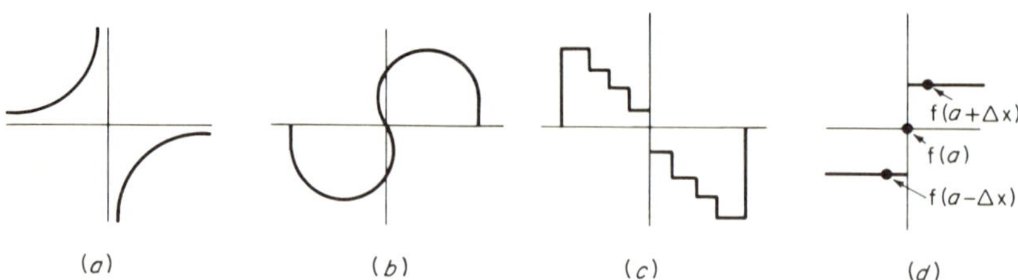

Fig. 9.1 The Dirichlet conditions.

each is finite in height. Figure 9.1*d* illustrates the evaluation of $f(x)$ at a discontinuity, as given by Eq. (9.2.1).

Let us attempt to represent $f(x)$ within the given interval by a second function $F(x)$, given by

$$F(x) = a_0 + a_1 \cos x + \cdots + a_n \cos nx + b_1 \sin x + \cdots + b_n \sin nx \tag{9.2.2}$$

or

$$F(x) = a_0 + \sum_{k=1}^{n} (a_k \cos kx + b_k \sin kx) \tag{9.2.3}$$

where k is a positive integer and the a's and b's are arbitrary real coefficients. We ask the following question: What choice of the $2n + 1$ coefficients will make $F(x)$ the best possible approximation to $f(x)$ in the interval?

A complete answer to this question would require a digression into approximation theory and error criteria,† which is beyond the scope of this text. Suffice it to say that the coefficients in Eq. (9.2.2) are customarily chosen to minimize the mean square error between $f(x)$ and $F(x)$, represented by

$$E = \frac{1}{2\pi} \int_{-\pi}^{\pi} [f(x) - F(x)]^2 \, dx \tag{9.2.4}$$

This minimization is accomplished by inserting $F(x)$ of Eq. (9.2.3) into Eq. (9.2.4), taking the partial derivatives of E with respect to the coefficients in $F(x)$, setting these partial derivatives equal to zero, and solving for the coefficients (see Prob. 9.18). The result is that to minimize E it is necessary that

$$\int_{-\pi}^{\pi} F(x)\, dx = \int_{-\pi}^{\pi} f(x)\, dx \tag{9.2.5}$$

$$\int_{-\pi}^{\pi} F(x) \cos lx \, dx = \int_{-\pi}^{\pi} f(x) \cos lx \, dx \tag{9.2.6}$$

$$\int_{-\pi}^{\pi} F(x) \sin lx \, dx = \int_{-\pi}^{\pi} f(x) \sin lx \, dx \tag{9.2.7}$$

where l is an integer which takes on all values from 1 through n.

The integral expressions in Eqs. (9.2.5) to (9.2.7) appear difficult to handle, but actually they are not because of the orthogonality‡ property of sinusoidal

† See, for example, N. Balabanian, "Network Synthesis," chap. 9, Prentice-Hall, Inc., Englewood Cliffs, N.J., 1958.

‡ Two functions $G(x)$ and $H(x)$ are orthogonal over an interval if $\int G(x)H(x)\, dx = 0$ over that interval. For a discussion of orthogonal functions, see F. B. Hildebrand, "Advanced Calculus for Engineers," chap. 5, Prentice-Hall, Inc., Englewood Cliffs, N.J., 1949.

and cosinusoidal functions. This property makes it possible for us to write

$$\int_{-\pi}^{\pi} \cos kx \cos lx \, dx = \pi\delta_{kl} \tag{9.2.8}$$

$$\int_{-\pi}^{\pi} \sin kx \sin lx \, dx = \pi\delta_{kl} \tag{9.2.9}$$

$$\int_{-\pi}^{\pi} \sin kx \cos lx \, dx = 0 \tag{9.2.10}$$

where δ_{kl} is the Kronecker delta, whose values are

$$\delta_{kl} = \begin{cases} 0 & l \neq k \\ 1 & l = k \end{cases} \tag{9.2.11}$$

If we insert $F(x)$ of Eq. (9.2.3) into Eqs. (9.2.5) to (9.2.7), employ the three orthogonality relations, and solve for a_0, a_k, and b_k, we obtain

$$a_0 = \frac{1}{2\pi} \int_{-\pi}^{\pi} f(x) \, dx \tag{9.2.12}$$

$$a_k = \frac{1}{\pi} \int_{-\pi}^{\pi} f(x) \cos kx \, dx \tag{9.2.13}$$

$$b_k = \frac{1}{\pi} \int_{-\pi}^{\pi} f(x) \sin kx \, dx \tag{9.2.14}$$

The expression for $F(x)$ in Eq. (9.2.3) and the coefficients in Eqs. (9.2.12) to (9.2.14) can be expressed in more concise form. To this end, we insert the identities

$$\cos kx = \frac{e^{jkx} + e^{-jkx}}{2} \tag{9.2.15}$$

and

$$\sin kx = \frac{e^{jkx} - e^{-jkx}}{2j} \tag{9.2.16}$$

into Eq. (9.2.3) and obtain

$$F(x) = a_0 + \sum_{1}^{n} \left(\frac{a_k - jb_k}{2} e^{jkx} + \frac{a_k + jb_k}{2} e^{-jkx} \right) \tag{9.2.17}$$

Now let

$$c_k = \frac{a_k - jb_k}{2} \tag{9.2.18}$$

Then

$$F(x) = a_0 + \sum_{1}^{n} (c_k e^{jkx} + c_k^* e^{-jkx}) \tag{9.2.19}$$

and from Eqs. (9.2.13) and (9.2.14)

$$c_k = \frac{1}{2\pi} \int_{-\pi}^{\pi} f(x)(\cos kx - j \sin kx)\, dx \tag{9.2.20}$$

which, from Eqs. (9.2.15) and (9.2.16), can be written as

$$c_k = \frac{1}{2\pi} \int_{-\pi}^{\pi} f(x) e^{-jkx}\, dx \tag{9.2.21}$$

We now have the coefficients in concise form in Eq. (9.2.21), but $F(x)$ in Eq. (9.2.19) is almost as spread out as it was in Eq. (9.2.3). If we could show that $c_k^* = c_{-k}$, we could let k run from $-n$ to n, instead of from 1 to n, and pick up both summations in Eq. (9.2.19), as well as the a_0 term, with a single summation. To this end, note that the coefficient a_k is an even function of x whereas b_k is an odd function of x. Thus

$$a_{-k} = a_k \qquad b_{-k} = -b_k \tag{9.2.22}$$

and we can write, using Eq. (9.2.18),

$$c_k^* = \frac{a_k + jb_k}{2} = \frac{a_{-k} - jb_{-k}}{2} = c_{-k} \tag{9.2.23}$$

Now Eq. (9.2.19) becomes

$$F(x) = \sum_{-n}^{n} c_k e^{jkx} \tag{9.2.24}$$

As k runs from $-n$ to -1, we develop the second summation in Eq. (9.2.19); as k runs from 1 to n, we develop the first summation. For $k = 0$, c_0 is obtained, which, from a comparison of Eqs. (9.2.12) and (9.2.21), equals a_0, as desired. [One is tempted to use Eq. (9.2.18) and say that $c_0 = a_0/2$. This is incorrect, for k cannot equal 0 in that equation.]

Nowhere has the word "periodic" been mentioned in the discussion. All we have done thus far is to match $F(x)$, the approximating function, to $f(x)$, the given function, in the interval $-\pi$ to π, by choosing the coefficients of $F(x)$ to minimize the mean square error.

The function $F(x)$ in Eqs. (9.2.3) and (9.2.24) is periodic of period 2π; that is,

$$F(x + 2m\pi) = F(x) \tag{9.2.25}$$

where m is any plus or minus integer. If $f(x)$ is also periodic with a period of 2π, then $F(x)$ matches $f(x)$ for all x, rather than just within the original interval. Here is where the major advantage of the Fourier series lies; it permits a periodic function to be represented by a series of sinusoidal and cosinusoidal, or exponential, terms.

Not many periodic functions of interest have periods of 2π rad. We are concerned instead with time functions, whose periods are given in seconds (or milliseconds or microseconds). It is necessary to change variables in Eqs. (9.2.21) and (9.2.24). In these two equations, let

$$x = \frac{2\pi}{T}t \tag{9.2.26}$$

When $x = \pi$, $t = T/2$, and when $x = -\pi$, $t = -T/2$. Thus the period is now T sec instead of 2π rad. Now Eqs. (9.2.21) and (9.2.24) become

$$c_k = \frac{1}{T}\int_{-T/2}^{T/2} f(t)e^{-jk\omega_1 t}\,dt \tag{9.2.27}$$

and

$$F(t) = \sum_{-n}^{n} c_k e^{jk\omega_1} \tag{9.2.28}$$

where

$$\omega_1 = \frac{2\pi}{T} \tag{9.2.29}$$

is the fundamental angular frequency and $k\omega_1$, for each value of k, is the kth harmonic of the fundamental.

With the inclusion of ω_1, Eq. (9.2.27) becomes a transformation equation from the time to the frequency domain. Likewise Eq. (9.2.28) becomes a transformation equation from the frequency to the time domain.

We call Eqs. (9.2.27) and (9.2.28) *Fourier series pairs.* They will be used later in the chapter as a starting point in the development of Fourier transform and Laplace transform pairs, which also relate the time and frequency domains.

Example 9.1 A portion of a rectangular voltage pulse train of period T sec and pulse width τ sec is shown in Fig. 9.2.

(*a*) Find its Fourier series coefficients from Eqs. (9.2.12) to (9.2.14) (with a proper change of variable) and from Eq. (9.2.27).

(*b*) Using the results of part (*a*), find the Fourier series for the pulse train when $\tau = T/2$ and when $\tau = T/6$.

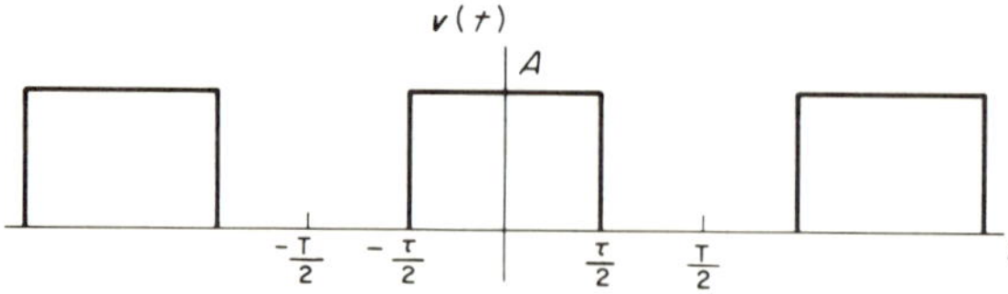

Fig. 9.2

(*a*) The proper change in variable is given by Eq. (9.2.26). Insertion in Eqs. (9.2.12) to (9.2.14) gives

$$a_0 = \frac{1}{T}\int_{-T/2}^{T/2} v(t)\,dt \qquad (9.2.30)$$

$$a_k = \frac{2}{T}\int_{-T/2}^{T/2} v(t)\cos k\frac{2\pi}{T}t\,dt \qquad (9.2.31)$$

$$b_k = \frac{2}{T}\int_{-T/2}^{T/2} v(t)\sin k\frac{2\pi}{T}t\,dt \qquad (9.2.32)$$

The given $v(t)$ is an even function; that is, $v(t) = v(-t)$. For this reason, every b_k is zero. Then

$$a_0 = \frac{1}{T}\int_{-\tau/2}^{\tau/2} A\,dt = \frac{A\tau}{T}$$

and

$$a_k = \frac{2}{T}\int_{-\tau/2}^{\tau/2} A\cos k\frac{2\pi}{T}t\,dt = \frac{2A\tau}{T}\frac{\sin(k\pi\tau/T)}{k\pi\tau/T}$$

If Eq. (9.2.27) is used instead, we have

$$c_k = \frac{1}{T}\int_{-\tau/2}^{\tau/2} Ae^{-jk(2\pi/T)t}\,dt = \frac{A\tau}{T}\frac{\sin(k\pi\tau/T)}{k\pi\tau/T}$$

When $k = 0$, use of L'Hospital's rule on the expression for c_k gives

$$c_0 = \lim_{k\to 0}\frac{A\tau}{T}\frac{\sin(k\pi\tau/T)}{k\pi\tau/T} = \frac{A\tau}{T}$$

Thus $c_0 = a_0$. For all k except $k = 0$, $a_k = 2c_k = 2c_{-k}$, the latter equality resulting because c_k is an even function of k.

(*b*) For $\tau = T/2$,

$$a_0 = \frac{A}{2} \qquad \text{and} \qquad a_k = A\frac{\sin(k\pi/2)}{k\pi/2}$$

giving $a_1 = 2A/\pi$, $a_2 = 0$, $a_3 = -2A/3\pi$, $a_4 = 0$, $a_5 = 2A/5\pi, \ldots$. Thus, from Eq. (9.2.3), with the variable change,

$$V(t) = \frac{A}{2} + \frac{2A}{\pi}(\cos\omega_1 t - \tfrac{1}{3}\cos 3\omega_1 t + \tfrac{1}{5}\cos 5\omega_1 t - \cdots)$$

Alternatively, the c_k's are

$$c_0 = \frac{A}{2} \qquad c_1 = c_{-1} = \frac{A}{\pi} \qquad c_2 = c_{-2} = 0$$

$$c_3 = c_{-3} = -\frac{A}{3\pi} \qquad c_4 = c_{-4} = 0 \qquad c_5 = c_{-5} = \frac{A}{5\pi}$$

Using these in Eq. (9.2.28),

$$V(t) = \frac{A}{2} + \frac{A}{\pi} e^{j\omega_1 t} + \frac{A}{\pi} e^{-j\omega_1 t} - \frac{A}{3\pi} e^{j3\omega_1 t} - \frac{A}{3\pi} e^{-j3\omega_1 t}$$

$$+ \frac{A}{5\pi} e^{j5\omega_1} + \frac{A}{5\pi} e^{-j5\omega_1 t} - \cdots$$

$$= \frac{A}{2} + \frac{2A}{\pi} (\cos \omega_1 t - \tfrac{1}{3} \cos 3\omega_1 t + \tfrac{1}{5} \cos 5\omega_1 t - \cdots)$$

which checks the previous result. Note that each c_k is only one-half of the corresponding a_k but that both c_k and c_{-k} are used to obtain the coefficient of the kth harmonic.

For $\tau = T/6$,

$$c_0 = \frac{A}{6} \qquad c_1 = c_{-1} = \frac{A}{2\pi} \qquad c_2 = c_{-2} = \frac{\sqrt{3}A}{4\pi}$$

$$c_3 = c_{-3} = \frac{A}{3\pi} \qquad c_4 = c_{-4} = \frac{\sqrt{3}A}{8\pi} \qquad c_5 = c_{-5} = \frac{A}{10\pi} \qquad c_6 = c_{-6} = 0$$

Doubling each c_k except c_0 gives

$$V(t) = \frac{A}{6} + \frac{2A}{\pi}\left(\frac{1}{2} \cos \omega_1 t + \frac{\sqrt{3}}{4} \cos 2\omega_1 t + \frac{1}{3} \cos 3\omega_1 t + \frac{\sqrt{3}}{8} \cos 4\omega_1 t + \frac{1}{10} \cos 5\omega_1 t + \cdots\right)$$

A physical interpretation can be given to the $V(t)$ expressions obtained in Example 9.1. In Fig. 9.3a, the Fourier series for the $\tau = T/2$ pulse train is

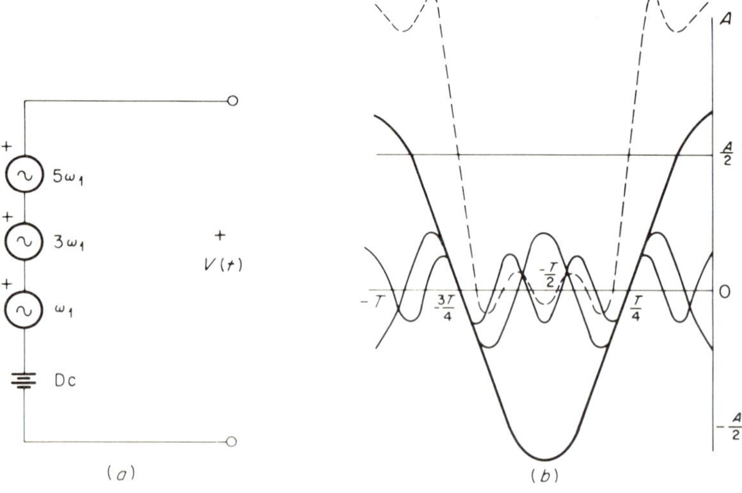

Fig. 9.3 Physical interpretation of a Fourier series.

represented by a battery and three cosine generators connected in series, terms higher than the fifth harmonic being neglected. Each generator and the battery are assumed to have been operating for a finite (very long) time prior to $t = 0$. In Fig. 9.3*b* the graphical addition of these four voltages up to $t = 0$ is shown. The dashed-line result is only an approximation to the given pulse train, but an amazingly good one considering that only three, rather than an infinite number, of harmonics have been included.

9.3 FOURIER SPECTRA

The reader will note that the a_k and b_k coefficients for the pulse train of Example 9.1 are presented in the general form $K(\sin x)/x$. This function occurs frequently in Fourier series work and is plotted, for the case where x is a continuous variable, in Fig. 9.4. We comment that it is an even function of x, that it has its maximum (of value K) at $x = 0$, that it oscillates through positive and negative values as the magnitude of x increases, and that it approaches zero as x approaches plus or minus infinity.

In Example 9.1, $x = k\pi\tau/T$ and is a discrete (rather than a continuous) variable, taking on the values 0, $\pm\pi\tau/T$, $\pm 2\pi\tau/T, \ldots$. Thus if we plot c_k vs. k, we shall obtain a discrete, or line, spectrum, rather than the continuous spectrum of Fig. 9.4. This is done in Fig. 9.5 for the two cases in Example 9.1*b*. The dashed envelope has, in each case, the form of the plot in Fig. 9.4.

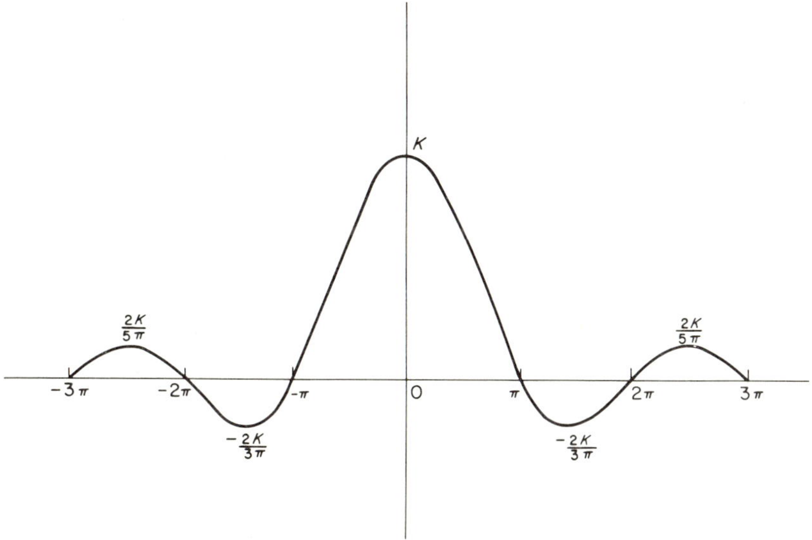

Fig. 9.4 Plot of $K(\sin x)/x$ *versus* x.

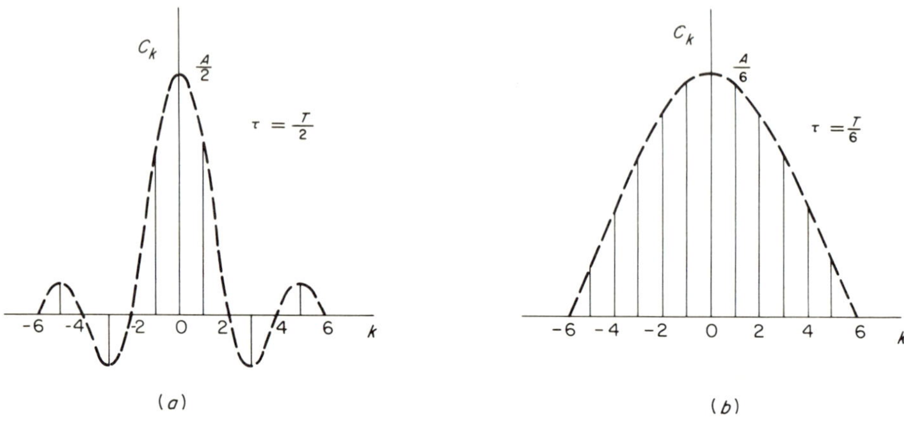

Fig. 9.5 Fourier line spectra for the pulse train in Fig. 9.2.

In each of these plots, a more useful abscissa is the frequency, or reciprocal period. Thus $k = 1$ corresponds to a fundamental frequency $f_1 = 1/T$, $k = 2$ to a second-harmonic frequency $f_2 = 2/T$, and so on. On this basis, the spacing between lines in each plot is

$$\Delta f = f_n - f_{n-1} = f_1 = \frac{1}{T} \tag{9.3.1}$$

The frequency of the first null is $2/T$ for $\tau = T/2$ and $6/T$ for $\tau = T/6$. Thus in general it is

$$f_c = \frac{1}{\tau} \tag{9.3.2}$$

where the c subscript denotes cutoff. Note that if τ is small compared with T, the rectangular pulse train is quite accurately represented by a Fourier series containing just those terms whose frequency is less than f_c (see Prob. 9.22). In this case f_c is effectively the bandwidth of the pulse.

The number of ac components out to the first null, including the first null component, is 2 for $\tau = T/2$ and 6 for $\tau = T/6$. In general, this number is

$$n = \frac{T}{\tau} \tag{9.3.3}$$

Finally, the maximum amplitude of the spectrum envelope is

$$c_0 = A\frac{\tau}{T} \tag{9.3.4}$$

Thus we see that the fundamental frequency and the frequency spacing between lines are functions of the pulse period only, the cutoff frequency is a function of the pulse width only, and the number of components within the bandwidth and the maximum amplitude of the spectrum envelope are functions of both the pulse period and pulse width.

Suppose a rectangular pulse train is given, like that in Fig. 9.2, whose period and pulse width are adjustable. If we hold T constant while decreasing τ, the fundamental frequency and the spacing between frequencies remain constant. The cutoff frequency increases, and there are more frequency components prior to it. The amplitudes of the frequency components decrease. The increase in f_c means that a greater system bandwidth will be required to pass the pulses with fidelity; i.e., the narrower the pulses, the greater the bandwidth.

As τ approaches 0, the cutoff frequency becomes the bandwidth and approaches infinity, and an infinite number of finite-spaced frequency components of vanishingly small amplitude is required in the Fourier series. If, however, the height of the pulse train can be made to vary inversely with pulse width (let $A = 1/\tau$), then as τ decreases and the pulses become narrower, they also become higher but the area under each pulse remains constant (equal to 1). In the limit, as τ approaches zero, the pulse train becomes an impulse train, where each impulse has zero width, infinite height, and unit area. The impulse-train spectrum has infinite bandwidth and an infinite number of finite-spaced finite-height frequency components. More will be said about impulses later, when singularity excitation functions are considered in Sec. 9.6.

Now, using the original adjustable pulse train, keep τ constant and let T increase. In this case, the cutoff frequency is constant, but the fundamental frequency and the frequency spacing decrease. As before, the number of frequency components prior to f_c increases, and the amplitudes of the components decrease. In the limit, as T approaches infinity, we have one nonrecurring pulse. Its spectrum becomes continuous; i.e., the fundamental frequency and the frequency spacing approach zero. The cutoff frequency f_c is fixed and is the pulse bandwidth, and the amplitude of each of the infinite number of frequency terms present approaches zero.

9.4 FOURIER TRANSFORM PAIRS

The attempt in the previous section to write Fourier series pairs for a single nonrecurring pulse resulted in an infinite number of frequency terms within a finite bandwidth, each with zero amplitude. This is a most unsatisfactory result for two reasons: (1) zero-amplitude coefficients are meaningless, and (2) even with nonzero coefficients, if we must add together an infinite number of frequency terms with infinitesimal spacing between each, we should prefer integrating to summing.

Let us tackle each of these problems in turn with an eye toward revamping Eqs. (9.2.27) and (9.2.28) to take care of nonrecurring (infinite-period) pulses. In Eq. (9.2.27), note that each c_k becomes zero as T becomes infinite, because of the $1/T$ factor. New coefficients that do not vanish as T becomes infinite can be created simply by defining

$$c'_k \equiv Tc_k = \frac{2\pi}{\omega_1} c_k \tag{9.4.1}$$

With this change, Eqs. (9.2.27) and (9.2.28) become

$$c'_k = \int_{-T/2}^{T/2} f(t) e^{-jk\omega_1 t}\, dt \tag{9.4.2}$$

and

$$F(t) = \sum_{-n}^{n} \frac{\omega_1}{2\pi} c'_k e^{jk\omega_1 t} \tag{9.4.3}$$

Now the second task is to convert the summation in Eq. (9.4.3) into an integral as T approaches infinity. Recall the following definition of a line integral as the limit of a sum:

$$\int_{-\infty}^{\infty} f(z)\, dz = \lim_{\substack{n\to\infty \\ \Delta z_k \to 0}} \sum_{k=-n}^{n} f(z_k)\, \Delta z_k \tag{9.4.4}$$

The mathematical process in Eq. (9.4.4) is explained physically by noting that the interval in question is broken up into $2n$ parts, not necessarily equal. Then the limit is taken as the number of these parts approaches infinity while the length of the largest part, and thus all others, approaches zero.

The summation in Eq. (9.4.3) resembles but does not exactly match that in Eq. (9.4.4). To obtain an exact match, recall that ω_1 is the fundamental frequency and also the frequency spacing. Also recall that $k\omega_1$ is the kth harmonic. Rename this kth harmonic ω_k and the frequency spacing $\Delta\omega_k$. Then Eqs. (9.4.2) and (9.4.3) can be rewritten as

$$c'(\omega_k) = \int_{-T/2}^{T/2} f(t) e^{-j\omega_k t}\, dt \tag{9.4.5}$$

and

$$F(t) = \frac{1}{2\pi} \sum_{-n}^{n} c'(\omega_k) e^{j\omega_k t}\, \Delta\omega_k \tag{9.4.6}$$

where the coefficients are now functions of ω_k rather than k. Now the summation in Eq. (9.4.6) does exactly match that in Eq. (9.4.4), with the side remark that whereas the Δz_k's can in general be different, all the $\Delta\omega_k$'s are the same. As $n \to \infty$, $F(t)$, the approximant, must approach $f(t)$, the given function. Also ω_k becomes ω and $\Delta\omega_k$ becomes $d\omega$. This forces T, which is inversely proportional to $\Delta\omega_k$, to approach infinity, the desired result. In the limit, Eqs. (9.4.5) and (9.4.6) become

$$F(\omega) = \int_{-\infty}^{\infty} f(t)e^{-j\omega t}\,dt \tag{9.4.7}$$

and

$$f(t) = \frac{1}{2\pi}\int_{-\infty}^{\infty} F(\omega)e^{j\omega t}\,d\omega \tag{9.4.8}$$

where the coefficients (now functions of ω rather than ω_k) have been renamed with the capital letter F to conform to conventional notation. $F(\omega)$ and $f(t)$ are the Fourier transform pairs which relate the time and frequency domains for nonperiodic (infinite-period) functions, just as the Fourier series pairs related these two domains for periodic (finite-period) functions.† Popularly, $F(\omega)$ in Eq. (9.4.7) is called the *forward Fourier transform*, or simply the Fourier transform, and $f(t)$ in Eq. (9.4.8) is called the *inverse Fourier transform*. The Fourier transform pairs are valid for any function $f(t)$ that obeys the Dirichlet conditions and, in addition, yields a finite value for $\int_{-\infty}^{\infty} |f(t)|\,dt$. This latter constraint is required to assure that the integral in Eq. (9.4.7) does not become infinite at either limit.

$F(\omega)$ in Eq. (9.4.7) is, in general, a complex function of ω because of the $e^{-j\omega t}$ present inside the integral. However, if $f(t)$ is even, or if $f(t)$ is odd, certain simplifications result. These will now be explored.

Any function $f(t)$ can be written in terms of its even and odd parts such that

$$f(t) = f_e(t) + f_o(t) \tag{9.4.9}$$

where e and o stand for even and odd. In Eq. (9.4.9), if t is replaced by $-t$, we have

$$f(-t) = f_e(-t) + f_o(-t) = f_e(t) - f_o(t) \tag{9.4.10}$$

Equations (9.4.9) and (9.4.10) can be solved for $f_e(t)$ and $f_o(t)$ to give

$$f_e(t) = \tfrac{1}{2}[f(t) + f(-t)] \tag{9.4.11}$$

$$f_o(t) = \tfrac{1}{2}[f(t) - f(-t)] \tag{9.4.12}$$

Thus, given $f(t)$, we can always find its even and odd parts. How to do this graphically is shown in Fig. 9.6. The given function appears in Fig. 9.6*a*. First $f(-t)$ is drawn in Fig. 9.6*b*. Then Eqs. (9.4.11) and (9.4.12) are used to obtain Fig. 9.6*c* and *d*.

† Some authors prefer to display the Fourier transform pairs with the 2π factor in Eq. (9.4.7); see, for example, Ref. 9.2, pp. 200–201.

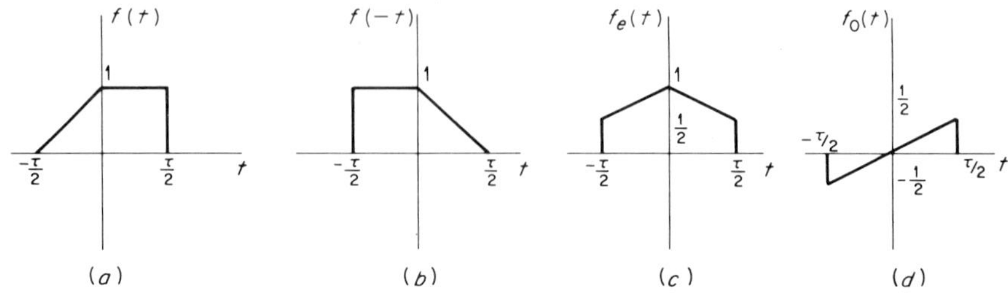

Fig. 9.6 Representation of a function by its even and odd parts.

Using Eq. (9.4.11), Eq. (9.4.7) can be rewritten in the form

$$F(\omega) = \int_{-\infty}^{\infty} [f_e(t) + f_o(t)](\cos \omega t - j \sin \omega t)\, dt$$

$$= \int_{-\infty}^{\infty} [f_e(t) \cos \omega t + f_o(t) \cos \omega t - jf_e(t) \sin \omega t - jf_o(t) \sin \omega t]\, dt \tag{9.4.13}$$

The second and third terms in Eq. (9.4.14) are odd functions of t (odd times even and even times odd yield odd) and thus vanish when the integrations are performed and the limits are inserted. The first and fourth terms are even functions of t (even times even and odd times odd yield even) and do not vanish. Thus we can write

$$F(\omega) = 2\int_{0}^{\infty} [f_e(t) \cos \omega t - jf_o(t) \sin \omega t]\, dt \tag{9.4.14}$$

Defining the real and imaginary parts of $F(\omega)$ by $F_r(\omega)$ and $F_i(\omega)$, Eq. (9.4.14) yields

$$F_r(\omega) = 2\int_{0}^{\infty} f_e(t) \cos \omega t\, dt \tag{9.4.15}$$

$$F_i(\omega) = -2\int_{0}^{\infty} f_o(t) \sin \omega t\, dt \tag{9.4.16}$$

It was noted earlier that $F(\omega)$ is, in general, a complex function of ω. But if $f(t)$ is an even function of time, Eq. (9.4.15) shows that $F(\omega)$ will be real and even in ω, whereas if $f(t)$ is odd in t, Eq. (9.4.16) shows that $F(\omega)$ will be imaginary and odd in ω.

A similar development can be carried out using the real and imaginary parts of $F(\omega)$ and Eq. (9.4.8). Inserting these quantities into that equation yields

$$f(t) = \frac{1}{2\pi}\int_{-\infty}^{\infty} [F_r(\omega) + jF_i(\omega)](\cos t\omega + j \sin t\omega)\, d\omega \tag{9.4.17}$$

which can be expanded to

$$f(t) = \frac{1}{2\pi}\int_{-\infty}^{\infty} [F_r(\omega)\cos t\omega - F_i(\omega)\sin t\omega + jF_r(\omega)\sin t\omega + jF_i(\omega)\cos t\omega]\,d\omega \qquad (9.4.18)$$

The third and fourth terms in Eq. (9.4.18) are odd functions of ω and hence vanish when the integrations are performed and the limits are inserted. The first and second terms are even in ω and do not vanish. Then Eq. (9.4.18) becomes

$$f(t) = \frac{1}{\pi}\int_{0}^{\infty} [F_r(\omega)\cos t\omega - F_i(\omega)\sin t\omega]\,d\omega \qquad (9.4.19)$$

The first term in Eq. (9.4.19) is even in time, the second odd. Thus

$$f_e(t) = \frac{1}{\pi}\int_{0}^{\infty} F_r(\omega)\cos t\omega\,d\omega \qquad (9.4.20)$$

$$f_o(t) = -\frac{1}{\pi}\int_{0}^{\infty} F_i(\omega)\sin t\omega\,d\omega \qquad (9.4.21)$$

Equations (9.4.20) and (9.4.21) provide a means of obtaining the even and odd parts of a time function directly from its real and imaginary parts in the ω domain.

A case of importance and practical interest is when $f(t) = 0$ for $t < 0$. From Eqs. (9.4.11) and (9.4.12) it follows that

$$f_e(t) = -f_o(t) \qquad \text{for } t < 0$$

$$f_e(t) = f_o(t) = \tfrac{1}{2}f(t) \qquad \text{for } t > 0 \qquad (9.4.22)$$

Equations (9.4.22) permit Eqs. (9.4.15) and (9.4.16) to be rewritten as

$$F_r(\omega) = \int_{0}^{\infty} f(t)\cos\omega t\,dt \qquad (9.4.23)$$

and

$$F_i(\omega) = -\int_{0}^{\infty} f(t)\sin\omega t\,dt \qquad (9.4.24)$$

and likewise permit Eqs. (9.4.20) and (9.4.21) to become

$$f(t) = 2f_e(t) = \frac{2}{\pi}\int_{0}^{\infty} F_r(\omega)\cos t\omega\,d\omega \qquad t > 0 \qquad (9.4.25)$$

and

$$f(t) = 2f_o(t) = -\frac{2}{\pi}\int_{0}^{\infty} F_i(\omega)\sin t\omega\,d\omega \qquad t > 0 \qquad (9.4.26)$$

The real integrals in Eqs. (9.4.25) and (9.4.26) are often easier to handle than the complex integral in Eq. (9.4.8). They show that $F_r(\omega)$ and $F_i(\omega)$ are sufficient in themselves to determine $f(t)$ if $f(t) = 0$ for $t < 0$.

Example 9.2 Referring to Fig. 9.2, assume that only the second of the three pulses is present. Find its Fourier transform and compare it with the equation for the c_k's in Example 9.1.

Using Eq. (9.4.7),

$$V(\omega) = \int_{-\infty}^{\infty} v(t)e^{-j\omega t}\,dt = \int_{-\tau/2}^{\tau/2} Ae^{-j\omega t}\,dt$$

$$= A\tau \frac{\sin(\omega\tau/2)}{\omega\tau/2}$$

is the Fourier transform. The c_k expression with which it is desired to compare this result is, from Example 9.1,

$$c_k = \frac{A\tau}{T}\frac{\sin(k\pi\tau/T)}{k\pi\tau/T} = \frac{A\tau}{T}\frac{\sin(k\omega_1\tau/2)}{k\omega_1\tau/2}$$

the second expression resulting from the use of Eq. (9.2.29).

The expressions for c_k and $V(\omega)$ are seen to be very similar, the two differences being the absence of the $1/T$ factor in the $V(\omega)$ expression, caused by Eq. (9.4.1), and the change from the discrete variable $k\omega_1$ to the continuous variable ω. We observe that for the periodic case it is not necessary to know τ in order to find the c_k's. All that is needed is the τ/T ratio. For the nonperiodic case, τ must be known.

The same physical interpretation for the nonperiodic case can be drawn as in Fig. 9.3*a* for the periodic case except that now with all frequencies present, the frequency of the kth generator differs only infinitesimally from that of the $(k + 1)$st generator. Many more generators would be required in this case to attain a degree of approximation to the single pulse comparable to that achieved for the pulse train by the three generators in Fig. 9.3*a*.

Example 9.3 Given $V(\omega) = A\tau \dfrac{\sin(\omega\tau/2)}{\omega\tau/2}$, find the inverse Fourier transform $v(t)$ and show that it is identical with the middle pulse in Fig. 9.2.

To obtain the inverse transform, we use Eq. (9.4.8) and write

$$v(t) = \frac{1}{2\pi}\int_{-\infty}^{\infty} A\tau \frac{\sin(\omega\tau/2)}{\omega\tau/2} e^{j\omega t}\,d\omega$$

which can be expanded to give

$$v(t) = \frac{A\tau}{2\pi}\int_{-\infty}^{\infty} \frac{\sin(\omega\tau/2)}{\omega\tau/2}(\cos\omega t + j\sin\omega t)\,d\omega$$

The second integrand is odd in ω, and thus its integral from $-\infty$ to $+\infty$ vanishes. The first integrand is even in ω. Therefore its integral is twice the integral from 0 to ∞.

This yields

$$v(t) = \frac{A\tau}{\pi}\int_0^\infty \frac{\sin(\omega\tau/2)}{\omega\tau/2}\cos\omega t\, d\omega$$

[Note that this expression for $v(t)$ could also have been obtained directly from Eq. (9.4.20), since $V_i(\omega) = 0$.]

Through trigonometric identities, $v(t)$ can be rewritten as

$$v(t) = \int_0^\infty \frac{A\tau}{2\pi}\frac{\sin\omega(t+\tau/2)}{\omega\tau/2}\,d\omega - \int_0^\infty \frac{A\tau}{2\pi}\frac{\sin\omega(t-\tau/2)}{\omega\tau/2}\,d\omega \qquad (9.4.27)$$

The integrals in Eq. (9.4.27) are known as *sine integrals* and cannot be evaluated in closed form. It is necessary to expand each into a power series and integrate term by term. Fortunately this work has been done and the results tabulated† for various values of x for the general form

$$\text{Si}\, x = \int_0^x \frac{\sin v}{v}\,dv \qquad (9.4.28)$$

A plot of Eq. (9.4.28) appears in Fig. 9.7. Note that $\text{Si}\,x$ is an odd function of x, approaching $\pi/2$ for large positive x and $-\pi/2$ for large negative x. It has peaks at $x = \pm(2n-1)\pi$ and valleys at $\pm 2n\pi$, where n is any integer.

The integrals in Eq. (9.4.27) are nearly in the general form of Eq. (9.4.28). If we change variables by letting $v = \omega(t+\tau/2)$ in the first of these and $v = \omega(t-\tau/2)$ in the second, each integral becomes $A/\pi\,\text{Si}\,x$. The value of x in the first integral is $+\infty$ for $t > -\tau/2$ and $-\infty$ for $t < -\tau/2$. Thus the first integral is $+A/2$ for $t > -\tau/2$ and $-A/2$ for $t < -\tau/2$. For $t = -\tau/2$, the first integral vanishes. In the second integral, x is $+\infty$ for $t > \tau/2$ and $-\infty$ for $t < \tau/2$. Thus the second integral is $+A/2$ for $t > \tau/2$ and $-A/2$ for $t < \tau/2$. It vanishes for $t = \tau/2$.

† See, for example, "CRC Standard Mathematical Tables," Chemical Rubber Publishing Co., Cleveland, Ohio, 1963, or E. Jahnke and I. F. Emde, "Tables of Functions," Dover Publications, Inc., New York, 1945.

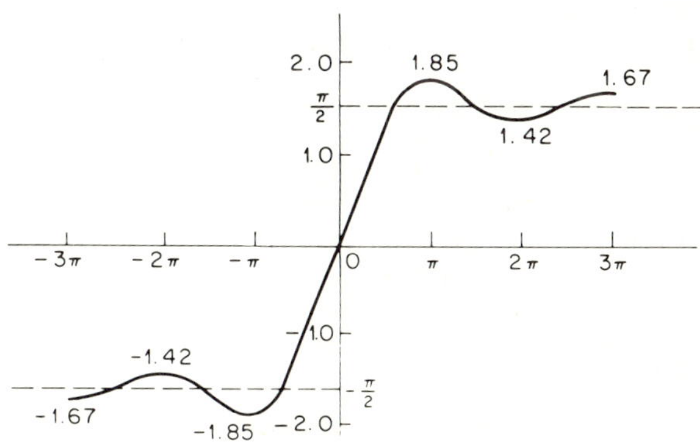

Fig. 9.7 Plot of Si *x versus x.*

Summarizing these results gives:

t	*First integral*	*Second integral*	$v(t)$
$-\frac{T}{2} < t < +\frac{T}{2}$	$\frac{A}{2}$	$-\frac{A}{2}$	A
$\frac{T}{2}$	$\frac{A}{2}$	0	$\frac{A}{2}$
$-\frac{T}{2}$	0	$-\frac{A}{2}$	$\frac{A}{2}$
$> \frac{T}{2}$	$\frac{A}{2}$	$\frac{A}{2}$	0
$< -\frac{T}{2}$	$-\frac{A}{2}$	$-\frac{A}{2}$	0

The resulting $v(t)$ is indeed the middle pulse in Fig. 9.2, as we set out to show.

This example has been included for two reasons: (1) it provided an opportunity to introduce the sine integral, a function that appears regularly in discussions of idealized filters, and (2) it demonstrated that it is often more difficult to obtain the inverse transform than it is the corresponding forward transform.

9.5 LAPLACE TRANSFORM PAIRS

Two very common excitations of electric networks are $e^{-\sigma t}$ and $u(t)$. The former is a decaying exponential, whereas the latter is the unit step function, having the value 0 for $t < 0$ and 1 for $t > 0$. If Eq. (9.4.7) is used to obtain the Fourier transform of each of these functions, immediate trouble is encountered because neither satisfies the condition that $\int_{-\infty}^{\infty} |f(t)|\,dt$ be finite. Thus these functions do not possess Fourier transforms.† However the Laplace transforms of these functions, and of many others that do not possess Fourier transforms, do exist. We now derive the Laplace transform pairs.

Define a function $g(t)$ by

$$g(t) = f(t)u(t)e^{-\sigma t} \tag{9.5.1}$$

where σ is a positive real number.

† Through the use of the impulse function, to be described in the next section, we can find a function of ω whose inverse transform is $u(t)$ (see Prob. 9.23).

The Fourier transform of $g(t)$ is, from Eq. (9.4.7),

$$G(\omega) = \int_{-\infty}^{\infty} f(t)u(t)e^{-(\sigma+j\omega)t}\, dt \tag{9.5.2}$$

The inverse transform of $G(\omega)$ is given by Eq. (9.4.8) as

$$g(t) = \frac{1}{2\pi}\int_{-\infty}^{\infty} G(\omega)e^{j\omega t}\, d\omega \tag{9.5.3}$$

The function $g(t)$ contains $e^{-\sigma t}$. Moving this quantity inside the integral on the right side of Eq. (9.5.3) changes that equation to

$$f(t)u(t) = \frac{1}{2\pi}\int_{-\infty}^{\infty} G(\omega)e^{(\sigma+j\omega)t}\, d\omega \tag{9.5.4}$$

Now, in Eqs. (9.5.2) and (9.5.4), let

$$s = \sigma + j\omega \qquad ds = j\, d\omega \tag{9.5.5}$$

The limits of $\pm\infty$ on ω become $\sigma \pm j\infty$ on s. The equations now read

$$G(s) = \int_{-\infty}^{\infty} f(t)u(t)e^{-st}\, dt \tag{9.5.6}$$

$$f(t)u(t) = \frac{1}{2\pi j}\int_{\sigma-j\infty}^{\sigma+j\infty} G(s)e^{st}\, ds \tag{9.5.7}$$

The $f(t)$ function in Eq. (9.5.1) was assumed to exist for all t, including negative time. If now we restrict $f(t)$ to be zero prior to $t = 0$, the $u(t)$ function associated with it is superfluous and Eqs. (9.5.6) and (9.5.7) can be rewritten to obtain

$$F(s) = \int_{0}^{\infty} f(t)e^{-st}\, dt \tag{9.5.8}$$

$$f(t) = \frac{1}{2\pi j}\int_{\sigma-j\infty}^{\sigma+j\infty} F(s)e^{st}\, ds \tag{9.5.9}$$

These are the Laplace transform pairs. The lower limit in Eq. (9.5.8) differs from that in Eq. (9.5.6) because of the new restriction imposed on $f(t)$. The name change from $G(s)$ to $F(s)$ is to use the same letter in each domain to represent the function, as in the Fourier transform case.

The Laplace transform pairs in Eqs. (9.5.8) and (9.5.9) differ only slightly from the Fourier transform pairs in Eqs. (9.4.7) and (9.4.8). Both transforms exist only for functions that satisfy the Dirichlet conditions. But, in addition, Fourier transformable functions yield a finite value for $\int_{-\infty}^{\infty} |f(t)|\, dt$, whereas Laplace transformable functions must be zero for $t < 0$ and must be of exponential order; i.e., for some positive value of σ (Re s) in Eq. (9.5.8), $|f(t)|\,e^{-\sigma t}$ must approach zero as t approaches infinity. Thus the function $f(t) = e^{t^2}$, for example, is not acceptable, whereas the function $f(t) = t^2$ is.

Other than these differences, the transforms are essentially the same, and each can be used in finding the time response of a network to an excitation. The Fourier transform is particularly useful when the magnitude and phase characteristics or the real and imaginary or even and odd parts of a network's transfer function are given or obtainable. On the other hand, the Laplace transform should be used when the transfer function itself or its poles, zeros, and gain constant are available.

Example 9.4 Find the Laplace transform of the pulse in Fig. 9.6 shifted so that it originates at $t = 0$.

It should be pointed out that the Laplace transform of the unshifted pulse does not exist. What would be obtained if Eq. (9.5.8) were used for this case is the Laplace transform of the right half of this pulse. When using the Laplace transform, we are in reality finding the transform of $f(t)u(t)$, as shown in Eq. (9.5.6).

Now, to find the transform of the shifted pulse, Eq. (9.5.8) gives

$$F(s) = \int_0^T Ae^{-st}\,dt = -\frac{A}{s}\,e^{-st}\Big|_0^T = \frac{A}{s}\,(1 - e^{-sT})$$

9.6 SINGULARITY FUNCTIONS

Before proceeding further with the development of the Laplace transform, it is pertinent to inquire what types of excitation functions are likely to be encountered in electric networks. In Chap. 8 we discussed the family of excitations obtainable from the general exponential excitation e^{st}. Included in this family were the sinusoidal excitation, the damped sinusoidal excitation, the decaying exponential, and, with $s = 0$, the unit step function. This latter function is also a member of a second family of excitations referred to as *singularity functions*, which we now discuss. This family includes the step, ramp, parabolic, impulse and doublet excitations, and combinations of them, some often delayed in time with respect to others, to produce pulses of various shapes through superposition.

For example, consider the two excitation pulses in Fig. 9.8 (the arrows indicate that the pulses continue forever without further change). In Fig. 9.8*a*, the triangular voltage pulse is the superposition of:

1. A ramp of slope $+1$ volt/sec starting at $t = 0$ sec
2. A ramp of slope -2 volts/sec starting at $t = 1$ sec
3. A ramp of slope $+2$ volts/sec starting at $t = 3$ sec
4. A ramp of slope -1 volt/sec starting at $t = 4$ sec

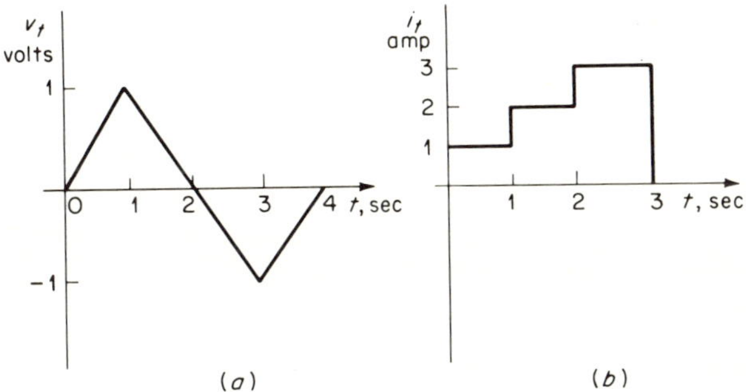

Fig. 9.8 Pulse excitations.

The staircase current pulse in Fig. 9.8*b* is the superposition of:

1. A +1-amp step starting at $t = 0$ sec
2. A +1-amp step starting at $t = 1$ sec
3. A +1-amp step starting at $t = 2$ sec
4. A −3-amp step starting at $t = 3$ sec

The first singularity function to be discussed is the delayed step function, described mathematically by

$$Au(t - a) = \begin{cases} A & t > a \\ 0 & t < a \end{cases} \tag{9.6.1}$$

where A is a real constant. Thus the step function is one which changes instantaneously from 0 to A at the point in time at which its argument $(t - a)$ is zero. When the argument is negative, the step function is zero; it is nonzero (of value A) when the argument is positive.

The quantity a is appropriately called a *delay factor*. If a is zero, there is no delay and the step begins at $t = 0$. Also A is the amplitude of the step, a unit step resulting when $A = 1$. To illustrate, the equation for the staircase pulse in Fig. 9.6*b* is

$$i(t) = u(t) + u(t - 1) + u(t - 2) - 3u(t - 3)$$

Consider the functions $-u(t - a)$, $u(t + a)$, and $u(a - t)$. The first is depicted in Fig. 9.9*a* as a negative unit step starting at $t = a$. The second (Fig. 9.9*b*) is a positive unit step beginning at $t = -a$. The third (Fig. 9.9*c*) is a positive unit step that exists *prior* to $t = a$ (the region in which its argument is positive) but is zero thereafter. It is perhaps clearer to write this last function as $u[-(t - a)]$. Thus a negative sign in front of the function

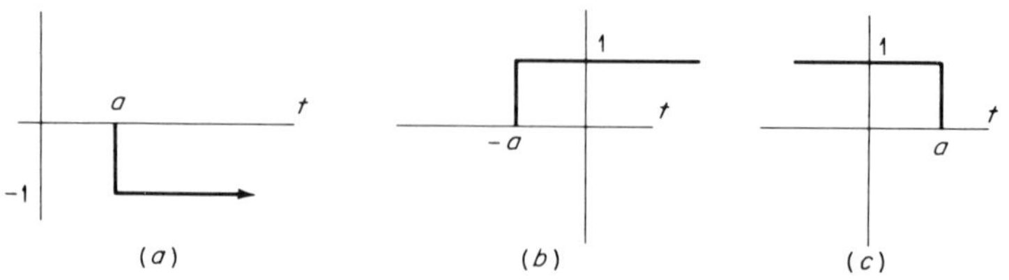

Fig. 9.9 *Step functions:* (a) $-u(t-a)$; (b) $u(t+a)$; (c) $u(a-t)$.

denotes a negative step, whereas a negative sign in front of t in the argument denotes the termination of a step.

The second singularity function to be considered is the delayed ramp function, defined by

$$Ar(t-a) = \begin{cases} A(t-a) & t > a \\ 0 & t < a \end{cases} \tag{9.6.2}$$

This function is zero prior to $t = a$, has a slope of A, and increases linearly with time thereafter. To illustrate, the triangular pulse in Fig. 9.8*a* is expressed in terms of delayed ramps by

$$v(t) = r(t) - 2r(t-1) + 2r(t-3) - r(t-4)$$

The delayed ramp function can also be defined in terms of the delayed step function. Thus

$$Ar(t-a) = A(t-a)u(t-a) \tag{9.6.3}$$

where the delayed step ensures that the ramp is zero prior to $t = a$. This function and two similar ones which do not define a ramp starting at $t = a$ are shown in Fig. 9.10.

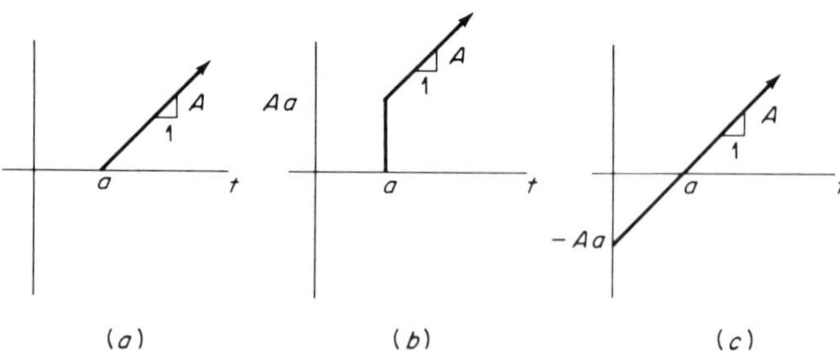

Fig. 9.10 *Ramp functions:* (a) $A(t-a)u(t-a)$; (b) $Atu(t-a)$; (c) $A(t-a)u(t)$.

Equation (9.6.3) suggests that the step and ramp functions are related through integration. Integrate the step in Eq. (9.6.1) to obtain

$$\int_0^t Au(t-a)\,dt = Au(t-a)\int_a^t dt = A(t-a)u(t-a) = Ar(t-a) \tag{9.6.4}$$

The second step in Eq. (9.6.4) is permitted because the integrand is 0 prior to $t = a$ and A thereafter, hence $Au(t-a)$ can be factored out and the lower limit changed.

From Eq. (9.6.4), we arrive at the important result that a delayed ramp is the definite integral from 0 to t of an equally delayed step. Conversely, the delayed step is the first time derivative of an equally delayed ramp.

The third in the family of singularity functions is the parabola, defined by

$$Ap(t-a) = \begin{cases} \dfrac{A}{2}(t-a)^2 & t > a \\ 0 & t < a \end{cases} \tag{9.6.5}$$

or, in terms of the step function, by

$$Ap(t-a) = \frac{A}{2}(t-a)^2 u(t-a) \tag{9.6.6}$$

where again the delayed unit step ensures that the function is zero prior to $t = a$.

The parabola can be related to the ramp through integration in the same fashion that we related the ramp to the step. This will also serve to explain the one-half in the definition. We have

$$\begin{aligned}\int_0^t Ar(t-a)\,dt &= Au(t-a)\int_a^t (t-a)\,dt = Au(t-a)\left[\frac{(t-a)^2}{2}\right]_a^t \\ &= Au(t-a)\frac{(t-a)^2}{2} = Ap(t-a)\end{aligned} \tag{9.6.7}$$

where the second step is justified as in Eq. (9.6.4) because the integrand must be 0 prior to $t = a$ and $A(t-a)$ thereafter.

The result in Eq. (9.6.7) shows that the delayed parabola is the definite integral from 0 to t of an equally delayed ramp and thus the second definite integral from 0 to t of an equally delayed step. Conversely, the ramp is the first derivative of the parabola; the step is the second derivative.

This development could be extended further to include the cubic as the definite integral of the parabola, and then on to higher-order functions, but they are so seldom used that it is of no practical importance to do so.

There are, however, two more members of interest to add to the family of singularity functions. One of these is the *impulse function*, also called the

Dirac delta function. Recall the discussion following Eq. (9.3.4). There we considered what happened to the periodic train of rectangular pulses, each of width τ and height $1/\tau$, as τ was allowed to become very small. In the limit, as τ approached zero, we noted that the pulse train became an impulse train, where each impulse had zero width, infinite height, and unit area.

Now consider the single rectangular pulse in Fig. 9.11*a*. It is like any of those in Fig. 9.2 except that its height is A/τ instead of A. Think of this pulse as being composed of (1) a step of height A/τ starting at $t = a$ and (2) a step of height $-A/\tau$ starting at $t = a + \tau$. This representation, suggested by the dashed lines in Fig. 9.11*a*, can be written mathematically as

$$f(t) = \frac{A}{\tau}\,[u(t - a) - u(t - a - \tau)] \tag{9.6.8}$$

Now take the limit as τ approaches zero to arrive at Fig. 9.11*b* and define a delayed impulse of amplitude A by

$$A\delta(t - a) = \lim_{\tau \to 0} \frac{A}{\tau}\,[u(t - a) - u(t - a - \tau)] \tag{9.6.9}$$

The limiting process is depicted in Fig. 9.12 for the values $A = a = 1$. Note that as the pulse width decreases, the pulse height increases; the center of the pulse moves closer to $t = a$, but the pulse area remains constant. In the limit, the pulse becomes a unit pulse of infinite height, zero width, and unit area located at $t = a$, which can be described by

$$\delta(t - a) = \begin{cases} 0 & t \neq a \\ \infty & t = a \end{cases} \tag{9.6.10}$$

$$\int_{-\infty}^{\infty} \delta(t - a)\, dt = 1$$

The first Eq. (9.6.10) shows that the unit impulse has a nonzero value

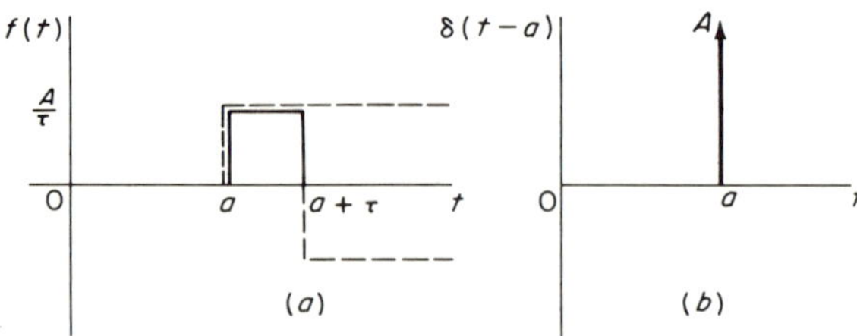

Fig. 9.11 Defining the impulse.

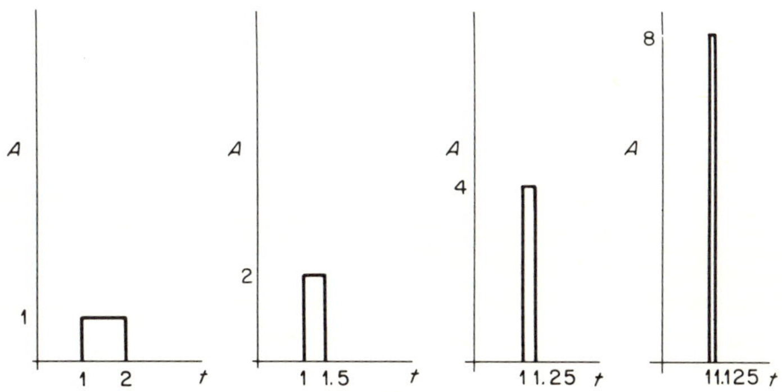

Fig. 9.12 Development of a delayed unit impulse.

only at $t = a$, and the second Eq. (9.6.10) indicates that the total area under its "curve" is unity. The second Eq. (9.6.10) also shows that the integral of a delayed unit impulse is an equally delayed unit step; for as we integrate along the time axis from $-\infty$ to ∞, we obtain 0 area until we reach $t = a$, at which time we suddenly acquire an area of 1 which then persists throughout the rest of the integration. Plotting this acquisition of area vs. time, we obtain the waveform of a unit step originating at $t = a$.

If the step is the integral of the impulse, it follows that the impulse is the derivative of the step. To verify this latter relationship directly, we are tempted to use Eq. (9.6.9), which is in the form of the definition of the derivative of $u(t - a)$. But recall that the definition of the derivative implies a continuous function at the point of differentiation, which the unit step function is not.

There are two ways of resolving this difficulty of differentiating the discontinuous unit step function. One is to define a continuous function having a parameter such that the characteristics of a unit step function are obtained as the parameter is allowed to approach a limit.† Taking the derivative prior to taking the limit yields the unit impulse function as the limit is taken. One such continuous function behaving in this fashion is

$$u(t) = \lim_{\alpha\to\infty} \alpha e^{-\alpha t} \qquad t > 0 \tag{9.6.11}$$

whose derivative is

$$\delta(t) = \lim_{\alpha\to 0} -\alpha^2 e^{-\alpha t} \qquad t > 0 \tag{9.6.12}$$

† See Ref. 9.5 for such a development using the function

$$f(t,\lambda) = \frac{1}{2} + \frac{1}{\pi}\tan^{-1}\lambda t$$

A second way of resolving the difficulty is to make use of the theory of distributions.†

Several interesting and useful properties of the impulse function should be considered. Consider a function $f(t)$ which is continuous at $t = a$. It follows that

$$\int_{-\infty}^{\infty} f(t)\delta(t - a)\, dt = f(a) \tag{9.6.13}$$

because, during the infinitesimal time interval at $t = a$ when the delayed impulse is nonzero, $f(t)$ has the constant value $f(a)$, which can be brought outside the integral.

The result in Eq. (9.6.13) lets us write

$$f(t)\delta(t - a) = f(a)\, \delta(t - a) \tag{9.6.14}$$

It also assists in obtaining the Fourier transform of $\delta(t - a)$ from Eq. (9.4.7). Thus

$$\Delta(\omega) = \int_{-\infty}^{\infty} \delta(t - a)e^{-j\omega t}\, dt = e^{-j\omega a} \tag{9.6.15}$$

If the unit impulse in Eq. (9.6.15) has no time delay, the Fourier transform becomes

$$\Delta(\omega) = 1 \tag{9.6.16}$$

Thus the frequency spectrum of a nondelayed unit impulse is uniformly 1 at all frequencies. Physically we can consider $\delta(t)$ as an infinity of equal-amplitude generators covering the entire frequency spectrum. If such a "signal" is applied to a linear system, the output is the frequency response of the system itself. More will be said later about this application of impulse functions and their close approximants, narrow pulses.

The last singularity function to be considered is the delayed doublet. Recall that to define the impulse we used a limiting process on the difference of two steps. The doublet will be defined in a similar fashion. Consider the pulse pair in Fig. 9.13. It can be synthesized as:

1. A step of height A/τ^2 starting at $t = a$
2. A step of height $-2A/\tau^2$ starting at $t = a + \tau$
3. A step of height A/τ^2 starting at $t = a + 2\tau$

The dashed lines in Fig. 9.13 suggest this representation. The pulse pair can be expressed mathematically by

$$f(t) = \frac{A}{\tau^2}\, [u(t - a) - 2u(t - a - \tau) + u(t - a - 2\tau)] \tag{9.6.17}$$

† See Ref. 9.4, appendix I, for a discussion of the mathematical properties of impulse functions in terms of the theory of distributions.

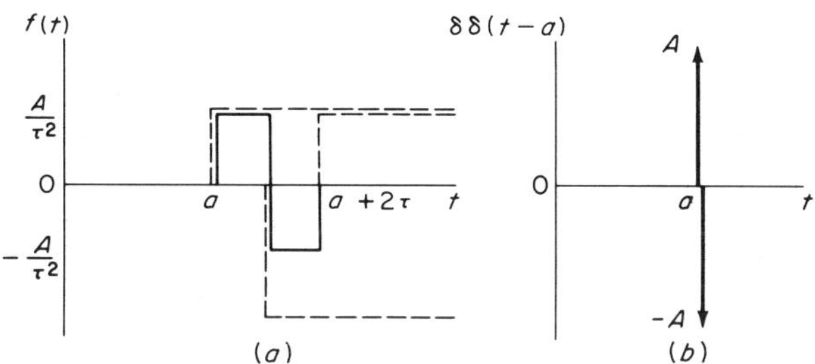

Fig. 9.13 Defining the doublet.

Taking the limit as τ approaches zero yields the delayed doublet

$$A\delta\delta(t-a) = \lim_{\tau\to 0} \frac{A}{\tau^2}\,[u(t-a) - 2u(t-a-\tau) + u(t-a-2\tau)] \tag{9.6.18}$$

As the limit is taken, we arrive at Fig. 9.13*b*. The pulse widths decrease to zero, the pulse heights increase to infinity, both pulses become centered at $t = a$, and the area under each pulse becomes infinite: it does not remain constant as it did for the impulse. However, the net area under the doublet is zero, since the two infinite areas cancel.

We summarize these observations by writing

$$\delta\delta(t-a) = \begin{cases} 0 & t \neq a \\ \pm\infty & t = a \end{cases} \tag{9.6.19}$$

$$\int_{-\infty}^{\infty} \delta\delta(t-a)\,dt = 0$$

The first Eq. (9.6.19) defines the pair of equal and opposite infinite spikes at $t = a$. The second Eq. (9.6.19) states that the net area under them is zero. This equation and the picture of the doublet in Fig. 9.13*b* show that the delayed impulse is the integral of the delayed doublet. For as we integrate along the real axis from $-\infty$ to $+\infty$ in Fig. 9.13*b*, we encounter first the positive spike at $t = a$ and acquire a positive infinite area, which, however, we immediately lose as the negative spike is passed. A plot of this acquisition of area vs. time yields the delayed impulse of Fig. 9.11*b*.

We have discussed five singularity functions and shown that they belong to a family. We have also indicated that any member of the family can be found by successive differentiations or integrations of any other member of the family. This latter attribute is illustrated in Table 9.1. Any number of additional members can be added to the family, the next two being the triplet

TABLE 9.1 Singularity function family

Function	*Symbol*	*Defining equation*
Unit doublet	$\delta\delta(t)$	(9.6.19)
Unit impulse	$\delta(t)$	(9.6.10)
Unit step	$u(t)$	(9.6.1)
Unit ramp	$r(t)$	(9.6.2)
Unit parabola	$p(t)$	(9.6.5)

and the cubic. Each member of the family is the derivative of the member below it in the table and the integral of the member above it.

The following three examples illustrate several of the ideas discussed in this section.

Example 9.5 Find the derivative and the integral of $f(t)$ in Fig. 9.14*a* and sketch each.

From Fig. 9.14*a* we write directly

$$f(t) = u(t) - 2u(t-1) + 2r(t-1) - 3r(t-2) + u(t-3) + r(t-3)$$

The derivative is obtained by differentiating $f(t)$ term by term. This yields

$$f'(t) = \delta(t) - 2\delta(t-1) + 2u(t-1) - 3u(t-2) + \delta(t-3) + u(t-3)$$

The sketch appears in Fig. 9.14*b*. Note the representation of the three impulses by arrows of lengths proportional to their areas.

The integral of $f(t)$ is

$$f^{-1}(t) = r(t) - 2r(t-1) + 2p(t-1) - 3p(t-2) + r(t-3) + p(t-3)$$

which appears in Fig. 9.14*c*. Any point on the curve can be obtained either from the expression for $f^{-1}(t)$ or by evaluating areas under the $f(t)$ plot in Fig. 9.14*a*.

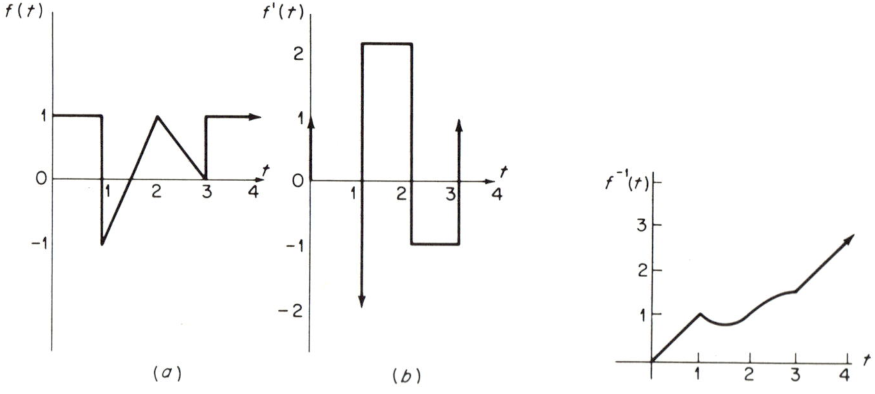

Fig. 9.14 Waveform differentiation and integration.

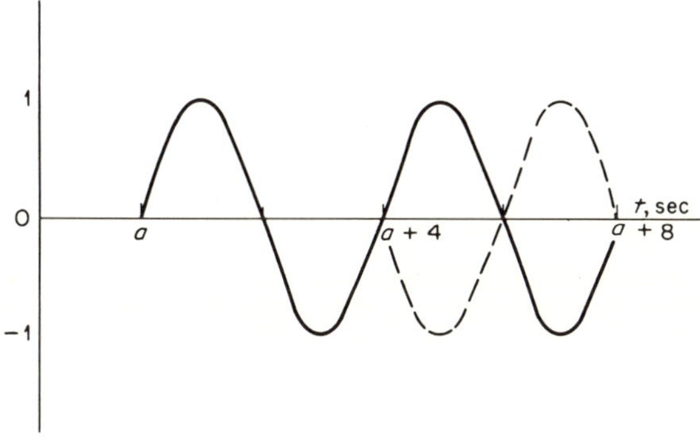

Fig. 9.15

Example 9.6 Write a mathematical expression for the single sine-wave pulse of voltage in Fig. 9.15.

The desired expression is

$$v(t) = \sin \frac{\pi}{2} (t - a)u(t - a) - \sin \frac{\pi}{2} (t - a)u(t - a - 4)$$

It is obtained as the difference of two sine waves, one starting at $t = a$, the other at $t = a + 4$, and each with a period of 4 sec. The cancellation of the component waves after $t = 4$ sec is illustrated in Fig. 9.15.

Example 9.7 The current of Fig. 9.16*b* flows in the inductor of Fig. 9.16*a* as shown. Find the source voltage required to sustain this current.

The current in the inductor is given by

$$i_L(t) = 2r(t) - 2r(t - 1) - 2u(t - 1)$$

Then the inductor voltage is

$$v_L(t) = L \frac{di_L}{dt} = 2u(t) - 2u(t - 1) - 2\delta(t - 1)$$

This voltage also appears across the resistor, so that the resistor current is

$$i_R(t) = 2u(t) - 2u(t - 1) - 2\delta(t - 1)$$

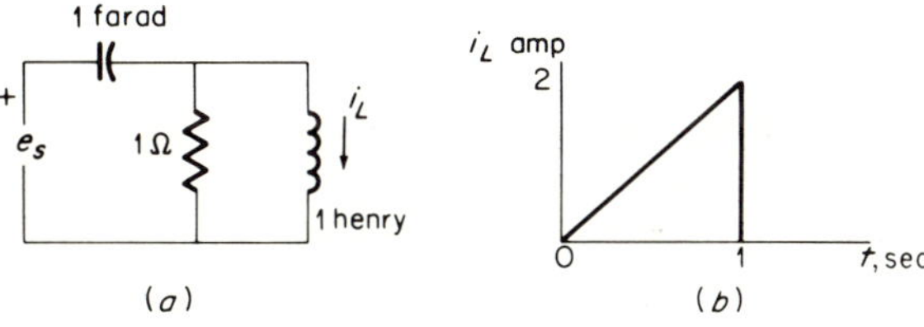

Fig. 9.16

and thus the total current is

$$i(t) = 2r(t) + 2u(t) - 2r(t-1) - 4u(t-1) - 2\delta(t-1)$$

Then the capacitor voltage is calculated as

$$v_c(t) = \frac{1}{C}\int i\,dt = 2p(t) + 2r(t) - 2p(t-1) - 4r(t-1) - 2u(t-1)$$

and the source voltage is

$$e_s(t) = 2p(t) + 2r(t) + 2u(t) - 2p(t-1) - 4r(t-1) - 4u(t-1) - 2\delta(t-1)$$

9.7 LAPLACE TRANSFORM THEOREMS

In the analysis of networks by Laplace transforms, the problem occurs of finding the forward transform [Eq. (9.5.8)] of a variety of functions. We could use Eq. (9.5.8) directly for each new excitation and/or system. But this would become tedious, and it would be helpful if we could tabulate the transforms of some of the simpler excitations and have a set of theorems that would enable us to find the transforms of more complicated functions from them. In this section, such a set of theorems is developed, and it is shown how they can be used in finding Laplace transforms. Tables listing the theorems and the transforms of common functions appear at the end of the section.

Theorem 1 Linearity Let k_1 and k_2 be constants and let $f_1(t)$ and $f_2(t)$ be Laplace transformable functions. Then

$$\begin{aligned}\mathscr{L}[k_1f_1(t) + k_2f_2(t)] &= k_1\mathscr{L}[f_1(t)] + k_2\mathscr{L}[f_2(t)]\\ &= k_1F_1(s) + k_2F_2(s) \qquad (9.7.1)\end{aligned}$$

where $\mathscr{L}$ means Laplace transform of.

The result in Eq. (9.7.1), which can be verified by direct substitution of $k_1f_1(t) + k_2f_2(t)$ into Eq. (9.5.8), is very useful. It permits the Laplace transform of, say, a summation of voltages around a closed loop to be found by finding the Laplace transform of each voltage and then summing the transforms.

Theorem 2 Real (t-domain) differentiation If $f(t)$ and its first n derivatives exist and are Laplace transformable (if the nth derivative is Laplace transformable, they all are), then

$$\mathscr{L}\left[\frac{d^nf}{dt^n}\right] = s^nF(s) - s^{n-1}f_+ - s^{n-2}\left.\frac{df}{dt}\right|_+ - \cdots - \left.\frac{d^{n-1}f}{dt^{n-1}}\right|_+ \qquad (9.7.2)$$

where f_+, $df/dt\,|_+, \ldots, d^{n-1}f/dt^{n-1}\,|_+$ are the values of $f(t)$ and its first $n - 1$ derivatives at $t = 0_+$.

We shall prove Eq. (9.7.2) by finding the Laplace transform of the first derivative and then extending the result to derivatives of higher order.

From Eq. (9.5.8),

$$\mathscr{L}\left[\frac{df}{dt}\right] = \int_0^\infty \frac{df}{dt} e^{-st}\,dt \tag{9.7.3}$$

The integration is performed by parts from the familiar equation

$$\int_0^\infty u\,dv = uv\Big|_0^\infty - \int_0^\infty v\,du \tag{9.7.4}$$

Let

$$u = e^{-st} \qquad \text{and} \qquad dv = df(t) \tag{9.7.5}$$

Then

$$du = -se^{-st} \qquad \text{and} \qquad v = f(t) \tag{9.7.6}$$

Inserting Eqs. (9.7.6) into Eq. (9.7.3) gives

$$\mathscr{L}\left[\frac{df}{dt}\right] = f(t)e^{-st}\Big|_0^\infty + s\int_0^\infty f(t)e^{-st}\,dt \tag{9.7.7}$$

Since $f(t)$ is Laplace transformable, the first term on the right side of Eq. (9.7.6) must vanish at the upper limit, from the discussion following Eq. (9.5.9). Thus we are left with

$$\mathscr{L}\left[\frac{df}{dt}\right] = sF(s) - f_+ \tag{9.7.8}$$

where we have written f_+, rather than f_0, to emphasize that the evaluation of $f(t)$ is to take place just after $t = 0$.

The result of Eq. (9.7.8) can be used to find the Laplace transform of the second derivative, and from it the third and higher derivatives. To this end we note that

$$\frac{d^2f}{dt^2} = \frac{d}{dt}\frac{df}{dt} \tag{9.7.9}$$

Thus all we need do is replace $f(t)$ by df/dt in Eq. (9.7.7) to obtain

$$\mathscr{L}\left[\frac{d^2f}{dt^2}\right] = \frac{df}{dt} e^{-st}\Big|_0^\infty + s\int_0^\infty \frac{df}{dt} e^{-st}\,dt \tag{9.7.10}$$

Assuming that df/dt is Laplace transformable, the first term on the right vanishes at the upper limit. The second term can be evaluated through

Eq. (9.7.7). The overall result is

$$\mathscr{L}\left[\frac{d^2f}{dt^2}\right] = s^2F(s) - sf_+ - \left.\frac{df}{dt}\right|_+ \tag{9.7.11}$$

The extension to higher-order derivatives and Eq. (9.7.2) follows directly.

The value of the result in Eq. (9.7.2) should not be lost in the mathematical manipulations leading to it. What this result shows is that the transform of $f(t)$ is preserved in the s domain during differentiation in the time domain; i.e., the Laplace transforms of $f(t)$ and df/dt both contain $F(s)$. It is this property, and a similar one for integration, that permits us to replace a differential equation in the t domain by an algebraic equation in the s domain when we Laplace transform the former.

Example 9.8 Find the Laplace transform of $\sin \omega t$.

We could, of course, use Eq. (9.5.8) directly to find the required transform. But in keeping with the theme of this section, we shall use the differentiation theorem instead.

Given $f(t) = \sin \omega t$, it follows that $df/dt = \omega \cos \omega t$ and $d^2f/dt^2 = -\omega^2 \sin \omega t = -\omega^2 f(t)$. Now from Eq. (9.7.11)

$$\mathscr{L}\frac{d^2f}{dt^2} = s^2\mathscr{L}[f(t)] - sf_+ - \left.\frac{df}{dt}\right|_+$$

$$f_+ = 0 \qquad \left.\frac{df}{dt}\right|_+ = \omega \qquad \mathscr{L}\frac{d^2f}{dt^2} = -\omega^2\mathscr{L}[f(t)]$$

Thus we have

$$-\omega^2\mathscr{L}[f(t)] = s^2\mathscr{L}[f(t)] - \omega$$

and the desired transform is

$$\mathscr{L}[\sin \omega t] = \frac{\omega}{s^2 + \omega^2}$$

The conclusion from Example 9.8 is that the Laplace transform of any function whose kth derivative is a constant times the original function can be found from the differentiation theorem in Eq. (9.7.2). For example, by this technique the Laplace transforms of $\cos \omega t$, $\sinh \beta t$, $\cosh \alpha t$, and $e^{s_0 t}$ can be found. This last function introduces an additional problem, which the following example and its discussion will illustrate.

Example 9.9 Find the Laplace transform of $e^{s_0 t}$.

From Eq. (9.7.8), we write

$$\mathscr{L}(s_0 e^{s_0 t}) = s_0\mathscr{L}(e^{s_0 t}) = s\mathscr{L}(e^{s_0 t}) - 1$$

Thus

$$\mathscr{L}(e^{s_0 t}) = \frac{1}{s - s_0}$$

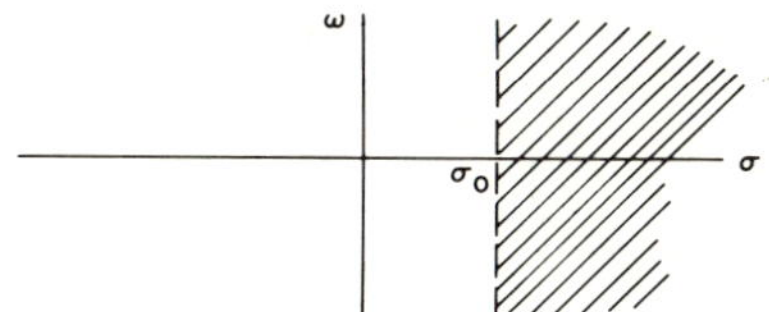

Fig. 9.17 Abscissa of convergence.

If $s_0 = 0$, $e^{s_0 t}$ becomes the unit step and we have

$$\mathscr{L}[u(t)] = \frac{1}{s}$$

Are there any constraints on the transforms in Example 9.9? The only constraint should be that $f(t) = e^{s_0 t}$ must be Laplace transformable. What does being Laplace transformable mean in this case? To answer this question, reexamine the first term on the right side of Eq. (9.7.7). For the function of Example 9.9, that term is

$$f(t)e^{-st}\Big|_0^\infty = e^{s_0 t}e^{-st}\Big|_0^\infty = e^{(\sigma_0-\sigma)t}e^{j(\omega_0-\omega)t}\Big|_0^\infty \tag{9.7.12}$$

where

$$s = \sigma + j\omega \qquad s_0 = \sigma_0 + j\omega_0 \tag{9.7.13}$$

If this term is to vanish at its upper limit, it is necessary and sufficient that

$$\sigma > \sigma_0 \tag{9.7.14}$$

Thus the Laplace transform of $e^{s_0 t}$ exists only to the right of the dashed line (within the shaded area) in the s plane of Fig. 9.17. This dashed line defines the *abscissa of convergence* σ_0 of the given function. If σ_0 is 0 or negative, the transform exists in the entire right half of the s plane; otherwise a portion of the right half of the s plane is excluded.

Theorem 3 Real (t-domain) integration If $f(t)$ is Laplace transformable, then

$$\mathscr{L}[f^{-n}(t)] = \frac{1}{s^n}F(s) + \frac{1}{s^n}f_+^{-1}\frac{1}{s^{n-1}}f_+^{-2} + \cdots + \frac{1}{s}f_+^{-n} \tag{9.7.15}$$

where $f^{-n}(t)$ is the nth indefinite integral of $f(t)$ [the nth integral of $f(t)$ between the limits of $-\infty$ and t] and f_+^{-1} through f_+^{-n} are the values of the first through the nth indefinite integrals of $f(t)$ at $t = 0_+$.

As in the derivative case, we shall prove Eq. (9.7.15) by finding the Laplace transform of the first indefinite integral and extending the result. We have

$$\mathscr{L}[f^{-1}(t)] = \int_0^\infty e^{-st}\left[\int_{-\infty}^t f(t)\,dt\right]dt \tag{9.7.16}$$

As before, the integration is by parts. Let

$$u = \int_{-\infty}^{t} f(t)\,dt \qquad \text{and} \qquad dv = e^{-st}\,dt \tag{9.7.17}$$

Then

$$du = f(t)\,dt \qquad \text{and} \qquad v = -\frac{1}{s}\,e^{-st} \tag{9.7.18}$$

Inserting these quantities into Eq. (9.7.16) gives

$$\mathscr{L}[f^{-1}(t)] = -\left[\int_{-\infty}^{t} f(t)\,dt\right]\frac{1}{s}\,e^{-st}\Big|_0^\infty + \frac{1}{s}\int_0^\infty f(t)e^{-st}\,dt \tag{9.7.19}$$

Since $f(t)$ is Laplace transformable, the first term on the right side vanishes at its upper limit, again from the discussion following Eq. (9.5.9). Thus we have

$$\mathscr{L}[f^{-1}(t)] = \frac{1}{s}\,f_+^{-1} + \frac{1}{s}\,F(s) \tag{9.7.20}$$

The first term on the right side of Eq. (9.7.20) occurs because we have found the transform of the indefinite integral $\int_{-\infty}^{t} f(t)\,dt$ rather than of the definite integral $\int_0^t f(t)\,dt$. The latter transform is simply

$$\mathscr{L}\left[\int_0^t f(t)\right]dt = \frac{1}{s}\,F(s) \tag{9.7.21}$$

Thus $\int_{-\infty}^{0} f(t)\,dt$ is really an initial-condition producer: it represents the "accumulation" for all time up to $t = 0$. If $f(t)$ is a capacitor current, the initial condition term is

$$\frac{1}{s}\int_{-\infty}^{0} i(t)\,dt = \frac{q_+}{s} = \frac{Cv_{c+}}{s} \tag{9.7.22}$$

whereas if $f(t)$ is an inductor voltage, there results

$$\frac{1}{s}\int_{-\infty}^{0} v(t)\,dt = \frac{\psi_+}{s} = \frac{Li_{L+}}{s} \tag{9.7.23}$$

where q_+ and ψ_+ are the values of charge and flux linkages, respectively, at $t = 0_+$ and v_{c+} and i_{L+} are the corresponding values of capacitor voltage and inductor current.

The result in Eq. (9.7.20) can be used to find the Laplace transform of the second indefinite integral of $f(t)$ in much the same fashion as we used Eq. (9.7.8) to obtain the Laplace transform of the second derivative of $f(t)$.

Thus

$$f^{-2}(t) = \int_{-\infty}^{t} f^{-1}(t)\,dt \tag{9.7.24}$$

and

$$\mathscr{L}[f^{-2}(t)] = -f^{-2}(t)\frac{1}{s}e^{-st}\Big|_0^\infty + \frac{1}{s}\mathscr{L}[f^{-1}(t)] \tag{9.7.25}$$

which becomes

$$\mathscr{L}[f^{-2}(t)] = \frac{1}{s}f_+^{-2} + \frac{1}{s^2}f_+^{-1} + \frac{1}{s^2}F(s) \tag{9.7.26}$$

The extension to higher-order integrals and to Eq. (9.7.15) follows directly.

Theorem 4 Complex (s-domain) differentiation If $f(t)$ is Laplace transformable, then $t^n f(t)$ is Laplace transformable and

$$\mathscr{L}[(-t)^n f(t)] = \frac{d^n F(s)}{ds^n} \tag{9.7.27}$$

To prove Theorem 4, let us differentiate $F(s)$ in Eq. (9.5.8) with respect to s. We obtain

$$\frac{dF(s)}{ds} = \frac{d}{ds}\left[\int_0^\infty f(t)e^{-st}\,dt\right] = \int_0^\infty f(t)\frac{d}{ds}e^{-st}\,dt$$

$$= \int_0^\infty -tf(t)e^{-st}\,dt = -\mathscr{L}[tf(t)] \tag{9.7.28}$$

Thus multiplication by $-t$ in the time domain corresponds to differentiation with respect to s in the frequency domain. It follows directly that

$$\frac{d^2F(s)}{dt^2} = \mathscr{L}[t^2 f(t)] \tag{9.7.29}$$

and so on until Eq. (9.7.27) is reached.

Example 9.10 Find the Laplace transforms of (*a*) $t \sin \omega t$ and (*b*) t^n.

(*a*) The transform of $\sin \omega t$ was found in Example 9.8. Differentiating once with respect to s yields

$$\mathscr{L}(t \sin \omega t) = \frac{2\omega s}{(s^2 + \omega^2)^2}$$

(*b*) The transform of $u(t)$ was found in Example 9.9 to be $1/s$. It follows directly that

$$\mathscr{L}[t^n u(t)] - (-1)^n \frac{d^n(1/s)}{ds^n} \quad \frac{n!}{s^{n+1}}$$

If $n = 1$, we have the unit ramp function. From the general result, its transform is given by

$$\mathscr{L}[tu(t)] = \frac{1}{s^2}$$

When $n = 2$, we get

$$\mathscr{L}[t^2u(t)] = \frac{2}{s^3}$$

which is twice the transform of the unit parabola $[p(t) = \frac{1}{2}t^2u(t)]$.

By continuing, the transforms of the entire family of singularity functions (except the delta and doublet functions) can be developed from Theorem 4.

Theorem 5 Complex (s-domain) integration If $(1/t)f(t)$ is Laplace transformable, then

$$\mathscr{L}\frac{1}{t}f(t) = \int_s^\infty F(s)\,ds \tag{9.7.30}$$

where $F(s)$ is the Laplace transform of $f(t)$.

To prove the theorem, we write

$$\int_s^\infty F(s)\,ds = \int_s^\infty \left[\int_0^\infty f(t)e^{-st}\,dt\right] ds \tag{9.7.31}$$

Interchanging the orders of the t- and s-domain integrations, which is permissible since $f(t)$ is Laplace transformable if $(1/t)f(t)$ is, we obtain

$$\begin{aligned}\int_s^\infty F(s)\,ds &= \int_0^\infty f(t)\left(\int_s^\infty e^{-st}\,ds\right) dt \\ &= \int_0^\infty f(t)\left[\frac{e^{-st}}{-t}\right]_s^\infty dt \\ &= \int_0^\infty f(t)\frac{1}{t}e^{-st}\,dt \end{aligned} \tag{9.7.32}$$

Example 9.11 Find, if they exist, the Laplace transforms of (*a*) $(\sin \omega t)/\omega t$ and (*b*) $u(t)/t$.

(*a*) The function is Laplace transformable. Identifying $f(t) = (\sin \omega t)/\omega$ and using Eq. (9.7.30), we obtain

$$\begin{aligned}\mathscr{L}\left[\frac{\sin \omega t}{\omega t}\right] &= \int_s^\infty \frac{1}{s^2 + \omega^2}\,ds = \frac{1}{\omega}\tan^{-1}\frac{s}{\omega}\Big|_s^\infty \\ &= \frac{1}{\omega}\left(\frac{\pi}{2} - \tan^{-1}\frac{s}{\omega}\right) = \frac{1}{\omega}\tan^{-1}\frac{\omega}{s}\end{aligned}$$

(*b*) The Laplace transform of $u(t)$ is $1/s$, but the function $(1/t)u(t)$ is not Laplace transformable, as can be verified through Eq. (9.5.8). To this end

$$\int_0^\infty \frac{1}{t} u(t)e^{-st}\,dt = \int_0^\infty \frac{1}{t} e^{-st}\,dt$$

$$= \ln t - st + \frac{s^2t^2}{4} = - \frac{s^3t^3}{18} + \cdots \Big|_0^\infty$$

which gives an infinite result.

The reader should note that although $tu(t)$ is a unit ramp function, it does not follow that $(1/t)u(t)$ is a unit impulse function. These two functions are derived from the unit step function through differentiation and integration, respectively, not through multiplication and division, respectively, by t.

Theorem 6 Real (t-domain) translation If $f(t)u(t)$ is Laplace transformable with transform $F(s)$, then

$$\mathscr{L}[f(t-\tau)u(t-\tau)] = e^{-\tau s}F(s) \tag{9.7.33}$$

where τ is a nonnegative real number.

The proof proceeds from Eq. (9.5.8)

$$\mathscr{L}[f(t-\tau)u(t-\tau)] = \int_0^\infty f(t-\tau)u(t-\tau)e^{-st}\,dt$$

$$= \int_\tau^\infty f(t-\tau)e^{-st}\,dt \tag{9.7.34}$$

where the second form of Eq. (9.7.34) arises because $u(t-\tau)$ causes the integrand to be zero for $0 < t < \tau$. Now, letting $a = t - \tau$ and $da = dt$, Eq. (9.7.34) can be written

$$\mathscr{L}[f(t-\tau)u(t-\tau)] = \int_0^\infty f(a)e^{-s(a+\tau)}\,da$$

$$= -e^{-\tau s}\int_0^\infty f(a)e^{-sa}\,da$$

$$= -e^{-\tau s}F(s) \tag{9.7.35}$$

The reader should note that t in Eq. (9.5.8) and a in Eq. (9.7.35) are dummy variables; neither one appears in the final result once the integration has been performed and the limits inserted. Thus it does not matter whether we use t, a, or some other letter in the integrand; the final result is a function of s only.

Example 9.12 Find the Laplace transform of the unit impulse function $\delta(t)$.

$$\delta(t) = \lim_{\tau \to 0} \frac{1}{\tau} [u(t) - u(t - \tau)]$$

Then

$$\mathcal{L}[\delta(t)] = \lim_{\tau \to 0} \frac{1}{\tau} \mathcal{L}[u(t) - u(t - \tau)]$$

$$= \lim_{\tau \to 0} \frac{1}{\tau} \left(\frac{1}{s} - \frac{e^{-\tau s}}{s} \right)$$

$$= \lim_{\tau \to 0} \frac{1 - e^{-\tau s}}{\tau s}$$

$$= \lim_{\tau \to 0} \frac{s e^{-\tau s}}{s} = 1$$

where we have used L'Hospital's rule to evaluate the indeterminate form in the second to the last step.

We observe that the Laplace and Fourier transforms of the impulse function are identical, the latter having been derived in Eq. (9.6.16). The reader may at this time wish to review the comments following Eq. (9.6.16) about the frequency spectrum of the impulse function.

From Theorem 6, the Laplace transform of a unit impulse delayed τ units of time is

$$\mathcal{L}[\delta(t - \tau)] = e^{-\tau s}$$

and, from the derivation just completed, the Laplace transform of a delayed unit step is

$$\mathcal{L}[u(t - \tau)] = \frac{e^{-\tau s}}{s}$$

Example 9.13 Find a general expression for the Laplace transform of a periodic function $f(t)$ of period T and use the result to find the Laplace transform of a repetitive square wave of period T and amplitude A.

Using Eq. (9.5.8), we have

$$\mathcal{L}f(t) = \int_0^T f(t) e^{-st}\, dt + \int_T^{2T} f(t) e^{-st}\, dt + \cdots$$

In the second term on the right let

$$a = t - T \qquad da = dt$$

to obtain for that term

$$\int_T^{2T} f(t) e^{-st}\, dt = \int_0^T f(a + T) e^{-s(a+T)}\, da$$

Extracting e^{-sT}, replacing the dummy variable a by t, and noting that

$$f(a + T) = f(t)$$

because $f(t)$ is periodic, we have

$$\int_T^{2T} f(t)e^{-st}\,dt = e^{-sT}\int_0^T f(t)e^{-st}\,dt$$

Thus

$$\mathscr{L}[f(t)] = \int_0^T f(t)e^{-st}\,dt + e^{-sT}\int_0^T e^{-st}\,dt + \cdots$$

$$= (1 + e^{-sT} + \cdots)\int_0^T f(t)e^{-sT}\,dt$$

The series in parentheses can be identified as $(1 - e^{-sT})^{-1}$. Thus

$$\mathscr{L}[f(t)] = \frac{1}{1 - e^{-sT}}\int_0^T f(t)e^{-st}\,dt \tag{9.7.36}$$

We can represent the given square wave during its first period by

$$f_1(t) = A\left[u(t) - 2u\left(t - \frac{T}{2}\right) + u(t - T)\right] \qquad 0 < t < T$$

The Laplace transform of $f_1(t)$ is

$$\mathscr{L}[f_1(t)] = \frac{A}{s}(1 - 2e^{-sT/2} + e^{-sT})$$

From this result the Laplace transform of $f(t)$ is

$$\mathscr{L}[f(t)] = \frac{A(1 - 2e^{-sT/2} + e^{-sT})}{s(1 - e^{-sT})} = \frac{A}{s}\frac{1 - e^{-sT/2}}{1 + e^{-sT/2}}$$

Theorem 7 Complex (s-domain) translation If $f(t)$ is Laplace transformable with transform $F(s)$, then $e^{-at}f(t)$, with a real, is Laplace transformable and

$$\mathscr{L}[e^{-at}f(t)] = F(s + a) \tag{9.7.37}$$

The proof of the theorem is obtained directly by noting that

$$\mathscr{L}[e^{-at}f(t)] = \int_0^\infty e^{-at}f(t)e^{-st}\,dt$$

$$= \int_0^\infty f(t)e^{-(s+a)t}\,dt = F(s + a) \tag{9.7.38}$$

Example 9.14 Find the Laplace transforms of $e^{-at}\sin\omega t$, $e^{-at}t^{n-1}$, and $te^{-at}\sin\omega t$.

In Example 9.8, we showed that the

$$\mathscr{L}[\sin\omega t] = \frac{\omega}{s^2 + \omega^2}$$

and, from the result in part (*b*) of Example 9.10, we deduce that

$$\mathscr{L}[t^{n-1}] = \frac{(n-1)!}{s^n}$$

Now, using Theorem 7, we obtain

$$\mathscr{L}[e^{-at} \sin \omega t] = \frac{\omega}{(s+a)^2 + \omega^2}$$

and

$$\mathscr{L}[e^{-at}t^{n-1}] = \frac{(n-1)!}{(s+a)^n}$$

Finally, from Theorem 4 and Eq. (9.7.28), we have

$$\mathscr{L}[t(e^{-at} \sin \omega t)] = -\frac{d}{ds}\frac{\omega}{(s+a)^2 + \omega^2}$$

$$= \frac{2\omega(s+a)}{[(s+a)^2 + \omega^2]^2}$$

These three transforms just developed will be of considerable use. The first shows that the time-domain representation of a pair of complex conjugate s-domain poles is a damped sinusoid; the second shows that a real nth-degree pole in the s domain transforms into an exponential multiplied by t^{n-1} in the time domain; the third, a combination of the other two, gives the time-domain form of a pair of second-degree complex conjugate poles having a zero equal to the real part of the poles as a damped sinusoid multiplied by t. Thus we see that increasing the degree of a pole by 1 in the s domain corresponds to multiplication by t in the time domain.

Theorem 8 Initial value If $f(t)$ and df/dt are Laplace transformable with the former having a transform $F(s)$, then

$$\lim_{s\to\infty} sF(s) = \lim_{\substack{t\to 0_+ \\ \text{from } t>0}} f(t) = f_+ \tag{9.7.39}$$

To prove Theorem 8, we make use of Eq. (9.7.8), rewriting that equation in the form

$$sF(s) = \int_0^\infty \frac{df}{dt} e^{-st}\, dt + f_+ \tag{9.7.40}$$

Taking the limit of the integral portion of the right side of Eq. (9.7.40) as s approaches infinity, we obtain

$$\lim_{s\to\infty} \int_0^\infty \frac{df}{dt} e^{-st}\, dt = \int_0^\infty \lim_{s\to\infty} \frac{df}{dt} e^{-st}\, dt = 0 \tag{9.7.41}$$

Now taking the same limit on the left side of Eq. (9.7.40), we obtain the result in Eq. (9.7.39).

In this derivation, the limit and integral operations in Eq. (9.7.41) have been interchanged. This is always permissible if df/dt is Laplace transformable. It has also been presumed that $\lim_{s\to\infty} sF(s)$ exists. If $F(s)$ is a ratio of polynomials, this limit will exist only if the largest power of s in the numerator of $F(s)$ is at least 1 less than the largest power of s in the denominator of $F(s)$. This will make the largest powers of s in the numerator and denominator of $sF(s)$ at worst equal and will result in a finite limit as s approaches infinity.

Example 9.15 A function $f(t)$ has a Laplace transform given by

$$F(s) = \frac{s+3}{s^3+4s^2+6s+4}$$

Find the values of $f(t)$ and its first and second derivatives at $t = 0_+$.

From Eq. (9.7.39), we have

$$\lim_{s\to\infty} \frac{s^2+3s}{s^3+4s^2+6s+4} = 0 = f_+$$

Now, treating df/dt as the function of interest, calling its Laplace transform $F_1(s)$, and following the form of Eq. (9.7.39), we can write

$$\lim_{s\to\infty} sF_1(s) = \left.\frac{df}{dt}\right|_+ \tag{9.7.42}$$

But, from Eq. (9.7.8), Eq. (9.7.42) becomes

$$\lim_{s\to\infty} s[sF(s) - f_+] = \lim_{s\to\infty} \frac{s^3+3s^2}{s^3+4s^2+6s+4} = 1 = \left.\frac{df}{dt}\right|_+$$

Proceeding in a similar fashion we consider d^2f/dt^2, with transform $F_2(s)$, as the function of interest and write

$$\lim_{s\to\infty} sF_2(s) = \left.\frac{d^2f}{dt^2}\right|_+ \tag{9.7.43}$$

Again using Eq. (9.7.8), Eq. (9.7.43) can be expanded to yield

$$\lim_{s\to\infty} s\left[sF_1(s) - \left.\frac{df}{dt}\right|_+\right] = \lim_{s\to\infty} s\left[s^2F(s) - \left.\frac{df}{dt}\right|_+\right]$$

$$= \lim_{s\to\infty} \frac{-s^3-6s^2-4s}{s^3+4s^2+6s+4} = -1 = \frac{d^2f}{dt^2}$$

We now have an additional method of evaluating initial conditions to go with those of Sec. 8.7, which were time-domain methods. The present approach of using the initial-value theorem is an s-domain approach.

Theorem 9 Final value If $f(t)$ and df/dt are Laplace transformable, with the former having a transform $F(s)$, and if the poles of $sF(s)$ are restricted to the left half of the s plane (excluding the $j\omega$ axis) then

$$\lim_{s\to 0} sF(s) = \lim_{t\to\infty} f(t) \tag{9.7.44}$$

We prove Theorem 9 in much the same way that we did Theorem 8, again using Eq. (9.7.40). Taking the limit of the integral portion of that equation as $s \to 0$, we have

$$\lim_{s\to 0}\int_0^\infty \frac{df}{dt}e^{-st}\,dt = \int_0^\infty \lim_{s\to 0}\frac{df}{dt}e^{-st}\,dt = \int_0^\infty \frac{df}{dt}\,dt = \lim_{t\to\infty} f(t) - f_+ \tag{9.7.45}$$

Now taking the limit as s approaches 0 in Eq. (9.7.40) and using the result of Eq. (9.7.45), we obtain Eq. (9.7.44).

We shall explain the necessity for the restriction on pole locations of $sF(s)$ in the statement of Theorem 9 through an example.

Example 9.16 Show why the final-value theorem is not applicable to the function $f(t) = \sin t$ but is applicable to the function $g(t) = u(t)$.

We note from Example 9.8 that $F(s) = 1/(s^2 + 1)$ for the sine function. Thus $sF(s)$ has a pair of poles on the $j\omega$ axis, and the final-value theorem does not apply. To show why this is so, let us compare the results obtained from the first and second terms in Eq. (9.7.45). The first term yields

$$\lim_{s\to 0}\int_0^\infty \cos te^{-st}\,dt = \lim_{s\to 0}\frac{s}{s^2+1} = 0$$

The second term gives

$$\int_0^\infty \lim_{s\to 0}\cos te^{-st}\,dt = \int_0^\infty \cos t\,dt$$

which is undefined at the upper limit.

Thus we see that unless the poles of $sF(s)$ lie in the left half of the s plane, we cannot interchange the limit and integral operations in Eq. (9.7.45).

The final-value theorem does apply to the unit step function $g(t) = u(t)$, whose transform is $G(s) = 1/s$. The pole of $G(s)$ is not located in the left half of the s plane, but it is the function $sG(s)$ that we must examine, and the latter has no poles. Thus the final-value theorem gives

$$\lim_{s\to 0} s\frac{1}{s} = 1 = \lim_{t\to\infty} u(t)$$

The final-value theorem is very useful in obtaining the steady-state value of $f(t)$, that is, the value of $f(t)$ after all transients have decayed to zero. If only the final value of $f(t)$ is of interest, the final-value theorem can be used in lieu of the more involved process of inverse transforming $F(s)$ to get $f(t)$ and then taking the limit as t approaches infinity.

The theorems we have developed in this section are assembled in Table 9.2. Transforms of commonly encountered functions, many of which we have derived in this section, appear in Table 9.3.

9.8 CONVOLUTION

Two additional Laplace transform theorems could have been included in Sec. 9.7, but they are so important in their own right that a separate section is devoted to their discussion.

Theorem 10 Real convolution Assume $g(t)$ and $h(t)$ have Laplace transforms $G(s)$ and $H(s)$, respectively. Then the inverse transform of $F(s) = G(s)H(s)$ is given by

$$f(t) = \int_0^t g(\tau)h(t - \tau)\, d\tau = \int_0^t h(\tau)g(t - \tau)\, d\tau \tag{9.8.1}$$

which is customarily written

$$f(t) = g(t) * h(t) = h(t) * g(t) \tag{9.8.2}$$

with the $*$ denoting convolved with.

TABLE 9.2 Laplace transform theorems

Theorem	*t domain*	*s domain*
Linearity	$k_1f_1(t) + k_2f_2(t)$	$k_1F_1(s) + k_2F_2(s)$
Real differentiation	$f^n(t)$	$s^nF(s) - s^{n-1}f_+ - \cdots - \left.\dfrac{d^{n-1}f}{dt^{n-1}}\right\rvert_+$
Real integration	$f^{-n}(t)$	$\dfrac{1}{s^n}F(s) + \dfrac{1}{s^n}f_+^{-1} + \cdots + \dfrac{1}{s}f_+^{-n}$
Complex differentiation	$(-t)^nf(t)$	$\dfrac{d^nF(s)}{ds^n}$
Complex integration	$\dfrac{1}{t}f(t)$	$\int_s^\infty F(s)\, ds$
Real translation	$f(t - \tau)u(t - \tau)$	$e^{-\tau s}F(s)$
Complex translation	$e^{-at}f(t)$	$F(s + a)$
Initial value	$\lim_{t\to 0} f(t)$	$\lim_{s\to\infty} sF(s)$
Final value	$\lim_{t\to\infty} f(t)$	$\lim_{s\to 0} sF(s)$

TABLE 9.3 Laplace transforms of common functions

No.	*t domain*	*s domain*
1	$\delta(t)$	1
2	$u(t)$	$\frac{1}{s}$
3	$r(t)$	$\frac{1}{s^2}$
4	$p(t)$	$\frac{1}{s^3}$
5	t^{n-1}	$\frac{(n-1)!}{s^n}$
6	e^{-at}	$\frac{1}{s+a}$
7	$\sin \omega t$	$\frac{\omega}{s^2+\omega^2}$
8	$\cos \omega t$	$\frac{s}{s^2+\omega^2}$
9	$\sin (\omega t + \theta)$	$\frac{s \sin \theta + \omega \cos \theta}{s^2+\omega^2}$
10	$\cos (\omega t + \theta)$	$\frac{s \cos \theta - \omega \sin \theta}{s^2+\omega^2}$
11	$\sin^2 \omega t$	$\frac{2\omega^2}{s(s^2+4\omega^2)}$
12	$\cos^2 \omega t$	$\frac{s^2+2\omega^2}{s(s^2+4\omega^2)}$
13	$\sinh at$	$\frac{a}{s^2-a^2}$
14	$\cosh at$	$\frac{s}{s^2-a^2}$
15	$t \sin \omega t$	$\frac{2\omega s}{(s^2+\omega^2)^2}$
16	$t \cos \omega t$	$\frac{s^2-\omega^2}{(s^2+\omega^2)^2}$
17	$\frac{1}{\omega t} \sin \omega t$	$\frac{1}{\omega} \tan^{-1} \frac{\omega}{s}$
18	$e^{-at} t^{n-1}$	$\frac{(n-1)!}{(s+a)^n}$
19	$e^{-at} \sin \omega t$	$\frac{\omega}{(s+a)^2+\omega^2}$
20	$e^{-at} \cos \omega t$	$\frac{s+a}{(s+a)^2+\omega^2}$

TABLE 9.3 Continued

No.	*t domain*	*s domain*
21	$te^{-at}\sin\omega t$	$\dfrac{-2\omega(s+a)}{[(s+a)^2+\omega^2]^2}$
22	$te^{-at}\cos\omega t$	$\dfrac{(s+a)^2-\omega^2}{[(s+a)^2+\omega^2]^2}$
23	$e^{-at}\sin(\omega t+\theta)$	$\dfrac{(s+a)\sin\theta+\omega\cos\theta}{(s+a)^2+\omega^2}$
24	$\delta(t-\tau)$	$e^{-\tau s}$
25	$u(t-\tau)$	$\dfrac{1}{s}e^{-\tau s}$
26	$r(t-\tau)$	$\dfrac{1}{s^2}e^{-\tau s}$
27	$u(t)-u(t-\tau)$ (rectangular pulse)	$\dfrac{1-e^{-\tau s}}{s}$
28	$r(t)-2r(t-\tau)-r(t-2\tau)$ (triangular pulse)	$\dfrac{(1-e^{-\tau s})^2}{s^2}$
29	$r(t)-r(t-\tau)-u(t-\tau)$ (sawtooth pulse)	$\dfrac{1-e^{-\tau s}-se^{-\tau s}}{s^2}$
30	$\sin\omega t\left[u(t)-u\left(t-\dfrac{T}{2}\right)\right]$ (half-sine pulse)	$\dfrac{1-e^{-sT/2}}{s^2+\omega^2}$
31	$\dfrac{1}{a}(1-e^{-at})$	$\dfrac{1}{s(s+a)}$
32	$\dfrac{1}{a^2}(at-1+e^{-at})$	$\dfrac{1}{s^2(s+a)}$
33	$\dfrac{e^{-at}-e^{-bt}}{b-a}$	$\dfrac{1}{(s+a)(s+b)}$
34	$\dfrac{(c-a)e^{-at}-(c-b)e^{-bt}}{b-a} =$	$\dfrac{s+c}{(s+a)(s+b)}$

To prove Eq. (9.8.1), we use the inverse Laplace transform Eq. (9.5.9) and write

$$f(t)=\frac{1}{2\pi j}\int_{\sigma-j\infty}^{\sigma+j\infty}G(s)H(s)e^{st}\,ds \tag{9.8.3}$$

Either $G(s)$ or $H(s)$ can be expressed in terms of the forward Laplace transform Eq. (9.5.8). Choosing $G(s)$ yields

$$f(t)=\frac{1}{2\pi j}\int_{\sigma-j\infty}^{\sigma+j\infty}\left[\int_0^{\infty}g(\tau)e^{-s\tau}\,d\tau\right]H(s)e^{st}\,ds \tag{9.8.4}$$

where the dummy variable τ has been used in the inner integral to distinguish what is integrated with respect to t from what is not. Now, assuming convergence of each integral, we interchange the order of integration and obtain

$$f(t) = \int_0^\infty \left[\frac{1}{2\pi j}\int_{\sigma-j\infty}^{\sigma+j\infty} H(s)e^{st}e^{-s\tau}\,ds\right]g(\tau)u(\tau)\,d\tau \tag{9.8.5}$$

Note that all quantities containing s have been placed within the inner integral. Also note that we have multiplied $g(\tau)$ by $u(\tau)$ to stress the fact that $g(t)$ is zero for $t < 0$. Now the inner integral is recognizable as $h(t-\tau)u(t-\tau)$ and thus

$$f(t) = \int_0^\infty h(t-\tau)u(t-\tau)g(\tau)u(\tau)\,d\tau \tag{9.8.6}$$

$u(\tau)$ is a step beginning at $\tau = 0$, whereas $u(t-\tau)$ is a step ending at $\tau = t$ (review Fig. 9.9c). Hence we can change the upper limit in Eq. (9.8.6) from ∞ to t, eliminate the two unit step functions, and obtain the first form of Eq. (9.8.1). Replacing $H(s)$ in Eq. (9.8.3) by its transform integral yields the second form of Eq. (9.8.1).

There is a similar real-convolution theorem for Fourier transforms. If $g(t)$ and $h(t)$ have Fourier transforms $G(\omega)$ and $H(\omega)$, respectively, the inverse Fourier transform of $F(\omega) = G(\omega)H(\omega)$ is

$$f(t) = \int_{-\infty}^\infty g(\tau)h(t-\tau)\,d\tau = \int_{-\infty}^\infty h(\tau)g(t-\tau)\,d\tau \tag{9.8.7}$$

The only difference in Eqs. (9.8.1) and (9.8.7) is in the limits of integration. The proof of Eq. (9.8.7) is requested in the problems.

Theorem 11 Complex convolution Assume that $g(t)$ and $h(t)$ have Laplace transforms $G(s)$ and $H(s)$, respectively. Then the function $f(t) = g(t)h(t)$ will have a Laplace transform given by

$$F(s) = \frac{1}{2\pi j}\int_{\sigma-j\infty}^{\sigma+j\infty} G(\lambda)H(s-\lambda)\,d\lambda = \frac{1}{2\pi j}\int_{\sigma-j\infty}^{\sigma+j\infty} H(\lambda)G(s-\lambda)\,d\lambda \tag{9.8.8}$$

The proof of Theorem 11 is very similar to that of Theorem 10. We write

$$\begin{aligned}
F(s) &= \int_0^\infty g(t)h(t)e^{-st}\,dt \\
&= \int_0^\infty g(t)\left[\frac{1}{2\pi j}\int_{\sigma-j\infty}^{\sigma+j\infty} H(\lambda)e^{\lambda t}\,d\lambda\right]e^{-st}\,dt \\
&= \frac{1}{2\pi j}\int_{\sigma-j\infty}^{\sigma+j\infty}\left[\int_0^\infty g(t)e^{-st}e^{\lambda t}\,dt\right]H(\lambda)\,d\lambda \\
&= \frac{1}{2\pi j}\int_{\sigma-j\infty}^{\sigma+j\infty} G(s-\lambda)H(\lambda)\,d\lambda
\end{aligned} \tag{9.8.9}$$

As in the case of real convolution, a complex-convolution integral for Fourier transforms exists and is given by

$$F(\omega) = \frac{1}{2\pi}\int_{-\infty}^{\infty} G(\omega - \lambda)H(\lambda)\, d\lambda \tag{9.8.10}$$

The proof is requested in the problems.

Because the complex-convolution integrals require contour integration for their evaluations, we shall not consider them further until we have discussed the theory of complex variables in Chap. 12. We devote the remainder of this section to the use of the real-convolution integral in the analysis of linear systems.

Consider first the use of the convolution integral in finding inverse transforms. Recall from Eq. (9.7.21) that the inverse Laplace transform of the quantity $G(s)/s$ is the first definite integral from 0 to t of $g(t)$. Can this inverse transform be expressed in an alternate form using the convolution integral?

Let $f(t)$ be the desired function. Its transform is to be $F(s) = G(s)H(s)$. Let $H(s)$ be $1/s$, giving $h(t) = u(t)$. Then from Eq. (9.8.1),

$$f(t) = \int_0^t g(\tau)u(t-\tau)\, d\tau = \int_0^t g(t-\tau)u(\tau)\, d\tau \tag{9.8.11}$$

Next, if $H(s) = 1/s^2$, $h(t) = t$, and

$$f(t) = \int_0^t g(\tau)(t-\tau)\, d\tau = \int_0^t g(t-\tau)\tau\, d\tau \tag{9.8.12}$$

Finally, for $H(s) = 1/s^{n+1}$, $h(t) = t^n/n!$ and

$$f(t) = \int_0^t g(\tau)\frac{(t-\tau)^n}{n!}\, d\tau = \int_0^t g(t-\tau)\frac{t^n}{n!}\, d\tau \tag{9.8.13}$$

Example 9.17 As an illustration of the use of this family of functions, find the inverse Laplace transform of

$$F(s) = \frac{1}{s^2(s+\alpha)^2}$$

Break $F(s)$ into

$$F(s) = G(s)H(s) = \frac{1}{(s+\alpha)^2}\frac{1}{s^2}$$

From Table 9.3, $g(t) = te^{-\alpha t}$. Then, using Eq. (9.8.12), we have

$$f(t) = \int_0^t \tau e^{-\alpha\tau}(t-\tau)\, d\tau = \int_0^t (t-\tau)e^{-\alpha(t-\tau)}\, d\tau$$

Performing either integration yields

$$f(t) = \frac{t}{\alpha^2}(e^{-\alpha t}+1) + \frac{2}{\alpha^3}(e^{-\alpha t}-1)$$

Thus far the quantities $G(s)$ and $H(s)$, which have been multiplied together in the s domain and convolved in the time domain, have received no physical interpretation. To give them physical meaning, consider the network in Fig. 9.18a. Assuming the capacitor initially uncharged, the differential equation relating the output and input voltages is

$$e_i = RC\frac{de_0}{dt} + e_0 \tag{9.8.14}$$

Laplace transform each side of Eq. (9.8.4) to obtain

$$E_i(s) = (RCs + 1)E_0(s) - RCe_{0+} \tag{9.8.15}$$

Since $e_{0+} = 0$, we can write

$$E_0(s) = \frac{1/RC}{s + 1/RC}E_i(s) \tag{9.8.16}$$

Equation (9.8.16) should be compared with Eq. (8.6.5). Each contains a function of s relating the response of a system to its excitation. We have called that function $T(s)$, the system transfer function, in Sec. 8.6, and we retain that designation here. Thus

$$E_0(s) = T(s)E_i(s) \tag{9.8.17}$$

Equation (9.8.17) leads to the generalization shown in Fig. 9.18b, where, in the s domain, we relate output, system, and input through

$$F(s) = G(s)H(s) \tag{9.8.18}$$

where $h(t)$ and $H(s)$ are the system excitation and its transform, $G(s)$ is the *system transfer function*, whose inverse transform $g(t)$ is called the *system weighting function*, and $f(t)$ and $F(s)$ are the system output and its transform.

The product relationship in Eq. (9.8.18) means that we can use the real-convolution theorem to find the output of a system whose weighting function and input as functions of time are given without ever entering the s domain. An important example is when $h(t)$ in Fig. 9.18b is a unit impulse at $t = 0$.

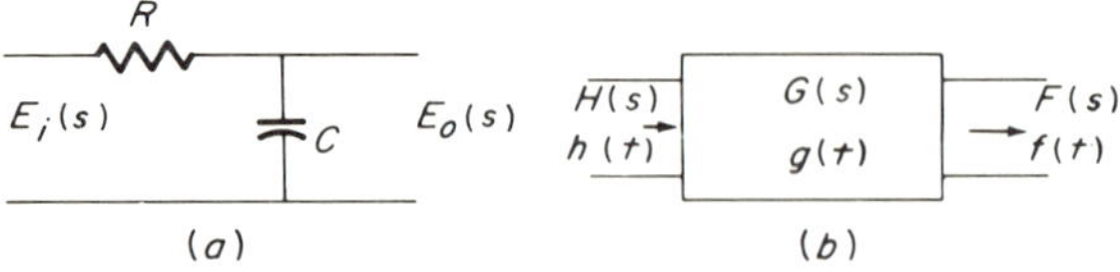

Fig. 9.18 *The transfer and weighting functions.*

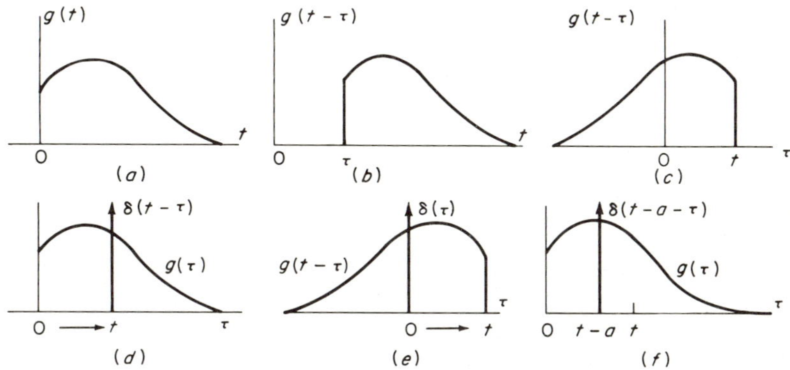

Fig. 9.19 Impulse response through convolution.

For this case, we have from Eq. (9.8.1)

$$f(t) = \int_0^t g(\tau)\delta(t - \tau)\, d\tau \qquad \text{or} \qquad f(t) = \int_0^t \delta(\tau)g(t - \tau)\, d\tau \tag{9.8.19}$$

The weighting function $g(t)$ is presumed zero for $t < 0$. An example of such a function is shown in Fig. 9.19*a*. To draw $g(t - \tau)$ vs. t, we merely displace $g(t)$ τ units to the right, as shown in Fig. 9.19*b*. However, in Eq. (9.8.19), the integrations are with respect to τ, not t. We need $g(\tau)$ vs. τ and $g(t - \tau)$ vs. τ. The first is merely Fig. 9.19*a* with t replaced by τ. To obtain the second, we rotate the plot in Fig. 9.19*b* about the $t = \tau$ vertical centerline to obtain Fig. 9.19*c*. Thus $g(t - \tau)$ plotted vs. t is a function that begins at $t = \tau$, whereas $g(t - \tau) = g[-(\tau - t)]$ plotted vs. τ is a function that ends at $\tau = t$.

Consider now the two integrations in Eq. (9.8.19). In the first, we are dealing with Fig. 9.19*d*. As the integration is performed from 0 to t (horizontal arrow), we obtain a nonzero product of the two functions only at $\tau = t$, and thus from Eq. (9.6.13), we have $f(t) = g(t)$, the weighting function evaluated at the location of the unit impulse. In the second form of Eq. (9.8.19), Fig. 9.19*e* applies. A nonzero product of the two functions occurs only at $\tau = 0$, again giving $f(t) = g(t)$.

The important conclusion we reach from these results is that the time-domain response of a system to a unit impulse excitation originating at $t = 0$ (the impulse response of the system) is the system weighting function $g(t)$. It follows that if we know the impulse response of a system, we can find its response to any other excitation simply by specifying that excitation and employing the real-convolution theorem.

Suppose next that a delayed unit impulse excitation $\delta(t - a)$ is applied to the system whose impulse response is given by Fig. 9.19*a* and we want the system response. The situation is depicted in Fig. 9.19*f*. Instead of occurring at $\tau = t$, the impulse now occurs at $\tau = t - a$. The resulting integration, using Eq. (9.8.1), yields

$$f(t) = \int_0^t g(\tau)\delta(t - a - \tau)\, d\tau = g(t - a) \tag{9.8.20a}$$

or

$$f(t) = \int_0^t \delta(\tau - a)g(t - \tau)\, d\tau = g(t - a) \tag{9.8.20b}$$

Thus the response to a delayed unit impulse is the weighting function equally delayed.

Example 9.18 A system excitation and its weighting function are shown in Fig. 9.20*a* and *b*. Find the system response by (*a*) Laplace transforms, (*b*) analytical convolution, and (*c*) graphical convolution.

(*a*) The given functions can be represented analytically by

$$h(t) = u(t) - u(t - 4)$$

$$g(t) = r(t) - \tfrac{3}{2}r(t - 1) + \tfrac{1}{2}r(t - 3)$$

Their Laplace transforms are

$$H(s) = \frac{1}{s} - \frac{e^{-4s}}{s}$$

$$G(s) = \frac{1}{s^2} - \frac{3e^{-s}}{2s^2} + \frac{e^{-3s}}{2s^2}$$

From Eq. (9.8.18), $F(s)$ is given by

$$F(s) = \frac{1}{s^3} - \frac{3e^{-s}}{2s^3} + \frac{e^{-3s}}{2s^3} - \frac{e^{-4s}}{s^3} + \frac{3e^{-5s}}{2s^3} - \frac{e^{-7s}}{2s^3}$$

and from Table 9.3

$$f(t) = \tfrac{1}{2}t^2u(t) - \tfrac{3}{4}(t - 1)^2u(t - 1) + \tfrac{1}{4}(t - 3)^2u(t - 3)$$
$$- \tfrac{1}{2}(t - 4)^2u(t - 4) + \tfrac{3}{4}(t - 5)^2u(t - 5) - \tfrac{1}{4}(t - 7)^2u(t - 7)$$

The response is plotted in Fig. 9.20c from the values in Table 9.4.

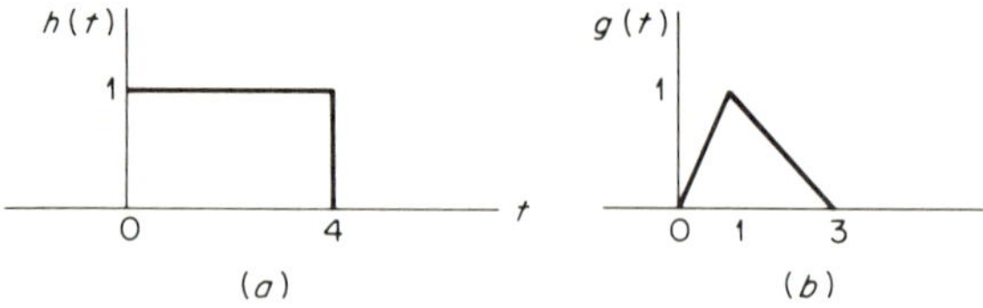

Fig. 9.20

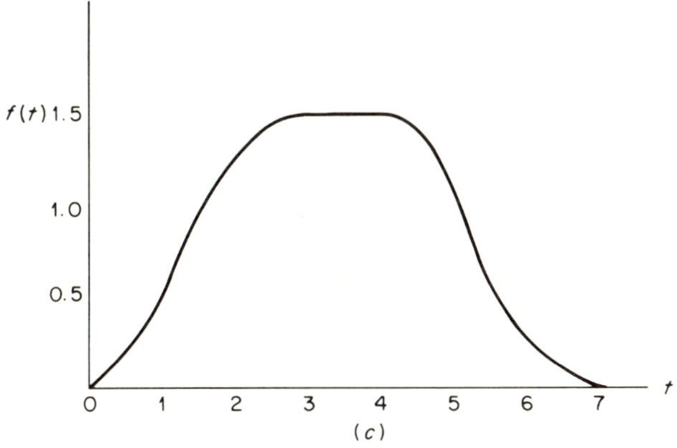

Fig. 9.20 (continued)

(*b*) Choosing the first form of Eq. (9.8.1), we have

$$\begin{aligned}
f(t) &= \int_0^t [r(\tau) - \tfrac{3}{2}r(\tau-1) + \tfrac{1}{2}r(\tau-3)][u(t-\tau) - u(t-4-\tau)]\,d\tau \\
&= \int_0^t [r(\tau) - \tfrac{3}{2}r(\tau-1) + \tfrac{1}{2}r(\tau-3)]\,d\tau \\
&\quad - \int_0^{t-4} [r(\tau) - \tfrac{3}{2}r(\tau-1) + \tfrac{1}{2}r(\tau-3)]\,d\tau \\
&= \int_0^t \tau\,d\tau - \frac{3}{2}\int_1^t (\tau-1)\,d\tau + \frac{1}{2}\int_3^t (\tau-3)\,d\tau \\
&\quad - \int_0^{t-4} \tau\,d\tau + \frac{3}{2}\int_1^{t-4} (\tau-1)\,d\tau - \frac{1}{2}\int_3^{t-4} (\tau-3)\,d\tau \\
&= \tfrac{1}{2}t^2u(t) - \tfrac{3}{4}(t-1)^2u(t-1) + \tfrac{1}{4}(t-3)^2u(t-3) \\
&\quad - \tfrac{1}{2}(t-4)^2u(t-4) + \tfrac{3}{4}(t-5)^2u(t-5) - \tfrac{1}{4}(t-7)^2u(t-7)
\end{aligned}$$

TABLE 9.4

t	$f(t)$
0	0
1	$\frac{1}{2}$
2	$2 - \frac{3}{4} = \frac{5}{4}$
3	$\frac{9}{2} - 3 = \frac{3}{2}$
4	$8 - \frac{27}{4} + \frac{1}{4} = \frac{3}{2}$
5	$\frac{25}{2} - 12 + 1 - \frac{1}{2} = 1$
6	$18 - \frac{75}{4} + \frac{9}{4} - 2 + \frac{3}{4} = \frac{1}{4}$
7	$\frac{49}{2} - 27 + 4 - \frac{9}{2} + 3 = 0$
>7	0

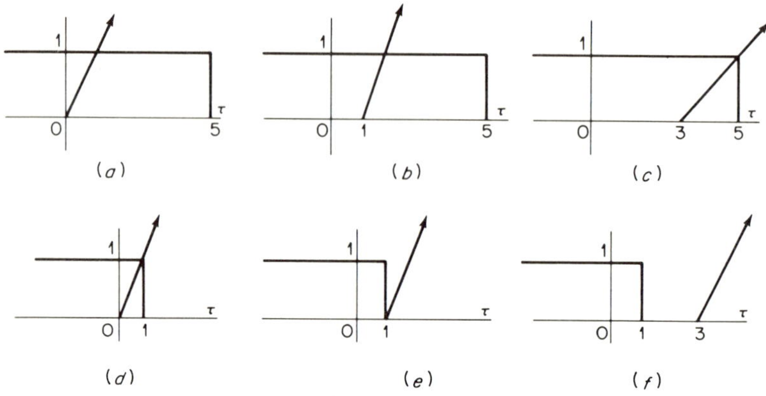

Fig. 9.21 *Upper and lower convolution integral limits: (a) $r(t)$ and $u(5-\tau)$; (b) $\frac{3}{2}r(\tau-1)$ and $u(5-\tau)$; (c) $\frac{1}{2}r(\tau-3)$ and $u(5-\tau)$; (d) $r(\tau)$ and $u(1-\tau)$; (e) $\frac{3}{2}r(\tau-1)$ and $u(1-\tau)$; (f) $\frac{1}{2}r(\tau-3)$ and $u(1-\tau)$.*

In the second step, the two unit steps are eliminated, and the integral is broken up into two integrals, with the upper limit of the second integral changed from t to $t-4$ to take into account the fact that $u(t-4-\tau)$ is zero for $\tau > t-4$. In the third step, the ramp functions are replaced by their equivalent representations in terms of τ. Six integrals are shown, with the lower limit of each equal to the value of τ at which the ramp commences. These upper and lower limits are illustrated in Fig. 9.21 for $t = 5$. In the fourth step each integration is performed and evaluated, and the appropriate delayed unit step is attached to each term in the response. For example the fifth response term has a $u(t-5)$ step since, with an upper limit of $t-4$ and a lower limit of 1, the fifth integral must be zero for $t < 5$. The total response is seen to be the same as in part (*a*).

(*c*) Here the purpose is to show how the integration in Eq. (9.8.1) can be carried out graphically. We choose the second form of Eq. (9.8.1), which means that we are dealing with $h(\tau)$ and $g(t-\tau)$. Now as t increases, $h(\tau)$ is fixed but $g(t-\tau)$ slides to the right from its position at $t = 0$. This initial position of $g(t-\tau)$ is displayed in the first Fig. 9.22. Its position at other integer intervals of time is shown in the other diagrams in that figure. The regions of overlap in each diagram (shaded regions in Fig. 9.22) are the regions of nonzero value for the second integral in Eq. (9.8.1). For example, for $t = 5$, the triangle and rectangle overlap only between $t = 2$ and $t = 4$. The area under the product of the two graphs is simply the shaded portion of the triangle, since $h(\tau) = 1$ from $t = 2$ to $t = 4$.

The sliding block approach in Fig. 9.22 permits us to determine easily when the response ends. The response cutoff time is simply the sum of 4 and 3, the cutoff times for $h(t)$ and $g(t)$, respectively. It also tells us that the response reaches a maximum at $t = 3$, which it holds until $t = 4$, this being the time period during which the rectangle completely overlaps the triangle.

It is seldom that the weighting function possesses the simple straight-line graph of $g(t)$ in Fig. 9.20*b*. More often it will have an irregular shape such

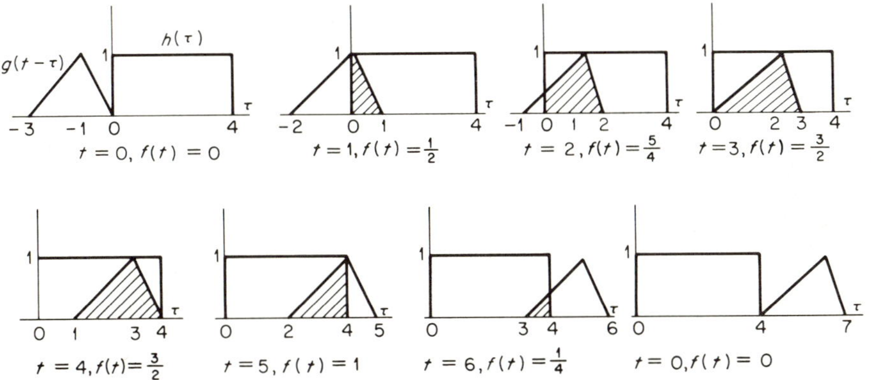

Fig. 9.22 Graphical convolution.

as that of $g(t)$ in Fig. 9.19*a*. Also the excitation is often more complicated than that in Fig. 9.20*a*. In such cases there are at least three ways in which we can proceed. One is to approximate the given $g(t)$ or $h(t)$ by a series of delayed ramp functions (parabolic functions can be used if greater accuracy is required). Then any of the techniques used in Example 9.18 can be employed to obtain the response. A second procedure we can follow stems from Eq. (9.8.18), which can be displayed in the following alternate forms:

$$
\begin{aligned}
F(s) &= G(s)H(s) = sG(s)\,\frac{H(s)}{s} = \frac{G(s)}{s}\,sH(s) \\
&= s^2G(s)\,\frac{H(s)}{s^2} = \cdots
\end{aligned}
\tag{9.8.21}
$$

Now, assuming that all initial conditions in the system are zero, we can write, from Eqs. (9.7.2) and (9.7.15),

$$
\begin{aligned}
f(t) &= g(t) * h(t) = \frac{dg}{dt} * h^{-1}(t) = g^{-1}(t) * \frac{dh}{dt} \\
&= \frac{d^2g}{dt^2} * h^{-2}(t) = \cdots
\end{aligned}
\tag{9.8.22}
$$

We shall illustrate the use of Eq. (9.8.22) by two examples.

Example 9.19 Find the response for $h(t)$ and $g(t)$ of Fig. 9.20 by convolving dh/dt with $g^{-1}(t)$.

Using the first form of Eq. (9.8.1),

$$
f(t) = \int_0^t g^{-1}(\tau)\,\frac{dh(t-\tau)}{d(t-\tau)}\,d\tau \tag{9.8.23}
$$

We show dh/dt in Fig. 9.23*a* as two unit impulses. In Fig. 9.23*b*, $g^{-1}(t) = \int_0^t g(t)\,dt$ is displayed. To obtain $dh(t-\tau)/d(t-\tau)$, dh/dt in Fig. 9.23*a* is shifted τ units to the right and rotated about the $t = \tau$ vertical centerline. This function, for $t = 2$, and $g^{-1}(\tau)$ appear in Fig. 9.23*c*.

Now, for $t = 0$, the positive impulse in Fig. 9.23*c* lies at the origin, the negative impulse at $t = -4$. Neither contributes anything to $f(t)$ in Eq. (9.8.21). At $t = 1$, the positive impulse convolved with $g^{-1}(1)$ gives ½ and the negative impulse gives nothing. After t reaches 4, the negative impulse begins to contribute. At $t = 5$, the negative impulse and $g^{-1}(1)$ give $-½$, and the positive impulse and $g^{-1}(5)$ give $\frac{3}{2}$, for a net of 1. At $t = 7$ and beyond, the negative impulse gives $-\frac{3}{2}$ and the positive impulse $\frac{3}{2}$, for a net of 0. The overall result is the plot in Fig. 9.20*c*.

From Example 9.19 we see that, through Eq. (9.8.22), we can convert a system and its excitation to a new "system" and an impulse-train excitation. The response of this new system to the impulse-train excitation is identical to the response of the original system to the given excitation. And since it is much easier to convolve with impulses than with other functions, we can obtain the response almost by inspection when the weighting function of the new system has been found. For example, the leading edge of $f(t)$ in Fig. 9.20*c* is identical to the leading edge of $g^{-1}(t)$ in Fig. 9.23*b*, and the trailing edge of $f(t)$ is given by 1.5 minus the leading edge of $g^{-1}(t)$.

Example 9.20 A system whose weighting function is $g(t) = e^{-t}$ is excited by a unit ramp function at $t = 0$. Find the system response by (*a*) analytical convolution and (*b*) impulse-train convolution.

(*a*) $f(t) = \int_0^t h(\tau)g(t-\tau)\,d\tau = \int_0^t \tau u(\tau)e^{-(t-\tau)}u(t-\tau)\,d\tau = \int_0^t \tau e^{-(t-\tau)}\,d\tau$ since the two unit steps are redundant in terms of the limits of integration. We place e^{-t} outside the integral and perform the integration by parts by letting

$$u = \tau \qquad dv = e^{\tau}\,d\tau$$

$$du = d\tau \qquad v = e^{\tau}$$

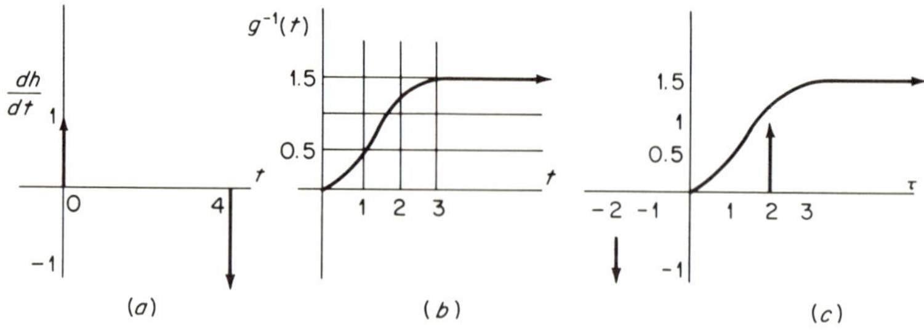

Fig. 9.23 Impulse-train convolution.

Then

$$f(t) = e^{-t}\tau e^{\tau}\big|_0^t - e^{-t}\int_0^t e^{\tau}\,d\tau$$

$$= t - e^{-t}(e^t - 1) = (t - 1 + e^{-t})u(t)$$

where the unit step has been reinstated to show the absence of a response prior to $t = 0$.

(*b*) The excitation is a unit ramp at $t = 0$. Its second derivative is a unit impulse, also at $t = 0$. We require the second integral of $g(t)$.

$$g^{-1}(t) = \int_0^t e^{-t}\,dt = 1 - e^{-t}$$

$$g^{-2}(t) = \int_0^t (1 - e^{-t})\,dt = t - 1 + e^{-t}$$

Then

$$f(t) = g^{-2}(t) * \frac{d^2h}{dt^2}$$

$$= (t - 1 + e^{-t}) * \delta(t)$$

$$= (t - 1 + e^{-t})u(t)$$

The response of the system to a unit step excitation appears as a bonus in this second procedure. It is merely $g^{-1}(t)u(t)$. Or we could differentiate the ramp response and obtain the step response. Either way, the step response is

$$f(t) = (1 - e^{-t})u(t)$$

We have discussed two ways of handling irregularly shaped weighting or excitation functions. A third way is to approximate the given $g(t)$ or $h(t)$ by an assemblage of narrow vertical pulses. We then consider each pulse to be an impulse whose amplitude is the area of the pulse and whose location is at the center of the pulse. We then convolve this impulse train with the nonapproximated time function [$g(t)$ or $h(t)$] to obtain an approximate response.

To represent this procedure mathematically, assume we are approximating $h(t)$. Let each pulse have a width $\Delta\tau$. Then the ith pulse will start at $\tau = (i - 1)\,\Delta\tau$, and the ith impulse function representing it will occur at

$$\tau = (i - 1)\,\Delta\tau + \frac{\Delta\tau}{2} = \frac{2i - 1}{2}\,\Delta\tau \tag{9.8.24}$$

The amplitude of the impulse is the area of the ith pulse, which is

$$A_i = h\left(\frac{2i - 1}{2}\,\Delta\tau\right)\Delta\tau \tag{9.8.25}$$

Now this delayed impulse is to be convolved with $g(t)$. The result is the value of $g(t)$ at the time the impulse strikes times the amplitude of the impulse and is given by

$$f_i(t) = g\left(t - \frac{2i-1}{2}\Delta\tau\right)A_i \tag{9.8.26}$$

The total response is then given by the summation

$$f(t) = \sum_{i=1}^{t/\Delta\tau} g\left(t - \frac{2i-1}{2}\Delta\tau\right)h\left(\frac{2i-1}{2}\Delta\tau\right)\Delta\tau \tag{9.8.27}$$

which when expanded gives

$$f(t) = \Delta\tau\left[g\left(t - \frac{\Delta\tau}{2}\right)h\left(\frac{\Delta\tau}{2}\right) + \cdots + g\left(\frac{\Delta\tau}{2}\right)h\left(t - \frac{\Delta\tau}{2}\right)\right] \tag{9.8.28}$$

where, in the last term, $i = t/\Delta\tau$.

Example 9.21 Represent $h(t)$ in Fig. 9.20*a* by a series of pulses of width $\Delta\tau = 0.5$. Then represent each pulse by an impulse at the center of the pulse and convolve the impulse train with $g(t)$ in Fig. 9.20*b*.

The representation of $h(t)$ by first pulses and then impulses appears in Fig. 9.24. Now, from Eqs. (9.8.27) and (9.8.28),

$$\begin{aligned} f(t) = \tfrac{1}{2}[f(t - \tfrac{1}{4})h(\tfrac{1}{4}) + g(t - \tfrac{3}{4})h(\tfrac{3}{4}) + \cdots \\ + g(\tfrac{3}{4})h(t - \tfrac{3}{4}) + g(\tfrac{1}{4})h(t - \tfrac{1}{4})] \end{aligned}$$

Inserting integer values of time and plotting, we obtain Fig. 9.20*c*. In this case the $f(t)$ obtained is an exact, rather than an approximate, response. This is the exception rather than the rule and occurs because $h(t)$ is constant and $g(t)$ is linear, to keep the example simple algebraically.

REFERENCES

9.1. Aseltine, J. A.: "Transform Method in Linear System Analysis," McGraw-Hill Book Company, New York, 1958. Presents a readable and thorough treatment of the topics covered in this chapter, including an excellent discussion of the impulse function.

9.2. Chen, W. H.: "The Analysis of Linear Systems," McGraw-Hill Book Company, New York, 1963. Chapter 9 is devoted to the use of Fourier transforms in finding network responses.

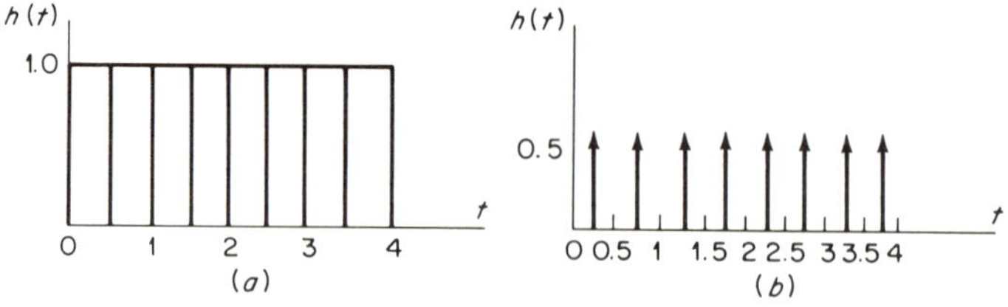

Fig. 9.24 Representing h(t) by pulses and impulses.

9.3. LePage, W. R.: "Complex Variables and the Laplace Transform for Engineers," McGraw-Hill Book Company, New York, 1961. Fourier and Laplace transforms and convolution are discussed in depth following several chapters on complex-variable theory.

9.4. Papoulis, A.: "The Fourier Integral and Its Applications," McGraw-Hill Book Company, New York, 1962. Most of the topics of the present chapter are included in this advanced-level text. Fourier transforms are accented, and the impulse function is developed from the theory of distributions.

9.5. Reza, F., and S. Seely: "Modern Network Analysis," McGraw-Hill Book Company, New York, 1959. Chapter 10 deals with singularity functions, and the Laplace transform is discussed in Chapter 11. The step and impulse are defined as limiting cases of a family of continuous functions.

9.6. Savant, C. J.: "Fundamentals of the Laplace Transformation," McGraw-Hill Book Company, New York, 1962. Chapter 7 contains an excellent discussion of Laplace transform theorems.

9.7. Schwarz, M.: "Information Transmission, Modulation, and Noise," McGraw-Hill Book Company, New York, 1959. Chapter 2 of this well-written book discusses Fourier series and transform theory with particular emphasis on application to pulse excitations.

9.8. Van Valkenburg, M. E.: "Network Analysis," Prentice-Hall, Inc., Englewood Cliffs, N.J., 1964. Written as an undergraduate text, the book presents a very clear and understandable development of Laplace transforms, singularity functions, and convolution.

PROBLEMS

Drill

9.1. Find the Fourier series of

$$f(t) = \begin{cases} \cos \omega_c t & -\frac{\tau}{2} < t < \frac{\tau}{2} \\ 0 & -\frac{\tau}{2} < t < -\frac{\tau}{2} \\ 0 & \frac{\tau}{2} < t < \frac{T}{2} \end{cases}$$

$f(t)$ periodic

$$\omega_c \gg \frac{2\pi}{T}$$

9.2. Find the Fourier series of the periodic waveform shown. Use the result to find values for $\pi/4$ and $\pi^2/8$ as sums of series.

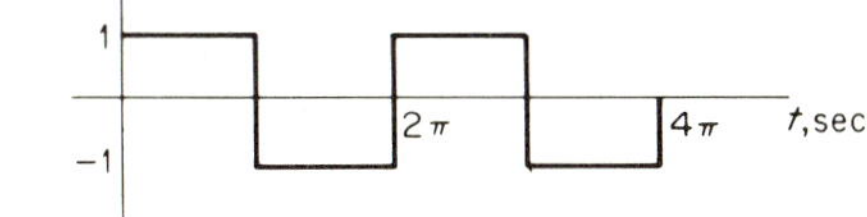

Fig. P9.2

9.3. Find the complex form of the Fourier series for the given periodic waveform. Find its Fourier transform, assuming now that $i(t) = 0$ for $t > 2\pi$.

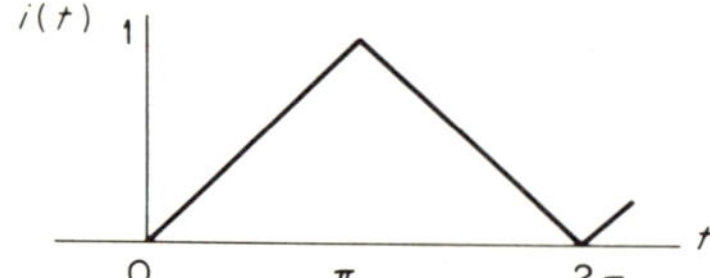

Fig. P9.3

9.4. Consider a periodic pulse train of period T and pulse width $T/6$. Four types of pulses, each with the same area, are shown. Plot the amplitude spectrum for each and compare.

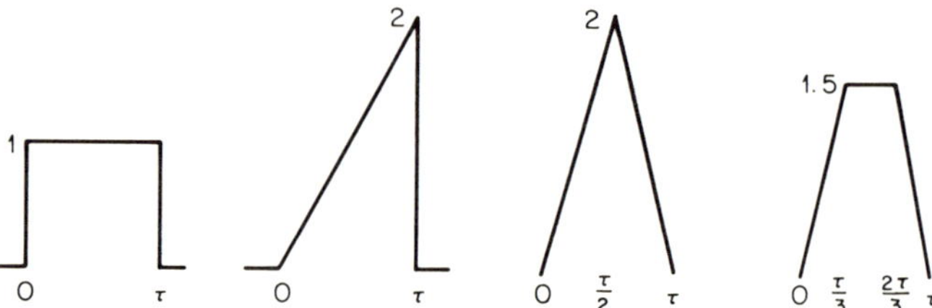

Fig. P9.4

9.5. Using the results of Example 9.1 and the principle of superposition, find the complex form of the Fourier series of the given waveform.

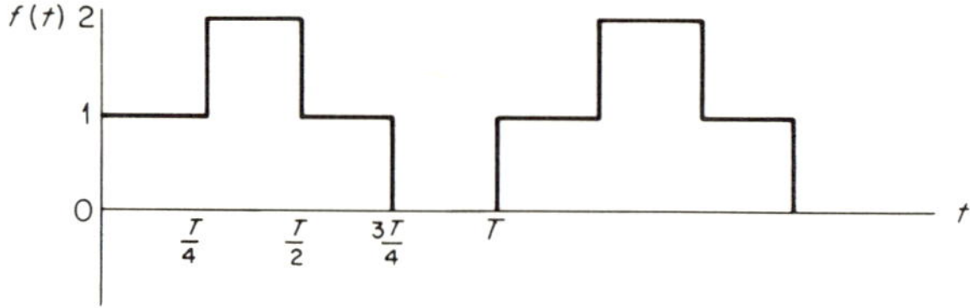

Fig. P9.5

9.6. Find the Fourier transform of

$$f(t) = \begin{cases} \cos 2\pi t & -\frac{\tau}{2} < t < \frac{\tau}{2} \\ 0 & \text{elsewhere} \end{cases}$$

Sketch $F(\omega)$ vs. ω for $\tau = 0.5, 1.0, 2.0, \infty$.

9.7. Find the Fourier transform of the given nonperiodic 3-pulse signal.

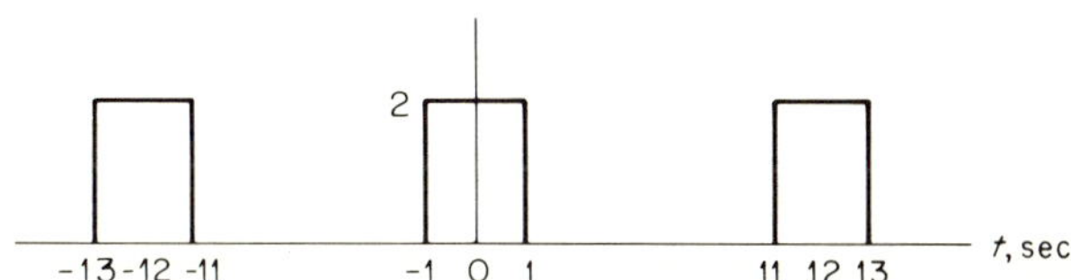

Fig. P9.7

9.8. The weighting and excitation functions of a system are such that

$$g(t) * h(t) = 1 - h(t) \qquad t \geqslant 0$$

If $g(t) = te^{-\alpha t}$, find $H(s)$.

9.9. Without using Eq. (9.5.8), find the Laplace transforms of

(*a*) $tu(t - \tau)$

(*b*) $t^2e^{-at} \cos \omega t$

(*c*) $\frac{1}{t} \sin^2 \omega t$

(*d*) $\int_0^t (t - \tau)u(\tau)\, d\tau$

(*e*)

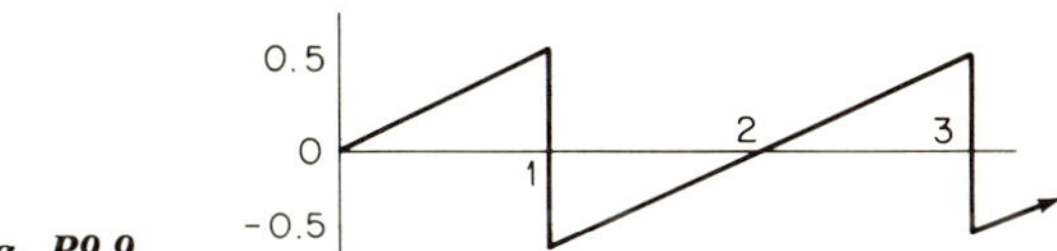

Fig. P9.9

9.10. Find the Laplace transform of $f(t)$ and of its derivative and integral. Sketch each.

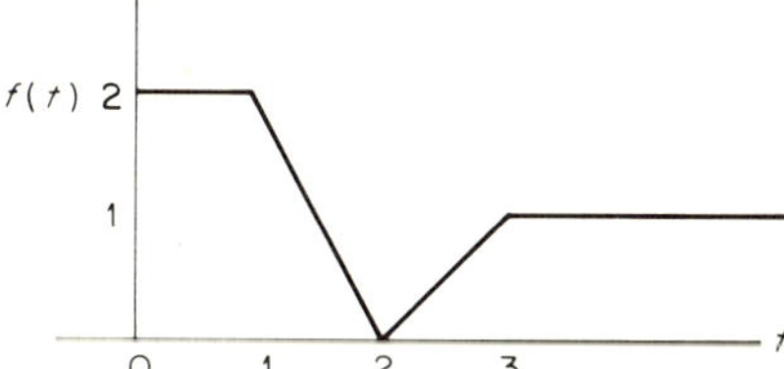

Fig. P9.10

9.11. The time response of a system to a unit-step-function excitation is

$$y(t) = 1 - 0.5e^{-0.2t} - 0.5e^{-3.5t}(0.7 \sin 5t + \cos 5t)$$

Find the system weighting and transfer functions.

9.12. Given:

$$F(s) = \frac{K(s + a)}{s^3 + bs^2 + cs + d}$$

(*a*) Find the initial and final values of $f(t)$, where

$$f(t) = \mathscr{L}^{-1}[F(s)]$$

(*b*) Find the initial values of the first two derivatives of $f(t)$.

9.13. An *RC* filter is shown.

(*a*) For $h(t) = tu(t)$ find $f(t)$ by convolving $h(t)$ with $g(t)$ and by convolving d^2h/dt^2 with $g^{-2}(t)$.

(*b*) Find the response of the network to unit step and unit impulse excitations.

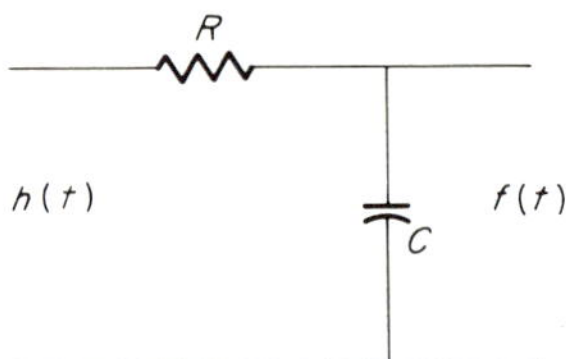

Fig. P9.13

9.14. Find and sketch the waveform of a voltage source $e(t)$ that will produce the current $i(t)$ shown in a series *RLC* circuit where $R = 3$ ohms, $L = 2$ henrys, $C = 1$ farad.

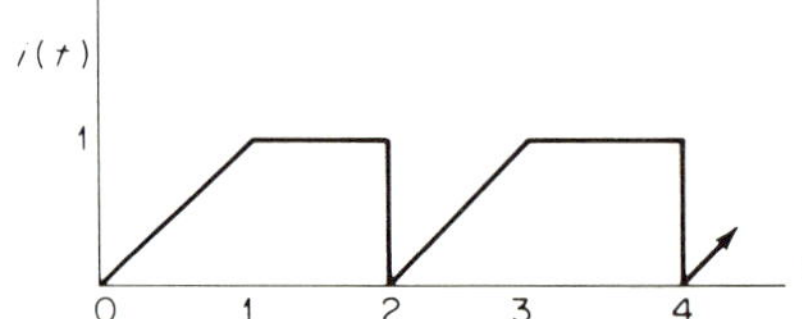

Fig. P9.14

Theory and proofs

9.15. The real part of the Fourier transform of nonrecurring signal is shown.

(*a*) Find the time-domain representation of the signal, assuming $h(t) = 0$ for $t < 0$.

(*b*) Prove that $h(t)$ can also be found from

$$h(t) = \frac{2}{(-jt)^2\pi} \int_0^\infty \frac{d^2H_r}{d\omega^2} e^{j\omega t}\, d\omega$$

Show that the integral reduces to a summation of two terms for the given H_r.

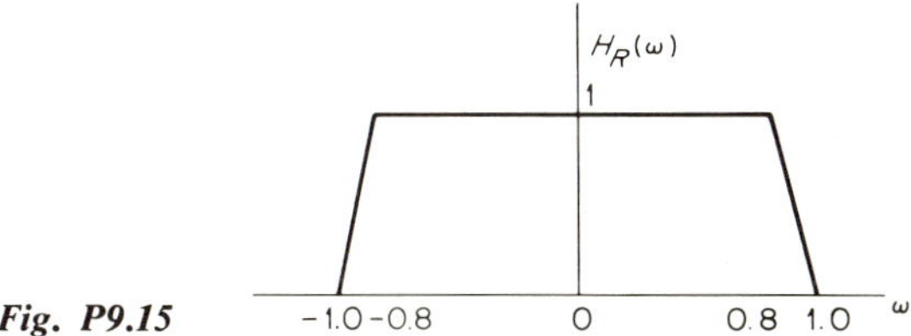

Fig. P9.15

9.16. Verify the orthogonality relations in Eqs. (9.2.8) to (9.2.10).

9.17. Simplify Eqs. (9.2.12) to (9.2.14) and (9.2.21) for (*a*) odd functions of x, (*b*) even functions of x, (*c*) functions of x with only even harmonics, and (*d*) functions of x with only odd harmonics.

9.18. Let $F(x) = \Sigma\, b_k \sin kx$ be the Fourier series approximation to a function $f(x)$ in the interval $-\pi < x < \pi$. Derive an expression for the b_k coefficients such that the mean square error between $f(x)$ and $F(x)$ is minimized.

9.19. Derive Eqs. (9.8.7) and (9.8.10). Also derive Eq. (9.8.1) from Eq. (9.8.7) by including appropriate step functions in the latter equation.

9.20. Let the pulse in Fig. 9.2 be displaced $\tau/2$ sec to the right. Show that this does not change the Fourier spectrum of the pulse but does introduce a phase shift in each term of the Fourier series.

9.21. Knowing the Fourier series for a pulse train of period T and pulse width τ, devise a method of writing down by inspection the Fourier series for a pulse train of period T and width $T - \tau$.

9.22. Verify the statement following Eq. (9.3.2) by plotting $F(t)$ for the pulse train in Example 9.1 for $\tau = T/2$ and $T/6$, using only those terms whose frequencies are less than f_c.

9.23. Let a Fourier transform be given by

$$F(\omega) = \pi\delta(\omega) + \frac{1}{j\omega}$$

Show that the inverse Fourier transform of $F(\omega)$ is a unit step starting at $t = 0$.

9.24. Define a function

$$u(t, \lambda) = \frac{1}{2} + \frac{1}{\pi}\tan^{-1}\lambda t \qquad \lambda > 0$$

(*a*) Show that this function approaches a unit step starting at $t = 0$ as $\lambda \to \infty$.

(*b*) Find the derivative of the given function. Show that the derivative approaches a unit impulse at $t = 0$ as $\lambda \to \infty$.

Application

9.25. Given that the Laplace transform of $f(t)$ is $F(s)$, prove that the Laplace transform of $f(t/a)$ is $aF(as)$. Use this result to find $F(s) = \mathscr{L}[e^{-50t}\cos 100t]$, t in seconds, such that when $F(s)$ is inverse-transformed, t will be in milliseconds.

9.26. An ideal low-pass filter and its excitation are shown. Find and plot the output of the filter as a function of time for $\omega_c = \pi/3\tau$, $2\pi/\tau$, and $12\pi/\tau$.

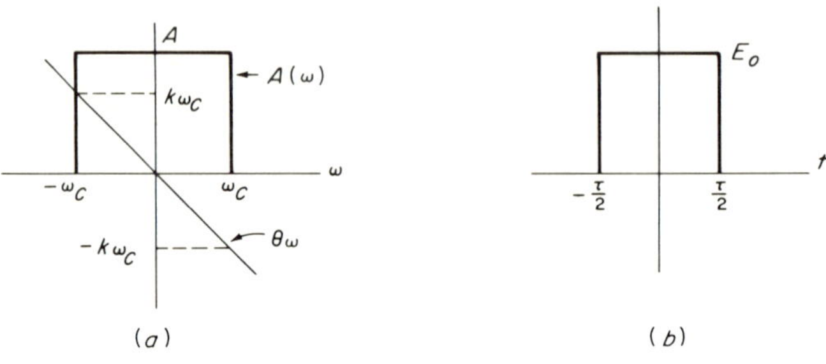

Fig. P9.26

9.27. The system shown has a weighting function $g(t)$. Find its response to the three pulse and the impulse excitations by (*a*) analytical convolution, (*b*) graphical convolution, and (*c*) derivative-integral convolution. Plot the four responses and compare them.

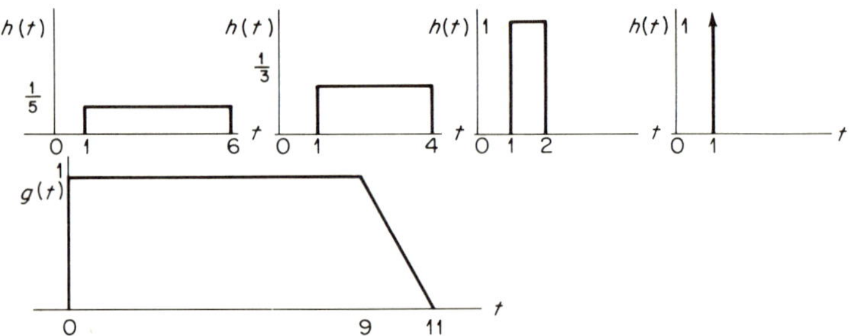

Fig. P9.27

9.28. The voltage and current at the terminals of a 1-port are given by $v = tu(t)$ and $i = e^{-t}u(t)$. Plot the Fourier spectrum of the power delivered to the 1-port.

9.29. The circuit shown is a two-section harmonic wave analyzer. Assume $e(t)$ is a voltage representable by a Fourier series. Also assume

$$Q_n = \frac{n\omega L}{R} \geq 10$$

for both coils over the range of frequencies of interest and

$$R_0 \geq 100RQ_n^2$$

for all n considered. Set both C's for resonance to the nth harmonic. Derive an expression for the ratio in the output of the amplitude of the nth harmonic to that of the mth harmonic (v_n/v_m) in terms of the ratio at the input (e_n/e_m).

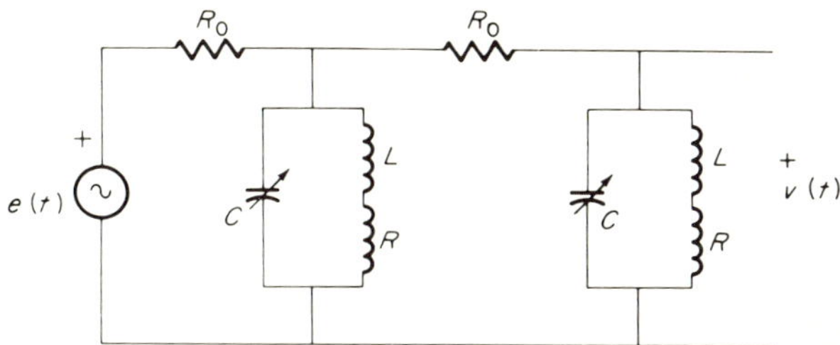

Fig. P9.29

9.30. The two boxes shown have identical weighting functions, and there is no loading of the second box on the first. The impulse response of the left box is te^{-t}. Find the impulse response of the overall system.

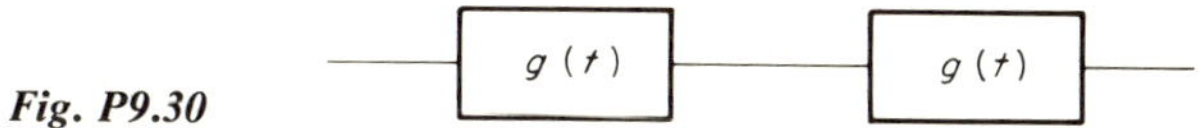

Fig. P9.30

9.31. A system has a transfer function given by $G(s) = 1/(s^2 + 1)$. Find its weighting function and then determine the system output for an input of $h(t) = tu(t)$ by (*a*) analytical convolution of $h(t)$ and $g(t)$, (*b*) analytical convolution of d^2h/dt^2 and $g^{-2}(t)$, (*c*) approximate graphical convolution of $h(t)$ and $g(t)$, using impulses for values of time $0 \leq t \leq 7$ and with increments of $\Delta t = 0.5$ sec, and (*d*) exact graphical convolution of $h(t)$ and $g(t)$ for values of time $0 \leq t \leq 7$ and with increments of $\Delta t = \pi/2$ sec. Plot approximate and exact responses and compare.

9.32. The excitation and weighting functions of a system are shown as $h(t)$ and $g(t)$, respectively. Find the system response by (*a*) analytical convolution of $h(t)$ and $g(t)$, (*b*) analytical convolution of dh/dt and $g^{-1}(t)$, (*c*) approximate graphical convolution of $h(t)$ and $g(t)$ for $0 \leq t \leq 6$ and time increments of $\Delta t = 0.5$ sec, and (*d*) exact graphical convolution of $h(t)$ and $g(t)$ for $0 \leq t \leq 6$ and time increments of $\Delta t = 0.5$ sec.

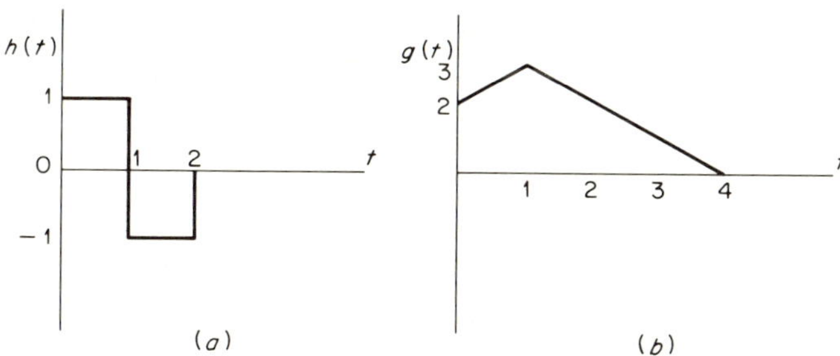

Fig. P9.32

10 FORCED AND STEADY-STATE RESPONSES

10.1 INTRODUCTION

In this chapter techniques will be developed for obtaining the forced and steady-state responses of any system that can be characterized by an nth-order linear differential equation with constant coefficients when such a system is simultaneously excited by external driving functions and internal initial conditions.

We have already considered one nth-order solution technique in Sec. 8.6, namely, the use of $T(s_0)$ in conjunction with an $e^{s_0 t}$ driving function to obtain the forced response. It was observed that this procedure had several inherent difficulties and was not particularly versatile when compared with a solution by Laplace transforms.

The present chapter begins with a review of the Laplace transformation of an nth-order differential equation, drawing on the Laplace transform theorems of Chap. 9. We are led to a further consideration of initial conditions and of certain difficulties that arise when impulses at $t = 0$ are present. In Sec. 10.3, the partial-fraction-expansion technique for obtaining inverse Laplace transforms is presented, and the reader is introduced to the residue-sum theorem, which considerably reduces the labor involved in computing partial-fraction-expansion coefficients.

With this groundwork completed, various classes of transient switching problems are presented. Solutions by Laplace theory and by exponential end-point analysis are developed, the latter being an almost-by-inspection technique when only one energy-storage element is present in the circuit. Following this, steady-state solutions to periodic excitations are considered. Four techniques are presented, namely, approximate solution by Fourier series, solution by end-point analysis, and open- and closed-form solutions by Laplace transforms.

The chapter concludes with two sections on frequency-domain response. In the first, break-point (Bode) analysis is developed, and several circuits with first-order factors of the form $1 + \tau s$ are considered. In the last section, a basic second-order system is considered in detail, with particular emphasis on a comparison of the frequency- and time-domain responses of such a system.

The chapter focuses on the analysis of an nth-order system from a single nth-order differential equation. The reader will recall from the discussion in Chap. 8 that an alternative approach is to analyze an nth-order system through a set of n first-order differential equations. This approach will be the subject of Chap. 11.

10.2 INITIAL CONDITIONS: II—IMPULSES

Reconsider the nth-order differential Eq. (8.2.1), repeated here for ease of reference

$$a_n \frac{d^n y}{dt^n} + a_{n-1} \frac{d^{n-1} y}{dt^{n-1}} + \cdots + a_1 \frac{dy}{dt} + a_0 y = b_m \frac{d^m x}{dt^m} + b_{m-1} \frac{d^{m-1} x}{dt^{m-1}} + \cdots + b_1 \frac{dx}{dt} + b_0 x \qquad (10.2.1)$$

where the a's and b's are all assumed to be constants. Laplace transforming Eq. (10.2.1) term by term, using Theorems 1 and 2 of Chap. 9 [Eqs. (9.7.1) and (9.7.2)], gives

$$Y(s)(a_n s^n + a_{n-1} s^{n-1} + \cdots + a_1 s + a_0) + IC_y = X(s)(b_m s^m + b_{m-1} s^{m-1} + \cdots + b_1 s + b_0) + IC_x \qquad (10.2.2)$$

In Eq. (10.2.2) $Y(s)$ and $X(s)$ are the Laplace transforms of $y(t)$ and $x(t)$, respectively, and IC_y and IC_x are the initial-condition terms in y and x, respectively, resulting from the application of Eq. (9.7.2), and are generally functions of s.

Letting the term in parentheses on the left in Eq. (10.2.2) be $Q(s)$, that on the right $P(s)$, and their ratio $P(s)/Q(s) = T(s)$ (as in Sec. 8.6), we have

$$Y(s)Q(s) + IC_y = X(s)P(s) + IC_x \tag{10.2.3}$$

Solving for $Y(s)$ gives

$$Y(s) = X(s)T(s) + \frac{IC_x - IC_y}{Q(s)} \tag{10.2.4}$$

In Eq. (10.2.4), the first term on the right is the transform of the forced, or zero-state, response. The second term is the transform of the free, or zero-input, response. It is this latter term that we wish to examine in greater detail now, particularly with reference to the handling of impulses at the origin.

In a passive system, m, the highest power of s in the numerator of $T(s)$ [in $P(s)$] can at most be 1 greater than n, the highest power of s in the denominator of $T(s)$ [in $Q(s)$]. This constraint follows from the discussion following Eq. (8.5.2). What will the time-domain response be like if m is 1 greater than n, that is, if $T(s)$ has a pole at infinity? Consider the following equation

$$\frac{dy}{dt} + a_0 y = \frac{d^2x}{dt^2} + a_1 \frac{dx}{dt} + a_0 x \tag{10.2.5}$$

and suppose that $x(t)$ is a unit step originating at the origin. This means that dx/dt will be an impulse and d^2x/dt^2 will be a doublet. Then the left side of Eq. (10.2.5) must also contain a doublet, and it will appear in the dy/dt term. It follows that y will contain an impulse. Thus, although the excitation does not contain an impulse, the response must contain one because $m > n$ for the system. Now suppose the d^2x/dt^2 term is missing from Eq. (10.2.5) but that $x(t)$ is a unit impulse originating at the origin. Then dx/dt is a doublet, dy/dt again contains a doublet, and y again contains an impulse. Conversely, if $m \leq n$, and if x does not contain an impulse, then y will not.

The presence of impulses at the origin in y, either because $m > n$ or because the excitation contains an impulse at the origin, makes it necessary for us to consider in further detail the behavior of the Laplace transformation in the vicinity of $t = 0$. To indicate the type of problem that occurs there, consider the application of Eq. (9.7.8), the Laplace transform of the first derivative, when $f(t)$ is a unit step function. We have, from Eq. (9.7.8),

$$\mathscr{L}\left[\frac{d}{dt}u(t)\right] = s\frac{1}{s} - u_+ \tag{10.2.6}$$

Now the left side of Eq. (10.2.6) should yield $\mathscr{L}[\delta(t)]$, which we derived as 1 in Sec. 9.7. But, noting that $u_+ = 1$, the right side of Eq. (9.8.1) yields 0. Something is obviously incomplete in our understanding of Eq. (9.7.8) and thus of Eq. (9.5.8), the basic Laplace transform integral. In particular, we need to reexamine the lower limit in Eq. (9.5.8).

It has been understood throughout our discussion of Laplace transforms that by $t = 0$ we mean $t = 0_+$; that is, we approach $t = 0$ from the right but never quite get to it. All initial conditions have been interpreted in this light, as has the lower limit in Eq. (9.5.8). With this interpretation in mind, reconsider the unit impulse at $t = 0$. If this impulse is considered to be slightly to the left of the lower limit in Eq. (9.5.8), its transform is 0, not 1, because $\delta(t)$ is not included in the region of integration. On the other hand, if $\delta(t)$ is considered as being slightly to the right of the lower limit in Eq. (9.5.8), its transform is then 1. We indicate these two situations by writing

$$\mathscr{L}[\delta_-(t)] = 0 \qquad \text{and} \qquad \mathscr{L}[\delta_+(t)] = 1 \tag{10.2.7}$$

where $\delta_-(t)$ and $\delta_+(t)$ are impulses to the left and right, respectively, of the lower limit in Eq. (9.5.8).

Returning now to our initial problem with Eq. (10.2.6) and defining $u_-(t)$ and $u_+(t)$ in the same manner as we defined the two impulse functions, we see that

$$\mathscr{L}\left[\frac{d}{dt}u_-(t)\right] = s\frac{1}{s} - 1 = 0 \qquad \text{and} \qquad \mathscr{L}\left[\frac{d}{dt}u_+(t)\right] = s\frac{1}{s} - 0 = 1 \tag{10.2.8}$$

that is, $u_+ = 1$ if $u(t)$ is to the left of the lower Laplace limit and 0 if it is to the right.

To lend some physical feeling to what we have just done, we comment that considering $\delta(t)$ to lie to the left of the lower Laplace limit means that we are considering $\delta(t)$ as an initial-condition producer, i.e., as part of the past history of the system before an excitation is applied. On the other hand, placing it to the right of the lower Laplace limit means that we are considering $\delta(t)$ as an excitation, i.e., as part of the system's current history. It is permissible to consider it in either way, but we must be careful, in a given problem, not to consider it in both ways at the same time. If we consider it as an excitation with a transform of 1, it must contribute nothing to the initial conditions, and vice versa.

We shall illustrate these concepts by solving several simple differential equations by Laplace transforms, which will also permit a discussion of some of the subtleties of the Laplace transform technique.

Example 10.1 An excitation $x = u(t)$ is applied to three initially relaxed systems whose differential equations are

$$(a)\quad \frac{dy}{dt} + 4y = 2x$$

$$(b)\quad \frac{dy}{dt} + 4y = 3\frac{dx}{dt} + 2x$$

$$(c)\quad \frac{dy}{dt} + 4y = \frac{d^2x}{dt^2} + 3\frac{dx}{dt} + 2x$$

In each case, find the response $y(t)$.

The reader will note that $m < n$, $m = n$, and $m > n$ in the first, second, and third equations, respectively.

(*a*) Laplace transforming, we obtain

$$(s + 4)Y(s) - y_4 = 2x(s) = \frac{2}{s}$$

To find y_+, we use the first method of Sec. 8.7 and obtain

$$y_+ = y_- = 0$$

Thus, in Eq. (10.2.4) the initial condition term is zero and we have

$$Y(s) = X(s)T(s) = \frac{2}{s(s + 4)} = \frac{1}{2s} - \frac{1}{2(s + 4)}$$

From Table 9.3 we find

$$y(t) = \tfrac{1}{2}u(t) - \tfrac{1}{2}e^{-4t}$$

As a check, insert the $y(t)$ found into the differential equation. We have

$$\frac{dy}{dt} + 4y = \tfrac{1}{2}\delta(t) + 2e^{-4t} + 2u(t) - 2e^{-4t} \neq 2x = 2u(t)$$

The two sides of the equation are not equal; thus $y(t)$ is incorrect. What we have neglected to do is to ensure that the exponential term in $y(t)$ is zero for all negative time, a requirement of the Laplace transform method. If we write

$$y(t) = \tfrac{1}{2}u(t) - \tfrac{1}{2}e^{-4t}u(t)$$

and insert this $y(t)$ into the differential equation, the two sides will be equal, recognizing that $\tfrac{1}{2}e^{-4t}\delta(t) = \tfrac{1}{2}\delta(t)$.

(*b*) Laplace transforming gives

$$(s + 4)Y(s) - y_+ = (3s + 2)X(s) - 3x_+$$

Applying the technique of Sec. 8.7, we have

$$y_+ = 3x_+$$

which causes the initial condition term in Eq. (10.2.4) to vanish again. Thus

$$Y(s) = \frac{3s + 2}{s(s + 4)} = \frac{1}{2s} + \frac{5}{2(s + 4)}$$

which gives

$$y(t) = \tfrac{1}{2}u(t) + \tfrac{5}{2}e^{-4t}u(t)$$

The solution with $m = n$ differs from that with $m < n$ only in the coefficient of the free response term.

(c) Laplace transforming gives

$$(s + 4)Y(s) - y_+ = (s^2 + 3s + 2)X(s) - sx_+ - \left.\frac{dx}{dt}\right|_+ - 3x_+$$

We know from the discussion earlier in this section that $y(t)$ must contain an impulse even though $x(t)$ does not. Because of this impulse, we must be careful in using the method of Sec. 8.7 to find y_+. In particular, if the excitation begins to the left of the lower Laplace limit, $\int_-^+ y\,dt$ is not zero but has the value x_+, as can be shown by performing the double integration of each side of the differential equation between the $t = 0_-$ and $t = 0_+$ limits. We have

$$\int_-^+\!\!\int \frac{dy}{dt}\,dt^2 + 4\int_-^+\!\!\int y\,dt^2 = \int_-^+\!\!\int \frac{d^2x}{dt^2}\,dt^2 + 3\int_-^+\!\!\int \frac{dx}{dt}\,dt^2 + 2\int_-^+ x\,dt^2$$

which simplifies to

$$\int_-^+ y\,dt + 0 = x_+ + 0 + 0 \qquad \text{or} \qquad x_+ = \int_-^+ y\,dt$$

Then a single integration of the differential equation between the same limits yields y_+

$$\int_-^+ \frac{dy}{dt}\,dt + 4\int_-^+ y\,dt = \int_-^+ \frac{d^2x}{dt^2}\,dt + 3\int_-^+ \frac{dx}{dt}\,dt + 2\int_-^+ x\,dt$$

$$y_+ + 4x_+ = \left.\frac{dx}{dt}\right|_+ + 3x_+$$

$$y_+ = \left.\frac{dx}{dt}\right|_+ - x_+$$

Proceeding now to form Eq. (10.2.4), we have

$$Y(s) = \frac{s^2 + 3s + 2}{s(s + 4)} - x_+\frac{s + 4}{s + 4} = 1 + \frac{1}{2s} - \frac{3}{2(s + 4)} - 1$$

The initial condition term is not zero; it has the value $-x_+ = -1$.

Finding the inverse transform using Table 9.3 gives

$$y(t) = \tfrac{1}{2}u(t) - \tfrac{3}{2}e^{-4t}u(t)$$

An examination of the result shows that we do not have the required impulse at the origin in the final answer. We have its transform in both $X(s)T(s)$ and in $(IC_x - IC_y)/Q(s)$ but not in their sum. To explain what has happened, consider Fig. 10.1. In Fig. 10.1*a*, we show the excitation $x(t)$ and in Fig. 10.1*b* the response $y(t)$, including the necessary unit impulse at the origin to the left of the Laplace limit. Thus the transform of the impulse should not appear in $T(s)$. We have two choices. We can

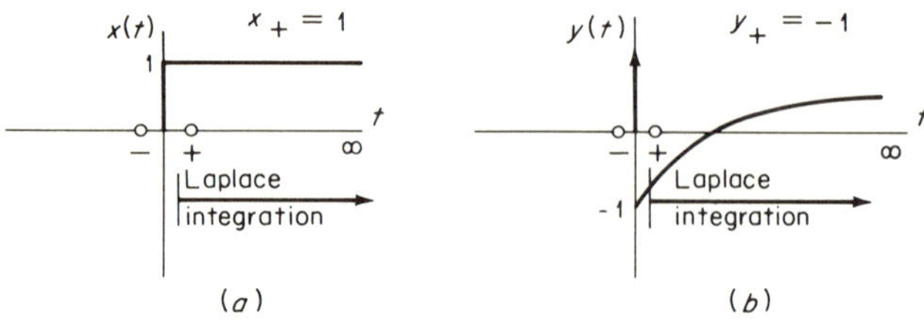

Fig. 10.1 Relating a jump at the origin to the lower Laplace limit.

work the problem as we have done with the excitation beginning to the left of the lower Laplace limit and then add the impulse to $y(t)$ at the end to obtain the correct response as

$$y(t) = \delta(t) + \tfrac{1}{2}u(t) - \tfrac{3}{2}e^{-4t}u(t)$$

or we can push the step excitation slightly to the right of the lower Laplace limit, which will make all initial conditions zero at that limit and thus cause the initial-condition term in Eq. (10.2.4) to vanish. It will also reinstate the transform of the impulse in $T(s)$, giving

$$Y(s) = \frac{s^2 + 3s + 2}{s(s + 4)} = 1 + \frac{1}{2s} - \frac{3}{2(s + 4)}$$

which, when inverse-transformed, also yields the correct $y(t)$.

We shall use this latter method in problems having $m > n$.

Example 10.2 An initially relaxed system described by

$$\frac{d^2v}{dt^2} + 3\frac{dv}{dt} + 2y = \frac{dx}{dt} + 4x$$

is excited by a unit impulse applied at $t = 0$. Find the response.

We shall work this problem in two ways: (*a*) assuming that $\delta(t)$ is to the left of the lower Laplace limit and hence is an initial-condition producer and (*b*) by assuming that $\delta(t)$ is to the right of the lower Laplace limit and hence is a driving function.

(*a*) Here $\delta(t)$ is outside the limits of the Laplace integration and thus its transform is zero. By the method of Sec. 8.9, a double integration of both sides of the differential equation between the limits of $t = 0_-$ and 0_+ yields

$$y_+ = \int_-^+ x\,dt = 1$$

Then a single integration gives

$$\left.\frac{dy}{dt}\right|_+ + 3y_+ = 4\int_-^+ x\,dt = 4 \qquad \left.\frac{dy}{dt}\right|_+ = 1$$

Now, transforming the differential equation, we obtain.

$$Y(s)(s^2 + 3s + 2) - sy_+ - \left.\frac{dy}{dt}\right|_+ - 3y_+ = X(s)(s + 4) - x_+$$

Both $X(s)$ and x_+ are zero, so that

$$Y(s) = \frac{s+4}{s^2+3s+2} = \frac{3}{s+1} - \frac{2}{s+2}$$

From Table 9.3

$$y(t) = 3e^{-t}u(t) - 2e^{-2t}u(t)$$

(*b*) Here $\delta(t)$ is within the Laplace limits and has a transform of 1. This means that

$$y_+ = \int_-^+ x\,dt = 0 \qquad \left.\frac{dy}{dt}\right|_+ = 0$$

and thus all initial conditions are zero. Transforming, we obtain

$$Y(s)(s^2 + 3s + 2) - 0 = X(s)(s + 4) - 0$$

and with $X(s) = 1$, we obtain the same $Y(s)$ and therefore $y(t)$ as before.

The difference between Example 10.2 and part (*c*) of Example 10.1 should be stressed. In the latter, the excitation contained an impulse but the response did not. In the former, the opposite was true. In the latter, it was easier to consider the excitation as lying to the right of the lower Laplace limit. This is also true in the former. Thus, in future problems involving either $m > n$ or an impulse excitation with the system initially relaxed, we shall consider initial conditions due to impulses to be zero and the transforms of impulses to be nonzero.

10.3 PARTIAL-FRACTION EXPANSION

In Examples 10.1 and 10.2, we performed simple partial-fraction expansions in order to expand $Y(s)$ into a sum of terms each of whose inverse transforms could be obtained directly from Table 9.3. It is the purpose of this section to present a general development of the partial-fraction-expansion technique for obtaining an inverse Laplace transform.

We begin with Eq. (10.2.4), rewritten in the form

$$Y(s) = \frac{X(s)P(s) + IC_x - IC_y}{Q(s)} \tag{10.3.1}$$

The right side of Eq. (10.3.1) is a ratio of polynomials in s whose numerator degree m may be greater than, equal to, or less than its denominator degree n. The first step in obtaining a partial fraction expansion is to divide the denominator into the numerator until the degree of the remainder numerator is one less than n. We indicate this process by writing

$$Y(s) = A_{m-n}s^{m-n} + \cdots + A_1 s + A_0 + \frac{R(s)}{Q(s)} \tag{10.3.2}$$

where $R(s)$ is of degree $n - 1$. As was noted in Sec. 10.2, m is generally not greater than $n + 1$; thus terms above $A_1 s$ are not likely to appear in Eq. (10.3.2).

Since $Q(s)$ is of degree n, it must have n roots, $s_1, s_2, \ldots, s_n$. Thus the expansion of $R(s)/Q(s)$ takes the general form

$$\frac{R(s)}{Q(s)} = \frac{K_1}{s - s_1} + \frac{K_2}{s - s_2} + \cdots + \frac{K_n}{s - s_n} \tag{10.3.3}$$

There are three types of roots that $Q(s)$ may have and thus three types of terms in Eq. (10.3.3). The roots of $Q(s)$ may be simple real, simple complex conjugates, and multiple. We shall consider each of these in turn.

SIMPLE REAL ROOTS

Assume that s_1 is a simple real root. Then K_1 must be real, and the time-domain response term is $K_1 e^{s_1 t}$. The value of K_1 is found by multiplying each side of Eq. (10.3.3) by $s - s_1$ and evaluating at $s = s_1$

$$\left.\frac{R(s)(s - s_1)}{Q(s)}\right|_{s_1} = \left.\frac{K_1(s - s_1)}{s - s_1}\right|_{s_1} + \left.\frac{K_2(s - s_1)}{s - s_2}\right|_{s_1} + \cdots + \left.\frac{K_n(s - s_1)}{s - s_n}\right|_{s_1} \tag{10.3.4}$$

Now, as $s \to s_1$, all terms on the right side vanish except the K_1 term. The left side does not vanish because $Q(s)$ contains $s - s_1$. Thus

$$K_1 = (s - s_1)\left.\frac{R(s)}{Q(s)}\right|_{s_1} \tag{10.3.5}$$

Alternatively, we can find K_1 from

$$K_1 = \lim_{s \to s_1} \frac{(s - s_1)R(s)}{Q(s)} = \lim_{s \to s_1} \frac{(s - s_1)R'(s) + R(s)}{Q'(s)} = \left.\frac{R(s)}{Q'(s)}\right|_{s_1} \tag{10.3.6}$$

where L'Hospital's rule has been used in the second step and the primes indicate differentiation with respect to s. The result in Eq. (10.3.6) is useful when $Q(s)$ is not in factored form.

SIMPLE COMPLEX-CONJUGATE ROOTS

Assume $s_2 = -\sigma_2 + j\omega_2$ and $s_3 = s_2^* = -\sigma_2 - j\omega_2$. That portion of Eq. (10.3.3) relating to these roots can be written as

$$\frac{K_2}{s - s_2} + \frac{K_3}{s - s_2^*} = \frac{(K_2 + K_3)s + (K_2 + K_3)\sigma_2 + j(K_2 - K_3)\omega_2}{(s + \sigma_2)^2 + {\omega_2}^2} \tag{10.3.7}$$

Now $R(s)$ has real coefficients; thus $K_2 + K_3$ must be real and $K_2 - K_3$ imaginary. It follows that

$$K_3 = K_2^* \tag{10.3.8}$$

Letting $K_2 = Ke^{j\theta}$, the time-domain response terms are

$$Ke^{j\theta}e^{(-\sigma_2+j\omega_2)t} + Ke^{-j\theta}e^{(-\sigma_2-j\omega_2)t} = 2Ke^{-\sigma_2 t}\cos(\omega_2 t + \theta) \tag{10.3.9}$$

The coefficient K_2 is found exactly as it was in the simple-real-root case by multiplying each side of Eq. (10.3.3) by $s - s_2$ and evaluating at $s = s_2$. Then K and θ are found from K_2. If s_2 is imaginary, σ_2 is 0 and the response is an undamped cosine wave given by $2K\cos(\omega_2 t + \theta)$.

MULTIPLE ROOTS

Assume that $Q(s)$ is composed of a multiple root at $s = s_4$ of degree r and the three simple roots previously discussed. Then Eq. (10.3.3) takes the form

$$\frac{R(s)}{Q(s)} = \frac{K_1}{s - s_1} + \frac{K_2}{s - s_2} + \frac{K_2^*}{s - s_2^*} + \frac{k_1}{s - s_4} + \frac{k_2}{(s - s_4)^2} + \cdots + \frac{k_r}{(s - s_4)^r} \tag{10.3.10}$$

and the time-domain response for the multiple-root portion, assuming s_4 to be a real root, is

$$k_1 e^{s_4 t} + \frac{1}{1!}k_2 t e^{s_4 t} + \cdots + \frac{1}{(r-1)!}k_r t^{r-1} e^{s_4 t} \tag{10.3.11}$$

If s_4 is complex, a root s_4^* of degree r is also present, the terms in (10.3.11) are added to a similar set containing s_4^*, and Eq. (10.3.9) is applied.

To evaluate k_r, the coefficient of the highest-power term in $s - s_4$, multiply each side of Eq. (10.3.10) by $(s - s_4)^r$

$$\frac{R(s)}{Q_1(s)} = \frac{K_1(s - s_4)^r}{s - s_1} + \frac{K_2(s - s_4)^r}{s - s_2} + \frac{K_2^*(s - s_4)^r}{s - s_2^*} + k_1(s - s_4)^{r-1} + k_2(s - s_4)^{r-2} + \cdots + k_r \tag{10.3.12}$$

where

$$Q_1(s) = (s - s_1)(s - s_2)(s - s_2^*) \tag{10.3.13}$$

Evaluating Eq. (10.3.12) at $s = s_4$ gives

$$k_r = \left.\frac{R(s)}{Q_1(s)}\right|_{s_4} = (s - s_4)^r \left.\frac{R(s)}{Q(s)}\right|_{s_4} \tag{10.3.14}$$

If we attempt the same sort of maneuver to find k_{r-1} by multiplying each side of Eq. (10.3.10) by $(s - s_4)^{r-1}$, trouble develops. The left side becomes infinite at $s = s_4$, as does the k_r term on the right side. We need another procedure to evaluate k_{r-1} through k_1.

The form of Eq. (10.3.12) suggests a method. If we take the derivative of that equation with respect to s, the k_r coefficient disappears and the k_{r-1} coefficient is left alone without an $s - s_4$ multiplier. All other terms on the right side have an $s - s_4$ to at least the first power. At $s = s_4$, these terms vanish, and we have

$$k_{r-1} = \frac{d}{ds}\frac{R(s)}{Q_1(s)}\bigg|_{s_4} = \frac{d}{ds}\left[(s - s_4)^r \frac{R(s)}{Q(s)}\right]\bigg|_{s_4} \tag{10.3.15}$$

If we differentiate a second time, k_{r-2} can be evaluated, and so on. In general, the ith k from the group of r k's is found from

$$k_i = \frac{1}{(r - i)!}\frac{d^{r-i}}{ds^{r-i}}\left[(s - s_4)^r \frac{R(s)}{Q(s)}\right]\bigg|_{s_4} \tag{10.3.16}$$

In Eq. (10.3.16), if $i = r$, Eq. (10.3.14) results; if $i = r - 1$, Eq. (10.3.15) results.

Before considering examples, we shall present a theorem which considerably reduces the labor involved in finding the coefficients of a partial-fraction expansion, expecially when $Q(s)$ has multiple roots. Define a *residue* as the coefficient of a first-degree term in s in a partial-fraction expansion. Thus, in Eq. (10.3.10), K_1, K_2, K_2^*, and k_1 are residues, but k_2 through k_r are not. Utilizing the forms of $P(s)$ and $Q(s)$ in Eq. (10.2.2),

$$\sum \text{Res} = \begin{cases} 0 & \text{if } n > m + 1 \\ \dfrac{b_m}{a_n} & \text{if } n = m + 1 \end{cases} \tag{10.3.17}$$

where Res stands for residue. This result, due to Hazony and Riley,† is known as the *residue-sum theorem*. Its proof requires a knowledge of complex-variable theory and is presented in Example 12.24.

Example 10.3 Perform partial-fraction expansions of the following functions without the use of differentiation:

(*a*) $\dfrac{s + 2}{(s + 1)^2(s + 3)}$

(*b*) $\dfrac{s + 2}{(s + 1)^3(s + 3)}$

† D. Hazony and J. Riley, Evaluating Residues and Coefficients of High Order Poles, *IRE Trans. Autom. Control*, vol. AC-4, Nov., 1959.

(c) $\dfrac{s+4}{(s+1)^2(s+3)^2}$

(a) $$\frac{s+2}{(s+1)^2(s+3)} = \frac{K_1}{(s+1)^2} + \frac{K_2}{s+1} + \frac{K_3}{s+3}$$

K_1 and K_3 are found directly by multiplying through by $(s+1)^2$ and $(s+3)$, respectively

$$K_1 = \left.\frac{s+2}{s+3}\right|_{-1} = \frac{1}{2}$$

$$K_3 = \left.\frac{s+2}{(s+1)^2}\right|_{-3} = -\frac{1}{4}$$

Now K_2 and K_3 are residues, but K_1 is not. With $n = 3$ and $m = 1$, the sum of the residues is zero, and

$$K_2 = -K_3 = \tfrac{1}{4}$$

(b) $$\frac{s+2}{(s+1)^3(s+3)} = \frac{K_1}{(s+1)^3} + \frac{K_2}{(s+1)^2} + \frac{K_3}{s+1} + \frac{K_4}{s+3}$$

K_1 and K_4 are found directly.

$$K_1 = \left.\frac{s+2}{s+3}\right|_{-1} = \frac{1}{2}$$

$$K_4 = \left.\frac{s+2}{(s+1)^3}\right|_{-3} = \frac{1}{8}$$

K_3 and K_4 are residues; K_1 and K_2 are not. The sum of residues is again zero. Thus $K_3 = -K_4 = -\tfrac{1}{8}$. To find K_2, multiply each side of the original expansion by $s + 1$ so as to make K_2 a residue

$$\frac{s+2}{(s+1)^2(s+3)} = \frac{K_1}{(s+1)^2} + \frac{K_2}{(s+1)} + K_3 + \frac{K_4(s+1)}{s+3}$$

$$= \frac{K_1}{(s+1)^2} + \frac{K_2}{s+1} + K_3 + K_4 - \frac{2K_4}{s+3}$$

The second step is necessary to convert the $s + 3$ term into partial-fraction-expansion form (note that $K_3 + K_4 = 0$). The sum of the residues of this new expansion is again zero, yielding

$$K_2 - 2K_4 = 0 \qquad K_2 = \tfrac{1}{4}$$

(c) $$\frac{s+4}{(s+1)^2(s+3)^2} = \frac{K_1}{(s+1)^2} + \frac{K_2}{s+1} + \frac{K_3}{(s+3)^2} + \frac{K_4}{s+3}$$

K_1 and K_3 are found directly

$$K_1 = \left.\frac{s+4}{(s+3)^2}\right|_{-1} = \frac{3}{4}$$

$$K_3 = \left.\frac{s+4}{(s+1)^2}\right|_{-3} = \frac{1}{4}$$

Next, each side of the expansion is multiplied by either $s + 1$ or $s + 3$. Choosing the former gives

$$\frac{s+4}{(s+1)(s+3)^2} = \frac{K_1}{s+1} + K_2 + \frac{K_3(s+1)}{(s+3)^2} + \frac{K_4(s+1)}{s+3}$$

The third and fourth terms are expanded to obtain

$$\frac{K_3(s+1)}{(s+3)^2} = \frac{K_5}{(s+3)^2} + \frac{K_6}{s+3}$$

$$K_5 = K_3(s+1)|_{-3} = -2K_3$$

$K_6 = K_3$, since the sum of the residues is K_3.

$$\frac{K_4(s+1)}{s+3} = K_4 - \frac{2K_4}{s+3}$$

Inserting these results in the expansion gives

$$\frac{s+4}{(s+1)(s+3)^2} = \frac{K_1}{s+1} + K_2 - \frac{2K_3}{(s+3)^2} + \frac{K_3}{s+3}$$
$$+ K_4 - \frac{2K_4}{s+3}$$

The sum of the residues is zero. Thus

$$K_1 + K_3 - 2K_4 = 0 \qquad K_4 = 1/2$$

In the original expansion, the sum of the residues is also zero, giving

$$K_2 = -K_4 = -1/2$$

Example 10.4 Find the impulse response of a system whose transfer function is given by

$$T(s) = \frac{2s^4 + s^3 + 3s^2 + s + 2}{(s+1)(s+2)^2(s^2+2s+2)}$$

The impulse response is simply the inverse Laplace transform of $T(s)$. The partial-fraction expansion takes the form

$$T(s) = \frac{K_1}{s+1} + \frac{K_2}{s+2} + \frac{K_3}{(s+2)^2} + \frac{K_4}{s+1-j1} + \frac{K_4^*}{s+1+j1}$$

K_1, K_3, and K_4 can be found directly.

$$K_1 = \left.\frac{2s^4 + s^3 + 3s^2 + s + 2}{(s+2)^2(s^2+2s+2)}\right|_{-1} = 5$$

$$K_3 = \left.\frac{2s^4 + s^3 + 3s^2 + s + 2}{(s+1)(s^2+2s+2)}\right|_{-2} = -18$$

$$K_4 = \left.\frac{2s^4 + s^3 + 3s^2 + s + 2}{(s+1)(s+2)^2(s+1+j1)}\right|_{-1+j1}$$

$$= \frac{-8 + 2 + j2 - j6 - 1 + j1 + 2}{(j1)\,(j2)\,(j2)} = \frac{3 - j5}{4} = 1.46 \underline{/-59^\circ}$$

Then

$$K_4^* = \frac{3 + j5}{4} = 1.46\,\underline{/59^\circ}$$

and, noting that the sum of the residues is 2,

$$K_2 = 2 - K_1 - K_4 - K_4^* = -9/2$$

Thus the impulse response is

$$T(t) = [5e^{-t} - 4.5e^{-2t} - 18te^{-2t} + 2.92e^{-t}\cos(t - 59^\circ)]u(t)$$

10.4 TRANSIENT SWITCHING PROBLEMS

A large class of electrical transient problems involves the closing or opening of switches in a circuit at sequential instants of time. Such problems can be solved:

1. Through Laplace transforms by considering each time interval individually and recognizing that the final conditions for time interval i are the initial conditions for time interval $i + 1$.

2. Through Laplace transforms by treating a switched excitation as a series of pulses, finding the transform of the pulse train, and obtaining one solution for all time, rather than individual solutions for each time interval. This technique is not applicable when components are switched in and out of a circuit.

3. Through exponential end-point analysis, where, by knowing the initial value of a voltage or current and its ultimate final value if no further switching occurs, the value of the voltage or current at all intermediate times can be written down by inspection. This technique is valid, in general, only if the circuit contains one energy-storage element.

We shall present these three techniques through examples.

Example 10.5 The switch in Fig. 10.2*a* has been closed for a long time. At $t = 0$ it opens, and it recloses at $t = 0.1$ msec. Find v_{ab} vs. t.

We shall solve this problem by Laplace transforms first, with the capacitor voltage as the final and initial condition linking the two time intervals. Then we shall show how the response can be written by inspection using exponential end-point analysis.

At $t = 0_-$, the capacitor voltage is 100 volts, as is v_{ab}. At $t = 0_+$, just after the switch opens, the capacitor voltage remains at 100 volts, but v_{ab} jumps to $v_{ab+} = 100 + \tfrac{1}{2} \times 200 = 200$ volts. From $t = 0$ to 0.1 msec, we write

$$300 = 2 \times 10^3 i + 2 \times 10^7 \int_{-\infty}^{t} i\,dt = 2 \times 10^3 i + 2 \times 10^7 \int_{0}^{t} i\,dt + v_{c0},$$

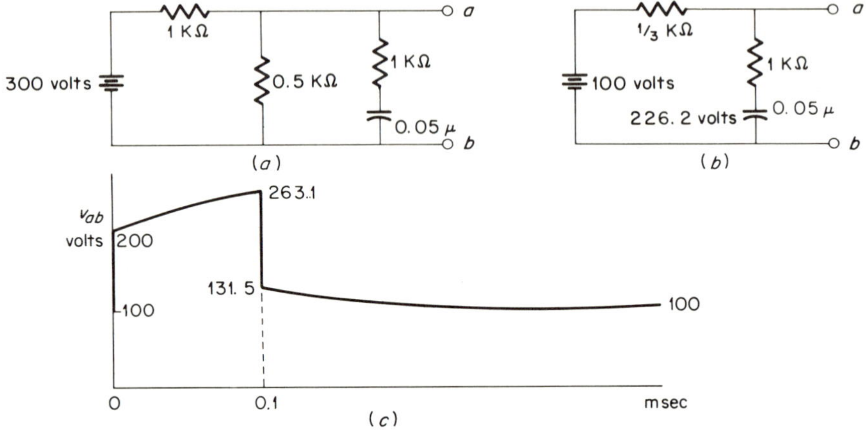

Fig. 10.2

where in the second form of the right side the lower limit on the capacitor integral has been changed and the initial capacitor voltage has been displayed explicitly.

Transforming either equation gives

$$\frac{300}{s} = 2 \times 10^3\left(1 + \frac{10^4}{s}\right)I(s) + \frac{100}{s}$$

yielding

$$I(s) = \frac{0.1}{s + 10^4}$$

Thus

$$i(t) = 0.1e^{-10^4 t} \qquad 0 < t < 0.1 \text{ msec}$$

and

$$\left.\begin{aligned} v_{ba} &= 300 - 10^3 i(t) = 300 - 100e^{-10^4 t} \\ v_c &= 300 - 2 \times 10^3 i(t) = 300 - 200e^{-10^4 t} \end{aligned}\right\} \qquad 0 < t < 0.1 \text{ msec}$$

At $t = 0.1$ msec, before the reclosing of the switch,

$$v_{ab} = 263.1 \text{ volts}, \qquad v_c = 226.2 \text{ volts}$$

The switch then recloses, and the circuit can be redrawn as shown in Fig. 10.2*b*, where the left portion of the original circuit has been replaced by its Thévenin equivalent. The voltage v_{ab} falls immediately to $226.2 - \frac{3}{4} \times 126.2 = 131.5$ volts. The new differential equation is

$$100 = \tfrac{4}{3} \times 10^3 i + 2 \times 10^7 \int_{-\infty}^{t'} i\,dt' = \tfrac{4}{3} \times 10^3 i + 2 \times 10^7 \int_0^{t'} i\,dt' + v_{c.1}$$

where $t' = t + 0.1$ msec. Transforming yields

$$\frac{100}{s} = 2 \times 10^3\left(\frac{2}{3} + \frac{10^4}{s}\right)I(s) + \frac{226.2}{s}$$

$$I(s) = \frac{-0.0946}{s + 1.5 \times 10^4}$$

and

$$i(t) = -0.0946e^{-1.5\times10^4 t'} = -0.0946e^{-1.5\times10^4(t-10^{-4})} \qquad t > 0.1 \text{ msec}$$

Thus

$$v_{ab} = 100 - \tfrac{1}{3} \times 10^3 i(t) = 100 + 31.5e^{-1.5\times10^4(t-10^{-4})} \qquad t > 0.1 \text{ msec}$$

A sketch of v_{ab} vs. t appears in Fig. 10.2*c*.

We have gone into considerable detail in the solution of this rather simple switching problem to illustrate several points. (1) The final capacitor voltage at the end of one time interval becomes the initial capacitor voltage at the beginning of the next time interval, because of the conservation of charge, but this is not true for other voltages in the circuit. For example, note the instantaneous changes in v_{ab} at $t = 0$ and 0.1 msec. (2) The use of Thévenin's theorem should be noted. In problems of this type it is often desirable to replace a portion of a circuit *not containing an energy-storage element* by its Thévenin or Norton equivalent. (3) The use of t' during the second time interval eliminates the need for a delay factor in the Laplace transforms for that time interval, thus simplifying the algebra. As a general rule, zero time should be reckoned from the start of each new time interval and appropriate delay introduced after the time response has been found. (4) Having worked this problem and having observed the form of the time-response expressions, we can use what we shall call *exponential end-point analysis* to write down these expressions almost by inspection. The argument here is that if we know the initial and final values of a voltage or current (the end-point values) and the time constant of the exponential path in a circuit containing one energy-storage element, we can write the equation for that path.

For example, let a parameter x (voltage or current) begin at x_i units at $t = 0$ and rise or fall toward an ultimate value of x_f units at $t = \infty$ in a circuit whose time constant is T_c (RC or L/R). The equation for x for $t > 0$ is

$$x = x_f + (x_i - x_f)e^{-t/T_c} \tag{10.4.1}$$

Note the logic of this equation. At $t = 0$, $x = x_i$; at $t = \infty$, $x = x_f$. In between, it is the difference between these two quantities that undergoes the exponential rise or decay.

Example 10.6 Solve Example 10.5 by exponential end-point analysis.

In the interval from 0 to 0.1 msec, the initial value of v_{ab} is 200 volts; the final value is 300 volts. The time constant is

$$RC = 2 \times 10^3 \times 5 \times 10^{-8} = 10^{-4} \text{ sec}$$

From Eq. (10.4.1)

$$v_{ab} = 300 - 100e^{-10^4 t} \text{ volts}$$

For t greater than 0.1 msec, v_i is 131.5 volts and v_f is 100 volts. The time constant is, from Fig. 10.2*b*,

$$RC = \tfrac{4}{3} \times 10^3 \times 5 \times 10^{-8} = \tfrac{2}{3} \times 10^{-4} \text{ sec}$$

From Eq. (10.4.1), with the appropriate time delay,

$$v_{ab} = 100 + 31.5e^{-1.5\times10^4(t-10^{-4})} \text{ volts}$$

Example 10.7 We have stated that, in general, exponential end-point analysis is valid only for circuitry containing one energy-storage element. An exception to that rule will now be illustrated.

The circuit in Fig. 10.3*a* is a conventional *RC* coupling network, with associated shunt capacitor, found in many electronic amplifiers. The network is excited by the single current pulse i, shown in Fig. 10.3*b*. This pulse is produced by having the switch in Fig. 10.3*a* close at $t = 0$ and open at $t = 30$ μsec. The output voltage v_{ab} as a function of time is desired.

This problem is very similar to the previous one, but here there are two capacitors, each with its own charge and discharge paths. Observe, though, that the coupling capacitor is much larger than the shunt capacitor. Hence, as a first approximation,

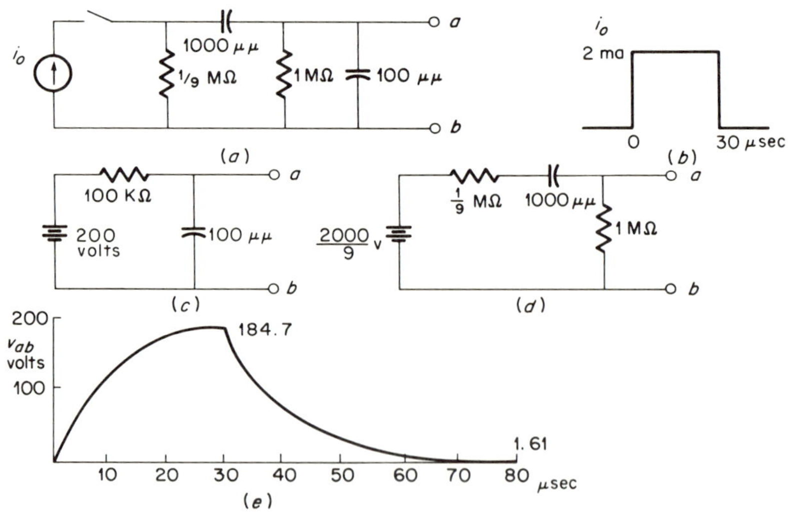

Fig. 10.3

assume that the coupling-capacitor voltage does not change during the charge period of the shunt capacitor. Thus we replace the coupling capacitor by a short circuit, combine the $\frac{1}{8}$- and 1-megohm resistors in parallel, and use Thévenin's theorem to arrive at the circuit in Fig. 10.3*c*. The shunt capacitor is initially uncharged. If allowed to charge fully, its voltage would be 200 volts. The time constant is 10^{-5} sec. From Eq. (10.4.1)

$$v_{ab} = 200(1 - e^{-10^5 t}) \qquad \text{volts} \qquad 0 < t < 30\ \mu\text{sec}$$

At 10 μsec, $v_{ab} = 126.2$ volts, and at the 30-μsec mark, where switching occurs, $v_{ab} =$ 190 volts.

Before considering the discharge cycle, let us determine the effect of the coupling capacitor on the 190-volt v_{ab} value at 30 μsec. Referring to Fig. 10.3*a*, we replace the shunt capacitor by an open circuit (actually an open circuit in series with a battery of 190 volts) and use Thévenin's theorem to obtain the circuit in Fig. 10.3*d*. The voltage across the coupling capacitor is initially zero and would eventually rise to $^{2000}\!/_{9}$ volts. The time constant is $\frac{1}{9} \times 10^{-2}$ sec. From Eq. (10.4.1)

$$v_{C_c} = {}^{2000}\!/_{9}\,(1 - e^{-900t}) \qquad \text{volts}$$

At 30 μsec

$$v_{C_c} = {}^{2000}\!/_{9}(1 - e^{-0.027}) = 5.8 \text{ volts}$$

Thus there is a 5.9-volt drop due to the coupling capacitor, nine-tenths of which is absorbed in v_{ab}. It follows that the corrected value for v_{ab} at 30 μsec is

$$v_{ab} = 190 - 5.3 = 184.7 \text{ volts}$$

During the discharge period, for t greater than 30 μsec, again temporarily neglecting the effect of the coupling capacitor, the circuit in Fig. 10.3*c* applies with the generator replaced by a short. The shunt capacitor has an initial voltage of 184.7 volts. Its final voltage is zero, and the time constant is 10^{-5} sec. From Eq. (10.4.1), with the appropriate delay factor,

$$v_{ab} = 184.7e^{-10^5(t-30\times10^{-6})} \qquad \text{volts} \qquad t > 30\ \mu\text{sec}$$

At $t = 40$ μsec (1 *RC*), $v_{ab} = 68.2$ volts, and at $t = 80$ μsec (5 *RC*), $v_{ab} = 1.25$ volts, essentially zero. Taking the coupling capacitor into account, the circuit of Fig. 10.3*d* is employed, with the battery replaced by a short. The coupling-capacitor voltage is initially 5.9 volts and finally 0 volts; the time constant is $\frac{1}{9} \times 10^{-2}$ sec. From Eq. (10.4.1), with appropriate delay,

$$v_{C_c} = 5.9e^{-900(t-30\times10^{-6})} \qquad \text{volts} \qquad t > 30\ \mu\text{sec}$$

At $t = 80$ μsec

$$v_{C_c} = 5.9e^{-0.045} = 5.64 \text{ volts}$$

Thus, at 80 μsec,

$$v_{ab} = 1.25 + 0.36 = 1.61 \text{ volts}$$

which is still approximately zero.

A plot of v_{ab} vs. t, including the effect of the coupling capacitor, appears in Fig. 10.3*a*.

Example 10.8 Find the exact expression for the output voltage V_{ab} in Fig. 10.3*a* when the excitation is the current step in Fig. 10.3*b*.

We shall find the exact response by Laplace transforms, as in Example 10.5, except that here we shall not consider each time interval separately by switching the excitation on and then off but treat the excitation as a current pulse whose transform is

$$I_0(s) = \frac{2 \times 10^{-3}}{s}(1 - e^{-\tau s})$$

where $\tau = 30$ μsec. Now

$$V_{ab}(s) = T(s)I_0(s)$$

where $T(s)$ is calculated from Fig. 10.3*a* to be

$$T(s) = \frac{10^{10}s}{s^2 + 1.01 \times 10^5 s + 9 \times 10^7}$$

Thus $V_{ab}(s) = \dfrac{2 \times 10^7(1 - \epsilon^{-\tau s})}{(s + 1.00101 \times 10^5)(s + 899)}$

Expanding by partial fractions gives

$$V_{ab}(s) = 201.61(1 - e^{-\tau s})\left(\frac{1}{s + 899} - \frac{1}{s + 1.00101 \times 10^5}\right)$$

Transforming to the time domain yields

$$\begin{aligned} v_{ab} &= 201.61(e^{-899t} - e^{-1.00101\times10^5 t})u(t) \\ &\quad - 201.61(e^{-899(t-\tau)} - e^{-1.00101\times10^5(t-\tau)})u(t - \tau) \end{aligned}$$

At $t = 10$ μsec, $v_{ab} = 125.0$ volts; at the switching point at 30 μsec, $v_{ab} = 186.0$ volts; at $t = 40$ μsec, $v_{ab} = 64.5$ volts; and at $t = 80$ μsec, $v_{ab} = -3.8$ volts. The corresponding values from the approximate analysis in Example 10.7 are 124.0, 184.7, 68.3, and 1.49 volts, not appreciably different from the exact values.

The actual time constant of the shunt capacitor is $1/1.00101 \times 10^5 = 9.989$ μsec; that of the coupling capacitor $1/899 = 1112$ μsec. In Example 10.7, the approximate values were 10 μsec for the shunt capacitor and 1111 μsec for the coupling capacitor, again in close agreement with the exact values.

When, in general, is it permissible to solve a transient switching problem by considering the excitation as a single pulse, as we did in Example 10.8? Stated another way, when must we resort to the technique of Example 10.5, where each time interval was considered separately? The answer is that the technique of Example 10.8 may not be used if the switching alters any of the time constants in the circuit. In Example 10.5, the switching did change the circuit time constants; in Example 10.8, it did not.

We conclude this section with one additional type of transient switching problem, namely, the case where the act of switching generates impulses of voltage or current in a circuit.

Example 10.9 The switch in Fig. 10.4*a* has been closed for a long time. At $t = 0$ it opens. Find the current i and the voltage v_{sw} across the switch as functions of time.

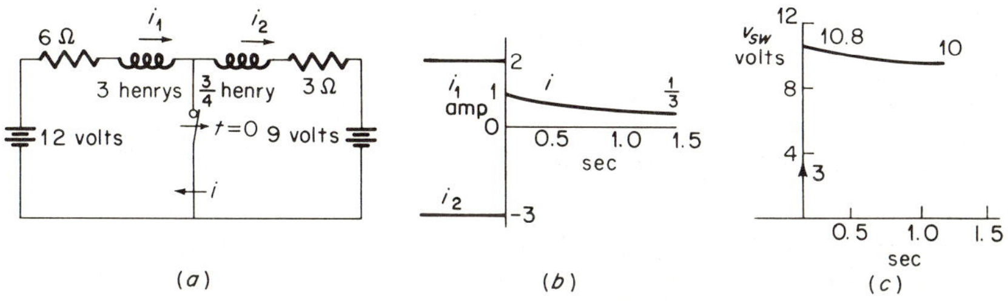

Fig. 10.4 A circuit with an inductive cut set.

The inductor currents at $t = 0_-$ are easily calculated to be

$$i_{1-} = 2a \qquad i_{2-} = -3a$$

At $t = 0_+$, after the switch closes, the following equations must hold:

$$i_{1+} = i_{2+} \equiv i_+$$

and

$$12 - 9 = 6i + 3\frac{di}{dt} + \frac{3}{4}\frac{di}{dt} + 3i$$

The major problem here is to determine i_+. We note immediately that

$$i_{1-} \neq i_{1+} \qquad i_{2-} \neq i_{2+}$$

which means that voltage impulses must occur across each inductor between $t = 0_-$ and $t = 0_+$. Now a unit impulse of voltage across an inductor at $t = 0$ will produce a change in current of

$$\Delta i = \frac{1}{L}\int_{-}^{+} \delta(t)\,dt = \frac{1}{L} \qquad \text{amp}$$

Thus the actual current changes in the circuit at $t = 0$ are

$i_+ - 2$ amp for 3-henry inductor
$i_+ + 3$ amp for $\frac{3}{4}$-henry inductor

Thus the impulses of voltage required are

$3(i_+ - 2)\delta(t)$ for 3-henry inductor
$\frac{3}{4}(i_+ + 3)\delta(t)$ for $\frac{3}{4}$-henry inductor

These two voltages must be equal and opposite in sign for Kirchhoff's voltage law to be satisfied when the switch is open. Therefore

$$3(i_+ - 2)\delta(t) = -\tfrac{3}{4}(i_+ + 3)\delta(t)$$

Solving for i_+ gives

$$i_+ = 1 \text{ amp}$$

We now Laplace transform the KVL equation written earlier and obtain

$$\frac{3}{s} = \left(9 + \frac{15s}{4}\right) I(s) - \frac{15}{4}$$

Solving for $I(s)$ and expanding by partial fractions gives

$$I(s) = \frac{5s + 4}{5s(s + 12/5)} = \frac{1}{3s} + \frac{2}{3(s + 12/5)}$$

yielding a time-domain current response of

$$i(t) = \tfrac{1}{3}u(t) + \tfrac{2}{3}e^{-12/5\,t}u(t)$$

The voltage across the switch is given by either of the following two expressions:

$$v_{sw}(t) = 12u(t) - 6i - 3\frac{di}{dt}$$

or

$$v_{sw}(t) = 9u(t) + 3i + \frac{3}{4}\frac{di}{dt}$$

In inserting $i(t)$ into these expressions, we must be very careful in our handling of di/dt. The expression for $i(t)$ says that $i(t) = 0$ for $t < 0$ and that $i(t)$ jumps from 0 to 1 amp at $t = 0$. In fact, the jump in current is from 2 to 1 amp in the 3-henry coil and from -3 to 1 amp in the $\frac{3}{4}$-henry coil at $t = 0$. Hence di/dt is not $\delta(t)$ for each coil, as the $i(t)$ expression implies, but $-\delta(t)$ for the 3-henry coil and $4\delta(t)$ for the $\frac{3}{4}$-henry coil. With this point in mind, the voltage across the switch is

$$\begin{aligned} v_{sw}(t) &= 12u(t) - 2u(t) - 4e^{-12/5\,t}u(t) \\ &\qquad - \delta(t) - 2e^{-12/5\,t}\delta(t) + \tfrac{24}{5}e^{-12/5\,t}u(t) + [6\delta(t)] \\ &= 10u(t) + \tfrac{4}{5}e^{-12/5\,t}u(t) + 3\delta(t) \end{aligned}$$

or

$$\begin{aligned} v_{sw} &= 9u(t) + u(t) + 2e^{-12/5\,t}u(t) \\ &\qquad + \tfrac{1}{4}\delta(t) + \tfrac{1}{2}e^{-12/5\,t}\delta(t) - \tfrac{6}{5}e^{-12/5\,t}u(t) + [\tfrac{9}{4}\delta(t)] \\ &= 10u(t) + \tfrac{4}{5}e^{-12/5\,t}u(t) + 3\delta(t) \end{aligned}$$

where the bracketed terms are the extra contributions to di/dt caused by the inductor currents not being zero for $t < 0$. For example, the bracketed $6\delta(t)$ term is included because the actual impulse voltage across the 3-henry coil is $+3\delta(t)$ volts rather than $-3\delta(t)$ volts.

Sketches of $i(t)$ vs. t and $v_{sw}(t)$ vs. t appear in Fig. 10.4*b* and *c*, respectively.

Example 10.9 illustrates one of two special circuit configurations in which impulses of voltage or current occur when switches are opened or closed. The case in Fig. 10.4*a* is that of a sudden change in current in a cut set of inductors caused by opening a switch. Such a situation will always require impulses of voltage across the inductors to satisfy Kirchhoff's voltage law around the loops of the network. The dual of the circuit of Fig. 10.4*a* is shown in Fig. 10.5 and illustrates another circuit configuration requiring

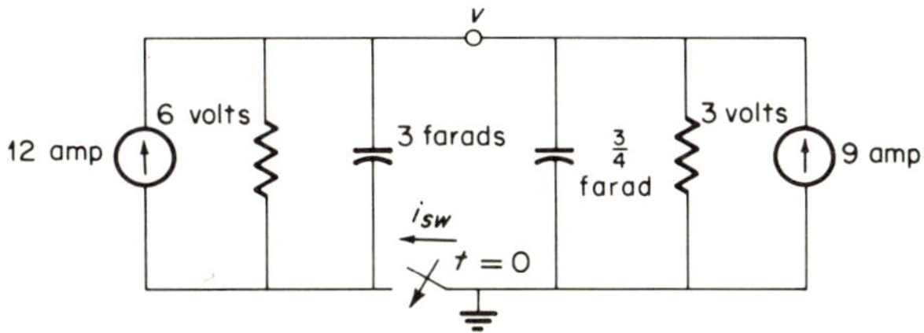

Fig. 10.5 Dual of Fig. 10.4a with capacitive tie-set constraint.

impulses. Here, a sudden change in voltage occurs in a tie set of capacitors when the switch is closed, requiring impulses of current in the capacitors to satisfy Kirchhoff's current law at the nodes of the network. Solution of this circuit is requested in a problem at the end of the chapter.

10.5 STEADY-STATE SOLUTIONS

In Sec. 10.4, our concern was in finding the total response of a circuit to changes occurring in the circuit at various points in time. These changes were the switching in and out of either excitations or components.

The total response of a circuit is not always the item of major concern to us. In the case of a periodic excitation switched on at $t = 0$, we may be more interested in the steady-state response of the circuit than we are in the wanderings that response takes before reaching its final value. In this section, several ways of obtaining the steady-state response of a circuit to a periodic excitation are investigated.

At this point we should be clear what we mean by steady state. In Eq. (10.3.1) the response of a system is expressed in terms of its transfer function, transformed excitation, and initial conditions. Assume that the initial conditions are zero, so that that equation becomes

$$Y(s) = X(s)\frac{P(s)}{Q(s)} \tag{10.5.1}$$

Now let $x(t)$ be a periodic excitation with period T and let $x_1(t)$ be the first cycle of $x(t)$. Thus

$$x_1(t) = \begin{cases} x(t) & 0 \leq t \leq T \\ 0 & t > T \end{cases} \tag{10.5.2}$$

In Example 9.13, it was shown that the Laplace transform of a periodic function can be simply expressed in terms of the Laplace transform of the first cycle of that function. We have

$$\mathscr{L}[x(t)] = \frac{1}{1 - e^{-sT}} \mathscr{L}[x_1(t)] \qquad \text{or} \qquad X(s) = \frac{1}{1 - e^{-sT}} X_1(s) \tag{10.5.3}$$

Inserting this result into Eq. (10.5.1) yields

$$Y(s) = \frac{X_1(s)}{1 - e^{-sT}} \frac{P(s)}{Q(s)} \tag{10.5.4}$$

The response $y(t)$ will be a summation of terms of the form $c_k e^{s_k t}$ where s_k can be real, complex, or imaginary. Each s_k is a pole of $Y(s)$, and thus either a zero of $Q(s)$ or a zero of $1 - e^{-sT}$. [If $x(t)$ is a bounded periodic function, $X_1(s)$ has no poles and thus contributes nothing to the s_k.†] Let us assume that the system transfer function $T(s)$ has no $j\omega$ axis poles. Then those terms in $y(t)$ resulting from the zeros of $Q(s)$ will decay with time and contribute nothing to the steady-state response. Thus the steady-state response must result from the roots of

$$1 - e^{-sT} = 0 \tag{10.5.5}$$

These roots lie on the $j\omega$ axis and are given by

$$s = j\frac{2k\pi}{T} \qquad k = 0, \pm 1, \pm 2, \ldots \tag{10.5.6}$$

We conclude that if $T(s)$ has no $j\omega$ axis poles, the forced response of a system to a periodic excitation is identical to its steady-state response, is periodic, and has the same period as the excitation.

With this background, we now want ways of finding the steady-state response to a periodic excitation without having to find the total response. Four techniques will be presented:

1. Approximate solution by Fourier series
2. Solution by end-point analysis
3. Open-form solution by Laplace transforms
4. Closed-form solution by Laplace transforms

APPROXIMATE SOLUTION BY FOURIER SERIES

The technique of obtaining an approximate steady-state response to a periodic excitation as a truncated Fourier series should be well known to the reader.

† To prove this, we need to show that the derivative of $X_1(s)$ exists throughout the finite s plane. Differentiation of complex functions is considered in the presentation of complex-variable theory in Chap. 12.

Using Eq. (9.2.28), we express the excitation as

$$x(t) = \sum_{-n}^{n} c(jk\omega_1)e^{jk\omega_1 t} \tag{10.5.7}$$

where $\omega_1 = 2\pi/T$ and from Eq. (9.2.27),

$$c(jk\omega_1) = \frac{1}{T}\int_0^T x(t)e^{-jk\omega_1 t}\,dt \tag{10.5.8}$$

To obtain $y(t)$, we simply multiply $c(jk\omega_1)$ by $T(jk\omega_1)$, the transfer function evaluated at each value of k, and write the steady-state response as

$$y(t) = \sum_{-n}^{n} c(jk\omega_1)T(jk\omega_1)e^{jk\omega_1 t} \tag{10.5.9}$$

The accuracy of $y(t)$ in Eq. (10.5.9) depends on how many terms of the Fourier series we retain, and how many we retain depends on how much the $c(jk\omega_1)$ coefficients of the excitation attenuate as k increases and how the values of $T(jk\omega_1)$ vary with k.

The technique will be illustrated through an example.

Example 10.10 A periodic square wave of amplitude 200 volts and period 1 msec is applied to the *RL* circuit shown in Fig. 10.6. It is desired to find the steady-state output voltage v across the inductor.

From Example 9.2, the Fourier series of the excitation is given by

$$e(t) = \sum_{-n}^{n} 100\,\frac{\sin(k\pi/2)}{k\pi/2}\,e^{-jk\omega_1 t}$$

or, in trigonometric form, by

$$e(t) = 100 + \frac{400}{\pi}\left(\cos\omega_1 t - \tfrac{1}{3}\cos 3\omega_1 t + \tfrac{1}{5}\cos 5\omega_1 t - \cdots\right)$$

where $\omega_1 = 2\pi \times 10^3$.

From Eq. (10.5.9), the output voltage is given by

$$v(t) = \sum_{-n}^{n} 100\,\frac{\sin(k\pi/2)}{k\pi/2}\,\frac{jk\omega_1 L}{R + jk\omega_1 L}\,e^{jk\omega_1 t}$$

To obtain the output voltage in trigonometric form, Table 10.1 is useful.

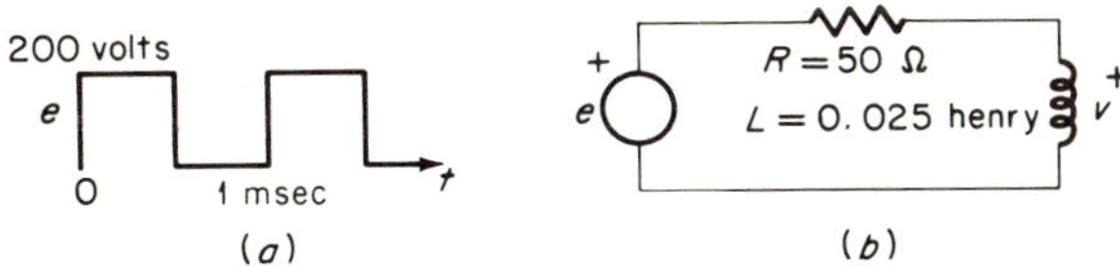

Fig. 10.6

TABLE 10.1

k	$E\ max$	$T(jk\omega_1) = \dfrac{jk\omega_1 L}{R + jk\omega_1 L}$	$V\ max$
0	100	0	0
1	$127.0\underline{/0^\circ}$	$0.955\underline{/17.66^\circ}$	$121.28\underline{/17.66^\circ}$
3	$42.4\underline{/-180^\circ}$	$0.993\underline{/6.05^\circ}$	$42.10\underline{/-173.95^\circ}$
5	$25.4\underline{/0^\circ}$	$0.998\underline{/3.66^\circ}$	$25.35\underline{/3.66^\circ}$
7	$18.1\underline{/-180^\circ}$	$0.998\underline{/2.60^\circ}$	$18.06\underline{/-177.40^\circ}$
9	$14.1\underline{/0^\circ}$	$0.999\underline{/2.00^\circ}$	$14.10\underline{/2.00^\circ}$

From the last column of Table 10.1 the trigonometric form of the output voltage is

$$v(t) = 121.28 \cos(\omega_1 t + 17.66^\circ) - 42.10 \cos(3\omega_1 t + 6.05^\circ) + 25.35 \cos(5\omega_1 t + 3.66^\circ) - 18.06 \cos(7\omega_1 t + 2.60^\circ) + 14.10 \cos(9\omega_1 t + 2.00^\circ)$$

The result is plotted in Fig. 10.7, along with the exact solution, which will be obtained subsequently from end-point analysis and by Laplace transforms. Terms through the ninth harmonic are included in the approximate result.

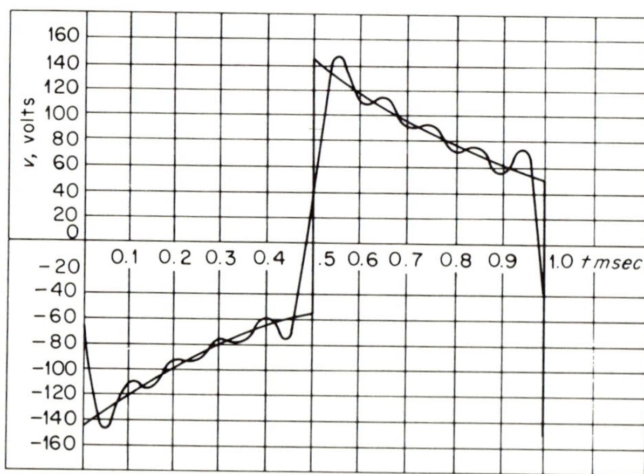

Fig. 10.7 *Approximate and exact responses of the circuit in Fig. 10.6b.*

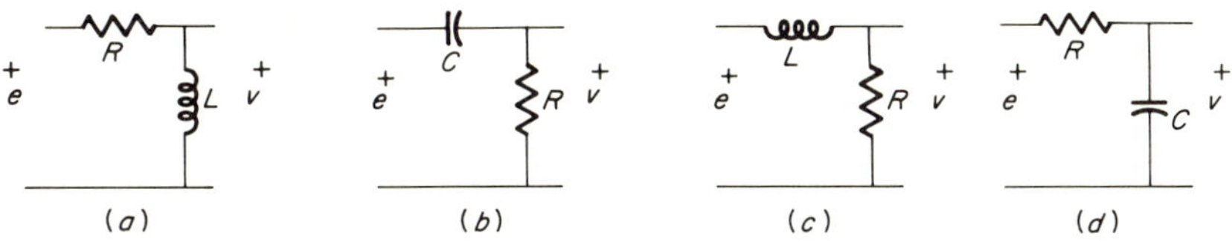

Fig. 10.8 Simple circuits whose steady-state responses can be found by exponential end-point analysis.

SOLUTION BY END-POINT ANALYSIS

Exponential end-point analysis is useful in finding the steady-state response to a periodic rectangular excitation in circuits containing one kind of energy-storage element. Let the excitation in Fig. 10.9*a* be applied to the circuits in Fig. 10.8, all assumed initially at rest. The output voltage of the first two of these appears in Fig. 10.9*b*; that of the second two in Fig. 10.9*c*. We see in Fig. 10.9*b* that the output voltage undergoes a transient decay over many cycles before reaching its steady-state condition with an average value of zero volts. In Fig. 10.9*c*, the transient is a buildup over many cycles to an ultimate steady-state voltage with an average value of $E_0 + \frac{1}{2}E$ volts.

It would save time to be able to find the voltages V_a, V_b, V_c, and V_d in Fig. 10.8*b* (the steady-state voltages) directly without having to find the transient portion of the output voltage first. This we can do with the aid of Eq. (10.4.1), repeated here for convenience with the *x*'s replaced by *v*'s

$$v = v_f + (v_i - v_f)e^{-t/T_c} \qquad (10.5.10)$$

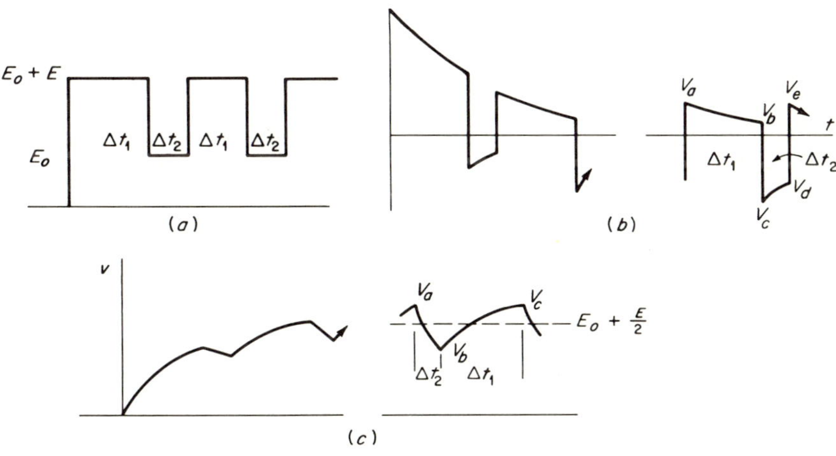

Fig. 10.9 Excitation and steady-state responses of circuits in Fig. 10.8.

All discharge curves in Fig. 10.9*b* tend ultimately to zero. Thus between *a* and *b* we have

$$v = V_a e^{-t'/T_c} \tag{10.5.11}$$

with time measured such that $v = V_a$ at $t' = 0$. Now at $t' = \Delta t_1$, $v = V_b$ and

$$V_b = V_a e^{-\Delta t_1/T_c} \tag{10.5.12}$$

The voltage then falls instantaneously by an amount E volts to point c, giving

$$V_c = V_a e^{-\Delta t_1/T_c} - E \tag{10.5.13}$$

Between c and d, with time now measured such that $v = V_c$ at $t'' = 0$, the voltage is given from Eq. (10.5.10) by

$$v = (V_a e^{-\Delta t_1/T_c} - E)e^{-t''/T_c} \tag{10.5.14}$$

and at point d

$$V_d = (V_a e^{-\Delta t_1/T_c} - E)e^{-\Delta t_2/T_c} \tag{10.5.15}$$

Now to go from d to e requires a jump of $+E$ volts. Thus

$$V_e = (V_a e^{-\Delta t_1/T_c} - E)e^{-\Delta t_2/T_c} + E = V_a \tag{10.5.16}$$

where the last equality holds because we are presuming that steady-state conditions have been reached. Letting T be the period and solving for V_a gives

$$V_a = E\frac{1 - e^{-\Delta t_2/T_c}}{1 - e^{-T/T_c}} \tag{10.5.17}$$

If the excitation is a square wave rather than a rectangular one, Eq. (10.5.17) reduces to

$$V_a = E\frac{1}{1 + e^{-T/2T_c}} \tag{10.5.18}$$

Knowing V_a in either expression, it is a simple matter to compute V_b, V_c, and V_d.

A similar analysis of the output voltage in Fig. 10.9*c* for the second pair of circuits can be performed. Here all charge curves tend ultimately to $E_0 + E$ volts in the steady state and all discharge curves to E_0 volts. The details of the derivation are left to the problems. The results are

$$V_a = E_0 + E\frac{1 - e^{-\Delta t_1/T_c}}{1 - e^{-T/T_c}} \tag{10.5.19}$$

for the rectangular wave and

$$V_a = E_0 + E\frac{1}{1 + e^{-T/2T}} \tag{10.5.20}$$

for the square wave.

Example 10.11 For the excitation in Fig. 10.6*a*, find the steady-state output voltage v of the circuit in Fig. 10.6*b* by end-point analysis.

Since the excitation is a square wave, Eq. (10.5.18) applies. The time constant is 0.5 msec. Thus

$$V_a = 200\frac{1}{1 + e^{-1}} = 146.2 \text{ volts}$$

Then from Eq. (10.5.12)

$$V_b = 146.2e^{-1} = 53.8 \text{ volts}$$

and from symmetry conditions

$$V_c = -146.2 \text{ volts} \qquad V_d = -53.8 \text{ volts}$$

The plot of the steady-state v vs. t is shown in Fig. 10.7.

OPEN-FORM SOLUTION BY LAPLACE TRANSFORMS

In Eq. (10.5.3), $X(s)$, the Laplace transform of the periodic excitation, is expressed in terms of $X_1(s)$, the Laplace transform of the first cycle of the excitation. The quantity $(1 - e^{-sT})^{-1}$ in that equation can be expanded by the binomial theorem into

$$\frac{1}{1 - e^{-sT}} = 1 + e^{-sT} + e^{-2sT} + \cdots \tag{10.5.21}$$

Incorporating Eq. (10.5.21) into Eq. (10.5.4) yields the Laplace transform of the total response as

$$Y(s) = \frac{P(s)}{Q(s)}X_1(s)(1 + e^{-sT} + e^{-2sT} + \cdots) \equiv \frac{P(s)}{Q(s)}X_1(s)D_1(s) \tag{10.5.22}$$

Before proceeding to obtain $y(t)$ by inverse transforming Eq. (10.5.22), we need to consider the makeup of $X_1(s)$ further. A typical $x_1(t)$ is made up of a time function to start the wave at $t = 0$ and at least one delayed time function to stop it before or at $t = T$. If $x_1(t)$ has a discontinuity at other times, $0 < t < T$, additional delayed functions will be needed. Consider, for example, the triangular excitation wave in Fig. 10.10. The equation of

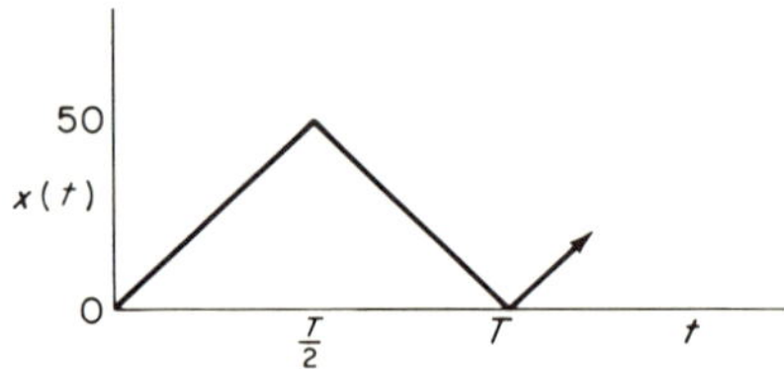

Fig. 10.10 A typical periodic excitation.

its first cycle is

$$x_1(t) = \frac{100}{T} r(t) - \frac{200}{T} r\left(t - \frac{T}{2}\right) + \frac{100}{T} r(t - T) \tag{10.5.23}$$

yielding a transform of

$$X_1(s) = \frac{100}{Ts^2} (1 - 2e^{-sT/2} + e^{-sT}) \equiv \frac{1}{R(s)} D_2(s) \tag{10.5.24}$$

Equation (10.5.24) illustrates the general form of $X_1(s)$, namely, a power of A denominator given by $R(s)$ and an exponential series numerator $D_2(s)$. Inserting this general form into Eq. (10.5.22) gives

$$Y(s) = \frac{P(s)}{Q(s)R(s)} D_1(s)D_2(s) \tag{10.5.25}$$

The product $D_1(s)D_2(s)$ yields a new exponential series. Its effect on $y(t)$ is to introduce delay. The quotient $P(s)/Q(s)R(s)$ can be expanded by partial fractions. Then the inverse transform of each term in the partial-fraction expansion can be obtained from Table 9.3 and multiplied by an appropriately delayed unit step to obtain each term in $y(t)$.

Example 10.12 Use the open-form Laplace method to find the steady-state output voltage of the circuit in Fig. 10.6*b* for the excitation in Fig. 10.6*a*.

The square excitation waveform is represented during its first cycle by

$$e_1(t) = 200u(t) - 200u(t - 0.5)$$

where the time units are milliseconds. Its transform is

$$E_1(s) = \frac{200}{s} (1 - e^{-0.5s})$$

The network transfer function is

$$T(s) = \frac{s}{s + 2}$$

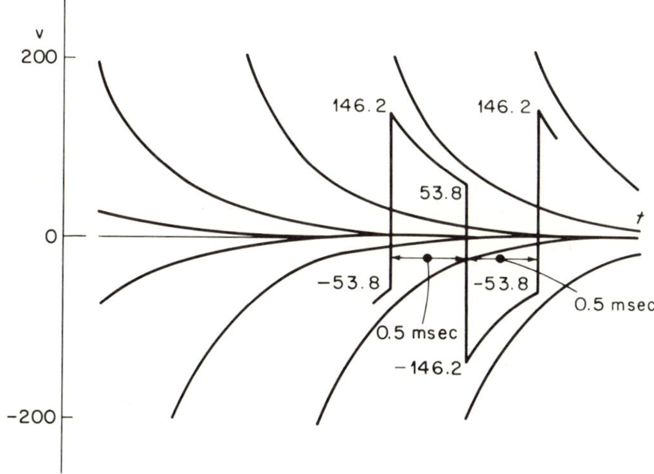

Fig. 10.11 *Plot of open-form Laplace solution.*

From Eq. (10.6.25)

$$V(s) = \frac{200}{s+2}(1 - e^{-0.5s})(1 + e^{-s} + e^{-2s} + \cdots)$$

$$= \frac{200}{s+2}(1 - e^{-0.5s} + e^{-s} - e^{-1.5s} + e^{-2s} - \cdots)$$

No partial-fraction expansion is necessary in this case and we write by inspection

$$v(t) = 200[e^{-2t}u(t) - e^{-2(t-0.5)}u(t - 0.5) + e^{-2(t-1)}u(t - 1) - e^{-2(t-1.5)}u(t - 1.5) + \cdots]$$

We have obtained the total response as a summation of exponential terms of equal time constant separated by 0.5 msec. The steady-state solution is quite easy to plot from this result. One merely makes a template of the exponential discharge curve, slides it horizontally to plot points, and then sums the contributions of each exponential curve at specific times to obtain the steady state $v(t)$. The results are shown in Fig. 10.11.

CLOSED-FORM SOLUTION BY LAPLACE TRANSFORMS

The open-form Laplace solution technique, though straightforward, is somewhat tedious. A better technique would be a closed-form (nonseries) Laplace solution yielding the steady-state response directly. Such a procedure will now be developed.

We assume, as before, that the system transfer function $T(s)$ has no $j\omega$ axis poles. We further assume that the excitation $x(t)$ is periodic, that it obeys the Dirichlet conditions, and that $|x(t)|\, e^{-\sigma t}$ approaches zero as t approaches infinity. Thus $x(t)$ is Laplace transformable.

Let us make the following identifications:

$y(t)$	total response
$y_t(t)$	transient response
$y_s(t)$	steady-state response
$y_1(t)$	total response during the first period
$y_{s1}(t)$	steady-state response during the first period
$x(t)$	periodic excitation
$x_1(t)$	excitation during the first period

What we wish to find is the steady-state response $y_s(t)$. Since $T(s)$ has no $j\omega$ axis poles, the free response eventually decays to zero and contributes nothing to the steady-state response. Also, because $x(t)$ is periodic, there is no transient component in the forced response. Thus the forced and steady-state responses are identical. Furthermore, the forced response to a periodic excitation during the first period is the same as the forced response during any other period. Thus, to find the steady-state response, we need merely to find the forced response during the first period. This is most easily obtained as the difference between the total response during the first period and the transient response. Formally we write

$$y_{s1}(t) = y_1(t) - y_t(t) \tag{10.5.26}$$

We can find the total response during the first period directly by inverse transforming Eq. (10.5.4) after deleting the $1 - e^{-sT}$ delay factor from the denominator of that expression. Thus

$$y_1(t) = \mathscr{L}^{-1}\left[X_1(s)\frac{P(s)}{Q(s)}\right] \tag{10.5.27}$$

From the discussion following Eq. (10.5.4), $X_1(s)$ has no poles. Thus all terms in $y_1(t)$ result from zeros of $Q(s)$, and $y_1(t)$ can be obtained quite easily through a routine partial-fraction expansion and use of Table 9.3.

To obtain the transient response $y_t(t)$, we formulate $y(t)$ from Eq. (10.5.4) as

$$y(t) = \mathscr{L}^{-1}\left[\frac{X_1(s)}{1 - e^{-sT}}\frac{P(s)}{Q(s)}\right] \tag{10.5.28}$$

and recall that the zeros of $1 - e^{-sT}$ determine the steady-state response; the zeros of $Q(s)$ the transient response. A partial-fraction expansion of the right side of Eq. (10.5.28) permits us to separate the transient terms from the steady-state terms, following which we use Table 9.3 to find $y_t(t)$.

Example 10.13 Use the closed-form Laplace method to find the steady-state output voltage of the circuit in Fig. 10.6*b* for the excitation in Fig. 10.6*a*.

From Example 10.12,

$$E_1(s) = \frac{200}{s}(1 - e^{-0.5s}) \qquad T(s) = \frac{s}{s+2}$$

Thus

$$V_1(s) = \frac{200}{s}(1 - e^{-0.5s})\frac{s}{s+2}$$

and

$$v_1(t) = 200[e^{-2t}u(t) - e^{-2(t-0.5)}u(t-0.5)]$$

Next, from Eq. (10.5.4),

$$V(s) = \frac{200(1 - e^{-0.5s})}{s(1 - e^{-s})}\frac{s}{s+2}$$

$$= V_t(s) + V_s(s)$$

where $V_t(s)$ is the transform of the transient response and $V_s(s)$ the transform of the steady-state response. In this case the transient response has only one term of the form $k/(s+2)$ with k given by

$$k = \left.\frac{200}{1 + e^{-0.5s}}\right|_{-2} = \frac{200}{1+e}$$

Thus

$$v_t(t) = \frac{200}{1+e}e^{-2t}u(t)$$

Finally the steady-state response during the first period is given by

$$v_{s1}(t) = v_1(t) - v_t(t)$$

$$= \frac{200}{1+e}e^{-2t}u(t) - 200e^{-2(t-0.5)}u(t-0.5)$$

The various responses for the first two cycles are shown in Fig. 10.12.

10.6 FREQUENCY RESPONSE THROUGH BREAK-POINT ANALYSIS

In determining the steady-state response of a network to a periodic excitation, we inevitably become concerned about the frequency-response characteristic of the network. This is so because the periodic excitation is often composed of a number of frequency components, and to understand how the network processes the excitation, we must understand how it processes each of the components.

There are many techniques for obtaining the frequency-response characteristic of a network. One of the most informative methods, which is also quite easy to apply, is break-point analysis.† This section develops this technique and applies it to a few interesting and useful networks.

In previous sections we have described the response of a network for a given excitation by

$$Y(s) = T(s)X(s) \tag{10.6.1}$$

where $Y(s)$, $T(s)$, and $X(s)$ are the Laplace transforms of the response, transfer function, and excitation, respectively. Our present concern is with periodic excitations. Thus it is useful for us to let $s = j\omega$ in Eq. (10.6.1) and write

$$Y(j\omega) = T(j\omega)X(j\omega) \tag{10.6.2}$$

where

$$T(j\omega) = A(\omega)e^{j\phi(\omega)} \tag{10.6.3}$$

in polar form and

$$T(j\omega) = T_r(\omega) + jT_i(\omega) \tag{10.6.4}$$

in cartesian form. In these expressions, $A(\omega)$ is $|T(j\omega)|$, $\phi(\omega)$ is Arg $T(j\omega)$,

† Originated by H. W. Bode and treated in his "Network Analysis and Feedback Amplifier Design," D. Van Nostrand Company, Inc., Princeton, N.J., 1947.

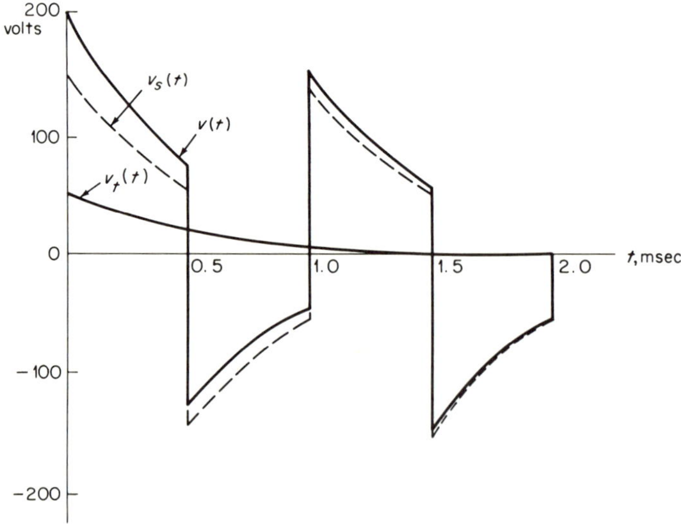

Fig. 10.12

and $T_r(\omega)$ and $T_i(\omega)$ are its real and imaginary parts. As was pointed out in Sec. 9.5, the quantities $A(\omega)$ and $\phi(\omega)$ can be measured and plotted, whereas $T(s)$ cannot; i.e., we can determine a network's magnitude and phase characteristics experimentally, but finding its poles and zeros experimentally is another matter. Thus it is quite important that we be able to analyze a network's performance in terms of $A(\omega)$ and $\phi(\omega)$.

In obtaining the frequency characteristic, it is possible, of course, to plot $T(j\omega)$ vs. ω directly as a phasor on the complex T plane. Such a presentation does have the advantage of presenting amplitude, phase, real-part, and imaginary-part information all on one diagram, but since ω values are not linearly spaced along the locus, interpretation of such a diagram is usually difficult. Also, it is no easy task to obtain the variation with frequency of the transfer function of two networks in cascade from their individual $T(j\omega)$-vs.-ω plots.

As an alternative procedure, separate plots of $A(\omega)$, $\phi(\omega)$, $T_r(\omega)$, and $T_i(\omega)$ vs. ω can be made. They are often easier to interpret than the $T(j\omega)$-vs.-ω plot. An added advantage is that for networks in cascade with no loading of one upon the others the magnitude and phase of the overall transfer function are found by multiplying the individual magnitudes and adding the individual phases at each frequency.

A still better technique is to plot $\phi(\omega)$ and the logarithm of $A(\omega)$ vs. ω on semilog paper. Now instead of multiplying transfer-function magnitudes of networks in cascade, one merely adds their logarithms. Also the semilog frequency scale gives more nearly equal emphasis to each portion of the frequency spectrum. A third advantage is the ease of constructing such plots through approximation by straight lines.

The most common procedure in making log magnitude plots is to use the base 10 and express the magnitude in decibels. From Eq. (10.6.3),

$$\text{Decibels} = 20 \log A(\omega) \tag{10.6.5}$$

A plot of Eq. (10.6.5), approximated by straight lines, vs. ω on semilog paper is called a *Bode plot*, and the analysis of such a plot is called *break-point analysis*.

To construct log magnitude and phase plots, we must first find $A(\omega)$ and $\phi(\omega)$. Recall that the network transfer function is a ratio of polynomials in s given in general form by

$$T(s) = \frac{P(s)}{Q(s)} = \frac{ks^r(s - s_a)(s - s_b)\cdots}{(s - s_1)(s - s_2)\cdots} \tag{10.6.6}$$

The roots in Eq. (10.6.6) are of three general types, namely, zero, real, and complex. If complex, they must appear as conjugates. Let us consider the following transfer function, which embodies all these root types:

$$T(s) = \frac{ks^2(s + a)}{(s + b)(s^2 + cs + d)} \tag{10.6.7}$$

where it is assumed that the roots of the quadratic term are complex.

The first step in performing a break-point analysis of Eq. (10.6.7) is to make the constant in each of the factors unity (so that the logarithm of each constant is zero for ease of plotting). We have

$$T(s) = \frac{Ks^2[(1/a)s + 1]}{[(1/b)s + 1][(1/d)s^2 + (c/d)s + 1]} \tag{10.6.8}$$

where $K = ka/bd$. Replacing s by $j\omega$ gives

$$T(j\omega) = \frac{-K\omega^2(1 + j\omega/a)}{(1 + j\omega/b)(1 + j\omega c/d - \omega^2/d)} \tag{10.6.9}$$

with a magnitude given by

$$A(\omega) = \frac{K\omega^2\sqrt{1 + \omega^2/a^2}}{\sqrt{1 + \omega^2/b^2}\sqrt{(1 - \omega^2/d)^2 + \omega^2c^2/d^2}} \tag{10.6.10}$$

and a phase of

$$\phi(\omega) = 180° - \tan^{-1}\frac{\omega}{b} + \tan^{-1}\frac{\omega}{a} - \tan^{-1}\frac{\omega c/d}{1 - \omega^2/d} \tag{10.6.11}$$

Converting Eq. (10.6.10) to decibels through Eq. (10.6.5) yields

$$\begin{aligned}\text{Decibels} &= 20 \log K + 40 \log \omega - 10 \log \left(1 + \frac{\omega^2}{b^2}\right) \\ &\quad + 10 \log \left(1 + \frac{\omega^2}{a^2}\right) - 10 \log \left[\left(1 - \frac{\omega^2}{d}\right)^2 + \frac{\omega^2 c^2}{d^2}\right]\end{aligned} \tag{10.6.12}$$

There are four types of terms in Eq. (10.6.12), namely, a constant, a second-order zero at the origin, two first-degree negative real roots, and one quadratic factor with complex roots. To find the slopes of the straight lines used to approximate these terms on a Bode plot, consider three frequencies, ω_1, ω_2, and ω_3, such that ω_2 is an octave above ω_1 ($\omega_2 = 2\omega_1$) and ω_3 is a decade above ω_1 ($\omega_3 = 10\omega_1$). The changes in decibels in going from ω_1 to ω_2 or ω_3 are given in general by

$$\begin{aligned}\Delta\text{db}_{1\text{-}2} &= 20 \log A(\omega_2) - 20 \log A(\omega_1) \\ &= 20 \log \frac{A(\omega_2)}{A(\omega_1)} \qquad \text{db/octave}\end{aligned} \tag{10.6.13}$$

and

$$\Delta db_{1\text{-}3} = 20 \log \frac{A(\omega_3)}{A(\omega_1)} \qquad \text{db/decade} \tag{10.6.14}$$

Now consider the case $A(\omega) = \omega^n$, n an integer, in Eqs. (10.6.13) and (10.6.14). There results

$$\Delta db_{1\text{-}2} = 20 \log \frac{(2\omega_1)^n}{\omega_1{}^n} = 20n \log 2 = 6n \qquad \text{db/octave} \tag{10.6.15}$$

and

$$\Delta db_{1\text{-}3} = 20 \log \frac{(10\omega_1)^n}{\omega_1{}^n} = 20n \log 10 = 20n \qquad \text{db/decade} \tag{10.6.16}$$

Thus the Bode plot slope for a first-degree term ($n = 1$) is 6 db/octave and 20 db/decade; for a second-degree term it is 12 db/octave and 40 db/decade; and so on.

With these results in mind, the Bode plot of Eq. (10.6.12) is drawn in Fig. 10.13*a*, where it is assumed that $1 \ll b \ll a \ll \sqrt{d}$. The 20 log K term is a horizontal line, independent of frequency. The 40 log ω term is a straight line of slope 12 db/octave ($6n$ with $n = 2$) whose intercept on the horizontal axis occurs at $\omega = 1$ (40 log 1 = 0). For frequencies much less than $\omega = a$, the 10 log $(1 + \omega^2/a^2)$ term yields a horizontal line of 0 db. For frequencies much greater than $\omega = a$, the term becomes 10 log (ω^2/a^2) = 20 log (ω/a), which is a straight line of slope +6 db/octave ($6n$ with $n = 1$) that intersects the horizontal axis at $\omega = a$. At $\omega = a$, the term becomes 10 log 2 $\approx$ 3 db. Thus the actual magnitude of the term is 3 db above the point of intersection of its two straight-line approximations. The -10 log $(1 + \omega^2/b^2)$ term is handled in the same way, except that its slope for $\omega \gg b$ is -6 db/octave.

The quadratic term is represented by a horizontal line of 0 db for $\omega \ll \sqrt{d}$. For $\omega \gg \sqrt{d}$ the term becomes -10 log $(\omega^4/d^2) = -40$ log $(\omega/\sqrt{d})$ with a slope of -12 db/octave. The decibel level of this term at $\omega = \sqrt{d}$ depends on the value of c. We shall explore this dependence in our comparison of the time- and frequency-domain responses of a second-order system in the next section.

The phase characteristic in Eq. (10.6.11) can be plotted directly on semilog paper. The result appears in Fig. 10.13*b*. The $-\omega^2$ term introduces a horizontal line of 180°. For $\omega \ll a$, $\tan^{-1}(\omega/a) \approx 0°$; for $\omega \gg a$, $\tan^{-1}(\omega/a) \approx 90°$. The angle is 45° at $\omega = a$. The $-\tan^{-1}(\omega/b)$ term is the same, except its angles are negative. The quadratic term has zero phase for $\omega \ll \sqrt{d}$ and approaches $-180°$ for $\omega \gg \sqrt{d}$. At $\omega = \sqrt{d}$, the angle is $-90°$. At other frequencies the angle depends on the value of c. This dependence will also be explored in the next section.

In plotting magnitude and phase for first-order factors, Table 10.2 is useful. It is constructed so that the decibel difference between the actual and approximate log magnitude plots can easily be ascertained.

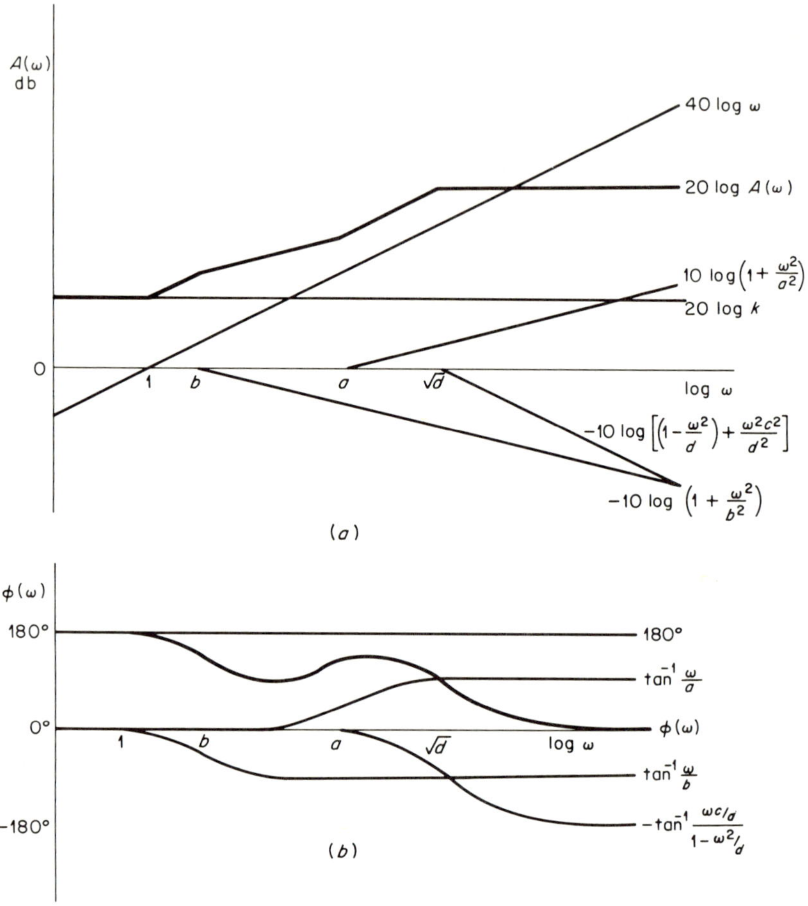

Fig. 10.13 *Bode log magnitude and phase plots for Eq. (10.6.9).*

The break-point analysis technique will now be used to analyze a network commonly encountered in control systems, namely, the lag-lead network shown in Fig. 10.14*a*. It is desired to obtain plots of log magnitude and phase vs. log ω of the voltage transfer function of this network.

To begin, the voltage ratio is given by

$$T(s) = \frac{V_0(s)}{V_i(s)} = \frac{R_2 + 1/sC_2}{R_2 + 1/sC_2 + (R_1/sC_1)/(R_1 + 1/sC_1)}$$

$$= \frac{(1 + sR_1C_1)(1 + sR_2C_2)}{1 + s(R_1C_1 + R_2C_2 + R_1C_2) + s^2R_1R_2C_1C} \qquad (10.6.17)$$

TABLE 10.2 Decibel and phase values for first-order factors of the form $1 + \tau s$

ω	*Actual decibels*	*Asymptotic decibels*	*Error*	*Phase*
$\frac{0.1}{\tau}$	≈ 0	0	≈ 0	5.7°
$\frac{0.51}{\tau}$	1.0	0	1.0	26.6°
$\frac{0.76}{\tau}$	2.0	0	2.0	37.4°
$\frac{1}{\tau}$	3.0	0	3.0	45.0°
$\frac{1.31}{\tau}$	4.34	2.34	2.0	52.6°
$\frac{1.96}{\tau}$	6.84	5.84	1.0	63.4°
$\frac{10}{\tau}$	≈ 20.0	20.0	≈ 0	84.3°

Note that $T(s)$ is written so that each of its factors has a unity constant term, as was done in Eq. (10.6.10).

We need to know whether the denominator roots are complex or real. They will be real if

$$(R_1C_1 + R_2C_2 + R_1C_2)^2 - 4R_1R_2C_1C_2 \geq 0 \qquad (10.6.18a)$$

which gives

$$(R_1C_1 - R_2C_2)^2 + R_1C_2(2R_1C_1 + 2R_2C_2 + R_1C_2) \geq 0 \qquad (10.6.18b)$$

The inequality in (10.6.18*b*) is satisfied with the greater-than sign for all non-zero values of the parameters. Hence the denominator roots of Eq. (10.6.17)

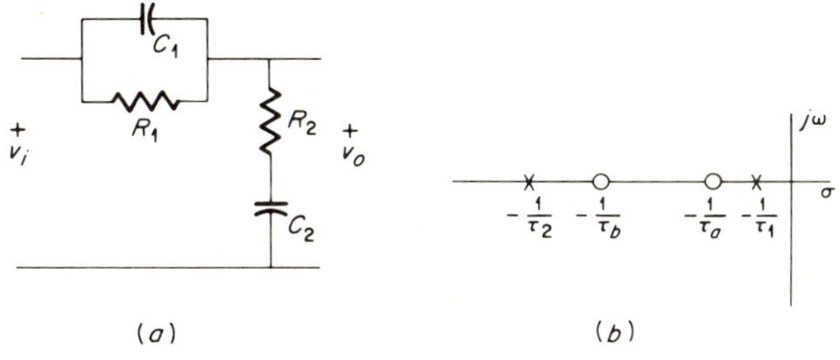

***Fig. 10.14** Lag-lead network and its pole-zero locations.*

are negative real and unequal, and $T(s)$ can be rewritten in the form

$$T(s) = \frac{(1 + s\tau_a)(1 + s\tau_b)}{(1 + s\tau_1)(1 + s\tau_2)} \tag{10.6.19}$$

where

$$\tau_a = R_1C_1 \qquad \tau_b = R_2C_2$$

$$\tau_1, \tau_2 = \frac{(R_1C_1 + R_2C_2 + R_1C_2) \pm \sqrt{(R_1C_1 + R_2C_2 + R_1C_2)^2 - 4R_1R_2C_1C_2}}{2} \tag{10.6.20}$$

The relative locations along the negative real axis of the poles and zeros of $T(s)$ must be determined next. From Eqs. (10.6.17) and (10.6.19),

$$\tau_1\tau_2 = \tau_a\tau_b \tag{10.6.21}$$

and

$$\tau_1 + \tau_2 = \tau_a + \tau_b + R_1C_2 \equiv \tau_a + \tau_b + \tau_c \tag{10.6.22}$$

Assume

$$\tau_1 = k\tau_a \tag{10.6.23}$$

Then, from Eq. (10.6.21),

$$k\tau_2 = \tau_b \tag{10.6.24}$$

and from Eqs. (10.6.21) to (10.6.24), with $k > 1$,

$$\frac{1}{\tau_1} < \frac{1}{\tau_b} \qquad \frac{1}{\tau_1} < \frac{1}{k\tau_2} \qquad \frac{1}{\tau_a} < \frac{1}{\tau_2} \qquad \frac{1}{\tau_a} < \frac{k}{\tau_b} \tag{10.6.25}$$

For example, for $k = 3/2$,

$$\tau_1 = 3/2\tau_a = \tau_b + 3\tau_c = 3/2\tau_2 + 3\tau_c$$

The inequalities in (10.6.25) fix the relative locations of the roots, except that $1/\tau_a$ may be either greater than or less than $1/\tau_b$. Choosing $1/\tau_a < 1/\tau_b$ yields the root-location plot in Fig. 10.14*b*, drawn for $k = 3/2$.

We are interested in the magnitude and phase of $T(s)$ for $s = j\omega$. These are given by

$$A(\omega) = \frac{\sqrt{1 + \omega^2\tau_a^2}\sqrt{1 + \omega^2\tau_b^2}}{\sqrt{1 + \omega^2\tau_1^2}\sqrt{1 + \omega^2\tau_2^2}} \tag{10.6.26}$$

and

$$\phi(\omega) = \tan^{-1} \omega\tau_a + \tan^{-1} \omega\tau_b - \tan^{-1} \omega\tau_1 - \tan^{-1} \omega\tau_2 \tag{10.6.27}$$

Expressing $A(\omega)$ in decibels yields

$$20 \log A(\omega) = 10 \log (1 + \omega^2\tau_a^2) + 10 \log (1 + \omega^2\tau_b^2) \\ - 10 \log (1 + \omega^2\tau_1^2) - 10 \log (1 + \omega^2\tau_2^2) \quad (10.6.28)$$

Sketches of $A(\omega)$ in decibels and $\phi(\omega)$ appear in Fig. 10.15. Reference to the log magnitude plot in Fig. 10.15*a* shows that $A(\omega)$ has an amplitude of 1 for small ω. Hence its decibel level there is 0. At $\omega = 1/\tau_1$, the $1 + s\tau_1$ factor introduces a slope of -6 db/octave. This is canceled after $\omega = 1/\tau_a$ by a slope of $+6$ db/octave from the $1 + s\tau_a$ term. At $\omega = 1/\tau_a$, the decibel level of the corner is

$$-10 \log \frac{\tau_1^2}{\tau_a^2} = -20 \log k \quad (10.6.29)$$

At $\omega = 1/\tau_b$, an additional $+6$ db/octave is inserted due to the $1 + s\tau_b$ term, and, finally, this is canceled after $\omega = 1/\tau_2$ by a -6 db/octave slope from the $1 + s\tau_2$ factor. The decibel level of the corner at $\omega = 1/\tau_2$ is

$$10 \log \frac{\tau_a^2}{\tau_2^2} + 10 \log \frac{\tau_b^2}{\tau_2^2} - \log \frac{\tau_1^2}{\tau_2^2} \\ = 20 \log \frac{\tau_a}{\tau_2}\frac{\tau_b}{\tau_2}\frac{\tau_2}{\tau_1} = 20 \log \frac{\tau_a\tau_b}{\tau_1\tau_2} = 0 \quad (10.6.30)$$

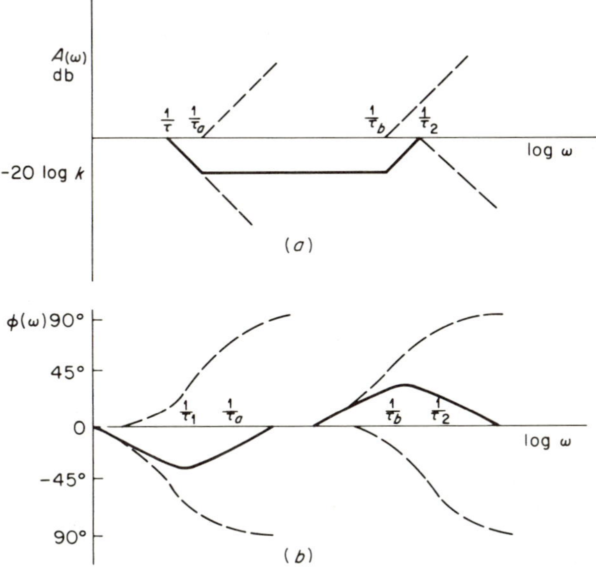

Fig. 10.15 Log magnitude and phase plots for a lag-lead network.

It is thus seen that the lag-lead network is essentially a band-elimination filter. At low frequencies, the capacitors appear as opens, making $v_o = v_i$ (0 db loss). At high frequencies they appear as shorts, again making $v_o = v_i$ (0 db loss). In a band of intermediate frequencies, they reduce the magnitude of v_o by 20 log k db.

The phase characteristic in Fig. 10.15b is, of course, the reason for the network's name. The $1 + s\tau_1$ and $1 + s\tau_a$ factors combine to produce a negative phase hump in the vicinity of $\omega = 1/\tau_1$ and $\omega = 1/\tau_a$. A similar positive phase hump occurs in the neighborhood of $\omega = 1/\tau_b$ and $\omega = 1/\tau_2$, due to the $1 + s\tau_b$ and $1 + s\tau_2$ factors. The first hump gives phase lag and the second phase lead. In between, there is a frequency at which $\phi(\omega)$ is zero. This occurs when

$$\tan^{-1} \omega\tau_a + \tan^{-1} \omega\tau_b = \tan^{-1} \omega\tau_1 + \tan^{-1} \omega\tau_2 \tag{10.6.31}$$

which requires that

$$\omega = \frac{1}{\sqrt{\tau_a\tau_b}} = \frac{1}{\sqrt{\tau_1\tau_2}} \equiv \omega_m \tag{10.6.32}$$

Equation (10.6.32) may be verified simply by inserting either expression for ω_m into Eq. (10.6.31), utilizing Eqs. (10.6.23) and (10.6.24), and noting that each side of Eq. (10.6.31) gives 90°.

If $1/\tau_b \gg 1/\tau_a$, the $1 + s\tau_b$ and $1 + s\tau_2$ factors have negligible influence on the negative phase hump. Then, in the region of $\omega = 1/\tau_1$ and $\omega = 1/\tau_a$, we can write

$$\phi(\omega) \approx \tan^{-1} \omega\tau_a - \tan^{-1} \omega\tau_1 \tag{10.6.33}$$

Differentiating and equating to zero gives the ω value at which $\phi(\omega)$ is a negative maximum as

$$\omega_- = \frac{1}{\sqrt{k}\tau_a} \tag{10.6.34}$$

Similarly, for the positive phase hump, again with $1/\tau_b \gg 1/\tau_a$,

$$\omega_+ = \sqrt{k}\,\frac{1}{\tau_b} \tag{10.6.35}$$

Combining Eqs. (10.6.32), (10.6.34), and (10.6.35) gives

$$\omega_m = \sqrt{\omega_+\omega_-} \tag{10.6.36}$$

showing that the frequency at which $\phi(\omega)$ is 0 is the geometric mean of the frequencies at which the magnitude of $\phi(\omega)$ is a maximum.

Several other basic networks can be derived from the lag-lead network. The network in Fig. 10.16a results if $C_1 = R_2 = 0$ in Fig. 10.14a. This new

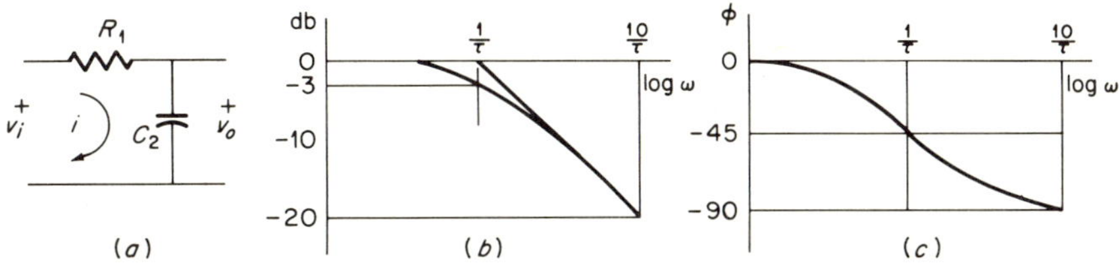

Fig. 10.16 Low-pass, phase-lag, or integrating network.

network is known variously as a low-pass filter, a phase-lag network, and an integrator, depending on the value of its time constant R_1C_2. The first two names are made obvious through the log magnitude and phase plots in Fig. 10.16*b* and *c*. To obtain these plots, we can form the voltage transfer function directly as

$$\frac{V_o(s)}{V_i(s)} = \frac{1}{1 + sR_1C_2} = \frac{1}{1 + \tau s} \tag{10.6.37}$$

and obtain

$$20 \log A(\omega) = -10 \log (1 + \tau^2\omega^2) \qquad \text{and} \qquad \phi(\omega) = -\tan^{-1} \tau\omega \tag{10.6.38}$$

Or we can note that, with $C_1 = R_2 = 0$, τ_a, τ_b, and τ_2 in Eq. (10.6.19) are zero and $\tau_1 = R_1C_2 \equiv \tau$. With these constraints, Eq. (10.6.19) and (10.6.37) become identical. What has happened is that $-1/\tau_2$, $-1/\tau_b$, and $-1/\tau_a$ in Fig. 10.14*b* have all migrated to $-\infty$, leaving only the root at $-1/\tau_1$.

To show that the network in Fig. 10.16*a* can also serve as an integrator, we require the differential equation

$$v_o = \frac{1}{C_2} \int i \, dt \tag{10.6.39}$$

If the input voltage v_i is periodic, and if the R_1C_2 time constant is large compared with the period, the capacitor will accumulate very little charge during the period. It follows that the current will remain approximately constant at its initial value of v_i/R_1 throughout the period, and we can write

$$v_o \approx \frac{1}{R_1C_2} \int v_i \, dt \tag{10.6.40}$$

which shows the output voltage to be a constant times the integral of the input voltage.

It is useful to compare the time-domain performance of the network in Fig. 10.16*a*, its ability to function as an integrator, with its frequency-domain performance, its ability to function as either a low-pass filter (amplitude selector) or a phase-lag network (phase selector). To correlate these performances, assume that the input voltage to the network is a rectangular pulse train of pulse width τ_p, pulse height V, and period T. Recall that in Sec. 9.3 we defined the cutoff frequency of a pulse train as the frequency of the first null in its Fourier series representation. From Eqs. (9.3.2) and (9.3.3), this frequency is given by

$$f_{\text{cp}} = \frac{1}{\tau_p} = \frac{n}{T} \tag{10.6.41}$$

where n is the number of ac frequency components out to and including the first null. It was noted that if τ_p were small compared with T (n large), the pulse train would be quite accurately represented by a Fourier series containing just those harmonics whose frequencies were less than f_{cp}.

Let us now define the cutoff frequency of the network in Fig. 10.16*a* as the frequency of its break point. Thus

$$f_{\text{cf}} = \frac{1}{2\pi R_1 C_2} \tag{10.6.42}$$

Four cases are to be considered:

$$f_{\text{cf}} = 10 f_{\text{cp}} \tag{10.6.43a}$$

$$f_{\text{cf}} = f_{\text{cp}} \tag{10.6.43b}$$

$$f_{\text{cf}} = \tfrac{1}{10} f_{\text{cp}} \tag{10.6.43c}$$

$$f_{\text{cf}} = \tfrac{1}{100} f_{\text{cp}} \tag{10.6.43d}$$

Equation (10.6.43*a*) requires that

$$R_1 C_2 = \frac{\tau_p}{20\pi} \approx 0.016\tau_p \tag{10.6.44}$$

Thus $5R_1C_2$, which is the time required for the capacitor to charge to 99 percent of its final value, is $0.08\tau_p$, and the output voltage reaches the value V early in the pulse on-time interval. It follows that the output voltage closely resembles the input voltage and the network is performing as a low-pass filter with a relatively high cutoff frequency, letting the first $10n$ frequency components pass essentially unattenuated and with nearly zero phase shift. The network is not performing as an integrator because the capacitor voltage changes drastically during the time the pulse is on and thus the current is not constant.

Turning next to Eq. (10.6.43*d*), we see that

$$R_1 C_2 = \frac{100\tau_p}{2\pi} \approx 16\tau_p \tag{10.6.45}$$

In this case, $5R_1C_2 = 80\tau_p$ and, in the time the pulse is on, the capacitor charges to

$$v_c = V(1 - e^{-0.063}) \approx 0.06 \text{ volts} \tag{10.6.46}$$

Now the current is essentially constant at V/R_1 throughout the on time of the pulse, and the output voltage during that time is a ramp, the integral of the input voltage step. The network is an extremely low-pass filter, severely and unequally attenuating the fundamental and all harmonics of the input voltage, unless $n > 100$ so that the fundamental frequency is less than f_{cf}. Also all frequencies are shifted by approximately $-90°$ in passing through the network, unless $n > 100$.

The cases represented by Eqs. (10.6.43*b*) and (10.6.43*c*) fall in between these two extremes. For $f_{cf} = f_{cp}$, $5R_1C_2 = 0.8\tau_p$, indicating that the capacitor reaches full charge about the time the pulse ends. This is neither integration nor faithful pulse reproduction. Frequencies less than f_{cf} in the input pulse train (there are n of these) are attenuated and shifted in phase very little; those above f_{cf}, a significant amount. For $f_{cf} = \frac{1}{10} f_{cp}$, $5R_1C_2 = 8\tau_p$ and the capacitor has a voltage of

$$v_c = V(1 - e^{-0.63}) = 0.47 \text{ volt}$$

at the end of the pulse. There is some integration effect here, but the current does change significantly during the pulse on time. Unless $n > 10$, all frequencies are attenuated and shifted in phase.

To summarize, the cutoff frequency of the network must be at least 10 times the cutoff frequency of the pulse if faithful reproduction of the pulse is to be achieved. If so, the network is performing as a low-pass filter and phase-lag device for the higher harmonics of the input voltage but is not performing at all as an integrator. For good integration, the cutoff frequency of the network should be no greater than 1 percent of the cutoff frequency of the pulse. Then the network is performing as a low-pass filter and $-90°$ phase shifter for all frequencies present, unless the pulse is very narrow compared with its period.

Three other basic networks derivable from the network in Fig. 10.14*a* are shown in Fig. 10.17. The network in Fig. 10.17*a* is very similar in performance to that in Fig. 10.16*a*, the difference being that the output voltage in Fig. 10.16 becomes zero as frequency becomes infinite whereas it becomes R_2v_i/R_1R_2 in Fig. 10.17*a*. This lesser reduction in high-frequency response is often necessary in low-pass-filter applications.

The network in Fig. 10.17*b* can be a high-pass filter, a phase-lead network, or a differentiator, depending on the value of its time constant R_2C_1. The network in Fig. 10.17*c* performs similarly to that in Fig. 10.17*b* except that it has a nonzero response to dc of R_2v_i/R_1R_2. Infinite low-frequency attenuation is often objectionable.

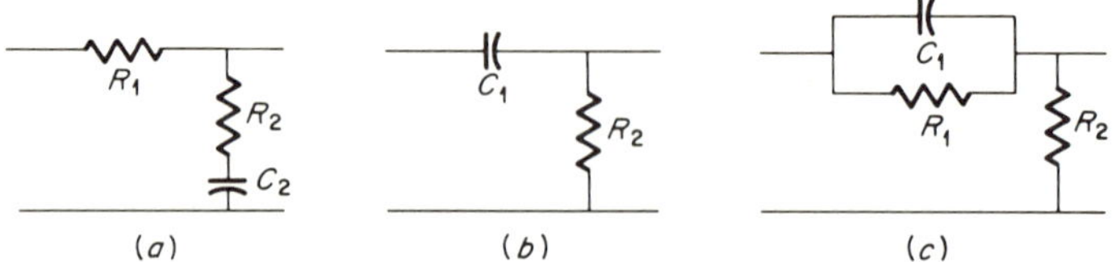

Fig. 10.17 Networks derivable from Fig. 10.14a.

10.7 SECOND-ORDER SYSTEM: COMPARISON OF s- AND t-DOMAIN RESPONSES

In the previous section, the quadratic factor $s^2 + cs + d$ was briefly discussed. Log magnitude and phase plots of this factor were given in Fig. 10.13, but left unanswered was how the magnitude and phase response varied in the vicinity of the break point at $\omega = \sqrt{d}$.

This section discusses the quadratic factor in detail, with particular emphasis on comparing and correlating the time- and frequency-domain responses of a second-order system.

The network in Fig. 10.18 will serve as a model. Its voltage transfer function is given by

$$T(s) = \frac{V(s)}{E(s)} = \frac{1/LC}{s^2 + (R/L)s + 1/LC} \tag{10.7.1}$$

The free response has the form

$$v(t) = k_1 e^{s_1 t} + k_2 e^{s_2 t} \tag{10.7.2}$$

where

$$s_1, s_2 = -\frac{R}{2L} \pm \sqrt{\frac{R^2}{4L^2} - \frac{1}{LC}} \tag{10.7.3}$$

The quantity under the radical in Eq. (10.7.3) can be positive real, zero, or negative real. If it is positive real, two unequal negative real natural frequencies result, and the free response is said to be *overdamped*. A zero under the radical yields two equal negative real roots and requires that the second term in Eq. (10.7.2) be replaced by $k_2 t e^{s_1 t}$. The circuit is then *critically damped*. *Underdamping* results if the quantity under the radical is negative

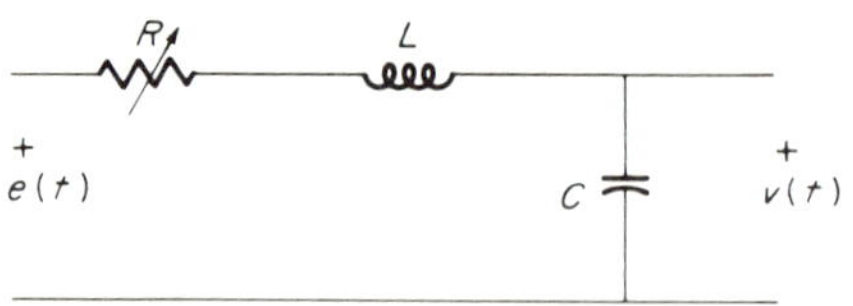

Fig. 10.18 Network whose T(s) contains a quadratic factor.

real. In this case the natural frequencies are complex conjugates with a non-positive real part. They become imaginary conjugates if no resistance is present. Then the circuit is underdamped to the extreme of being undamped.

There are two other, more descriptive, ways of representing the transfer function in Eq. (10.7.1). One is to express $T(s)$ in terms of the damping factor ζ and the undamped natural frequency ω_n. Such a representation is especially desirable when time-domain considerations are paramount. For example, the time response of the circuit in Fig. 10.18 to a unit step input voltage, measured in terms of rise time, overshoot, settling time, and ripple period, can be described effectively in terms of ζ and ω_n.

An alternative representation of $T(s)$ is useful when the input voltage is sinusoidal and the steady-state frequency-response characteristic of the circuit is of primary concern. In such cases it is descriptive to express $T(s)$ in terms of the circuit Q or its band width ω_B. We now develop each of these alternative representations in detail and compare them.

With R as the variable parameter (as indicated in Fig. 10.18), the resistance for critical damping (equal negative real roots) in Eq. (10.7.3) is

$$R_{CR} = 2\sqrt{\frac{L}{C}} \tag{10.7.4}$$

Define the damping factor ζ as the ratio of the actual resistance present to that required for critical damping. Thus

$$\zeta \equiv \frac{R}{R_{CR}} = \frac{R}{2}\sqrt{\frac{C}{L}} \tag{10.7.5}$$

The undamped natural frequency is obtained by setting $R = 0$ in Eq. (10.7.3). This yields

$$\omega_n = \frac{1}{\sqrt{LC}} \tag{10.7.6}$$

With these two defining relations, Eqs. (10.7.1) and (10.7.3) can be replaced by

$$T(s) = \frac{\omega_n^{\ 2}}{s^2 + 2\zeta\omega_n + \omega_n^{\ 2}} \tag{10.7.7}$$

and

$$s_1, s_2 = -\zeta\omega_n \pm \omega_n\sqrt{\zeta^2 - 1} \tag{10.7.8}$$

Note that $\zeta = 1$ gives the equal-root critically damped situation. With $\zeta > 1$, the circuit is overdamped; $\zeta < 1$ yields underdamping, and $\zeta = 0$ gives the undamped condition. The locus of the roots in Eq. (10.7.8), as $\zeta(R)$ varies from 0 to ∞, is shown in Fig. 10.19. The curved portion of the locus is circular of radius ω_n, as can be verified by squaring and adding the

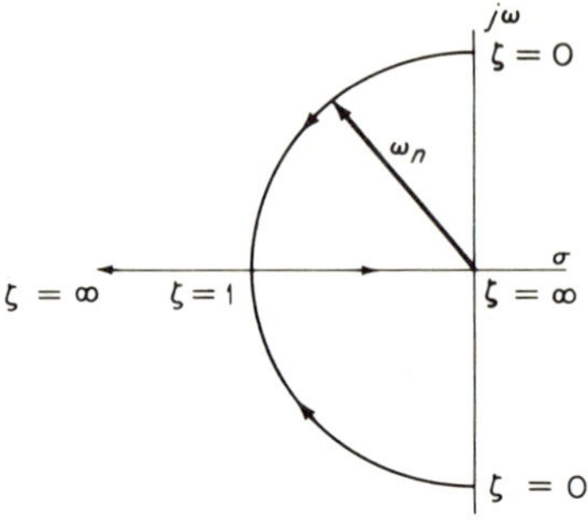

Fig. 10.19 Locus of roots of Eq. (10.7.8).

real and imaginary parts of Eq. (10.7.8). As ζ increases from 0, the roots move along the circle locus from the imaginary axis to a point of coalescence on the negative real axis at $\zeta = 1$. As ζ increases from 1, the roots migrate to $-\infty$ and 0 along the negative real axis.

As mentioned previously, ζ and ω_n are particularly useful in describing the transient response of a system. Assume $e(t)$ in Fig. 10.18 is a unit step $u(t)$. Then $V(s)$, the Laplace transform of the output voltage, is given by

$$V(s) = \frac{{\omega_n}^2}{s(s^2 + 2\zeta\omega_n + {\omega_n}^2)} \tag{10.7.9}$$

Performing a partial-fraction expansion yields

$$V(s) = \frac{K_0}{s} + \frac{K_1}{s + \zeta\omega_n - j\omega_d} + \frac{K_1^*}{s + \zeta\omega_n + j\omega_d} \tag{10.7.10}$$

where ω_d is the angular ripple frequency given by

$$\omega_d = \omega_n\sqrt{1 - \zeta^2} \tag{10.7.11}$$

and it is assumed that $\zeta < 1$. The constants in Eq. (10.7.10) are

$$K_0 = 1 \qquad \text{and} \qquad K_1 = \tfrac{1}{2}\left(-1 + j\zeta\frac{\omega_n}{\omega_d}\right) \tag{10.7.12}$$

and inverse transformation and algebraic simplification yield

$$v(t) = \left[1 - e^{-\zeta\omega_n t}\left(\cos\omega_d t + \zeta\frac{\omega_n}{\omega_d}\sin\omega_d t\right)\right]u(t) \tag{10.7.13}$$

or

$$v(t) = \left[1 - \frac{\omega_n}{\omega_d}e^{-\zeta\omega_n t}\sin\left(\omega_d t + \tan^{-1}\frac{\omega_d}{\zeta\omega_n}\right)\right]u(t) \tag{10.7.14}$$

Note that the solution in Eq. (10.7.14) is valid for $\zeta < 1$. If $\zeta = 0$, ω_d becomes equal to ω_n and

$$v(t) = [1 - \sin(\omega_n t + 90°)]u(t) = (1 - \cos\omega_n t)u(t) \tag{10.7.15}$$

If $\zeta = 1$, ω_d is zero and the sine term in Eq. (10.7.13) is indeterminate. Use of L'Hospital's rule gives

$$v(t) = [1 - e^{-\omega_n t}(1 + \omega_n t)]u(t) \tag{10.7.16}$$

This is the repeated-root case of critical damping. For the overdamped case, the partial-fraction expansion in Eq. (10.7.10) becomes

$$v(s) = \frac{K_0}{s} + \frac{K_2}{s + \zeta\omega_n - \omega_n\sqrt{\zeta^2 - 1}} + \frac{K_3}{s + \zeta\omega_n + \omega_n\sqrt{\zeta^2 - 1}} \tag{10.7.17}$$

The constants are

$$\begin{aligned} K_0 &= 1 \\ K_2 &= \frac{1}{2\sqrt{\zeta^2 - 1}(\sqrt{\zeta^2 - 1} - \zeta)} \\ K_3 &= \frac{1}{2\sqrt{\zeta^2 - 1}(\sqrt{\zeta^2 - 1} + \zeta)} \end{aligned} \tag{10.7.18}$$

Thus, for $\zeta > 1$, $v(t)$ is given by

$$\begin{aligned} v(t) = \Bigg\{1 &+ \frac{\frac{1}{2}\sqrt{\zeta^2 - 1}}{\sqrt{\zeta^2 - 1} - \zeta} \exp\left[-\omega_n(\zeta - \sqrt{\zeta^2 - 1})t\right] \\ &+ \frac{\frac{1}{2}\sqrt{\zeta^2 - 1}}{\sqrt{\zeta^2 - 1} + \zeta} \exp\left[-\omega_n(\zeta + \sqrt{\zeta^2 - 1})t\right]\Bigg\} u(t) \end{aligned} \tag{10.7.19}$$

For $\zeta = 1$, Eq. (10.7.19) should reduce to Eq. (10.7.16). Placing the exponential terms in Eq. (10.7.19) over a common denominator creates an indeterminate form to which L'Hospital's rule can be applied to yield Eq. (10.7.16).

A set of universal curves of $v(t)$ in Eqs. (10.7.14) to (10.7.16) and (10.7.19) vs. $\omega_n t$ appears in Fig. 10.20. These curves are universal in the sense that $\omega_n t$, rather than t, is the abscissa, and thus only one set of curves is needed for all natural frequencies. They show a range from sustained oscillation for $\zeta = 0$ to a monotonically increasing response for $\zeta = 2$.

In order to obtain quantitative measures of the dynamic performance of a system from these curves, we are led to consider the four previously mentioned quantities, namely, overshoot, rise time, settling time, and ripple period. These quantities have practical significance. If we can express each of them in terms of ζ and ω_n, we should be able to predict the dynamics of a given second-order system quite well.

Consider overshoot first and, in particular, the first, or peak, overshoot. To find the time at which it occurs, Eq. (10.7.13) is differentiated with respect

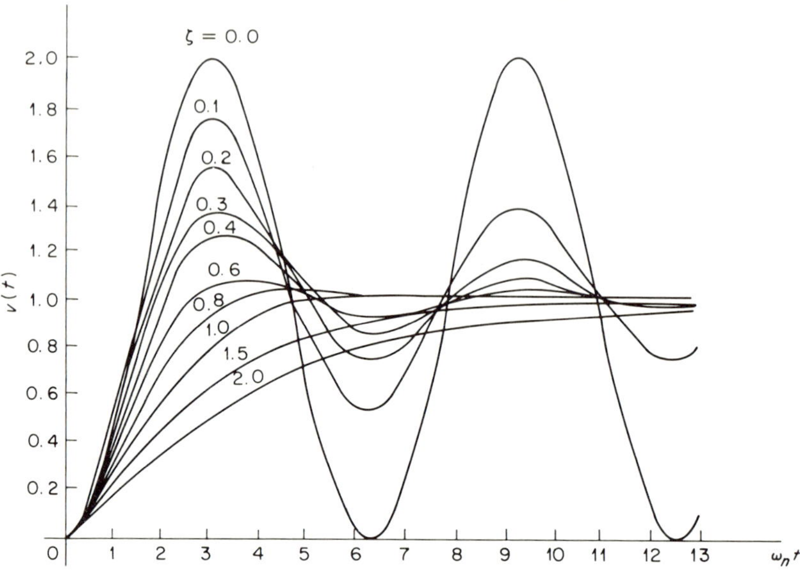

Fig. 10.20 *Second-order system unit step response.*

to time and equated to zero

$$\frac{dv(t)}{dt} = \zeta\omega_n e^{-\zeta\omega_n t}\left(\cos\omega_d t + \zeta\frac{\omega_n}{\omega_d}\sin\omega_d t\right)$$

$$- e^{-\zeta\omega_n t}(-\omega_d \sin\omega_d t + \zeta\omega_n \cos\omega_d t) \tag{10.7.20}$$

$$0 = \frac{\omega_n}{\sqrt{1-\zeta^2}} e^{-\zeta\omega_n t}\sin(\omega_n\sqrt{1-\zeta^2}\,t)$$

Calling the normalized time of peak overshoot $\omega_n t_0$ and letting the sine term in Eq. (10.7.20) equal $\pi/2$, we have

$$\omega_n t_0 = \frac{\pi}{\sqrt{1-\zeta^2}} \tag{10.7.21}$$

The maximum percent overshoot, denoted by 0_{max}, is obtained by inserting $\omega_n t_0$ into Eq. (10.7.13). The result is

$$0_{\text{max}} = 100\exp\left(-\frac{\zeta\pi}{\sqrt{1-\zeta^2}}\right) \qquad \text{percent} \tag{10.7.22}$$

We note that the peak overshoot is a function of ζ only. Thus specifying ζ fixes the peak overshoot; the larger the ζ, the smaller the peak overshoot.

For $\zeta \geq 1$, there is no overshoot, as can be shown by differentiating Eq. (10.7.19), equating to zero, and noting that $v(t)$ reaches its maximum at $t = \infty$. The variation of 0_{max} with ζ is displayed in Table 10.3.

The second factor of importance is rise time, denoted by t_r and customarily defined as the time for the response to rise from 10 to 90 percent of its final value. It is not possible to obtain an explicit expression for t_r using the 10 to 90 percent boundaries, but an explicit expression for the time to rise from 0 to 100 percent of final value can be obtained, and this expression serves nearly as well to show how rise time depends on ζ. Redefining t_r as the 0 to 100 percent time, and thus letting $v(t) = 1$ in Eq. (10.7.14), we obtain

$$\sin\left(\omega_d t_r + \tan^{-1}\frac{\omega_d}{\zeta\omega_n}\right) = \pi \tag{10.7.23}$$

which yields

$$\omega_n t_r = \frac{1}{\sqrt{1-\zeta^2}}\left(\pi - \tan^{-1}\frac{\sqrt{1-\zeta^2}}{\zeta}\right) \tag{10.7.24}$$

Values of normalized rise time $\omega_n t_r$ vs. ζ are shown in Table 10.3. As ζ increases, rise time increases, though not linearly. The values of $\omega_n t$ to go from 10 to 90 percent of final value are, as seen from Fig. 10.20, significantly less than the 0 to 100 percent figures in Table 10.3 only for the larger ζ values.

TABLE 10.3 Overshoot, rise time, settling time, and ripple period vs. ζ for a second-order system excited by a unit step

ζ	0_{max}, *percent*	$\omega_n t_r$, *rad*	$\omega_n t_s$, *rad*	$\omega_n T$, *rad*
0	100	1.57	∞	6.28
0.1	73	1.68	38.40	6.34
0.2	53	1.81	19.50	6.40
0.3	38	1.97	11.25	6.59
0.4	25	2.17	8.40	6.84
0.5	16	2.42	8.12	7.28
0.6	10	2.77	5.94	7.85
0.7	6	3.29	5.97	8.79
0.8	2	4.16	3.75	10.5
0.9	1	6.19	4.70	14.4
1.0	0	∞	5.80	∞

The third factor to consider is settling time. This will be denoted by t_s and defined as the time for the response to settle and remain within ± 2 percent of its final value. The settling time can be found from Eq. (10.7.14). The requirement is to find the smallest time such that

$$\frac{\omega_n}{\omega_d} e^{-\zeta\omega_n t} \sin\left(\omega_d t + \tan^{-1}\frac{\omega_d}{\zeta\omega_n}\right) = \pm 0.02 \tag{10.7.25}$$

Then, because of the decaying exponential, the response will remain within the ± 2 percent band from that time on.

It is difficult to tell which sign will apply in Eq. (10.7.25) because, especially for small ζ, there may be a number of overshoots and undershoots before the response settles within the prescribed tolerance. We do know, though, that if ζ is large enough, the overshoot will never exceed 2 percent and hence the response will settle into the ± 2 percent band from below. Setting $0_{max} = 2$ in Eq. (10.7.22) yields a value of 0.78 for ζ. Thus for any ζ of 0.78 or more, the plus sign should be used to obtain the settling time in Eq. (10.7.25). Values of normalized settling time, calculated by trial and error from Eq. (10.7.25), are tabulated in Table 10.3. As ζ increases, settling time decreases, but not in any smooth fashion, largely because of the oscillation introduced by the sine term

The last of the four descriptive factors is normalized ripple period, defined from Eq. (10.7.11) by

$$\omega_n T = \omega_n \frac{2\pi}{\omega_d} = \frac{2\pi}{\sqrt{1 - \zeta^2}} \tag{10.7.26}$$

This is the period of the sine term in Eq. (10.7.14) and gives the approximate spacing between successive overshoots in Eq. (10.7.14). As ζ increases from 0, $\omega_n T$ increases from 2π, approaching infinity as ζ approaches 1.

An examination of the universal curves in Fig. 10.20 shows that for a second-order system to have a unit step response with tolerable overshoot, rise time, and settling time, ζ should lie between 0.4 and 0.8. From Table 10.3 a ζ of 0.4 gives 25 percent overshoot, a rise time of 2.17 rad and a settling time of 8.40 rad. For $\zeta = 0.8$, these values are 2 percent, 4.18 rad, and 3.75 rad. For $\zeta < 0.4$, settling time becomes excessive and overshoot large: oscillations persist for too long a time. For $\zeta > 0.8$, rise time is too long and the system is sluggish.

We turn now to the second alternative representation of $T(s)$ in Eq. (10.7.1), which is useful when the frequency response of a system is of interest. As mentioned previously, it is more descriptive in this situation to express system performance in terms of ω_n and either the system Q or the bandwidth ω_B.

The Q of any system is defined as

$$Q = \frac{2\pi \text{ max stored energy}}{\text{energy dissipated/cycle}} \tag{10.7.27}$$

Assume for the circuit in Fig. 10.18 that the excitation voltage is sinusoidal with adjustable frequency. Resonance will occur when $\omega L = 1/\omega c$, giving

$\omega = 1/\sqrt{LC} = \omega_n$. Defining the Q at resonance as Q_n, we have from Eq. (10.7.27)

$$Q_n = 2\pi \frac{LI^2}{I^2 R 2\pi/\omega_n} = \frac{\omega_n L}{R} = \frac{1}{2\zeta} \tag{10.7.28}$$

where I is the rms current flowing.

At $\omega = \omega_n$, the circuit appears resistive (unity power factor), and the current is a maximum. The output voltage $v(t)$ is not a maximum at this frequency. Rather it reaches its peak at a frequency slightly less than the resonant frequency (as frequency is increased through resonance, the current increases and then decreases, whereas the capacitive reactance only decreases). The frequency separation between the output voltage maximum and the current maximum depends on Q_n, and is small if Q_n is large, as our subsequent development will show.

On each side of the resonant frequency, there is a frequency at which the current is down 3 db from its value at ω_n. Identifying these frequencies as ω_l (below resonance) and ω_n (above resonance), we have

$$\frac{1}{\omega_l C} - \omega_l L = R \qquad \text{and} \qquad \omega_h L - \frac{1}{\omega_h C} = R \tag{10.7.29}$$

which yield

$$\omega_l, \omega_h = \mp \frac{R}{2L} + \sqrt{\frac{R^2}{4L^2} + \frac{1}{LC}} \tag{10.7.30}$$

The difference between the 3-db frequencies is the circuit bandwidth

$$\omega_B = \omega_h - \omega_l = \frac{R}{L} = \frac{\omega_n}{Q_n} \tag{10.7.31}$$

where the last equality arises from Eq. (10.7.28). The product of the 3-db frequencies is the square of the resonant frequency

$$\omega_l \omega_h = \omega_n^2 \tag{10.7.32}$$

Using Eqs. (10.7.6) and (10.7.28), we can rewrite Eqs. (10.7.1), (10.7.3), and (10.7.30) in the form

$$T(s) = \frac{\omega_n^2}{s^2 + (\omega_n/Q_n)\, s + \omega_n^2} \tag{10.7.33}$$

$$s_1, s_2 = -\frac{\omega_n}{2Q_n} \pm \frac{\omega_n}{2Q_n}\sqrt{1 - 4Q_n^2} \tag{10.7.34a}$$

and

$$\omega_l, \omega_h = \mp \frac{\omega_n}{2Q_n} + \frac{\omega_n}{2Q_n}\sqrt{1 + 4Q_n^2} \tag{10.7.34b}$$

Note that $Q_n = \frac{1}{2}$ ($\omega_B = 2\omega_n$) gives critical damping; overdamping occurs for $Q_n < \frac{1}{2}$; underdamping for $Q_n > \frac{1}{2}$, with the undamped case requiring an infinite Q_n (zero bandwidth). Note further the similarity between the expressions for s_1, s_2 and ω_l, ω_h. It should be observed that the upper and lower 3-db frequencies can be found directly from the poles of $T(s)$; for the negative of the sum of the poles is the difference in the 3-db frequencies, and the product of the poles is their product.

A plot of the variation of these quantities with Q_n for fixed ω_n appears in Fig. 10.21. The locus of the s roots is identical to that in Fig. 10.19 and includes the semicircle and the negative σ axis. The locus of the ω roots is the positive ω axis. At $Q_n = 0$ (infinite bandwidth and infinite damping), s_1 and ω_l are zero, s_2 is $-\infty$, and ω_h is $+\infty$. At $Q_n = \infty$ (zero bandwidth and no damping), ω_l and $\omega_h = \omega_n$, $s_1 = j\omega_n$, and $s_2 = -j\omega_n$.

The stipulation previously made that ζ lie between 0.4 and 0.8 for tolerable overshoot, rise time, and settling time requires, from Eq. (10.7.28), a Q_n between 1.25 and 0.625 and, from Eq. (10.7.31), a bandwidth between $0.8\omega_n$ and $1.6\omega_n$. Contrast this with the situation encountered in selective resonant circuits, where $Q_n > 10$, giving $\zeta < 0.05$. Inserting this small ζ in Eqs. (10.7.22) and (10.7.24) to (10.7.26), we obtain the approximate expressions

$$0_{\text{max}} \approx 100 \text{ percent} \qquad \omega_n t_r \approx \frac{\pi}{2} \qquad \omega_n t_r \approx 8Q_n \qquad \omega_n T = 2\pi \tag{10.7.35}$$

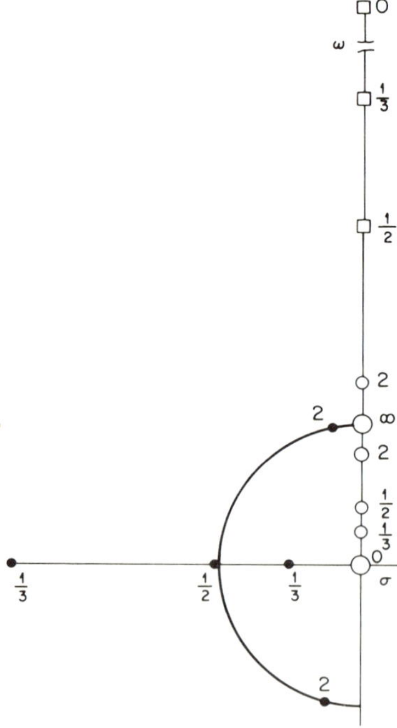

Fig. 10.21 s_1, s_2 *(dots),* ω_l *(circles), and* ω_n *(squares) versus* Q_n *(values numbered on plot) for the circuit of Fig. 10.18 as R is varied.*

The maximum overshoot is very large, but it is reached quite quickly. The settling time is long, and many cycles of ripple occur before it is reached. Thus highly selective frequency response means nearly oscillatory transient response.

Actually, to make a proper comparison between frequency- and time-domain responses, we should consider the same output quantity in each case. The unit-step-response curves in Fig. 10.20 display the output voltage $v(t)$ as a function of time, whereas up to now we have been considering the current in the circuit as a function of frequency. Let us now examine the output voltage as a function of frequency, or, what is the same for constant input voltage, $T(s)$ in Eq. (10.7.33) as a function of ω.

It was mentioned earlier that the output voltage and current do not reach their maxima at the same frequency. The current is maximum at $\omega = \omega_n$ and the output voltage at a frequency slightly less than ω_n—call it ω_0. To find ω_0, we require the magnitude of $T(j\omega)$, given from Eq. (10.7.33) by

$$|T(j\omega)| = \frac{\omega_n^{\,2}}{\sqrt{(\omega_n^{\,2} - \omega^2)^2 + (\omega_n^{\,2}/Q_n^{\,2})\omega^2}} \qquad (10.7.36)$$

Differentiating Eq. (10.7.36) and equating to zero yields

$$\omega_0 = \omega_n\sqrt{1 - 1/2Q_n^{\,2}} \qquad (10.7.37)$$

For $Q_n = 10$, $\omega_0 = 0.998\omega_n$. Thus for $Q_n \geq 10$, the difference between ω_0 and ω_n is negligible. For lesser Q_n's, this is not so. For example, for $Q_n = 2$, $\omega_0 = 0.935\omega_n$. For $Q_n \leq 0.707$, $\omega_0 = 0$ and there is no resonance peak.

The magnitude of the output voltage at $\omega = \omega_0$ is easily found by inserting Eq. (10.7.37) into Eq. (10.7.36). The result is

$$|T(j\omega)|_{\text{max}} = \frac{Q_n}{\sqrt{1 - 1/4Q_n^{\,2}}} \qquad (10.7.38)$$

For $Q_n = 10$, $|T(j\omega)|_{\text{max}} = 0.999Q_n$. Thus for $Q_n \geq 10$, the maximum magnitude is essentially Q_n.

A plot of $|T(j\omega)|$ vs. ω/ω_n for various values of Q_n is shown in Fig. 10.22. Note the shift of the resonant peak to lower frequencies as Q_n is decreased. Note also that, for a given Q_n, there are not always two frequencies at which the response is down 3 db from its maximum. The lower 3-db frequency is absent for small Q_n. To explore this latter matter quantitatively, we must find those frequencies at which $|T(j\omega)|$ is 0.707 of its maximum value. From Eqs. (10.7.36) and (10.7.38), the condition can be expressed as

$$\frac{\omega_n^{\,4}}{(\omega_n^{\,2} - \omega^2)^2 + (\omega_n^{\,2}/Q_n^{\,2})\omega^2} = \frac{Q_n^{\,2}}{2(1 - 1/4Q_n^{\,2})} \qquad (10.7.39)$$

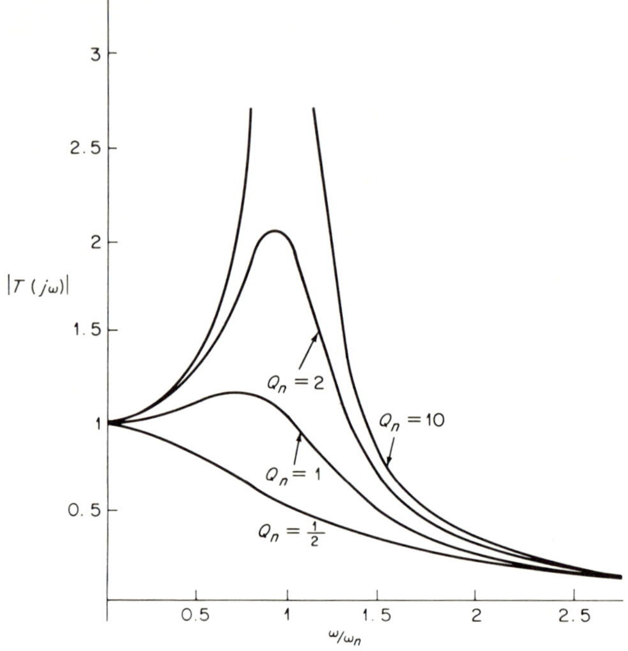

Fig. 10.22 Frequency response of the circuit in Fig. 10.18.

Solving for ω yields, after some algebraic manipulation,

$$\omega = \omega_n \sqrt{1 - \frac{1}{2Q_n^{\,2}} \pm \frac{1}{Q_n}\sqrt{1 - \frac{1}{4Q_n^{\,2}}}} \tag{10.7.40}$$

For large Q_n, Eq. (10.7.40) reduces to

$$\omega \approx \omega_n \sqrt{1 \pm \frac{1}{Q_n}} \approx \omega_n\left(1 \pm \frac{1}{2Q_n}\right) \tag{10.7.41}$$

which is identical to the values of ω_l and ω_h when Q_n is large in Eq. (10.7.34*b*). So for large Q_n, there are always two 3-db frequencies in the output voltage, and the output voltage bandwidth and the current bandwidth are essentially identical, with each centered at $\omega = \omega_n$.

The value of Q_n below which there is only one 3-db frequency is that value of Q_n in Eq. (10.7.40) which makes the term under the large radical negative when the minus sign is applied to the term with the small radical. Thus there will be only one 3-db frequency if

$$\frac{1}{2Q_n^{\,2}} + \frac{1}{Q_n}\sqrt{1 - \frac{1}{4Q_n^{\,2}}} > 1 \qquad \text{or} \qquad 2Q_n^{\,2} - 4Q_n^{\,2} + 1 < 0 \tag{10.7.42}$$

Solving for Q_n yields the condition

$$0.54 < Q_n < 1.31 \tag{10.7.43}$$

The lower constraint does not apply, since we know that if $Q_n < 0.707$, there is no resonant peak. Thus our conclusion is that Q_n must be greater than 1.31 for the output voltage to have two 3-db frequencies. From Eq. (10.7.28), this means that ζ must be less than 0.38. Here again is the trade off between good frequency-domain and good time-domain response. For the circuit to be selective in terms of its frequency response, its transient response must be sacrificed to the extent of considerable overshoot and settling time.

In the discussion in Sec. 10.6 of log modulus and phase plots, the quadratic factor was only briefly mentioned. Left unanswered was the manner in which these characteristics varied near the break frequency. The transfer function $T(s)$ in Eq. (10.7.33) is composed of a single quadratic factor. To obtain its

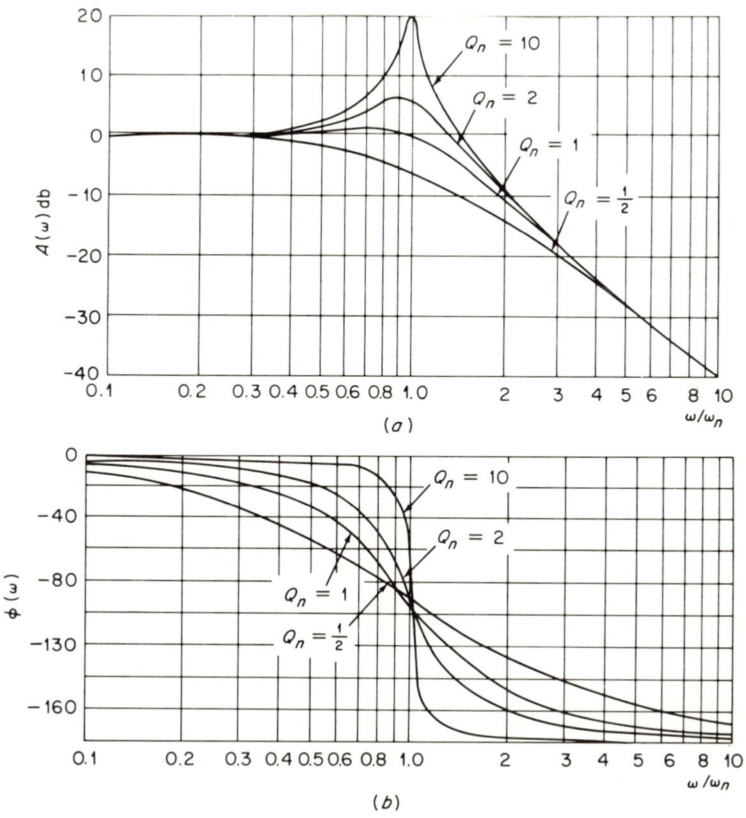

Fig. 10.23 Log magnitude and phase plots for Fig. 10.18.

Bode plot, we rewrite $T(s)$ in the form

$$T(s) = \frac{1}{(1/\omega_n^2)s^2 + (1/\omega_n Q_n)s + 1} \tag{10.7.44}$$

from which we obtain

$$A(\omega) = \frac{1}{\sqrt{(1 - \omega^2/\omega_n^2)^2 + \omega^2/\omega_n^2 Q_n^2}} \tag{10.7.45}$$

yielding

$$\text{Decibels} = -10 \log\left[\left(1 - \frac{\omega^2}{\omega_n^2}\right)^2 + \frac{\omega^2}{\omega_n^2 Q_n^2}\right] \tag{10.7.46}$$

and

$$\phi(\omega) = -\tan^{-1} \frac{\omega/\omega_n Q_n}{1 - \omega^2/\omega_n^2} \tag{10.7.47}$$

These two factors are plotted in Fig. 10.23 for several values of Q_n. The magnitude factor is represented by a horizontal line for $\omega \ll \omega_n$. For $\omega \gg \omega_n$, the factor becomes $-40 \log (\omega/\omega_n)$ with a slope of -12 db/octave. In between these two extremes, the factor depends on Q_n and at $\omega = \omega_n$ it is given by $-20 \log (1/Q_n)$. The phase characteristic is zero for $\omega \ll \omega_n$. At $\omega = \omega_n$, the phase is $-90°$ and for $\omega \gg \omega_n$, the phase approaches $-180°$. At other values of ω, the phase depends on Q_n, showing more rapid variation in the neighborhood of ω_n as Q_n is increased.

REFERENCES

10.1. Aseltine, J. A.: "Transform Method in Linear System Analysis," McGraw-Hill Book Company, New York, 1958. Chapter 3 contains an excellent discussion of the impulse function and problems at the origin. Chapter 7 reviews the partial-fraction-expansion technique for taking the inverse Laplace transform. Chapter 9 considers Bode plots.

10.2. Del Toro, V., and S. R. Parker: "Principles of Control System Engineering," McGraw-Hill Book Company, New York, 1960. Contains an excellent presentation of Bode plots and the analysis of first- and second-order systems.

10.3. Seshu, S., and N. Balabanian: "Linear Network Analysis," John Wiley & Sons, Inc., New York, 1959. Initial conditions are discussed in Chapter 4. Chapter 5 contains a detailed discussion of the steady-state response of a network to a general periodic excitation.

10.4. Van Valkenburg, M. E.: "Network Analysis," Prentice-Hall, Inc., Englewood Cliffs, N.J., 1964. Partial-fraction expansions are well covered in Chapter 7. Steady-state response to periodic signals is discussed in Chapter 15.

10.5. Von Tersch, L. W., and A. W. Swago: "Recurrent Electrical Transients," Prentice-Hall, Inc., Englewood Cliffs, N.J., 1953. Chapter 3 contains an excellent discussion of the response of basic circuits to simple waveforms, stressing end-point analysis.

PROBLEMS

Drill

10.1. Without differentiating, find the partial-fraction expansion of:

$$(a)\quad T(s) = \frac{s+4}{s(s+1)^2(s+2)^2(s+3)}$$

$$(b)\quad T(s) = \frac{s+2}{(s+1)^3(s^2+2s+2)^2}$$

10.2. (*a*) A circuit at rest is described by

$$\frac{di}{dt} + 2i = \frac{dv}{dt} + v$$

For an excitation $v = e^{-3t}u(t)$, determine the response $i(t)$. Is it correct, for the given excitation, to replace the entire right side of the differential equation by $-2e^{-3t}u(t)$ prior to solving for $i(t)$? Explain.
(*b*) Repeat the first portion of part (*a*) for

$$\frac{di}{dt} + 2i = \frac{d^2v}{dt^2} + 3\frac{dv}{dt} + v$$

The excitation remains the same.

10.3. For an excitation $v = u(t)$, solve the following differential equation for $i(t)$:

$$\frac{di}{dt} + 3i = \frac{d^2v}{dt^2} + 3\frac{dv}{dt} + 2v$$

10.4. A system is initially at rest and is described by

$$\frac{d^3y}{dt^3} + 3\frac{d^2y}{dt^2} + 4\frac{dy}{dt} + 2y = \frac{dx}{dt} + 2x$$

(*a*) Find y, dy/dt, and d^2y/dt^2 at $t = 0_+$ when x is a unit ramp, a unit step, a unit impulse.
(*b*) Determine $y(t)$ for each $x(t)$ in part (*a*).

10.5. The switch shuts at $t = 0$ and opens again at $t = 100$ msec. Find v_{ab} vs. t. Repeat the problem with the left 2-kilohm resistor replaced by a 1-henry inductor.

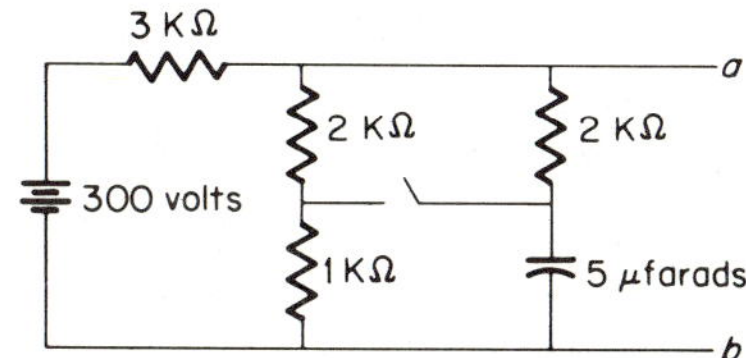

Fig. P10.5

10.6. C is uncharged for $t < 0$. Find $i(t)$ vs. t for $t > 0$ and sketch.

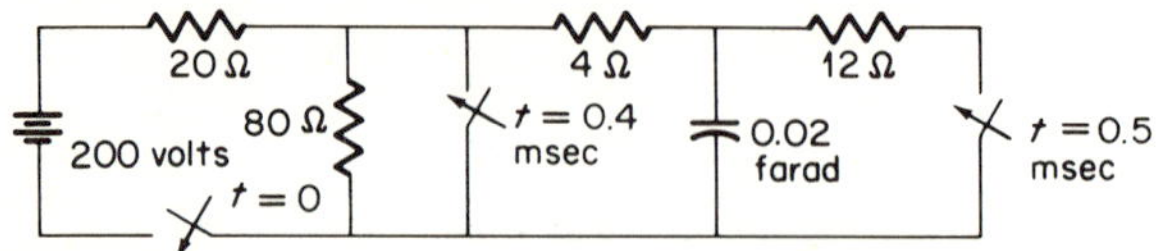

Fig. P10.6

10.7. S_2 and S_3 are closed. After a long time, S_2 and S_3 are opened and S_1 and S_4 are closed. Find v_{ab} vs. t for all time after the closure of S_1 and S_4.

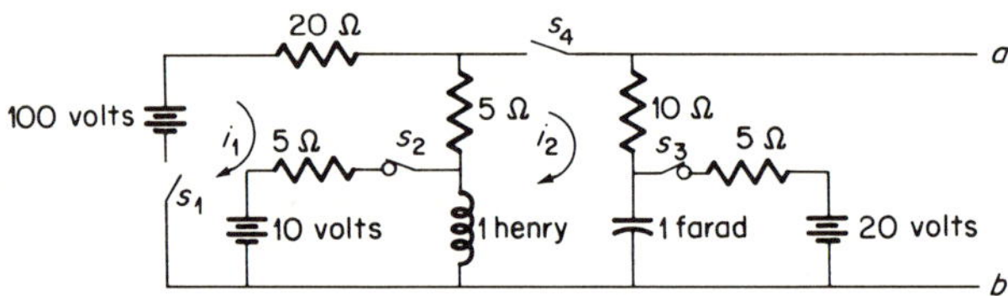

Fig. P10.7

10.8. The switch shuts at $t = 0$. Find v_{ab} and i_{SW} for $t > 0$.

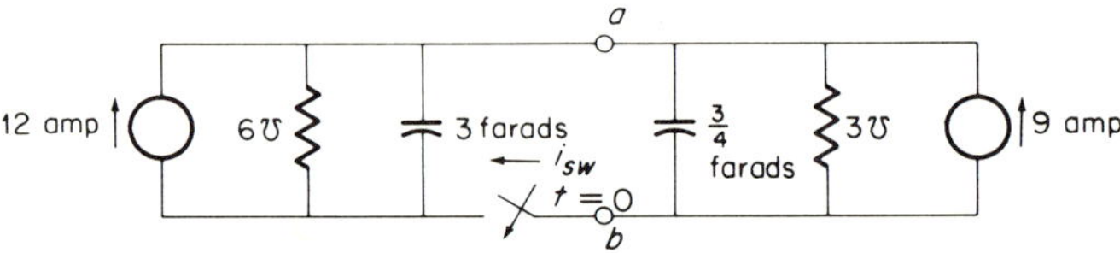

Fig. P10.8

10.9. The periodic voltage waveform shown is applied to the *RC* circuit. If the circuit time constant is 700 μsec, find the steady-state output voltage.

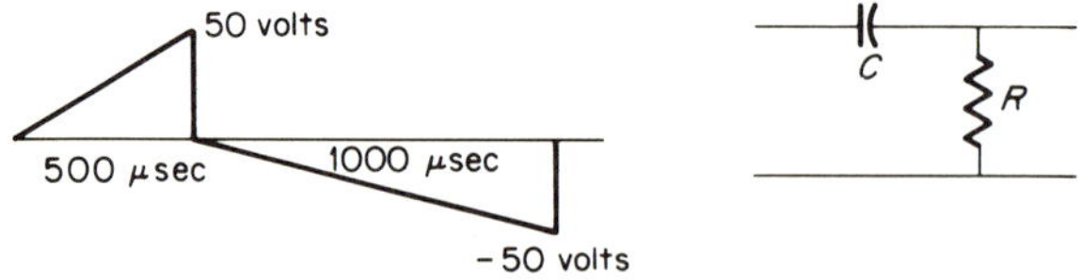

Fig. P10.9

10.10. For the circuit and input voltage waveform shown, find the output voltage across the 10-kilohm resistor (*a*) by an approximate analysis similar to that in Example 10.7 and (*b*) by an exact analysis similar to that in Example 10.8.

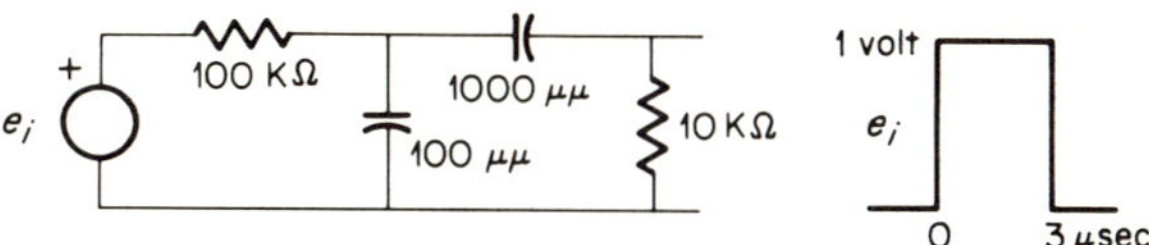

Fig. P10.10

10.11. The switch shuts at $t = {}^{2000}\!/_{377}$ sec. Find the initial conditions in the circuit just after the switch closes and find $i(t)$ for all time thereafter.

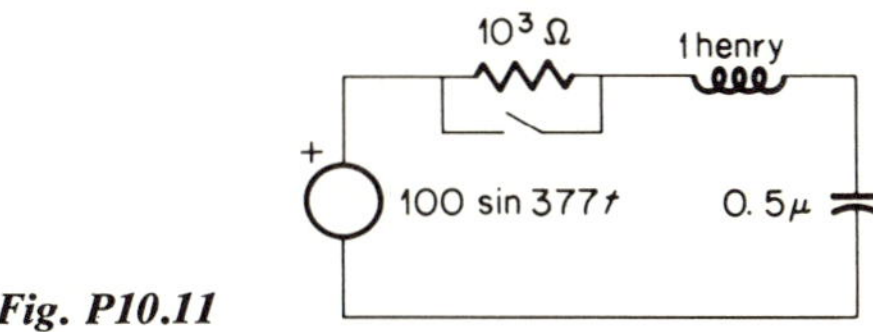

Fig. P10.11

10.12. For the given circuit and half-sinusoidal periodic excitation, find the steady-state output voltage (*a*) approximately as the first four terms of a Fourier series and (*b*) exactly by the open-form Laplace technique. Plot results and compare. How do these results change if the series resistance is zero?

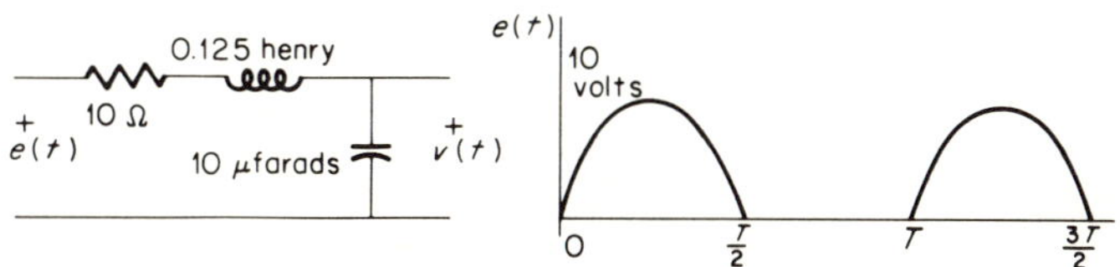

Fig. P10.12

10.13. Using the closed-form Laplace technique, find the steady-state output voltage in Fig. 10.6*b* for the periodic excitations whose representations during the first period are:

(*a*) $e_1(t) = 200u(t) - 100u(t - 0.5) - 100u(t - 1)$ volts

(*b*) $e_1(t) = 200tu(t) - 400(t - 0.5)u(t - 0.5) + 200(t - 1)u(t - 1)$ volts

where t is measured in milliseconds.

10.14. Repeat Examples 10.10 to 10.13 for $R = 100$ ohms. In each case, plot the output voltage vs. time.

10.15. Draw log magnitude and phase plots for the following transfer functions:

$$(a)\quad T(s) = \frac{s^2 + 5s + 4}{s^3 + 8s^2 + 37s + 50}$$

$$(b)\quad T(s) = \frac{(s + 2)^2}{(s + 1)(s^2 + 24s + 153)}$$

$$(c)\quad |T(j\omega)|^2 = \frac{1}{1 + \omega^6}$$

10.16. Find expressions for $\omega_n t_0$, $0_{\max}$, $\omega_n t_r$, $\omega_n t_s$, and $\omega_n T$ for the circuit shown when it is excited by a unit current step.

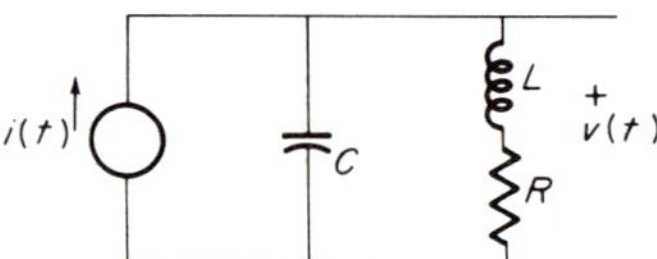

Fig. P10.16

10.17. For the circuit in Fig. P10.16 with $R = 2$ ohms, $L = 8$ henrys, and $C = 2$ farads, plot the output voltage as a function of ω assuming $i(t)$ is sinusoidal of variable frequency. Repeat for $R = 2$ ohms, $L = 2$ henrys, and $C = \frac{1}{2}$ farad.

10.18. The circuits shown are identical to that in Fig. 10.18 with the exception of the element across which the output voltage is taken. For $e(t) = u(t)$, plot $v(t)$ vs. $\omega_n t$ for several values of ζ for each circuit arrangement. Compare with the universal curves in Fig. 10.20.

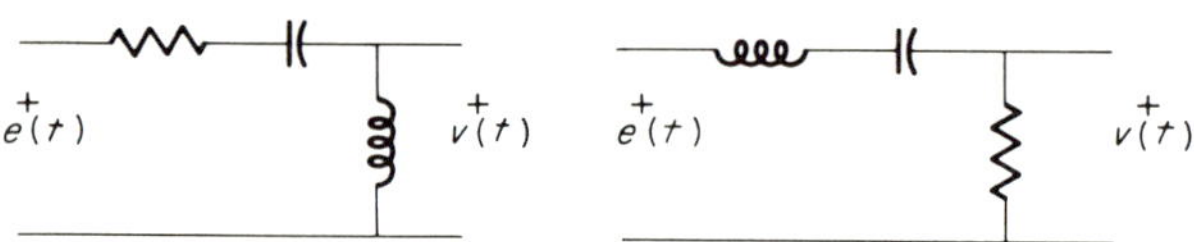

Fig. P10.18

10.19. For the circuit shown with $L = 2$ henrys, $C = \frac{1}{2}$ farad, and $R_2 = 1$ ohm:

(*a*) Find $v(t)$ vs. t when $e(t)$ is a unit step and $R_1 = \frac{5}{4}$ ohm. Repeat for $R_1 = 3$ ohms.

(*b*) Find $V(\omega)$ vs. ω when $e(t)$ is a unit sine wave of variable frequency and $R_1 = \frac{5}{4}$ ohm. Repeat for $R_1 = 3$ ohms.

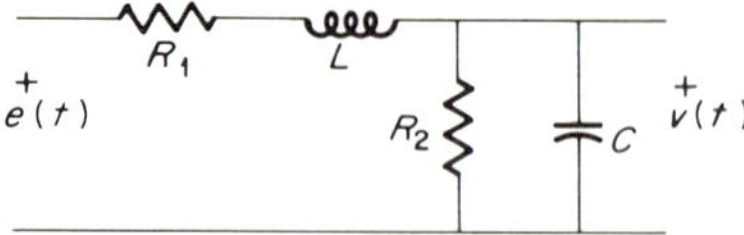

Fig. P10.19

Theory and Proofs

10.20. Verify Eqs. (10.5.19) and (10.5.20).

10.21. (*a*) Show that Eq. (10.5.1) gives correct expressions for V_c and V_{R2} in the circuit of Fig. P10.21*a*.

(*b*) Show that Eq. (10.4.1) gives correct expressions for V_L and V_{R2} in the circuit of Fig. P10.21*b*.

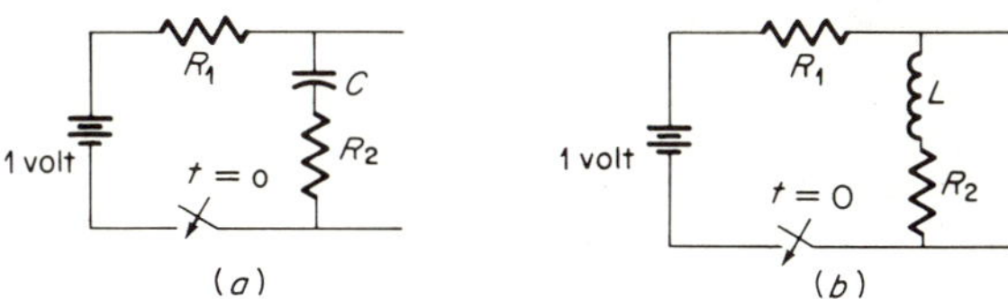

Fig. P10.21

10.22. A ratio of polynomials is given by

$$T(s) = \frac{a_n s^n + a_n s^{n-1} + \cdots + a_1 s + a_0}{b_m s^m + b_{m-1} s^{m-1} + \cdots + b_1 s + b_0}$$

Derive an expression for the sum of the residues of $T(s)$ when $m = n$. Check your expression by performing the expansion of

$$T(s) = \frac{(2s + 5)(s + 2)}{(s + 3)(s + 1)}$$

10.23. A transfer function with one second-degree pole can be represented by

$$T(s) = \frac{\Pi(s + a_i)}{(s + k)^2 \Pi(s + b_j)}$$

$$= \frac{k_1}{(s + k)^2} + \frac{k_2}{s + k} + \sum \frac{r_j}{s + b_j} \qquad i = 1, 2, \ldots, p$$

$$j = 1, 2, \ldots, q$$

$$p < q + 2$$

Find the general expressions for k_1, k_2, and the r_j's.

10.24. For the periodic excitation shown applied to the circuit in Fig. 10.8*b*, devise an exponential end-point analysis procedure, similar to that in Sec. 10.5, to find the steady-state output voltage. Sketch this voltage.

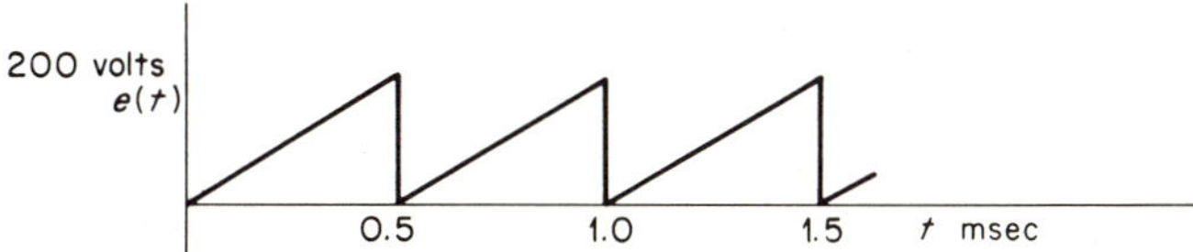

Fig. P10.24

Application

10.25. The s-domain response of a system to a unit step excitation is

$$T(s) = \frac{326(s+4)}{s(s^2+2s+11.6)(s^2+12.8s+113.2)}$$

Find $T(t)$, the inverse Laplace transform of $T(s)$, by a graphical evaluation of the coefficients in the partial-fraction expansion. Comment on the relative contributions of the various complex roots to the transient solution. Could $T(s)$ be simplified to a reasonable degree of accuracy?

10.26. Plot the given transfer-function data on semilog paper. Approximate the plot by straight-line asymptotes. From these, compose an analytical expression for $T(s)$. Obtain the time-domain response when a unit step excitation is applied to the system represented by $T(s)$.

ω, *rad/sec*	$\lvert T(j\omega)\rvert$, *db*	ω, *rad/sec*	$\lvert T(j\omega)\rvert$, *db*
0.1	−7.16	2.12	−16.82
0.5	−9.36	2.3	−16.00
0.707	−11.66	2.5	−15.04
1.0	−16.88	3.0	−14.24
1.2	−23.22	4.0	−14.52
1.414	−∞	10.0	−21.40
1.6	−26.00	100.0	−40.00
2.0	−17.84		

10.27. In the circuit shown, the switch closes at $t = 0$. Find E_m and θ such that the current response reaches its steady-state value immediately.

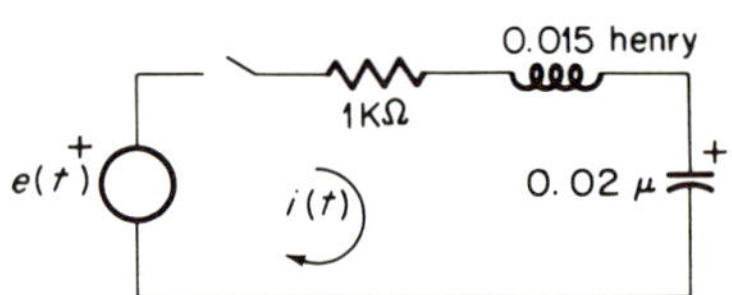

Fig. P10.27 $i_+ = 0$ *amp*, $V_{c+} = 100$ *volts*, $e = E_m \sin(\omega t + \theta)$ *volts, and* $\omega = 4 \times 10^4$ *rad/sec.*

10.28. The circuit shown is called a *clamping circuit*. It is excited by the given periodic voltage applied at $t = 0$. The diode has 500 ohms forward resistance and 500 kilohms back resistance. Compute and sketch the transient and steady-state output voltage v_{ab}.

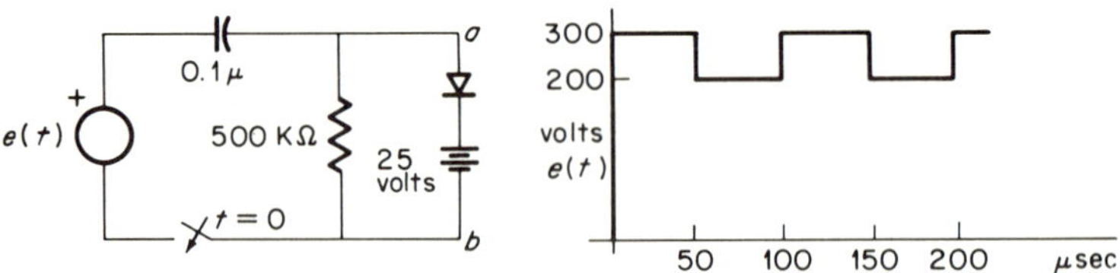

Fig. P10.28

10.29. A system has a frequency response T represented by the Bode plot shown. For the given square-wave input, find the steady-state output. Use a Fourier series approximation to the input function through the seventh harmonic. Plot one cycle of the approximate response.

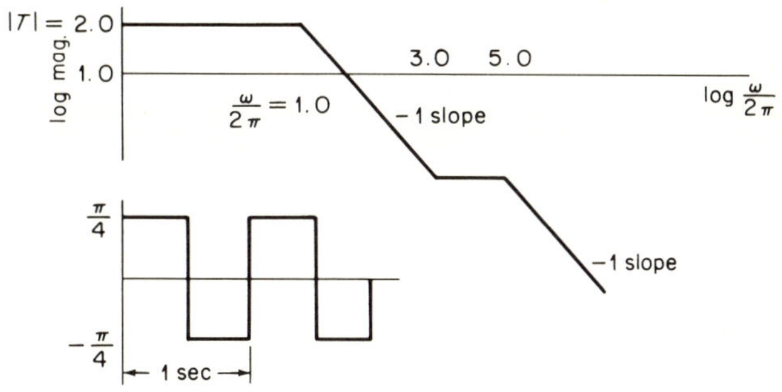

Fig. P10.29

10.30. Find the output voltage response of the RC circuit to an impulse voltage excitation of value 10^{-4} volt and compare it with the output voltage response of the circuit to the three pulse excitation voltages. From these results, comment on the conditions under which the pulsed response of a circuit may be considered equivalent to the impulse response.

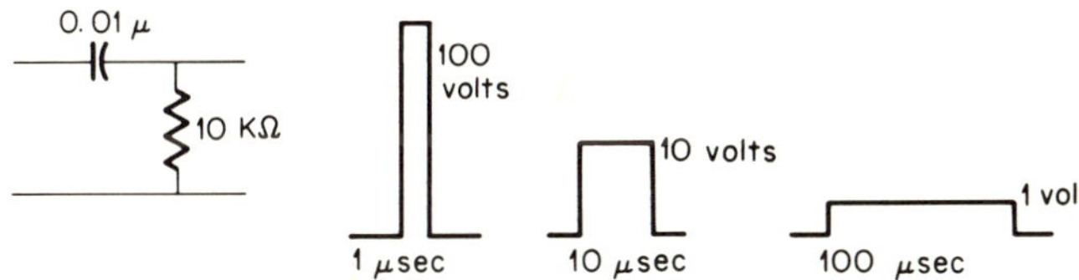

Fig. P10.30

11 STATE VARIABLES AND EQUATIONS

11.1 INTRODUCTION

The notion of state was introduced in Sec. 8.4, where a set of n first-order state equations was derived from the nth-order equation and the free response of the system was obtained from these n equations.

In this chapter the concept of state is explored in greater detail. The chapter is in two parts. In the first part, three methods of writing state equations directly from the network are presented: the by-inspection technique, the superposition method, and Bashkow's topological formulation. In the last part, s- and t-domain techniques for obtaining the total solution of a set of state equations are formulated, the latter involving a mathematical digression into functions of matrices.

Let us review some definitions, given earlier in connection with Eq. (8.4.4). The *state of a system* may be considered as the least amount of information that must be known about the system at a given time to determine its subsequent performance completely. It follows that the *states of a linear network* are a set of n independent initial conditions existing in the network at any time t_0. The *state variables* are the n dynamically independent variables whose values at time t_0 constitute the states at that time. The *state equations* are the n independent first-order differential equations in the state variables. If we know the values of the state variables at time t_0, and if we also know the external excitation from t_0 onward, the dynamic response of the network is determined for $t > t_0$.

The states of the network describe the energy stored in the network. Thus, as pointed out in Sec. 8.4, a set of independent inductor currents and capacitor voltages is a convenient set of state variables. But we also know that any nonsingular linear transformation of these, as indicated in Eq. (8.4.5), is also acceptable. Such a transformation permits the choice of a set of variables that may be easier to measure or control than those which directly specify the network's stored energy. In particular, sets of loop currents or node voltages can be used.

A simple example will illustrate the various choices of state variables.

Example 11.1 For the network in Fig. 11.1 with $L = 1$ henry, $C = \frac{1}{4}$ farad, $R_1 = 2$ ohms, and $R_2 = 6$ ohms, write homogeneous state equations in the form $\dot{Y} = AY$ using the following sets of variables: (*a*) i_1 and $\dot{i}_1$, (*b*) v_1 and $\dot{v}_1$, (*c*) i_1 and i_2, (*d*) v_1 and v_2, (*e*) v_C and i_L.

Cases (*a*) *and* (*b*)

These are cases where the n first-order equations contain a single variable and $n - 1$ of its derivatives, in this case one derivative. To arrive at the $\dot{Y} = AY$ form in these cases, it is easiest to proceed directly from the nth-order equation, as was done in Eqs. (8.4.1) and (8.4.2).

For case (*a*), equations are formulated on a loop basis

$$8i_1 + 4\int_0^t i_1\,dt + v_{C+} - 6i_2 = e$$

$$-6i_1 + 6i_2 + \frac{di_2}{dt} = 0$$

They are manipulated to obtain a second-order equation in i_1

$$\frac{d^2i_1}{dr^2} + 2\frac{di_1}{dt} + 3i_1 = \frac{1}{8}\frac{d^2e}{dt^2} + \frac{3}{4}\frac{de}{dt}$$

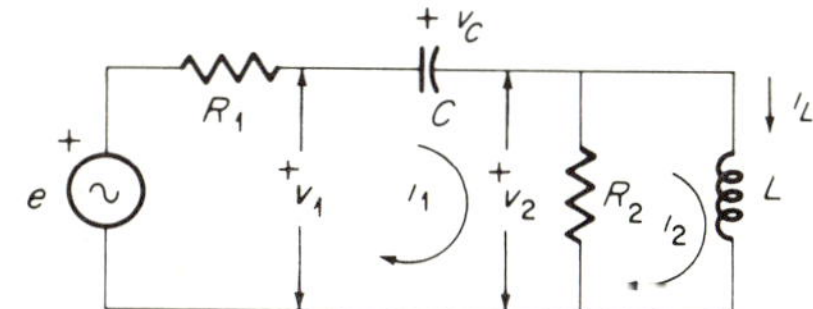

Fig. 11.1

The homogeneous equation is

$$\frac{d^2i_1}{dt^2} + 2\frac{di_1}{dt} + 3i_1 = 0$$

Letting $i_1 = y_1$ and $\dot{i}_1 = y_2$, we obtain

$$\begin{bmatrix} \dot{y}_1 \\ \dot{y}_2 \end{bmatrix} = \begin{bmatrix} 0 & 1 \\ -3 & -2 \end{bmatrix} \begin{bmatrix} y_1 \\ y_2 \end{bmatrix}$$

For case (*b*), equations are formulated from Fig. 11.1 on a node-to-datum basis

$$\tfrac{1}{2}(v_1 - e) + \frac{1}{4}\frac{d(v_1 - v_2)}{dt} = 0$$

$$\frac{1}{4}\frac{d(v_2 - v_1)}{dt} + \frac{1}{6}v_2 + \int_0^t v_2\,dt + i_{L+} = 0$$

Manipulating these equations gives

$$\frac{d^2v_1}{dt^2} + 2\frac{dv_1}{dt} + 3v_1 = \frac{4}{3}\frac{d^2e}{dt^2} + \frac{1}{2}\frac{de}{dt} + 3e$$

The homogeneous portion of this equation has the same coefficients as those previously obtained for the currents. With $v_1 = y_1$, and $\dot{v}_1 = y_2$, the identical A matrix results.

Cases (*c*), (*d*), *and* (*e*)

These are cases where the n first-order equations contain n variables not related through differentiation. To arrive at the $\dot{Y} = AY$ form in these cases, it is not necessary to derive the nth order differential equation. Instead, suitable loop, node-pair, node-to-datum, or other equations are written directly from an examination of the network, and they are manipulated into state-equation form. For case (*c*), we want to express $\dot{i}_1$ and $\dot{i}_2$ in terms of i_1 and i_2. With $e = 0$ (the homogeneous case), we differentiate the first loop equation and solve it and the second loop equation for $\dot{i}_1$ and $\dot{i}_2$. The result in matrix form is

$$\begin{bmatrix} \dot{i}_1 \\ \dot{i}_2 \end{bmatrix} = \begin{bmatrix} 4 & -9/2 \\ 6 & -6 \end{bmatrix} \begin{bmatrix} i_1 \\ i_2 \end{bmatrix}$$

For case (*d*), the node-to-datum equations for case (*b*), with $e = 0$, can be added and differentiated to yield

$$\tfrac{1}{2}\dot{v}_1 + \tfrac{1}{6}\dot{v}_2 + v_2 = 0$$

This equation and the first node-to-datum equation can then be solved for $\dot{v}_1$ and $\dot{v}_2$, giving

$$\begin{bmatrix} \dot{v}_1 \\ \dot{v}_2 \end{bmatrix} = \begin{bmatrix} -1/2 & -3/2 \\ -3/2 & -3/2 \end{bmatrix} \begin{bmatrix} v_1 \\ v_2 \end{bmatrix}$$

For case (*e*) the variables are the capacitor voltage and the inductor current. To obtain the state equations, we write a node equation at a node touched by the capacitor and a loop equation around a loop containing the inductor, each of these equations in terms of v_C and i_L

$$\tfrac{1}{4}\dot{v}_C + \frac{v_C + \dot{i}_L - e}{2} = 0$$

$$\dot{i}_L + (i_L - \tfrac{1}{4}\dot{v}_C)6 = 0$$

Solving for $\dot{v}_C$ and $\dot{i}_L$ with $e = 0$ yields

$$\begin{bmatrix} \dot{v}_C \\ \dot{i}_L \end{bmatrix} = \begin{bmatrix} -\tfrac{1}{2} & 3 \\ -\tfrac{3}{4} & -\tfrac{3}{2} \end{bmatrix} \begin{bmatrix} v_C \\ i_L \end{bmatrix}$$

11.2 GENERAL FORM OF STATE EQUATIONS

In Example 11.1, we have shown how to write homogeneous state equations for a simple network. Before proceeding further, we need to find how to include forcing functions. Reexamine case (*e*) of Example 11.1. With the voltage excitation included, the state equations become

$$\begin{bmatrix} \dot{v}_C \\ \dot{i}_L \end{bmatrix} = \begin{bmatrix} -\tfrac{1}{2} & 3 \\ -\tfrac{3}{4} & -\tfrac{3}{2} \end{bmatrix} \begin{bmatrix} v_C \\ i_L \end{bmatrix} + \begin{bmatrix} \tfrac{1}{2} \\ \tfrac{3}{4} \end{bmatrix} e \tag{11.2.1}$$

This result suggests the general form

$$\dot{Y} = AY + BX \tag{11.2.2}$$

For an *n*th-order system with *m* excitations, $\dot{Y}$ and Y will be $n \times 1$ column matrices, A an $n \times n$ square matrix, X an $m \times 1$ column matrix, and B an $n \times m$ matrix.

The reader should note that cases (*a*) to (*d*) in Example 11.1 cannot be put in the form of Eq. (11.2.2), in which X is an $m \times 1$ column matrix of excitations. It is necessary that derivatives of the excitations be present. For example, including the voltage excitation in case (*d*) yields

$$\begin{bmatrix} \dot{v}_1 \\ \dot{v}_2 \end{bmatrix} = \begin{bmatrix} -\tfrac{1}{2} & -\tfrac{3}{2} \\ \tfrac{3}{2} & -\tfrac{3}{2} \end{bmatrix} \begin{bmatrix} v_1 \\ v_2 \end{bmatrix} + \begin{bmatrix} \tfrac{1}{2} \\ -\tfrac{3}{2} \end{bmatrix} e + \begin{bmatrix} \tfrac{3}{4} \\ \tfrac{3}{4} \end{bmatrix} \dot{e} \tag{11.2.3}$$

in which both e and $\dot{e}$ are present.

When capacitor voltages and inductor currents are the state variables, it is always possible to write Eq. (11.2.2) (without derivatives of the excitations) simply because it is never necessary to differentiate the loop and node equations in order to assemble the state equations. When loop currents or node voltages are the state variables, it is necessary to have an $\dot{X}$ term as well as an X term in the equations, as Eq. (11.2.3) illustrates. To verify this latter point for the general case, let Y represent a set of capacitor-voltage, inductor-current state variables, and let Z be a set of loop-current or node-voltage variables. Assuming Z to be an acceptable set of state variables, Y and Z will be related by

$$Y = QZ + PX \tag{11.2.4}$$

where Q is an nonsingular $n \times n$ matrix and P is an $n \times m$ matrix. Solving for Z gives

$$Z = Q^{-1}(Y - PX) \tag{11.2.5}$$

which can be differentiated and rearranged to yield

$$Q^{-1}\dot{Y} = \dot{Z} + Q^{-1}P\dot{X} \tag{11.2.6}$$

Now premultiply Eq. (11.2.2) by Q^{-1} to obtain

$$Q^{-1}\dot{Y} = Q^{-1}AY + Q^{-1}BX \tag{11.2.7}$$

Substitution of Eqs. (11.2.4) and (11.2.6) into Eq. (11.2.7) yields, upon rearrangement,

$$\dot{Z} = Q^{-1}AQZ + Q^{-1}(B + AP)X - Q^{-1}P\dot{X} \tag{11.2.8}$$

which can be written in the abbreviated form

$$\dot{Z} = A'Z + B'X + C'\dot{X} \tag{11.2.9}$$

The result in Eq. (11.2.9) shows clearly that the first derivatives of the excitations, as well as the excitations themselves, are required in the state equations when loop currents or node voltages are the state variables.

Example 11.2 For the network in Fig. 11.1, obtain the state equations in Eq. (11.2.3) from the state equations in Eq. (11.2.1).

From Eq. (11.2.1)

$$A = \begin{bmatrix} -\frac{1}{2} & 3 \\ -\frac{3}{4} & -\frac{3}{2} \end{bmatrix} \qquad B = \begin{bmatrix} \frac{1}{2} \\ \frac{3}{4} \end{bmatrix}$$

To relate v_C and i_L to v_1, v_2, and e, we write by inspection of the circuit

$$v_C = v_1 - v_2$$

$$i_L = \tfrac{1}{2}(e - v_1) - \tfrac{1}{6}v_2$$

Thus

$$Q = \begin{bmatrix} 1 & -1 \\ -\frac{1}{2} & -\frac{1}{6} \end{bmatrix} \qquad Q^{-1} = \begin{bmatrix} \frac{1}{4} & -\frac{3}{2} \\ -\frac{3}{4} & -\frac{3}{2} \end{bmatrix} \qquad P = \begin{bmatrix} 0 \\ \frac{1}{2} \end{bmatrix}$$

and

$$A' = Q^{-1}AQ = \begin{bmatrix} -\frac{1}{2} & -\frac{3}{2} \\ \frac{3}{2} & -\frac{3}{2} \end{bmatrix}$$

$$B' = Q^{-1}(B + AP) = \begin{bmatrix} \frac{1}{2} \\ -\frac{3}{2} \end{bmatrix}$$

$$C' = -Q^{-1}P = \begin{bmatrix} \frac{3}{4} \\ \frac{3}{4} \end{bmatrix}$$

Assembling these quantities in Eq. (11.2.9) yields

$$\begin{bmatrix} \dot{v}_1 \\ \dot{v}_2 \end{bmatrix} = \begin{bmatrix} -\frac{1}{2} & -\frac{3}{2} \\ \frac{3}{2} & -\frac{3}{2} \end{bmatrix} \begin{bmatrix} v_1 \\ v_2 \end{bmatrix} + \begin{bmatrix} \frac{1}{2} \\ -\frac{3}{2} \end{bmatrix} e + \begin{bmatrix} \frac{3}{4} \\ \frac{3}{4} \end{bmatrix} \dot{e}$$

which is identical with Eq. (11.2.3).

In the remainder of this chapter, our concern will be with writing and solving state equations when the state variables are an independent set of capacitor voltages and inductor currents. Thus we shall be dealing with Eq. (11.2.2) rather than Eq. (11.2.9).

11.3 WRITING STATE EQUATIONS

In Example 11.1 it was shown how to write state equations for a simple network by juggling appropriate loop and node equations. It is necessary to develop more systematic procedures for the writing of such equations. Three techniques will be developed. In this section we shall examine the capacitor-node–inductor-loop method, which permits state equations to be written almost by inspection for simple networks, and the superposition method, which is applicable to more complicated networks. Then in Sec. 11.4 a completely general topological method will be presented.

In the capacitor-node–inductor-loop method, the procedure is as follows:

1. Write a Kirchhoff's current law (KCL) equation for each independent capacitor at a node it touches in terms of the independent capacitor voltages, independent inductor currents, and their first derivatives.
2. Write a Kirchhoff's voltage law (KVL) equation for each independent inductor around a loop containing it in terms of the independent capacitor voltages, independent inductor currents, and their first derivatives.
3. Solve each of the equations in steps 1 and 2 for the derivative of the voltage or current of the energy-storage element touching the node or contained in the loop.

4. Assemble the resulting equations in the matrix form of Eq. (11.2.2).

We shall illustrate the method and its limitations through two examples.

Example 11.3 Using the capacitor-node–inductor-loop method, write state equations for the network in Fig. 11.2.

At the two capacitor nodes, we write

$$i_s = C_1\dot{v}_1 + Gv_1 + i$$

$$\frac{e_s}{R} = \frac{v_2}{R} + C_2\dot{v}_2 - i$$

and around the inductor loop containing the two capacitors, we have

$$0 = -v_1 + L\dot{i} + v_2$$

Solving for the derivatives and assembling in matrix form gives

$$\begin{bmatrix} \dot{v}_1 \\ \dot{v}_2 \\ \dot{i} \end{bmatrix} = \begin{bmatrix} -\dfrac{G}{C_1} & 0 & -\dfrac{1}{C_1} \\ 0 & -\dfrac{1}{C_2R} & \dfrac{1}{C_2} \\ \dfrac{1}{L} & -\dfrac{1}{L} & 0 \end{bmatrix} \begin{bmatrix} v_1 \\ v_2 \\ i \end{bmatrix} + \begin{bmatrix} 0 & \dfrac{1}{C_1} \\ \dfrac{1}{C_2R} & 0 \\ 0 & 0 \end{bmatrix} \begin{bmatrix} e_s \\ i_s \end{bmatrix}$$

Example 11.4 Repeat Example 11.3 for the network in Fig. 11.3.

At a capacitor node we write

$$C\dot{v} = i_1 + i_2 + \frac{L_1}{R_2}\dot{i}_1$$

and around loops containing the inductors we obtain

$$e_s = R_1C\dot{v} + v + L_1\dot{i}_1$$

$$0 = -L_1\dot{i}_1 + L_2\dot{i}_2 + (i_2 + i_s)R_3$$

The first thing we note is that these equations are not in the convenient form of those in the previous example, with only one derivative in each equation. Considerable algebraic

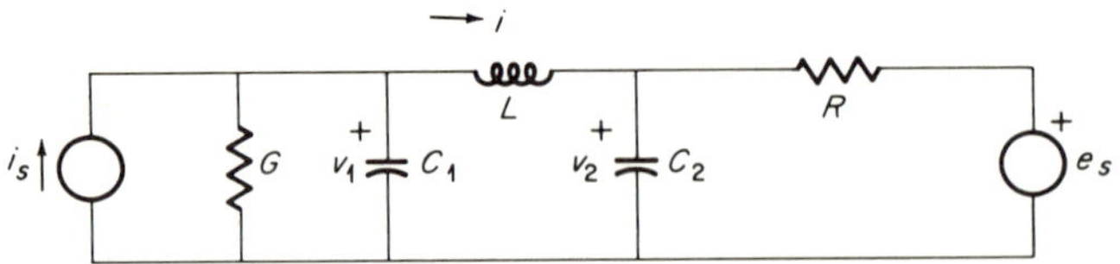

Fig. 11.2

juggling is necessary to reach state-equation form. Performing this juggling, we obtain

$$\begin{bmatrix} \dot{i}_1 \\ \dot{i}_2 \\ \dot{v} \end{bmatrix} = \begin{bmatrix} -\dfrac{R_1R_2}{L_1(R_1+R_2)} & -\dfrac{R_1R_2}{L_1(R_1+R_2)} & -\dfrac{R_2}{L_1(R_1+R_2)} \\ -\dfrac{R_1R_2}{L_1(R_1+R_2)} & -\dfrac{1}{L_2}\left(R_3+\dfrac{R_1R_2}{R_1+R_2}\right) & -\dfrac{R_2}{L_1(R_1+R_2)} \\ \dfrac{R_2}{C_1(R_1+R_2)} & \dfrac{R_2}{C_1(R_1+R_2)} & -\dfrac{1}{C_1(R_1+R_2)} \end{bmatrix} \begin{bmatrix} i_1 \\ i_2 \\ v \end{bmatrix}$$

$$+ \begin{bmatrix} 0 & \dfrac{R_2}{L_1(R_1+R_2)} \\ -\dfrac{R_3}{L_2} & \dfrac{R_2}{L_2(R_1+R_2)} \\ 0 & -\dfrac{R_2}{R_1C_1(R_1+R_2)} \end{bmatrix} \begin{bmatrix} i_s \\ e_s \end{bmatrix}$$

It is easy to see that the network would not have to become much more complicated than that in Fig. 11.3 for the algebra involved in writing the state equations to become prohibitive. Let us turn now to the superposition procedure, by which the state equations for the network in Fig. 11.3 can be written directly. Each capacitor current $C\dot{v}$ and inductor voltage $L\dot{i}$ is considered as the sum of $n + m$ currents and voltages, respectively, where n is the number of independent energy-storage elements and m is the number of excitations. Thus each of these $n + m$ currents and voltages is the result of an excitation, a capacitor voltage, or an inductor current acting individually as a source. When any excitation, capacitor, or inductor is activated, all other voltage sources and capacitors are treated as short circuits and all other current sources and inductors are treated as open circuits.

Example 11.5 Use the superposition procedure to write the state equations in Example 11.4.

We note that the first state equation, if multiplied through by L_1, expresses the voltage across L_1 as the sum of five voltages, three due to the state variables and two due to the excitations. Consider the contribution to $L\dot{i}_1$ due to i_1. With i_1 activated, we short e_s and C and open i_s and L_2, to obtain the network in Fig. 11.4*a*. The voltage across L_1 (across i_1) is $-[R_1R_2/(R_1 + R_2)]i_1$. Dividing by L_1 gives the first term in the state equation for $\dot{i}_1$.

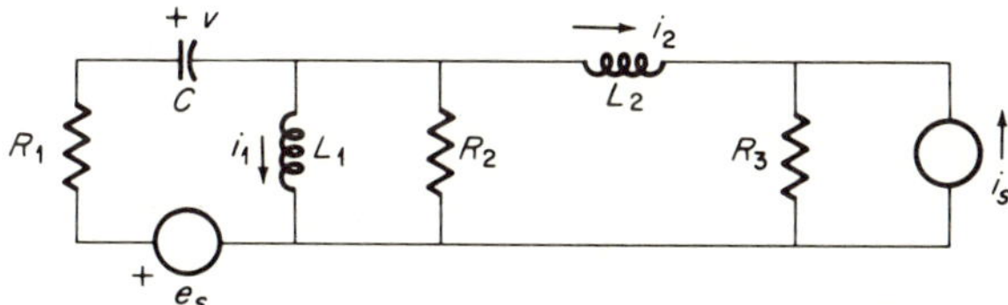

Fig. 11.3

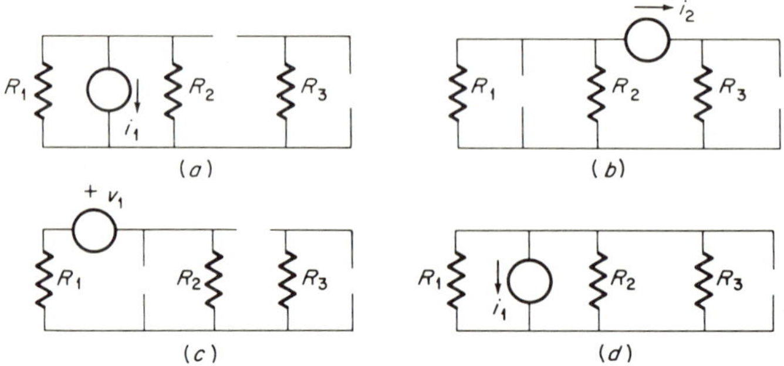

Fig. 11.4

To find the contribution of i_2 to $L\dot{i}_1$, the circuit in Fig. 11.4*b* is used. Note that e_s and C are shorted, i_s and L_1 opened. The voltage across L_1 is $-[R_1R_2/(R_1 + R_2)]i_2$. Dividing by L_1 gives the second term in the state equation for $\dot{i}_1$.

The contribution of v to $L\dot{i}_1$ is obtained from the circuit in Fig. 11.4*c*, in which e_s is shorted and i_s, L_1, and L_2 are opened. The voltage across L_1 is $-[R_2/(R_1 + R_2)]v$, which, when divided by L_1, gives the third term in the $\dot{i}_1$ state equation.

The other 12 terms in the state equations are obtained in a similar fashion. For example, the first term in the third state equation is obtained from the circuit in Fig. 11.4*d* as the current through the shorted capacitor.

The two procedures for writing state equations developed in this section are perfectly general and will handle the majority of networks the reader is likely to encounter. However, as network complexity increases, even the superposition procedure can become tedious to apply, especially if the network has a large number of resistors. The topological procedure developed in the next section is the cannon to be used when the pistol and the machine gun will not suffice, i.e., for unusually complex network structures and for networks containing L or C tie-set and cut-set constraints.

11.4 BASHKOW'S† FORMULATION OF STATE EQUATIONS

Assume initially that the network under consideration, although it may be complicated, contains no inductive or capacitive cut sets or tie sets. We shall discuss the reasons for this assumption and how it can be dispensed with after we have developed the procedure.

Define a *proper tree* of a network as one whose branches contain every capacitive element in the network, possibly some resistive elements, but no

† See Ref. 11.1.

inductive elements. If the network contains no capacitive tie sets or inductive cut sets, a proper tree can always be drawn for it; for if there are no capacitive tie sets, all capacitive elements can be included in a tree, and if there are no inductive cut sets, no inductive elements need appear in a tree.

To explain the method, consider the network in Fig. 11.5*a*. From previous symbology, note that $B = 7$, $N = 4$, and $L = 3$. The first step is to try to draw a proper tree. Since there are no capacitive tie sets or inductive cut sets present, such a tree can be formed, as in Fig. 11.5*b*. Next, after assigning branch arrows, the branches are systematically numbered so that the capacitive tree branches are 1, 2, and 3, the conductive tree branch 4, the inductive links 5 and 6, and the resistive link 7. Following this, the fundamental-tie-set and cut-set matrices for the network are formed. They are

$$\beta_F = \begin{matrix} 5 \\ 6 \\ 7 \end{matrix}\begin{bmatrix} 1 & 1 & 0 & 0 & 1 & 0 & 0 \\ 0 & 0 & 1 & 1 & 0 & 1 & 0 \\ 0 & 1 & 1 & 0 & 0 & 0 & 1 \end{bmatrix}$$

$$\alpha_F = \begin{matrix} 1 \\ 2 \\ 3 \\ 4 \end{matrix}\begin{bmatrix} 1 & 0 & 0 & 0 & -1 & 0 & 0 \\ 0 & 1 & 0 & 0 & -1 & 0 & -1 \\ 0 & 0 & 1 & 0 & 0 & -1 & -1 \\ 0 & 0 & 0 & 1 & 0 & -1 & 0 \end{bmatrix} \tag{11.4.1}$$

Note that because of the numbering scheme, β_F and α_F are in the forms

$$\beta_F = [\beta_{LN} \mid U_{LL}] \qquad \alpha_F = [U_{NN} \mid \alpha_{NL}] \tag{11.4.2}$$

where the subscripts, as previously, denote the orders of the submatrices and U_{LL} and U_{NN} are identity matrices.

Now the objective is to write state equations in the form of Eq. (11.2.2), using the three capacitor voltages and the two inductor currents as the state variables. To do this, use is made of Eqs. (4.7.2) and (4.8.2), here repeated

$$J_B = \beta_F{}^T I_L \qquad V_B = \alpha_F{}^T E_N \tag{11.4.3}$$

Recall that J_B and V_B are $B \times 1$ column matrices of branch currents and

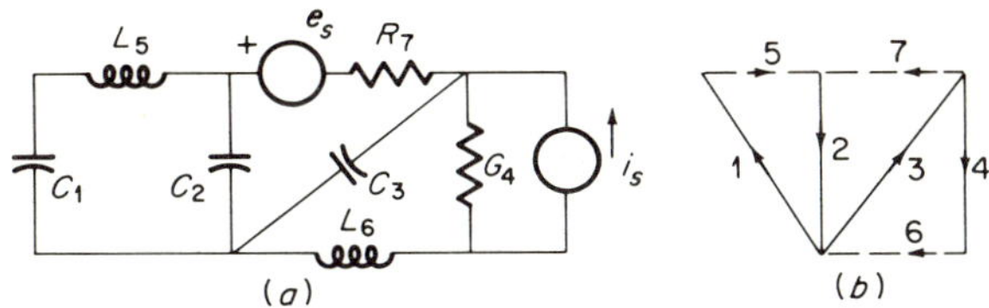

Fig. 11.5 Network and graph for Bushkow's formulation.

voltages, respectively, I_L is an $L \times 1$ current-variable matrix, and E_N is an $N \times 1$ voltage-variable matrix. Now using Eqs. (11.4.2), we partition Eqs. (11.4.3) as follows:

$$\begin{bmatrix} J_N \\ \hdashline J_L \end{bmatrix} = \begin{bmatrix} \beta_{LN}{}^T \\ \hdashline U_{LL} \end{bmatrix} I_L \qquad \begin{bmatrix} V_N \\ \hdashline V_L \end{bmatrix} = \begin{bmatrix} U_{NN} \\ \hdashline \alpha_{NL}{}^T \end{bmatrix} E_N \tag{11.4.4}$$

The lower portion of the first of Eqs. (11.4.4) and the upper portion of the second of Eqs. (11.4.4) are rejected (they are merely identities), leaving

$$J_N = \beta_{LN}{}^T I_L \qquad V_L = \alpha_{NL}{}^T E_N \tag{11.4.5}$$

For the network in Fig. 11.5*a* and the β_F and α_F matrices in Eqs. (11.4.1), Eqs. (11.4.5) take the form

$$\begin{bmatrix} j_1 \\ j_2 \\ j_3 \\ j_4 \end{bmatrix} = \begin{bmatrix} 1 & 0 & 0 \\ 1 & 0 & 1 \\ 0 & 1 & 1 \\ 0 & 1 & 0 \end{bmatrix} \begin{bmatrix} j_5 \\ j_6 \\ j_7 \end{bmatrix} \qquad \begin{bmatrix} v_5 \\ v_6 \\ v_7 \end{bmatrix} = \begin{bmatrix} -1 & -1 & 0 & 0 \\ 0 & 0 & -1 & -1 \\ 0 & -1 & -1 & 0 \end{bmatrix} \begin{bmatrix} v_1 \\ v_2 \\ v_3 \\ v_4 \end{bmatrix} \tag{11.4.6}$$

The next step is to combine the two equations (11.4.6) into a single equation, in which the variables are to be the capacitor voltages and inductor currents [the independent variables in Eqs. (11.4.6)] and their derivatives. To this end, an examination of the network leads to the branch relations

$$\begin{bmatrix} j_1 \\ j_2 \\ j_3 \\ j_4 \end{bmatrix} = \begin{bmatrix} C_1\dot{v}_1 \\ C_2\dot{v}_2 \\ C_3\dot{v}_3 \\ G_4v_4 - i_s \end{bmatrix} \qquad \begin{bmatrix} v_5 \\ v_6 \\ v_7 \end{bmatrix} = \begin{bmatrix} L_5\dot{j}_5 \\ L_6\dot{j}_6 \\ R_7j_7 - e_s \end{bmatrix} \tag{11.4.7}$$

and substitution of these relations into Eqs. (11.4.6) gives the combined form

$$\begin{bmatrix} C_1\dot{v}_1 \\ C_2\dot{v}_2 \\ C_3\dot{v}_3 \\ G_4v_4 - i_s \\ L_5\dot{j}_5 \\ L_6\dot{j}_6 \\ R_7j_7 - e_s \end{bmatrix} = \begin{bmatrix} & & & & 1 & 0 & 0 \\ & \mathbf{0} & & & 1 & 0 & 1 \\ & & & & 0 & 1 & 1 \\ & & & & 0 & 1 & 0 \\ -1 & -1 & 0 & 0 & & & \\ 0 & 0 & -1 & -1 & & \mathbf{0} & \\ 0 & -1 & -1 & 0 & & & \end{bmatrix} \begin{bmatrix} v_1 \\ v_2 \\ v_3 \\ v_4 \\ j_5 \\ j_6 \\ j_7 \end{bmatrix} \tag{11.4.8}$$

The square matrix in Eq. (11.4.8) has the general partitioned form

$$\begin{bmatrix} \mathbf{0}_{NN} & \beta_{LN}{}^T \\ \alpha_{NL}{}^T & \mathbf{0}_{LL} \end{bmatrix} = \begin{bmatrix} \mathbf{0}_{NN} & -\alpha_{NL} \\ \alpha_{NL}{}^T & \mathbf{0}_{LL} \end{bmatrix} \tag{11.4.9}$$

where the second result arises from the relationship in Eq. (4.12.11).

In Eq. (11.4.8), we have almost achieved the objective of writing state equations of the form of Eq. (11.2.2). It remains to eliminate those equations and variables pertaining to conductive and resistive branches, i.e., the non-state equations and variables. To do this, we remove the fourth and seventh equations from Eq. (11.4.8), then use them to obtain expressions for v_4 and j_7, respectively, in terms of the other variables, replace v_4 and j_7 by these expressions in the remaining five equations in Eq. (11.4.8), and combine terms to get a new square matrix of order 2 less than before.

To carry out the mechanics of this reduction, we obtain from the fourth and seventh equations in Eq. (11.4.8)

$$v_4 = \frac{1}{G_4} j_6 + \frac{1}{G_4} i_s \qquad j_7 = -\frac{1}{R_7} v_2 - \frac{1}{R_7} v_3 + \frac{1}{R_7} e_s \tag{11.4.10}$$

and incorporate these expressions into the remaining five equations in Eq. (11.4.8) to obtain

$$\begin{bmatrix} C_1\dot{v}_1 \\ C_2\dot{v}_2 \\ C_3\dot{v}_3 \\ L_5\dot{j}_5 \\ L_6\dot{j}_6 \end{bmatrix} = \begin{bmatrix} 0 & 0 & 0 & 1 & 0 \\ 0 & -\frac{1}{R_7} & -\frac{1}{R_7} & 1 & 0 \\ 0 & -\frac{1}{R_7} & -\frac{1}{R_7} & 0 & 1 \\ -1 & -1 & 0 & 0 & 0 \\ 0 & 0 & -1 & 0 & -\frac{1}{G_4} \end{bmatrix} \begin{bmatrix} v_1 \\ v_2 \\ v_3 \\ j_5 \\ j_6 \end{bmatrix} + \begin{bmatrix} 0 & 0 \\ \frac{1}{R_7} & 0 \\ \frac{1}{R_7} & 0 \\ 0 & 0 \\ 0 & -\frac{1}{G_4} \end{bmatrix} \begin{bmatrix} e_s \\ i_s \end{bmatrix} \tag{11.4.11}$$

Now, clearing the left column matrix of its coefficients yields the state equations in the desired form

$$\begin{bmatrix} \dot{v}_1 \\ \dot{v}_2 \\ \dot{v}_3 \\ \dot{j}_5 \\ \dot{j}_6 \end{bmatrix} = \begin{bmatrix} 0 & 0 & 0 & \frac{1}{C_1} & 0 \\ 0 & -\frac{1}{C_2R_7} & -\frac{1}{C_2R_7} & \frac{1}{C_2} & 0 \\ 0 & -\frac{1}{C_3R_7} & -\frac{1}{C_3R_7} & 0 & \frac{1}{C_3} \\ -\frac{1}{L_5} & -\frac{1}{L_5} & 0 & 0 & 0 \\ 0 & 0 & -\frac{1}{L_6} & 0 & -\frac{1}{L_6G_4} \end{bmatrix} \begin{bmatrix} v_1 \\ v_2 \\ v_3 \\ j_5 \\ j_6 \end{bmatrix} + \begin{bmatrix} 0 & 0 \\ \frac{1}{C_2R_7} & 0 \\ \frac{1}{C_3R_7} & 0 \\ 0 & 0 \\ 0 & -\frac{1}{G_4L_6} \end{bmatrix} \begin{bmatrix} e_s \\ i_s \end{bmatrix} \tag{11.4.12}$$

The reader will note that the sources in Fig. 11.5*a* are associated with resistive branches. If they should be associated with inductive or capacitive branches, no change in the procedure outlined is necessary, but it must be remembered that topological formulations are always made in terms of branch voltages and currents, not voltages across or currents through branch elements. Yet the state equations are always written in terms of element voltages and currents. For example, if the voltage source e_s and the current source i_s in Fig. 11.5*a* were in series with branch 1 and across branch 6 respectively, we would have in Eq. (11.4.7)

$$j_1 = C_1\dot{v}_{C_1} = C_1(\dot{v}_1 + \dot{e}_s)$$

and

$$v_6 = L_6\dot{i}_{L_6} = L_6(\dot{j}_6 + \dot{i}_s) \tag{11.4.13}$$

The state equations would be written in terms of $\dot{v}_{C_1}$, v_{C_1}, $\dot{i}_{L_6}$, and i_{L_6}, rather than in terms of $\dot{v}_1$, v_1, $\dot{j}_6$, and j_6.

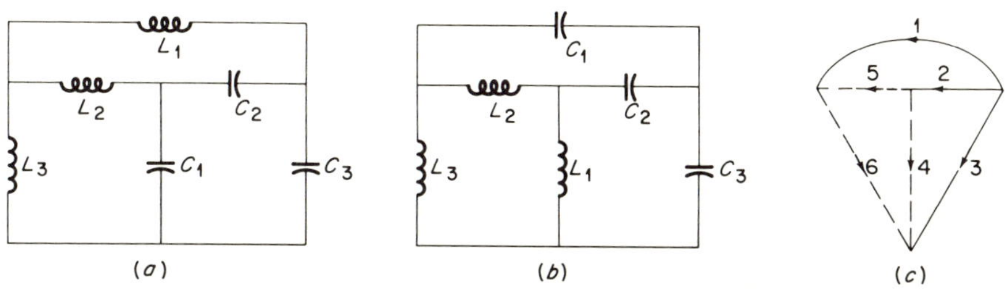

Fig. 11.6 Networks to illustrate the effects of dynamic constraints on state equations.

At the outset of the discussion of the Bashkow formulation, it was assumed that the network did not contain capacitive or inductive cut sets or tie sets. We now ask why it was necessary to make this assumption and whether a way can be devised of dispensing with it. These questions will be answered by considering two specific networks.

The network in Fig. 11.6*a* contains an inductive cut set and a capacitive tie set and that in Fig. 11.6*b* a capacitive cut set and an inductive tie set. To simplify the discussion, neither network contains resistances or sources; this does not alter the procedures to be developed.

From the previous discussion it is noted that a proper tree cannot be drawn for the network in Fig. 11.6*a*. Suppose we persist, however, and try to write equations in the form of Eq. (11.2.2), using as close to a proper tree as possible, as shown in Fig. 11.6*c*, namely, a tree with the maximum number of capacitive branches and the minimum number of inductive branches.

For this tree, the fundamental-tie-set and cut-set matrices are

$$\beta_F = \begin{bmatrix} 0 & 1 & -1 & 1 & 0 & 0 \\ -1 & 1 & 0 & 0 & 1 & 0 \\ 1 & 0 & -1 & 0 & 0 & 1 \end{bmatrix}$$

$$\alpha_F = \begin{bmatrix} 1 & 0 & 0 & 0 & 1 & -1 \\ 0 & 1 & 0 & -1 & -1 & 0 \\ 0 & 0 & 1 & 1 & 0 & 1 \end{bmatrix} \tag{11.4.14}$$

Writing Eqs. (11.4.5) yields

$$\begin{bmatrix} j_1 \\ j_2 \\ j_3 \end{bmatrix} = \begin{bmatrix} 0 & -1 & 1 \\ 1 & 1 & 0 \\ -1 & 0 & -1 \end{bmatrix} \begin{bmatrix} j_4 \\ j_5 \\ j_6 \end{bmatrix} \qquad \begin{bmatrix} v_4 \\ v_5 \\ v_6 \end{bmatrix} = \begin{bmatrix} 0 & -1 & 1 \\ 1 & -1 & 0 \\ -1 & 0 & 1 \end{bmatrix} \begin{bmatrix} v_1 \\ v_2 \\ v_3 \end{bmatrix} \tag{11.4.15}$$

Now we must write equations similar to Eqs. (11.4.7) and (11.4.8). For the network being considered, Eqs. (11.4.7) take the form

$$\begin{bmatrix} j_1 \\ j_2 \\ j_3 \end{bmatrix} = \begin{bmatrix} \dfrac{1}{L_1}\displaystyle\int v_1\,dt \\ c_2\dot{v}_2 \\ c_3\dot{v}_3 \end{bmatrix} \qquad \begin{bmatrix} v_4 \\ v_5 \\ v_6 \end{bmatrix} = \begin{bmatrix} \dfrac{1}{C_1}\displaystyle\int j_4\,dt \\ L_2\dot{j}_5 \\ L_3\dot{j}_6 \end{bmatrix} \tag{11.4.16}$$

It is here that trouble develops. For if we use the relations in Eqs. (11.4.16) to try to write an equation similar to Eq. (11.4.8), we end up with integrals, as well as derivatives, on the left side and obtain

$$\begin{bmatrix} \dfrac{1}{L_1}\displaystyle\int v_1\,dt \\ C_2\dot{v}_2 \\ C_3\dot{v}_3 \\ \dfrac{1}{C_1}\displaystyle\int j_4\,dt \\ L_2\dot{j}_5 \\ L_3\dot{j}_6 \end{bmatrix} = \begin{bmatrix} & \mathbf{0} & & 0 & -1 & 1 \\ & & & 1 & 1 & 0 \\ & & & -1 & 0 & -1 \\ 0 & -1 & 1 & & & \\ 1 & -1 & 0 & & \mathbf{0} & \\ -1 & 0 & 1 & & & \end{bmatrix} \begin{bmatrix} v_1 \\ v_2 \\ v_3 \\ j_4 \\ j_5 \\ j_6 \end{bmatrix} \tag{11.4.17}$$

From the discussion in Sec. 8.10, $n = 4$ for the network in Fig. 11.6*a*, there being six energy-storage elements and two dynamic constraints. Thus there must be four, not six, state equations. Two equations in Eq. (11.4.17) must be deleted. Actually we can delete any one of the first three equations and any one of the last three equations, but to arrive at the form of Eq. (11.2.2), we should delete the first and the fourth equations. A way also must be found to eliminate the variables v_1 and j_4 associated with these two equations from the remaining four equations. There is a method, adapted from Bryant,† that permits this to be done. Consider the elimination of j_4 from the second and third of Eqs. (11.4.17). Separating out these two equations and isolating the j_4 portions gives

$$\begin{bmatrix} C_2\dot{v}_2 \\ C_3\dot{v}_3 \end{bmatrix} = \begin{bmatrix} 1 \\ -1 \end{bmatrix} j_4 + \begin{bmatrix} 1 & 0 \\ 0 & -1 \end{bmatrix} \begin{bmatrix} j_5 \\ j_6 \end{bmatrix} \tag{11.4.18}$$

Now to eliminate j_4 from Eq. (11.4.18), we must express this current either in terms of $\dot{v}_2$ and $\dot{v}_3$ or in terms of j_5 and j_6. An inspection of the graph in Fig. 11.6*c* shows that the latter approach cannot be pursued, since the three

† P. R. Bryant, The Explicit Form of Bashkow's *A* Matrix, *IRE Professional Group on Circuit Theory*, Sept., 1962.

currents under consideration do not form a node. Pursuing the former approach, we note from the network that

$$j_4 = C_1\dot{v}_4 \tag{11.4.19}$$

and from the second of Eqs. (11.4.15) that

$$v_4 = [0 \quad -1 \quad 1]\begin{bmatrix} v_1 \\ v_2 \\ v_3 \end{bmatrix} = -v_2 + v_3 \tag{11.4.20}$$

Combining Eqs. (11.4.19) and (11.4.20) gives the desired relation

$$j_4 = [-C_1 \quad C_1]\begin{bmatrix} \dot{v}_2 \\ \dot{v}_3 \end{bmatrix} \tag{11.4.21}$$

Equation (11.4.20) is the key to the success of the procedure. The important thing to note in that equation is the zero in the row matrix. This zero means that v_4, the voltage across the link capacitor, is not a function of v_1, the voltage across the tree-branch inductor. This is a general result (in the general case a square matrix of zeros, instead of the single zero, would be present). With a tree containing the maximum number of capacitive branches and the minimum number of inductive branches, it is impossible for any capacitive-link voltage to depend on any inductive tree-branch voltage. This is true because each capacitive link forms a loop with one or more capacitive tree branches, and thus its voltage depends only on the voltages of those tree branches.

It remains now to substitute Eq. (11.4.21) into Eq. (11.4.18) and solve for $\dot{v}_2$ and $\dot{v}_3$ in terms of j_5 and j_6. The substitution yields

$$\begin{bmatrix} C_2\dot{v}_2 \\ C_3\dot{v}_3 \end{bmatrix} - \begin{bmatrix} 1 \\ -1 \end{bmatrix}[-C_1 \quad C_1]\begin{bmatrix} \dot{v}_2 \\ \dot{v}_3 \end{bmatrix} = \begin{bmatrix} 1 & 0 \\ 0 & -1 \end{bmatrix}\begin{bmatrix} j_5 \\ j_6 \end{bmatrix} \tag{11.4.22}$$

Expanding gives

$$\begin{bmatrix} C_1 + C_2 & -C_1 \\ -C_1 & C_1 + C_3 \end{bmatrix}\begin{bmatrix} \dot{v}_2 \\ \dot{v}_3 \end{bmatrix} = \begin{bmatrix} j_5 \\ -j_6 \end{bmatrix} \tag{11.4.23}$$

which, when solved, yields

$$\begin{bmatrix} \dot{v}_2 \\ \dot{v}_3 \end{bmatrix} = \begin{bmatrix} \dfrac{C_1 + C_3}{C_p} & -\dfrac{C_1}{C_p} \\ \dfrac{C_1}{C_p} & -\dfrac{C_1 + C_2}{C_p} \end{bmatrix}\begin{bmatrix} j_5 \\ j_6 \end{bmatrix} \tag{11.4.24}$$

where

$$C_p = C_1C_2 + C_2C_3 + C_1C_3 \tag{11.4.25}$$

A dual procedure is used to eliminate v_1 from the fifth and sixth equations in Eqs. (11.4.17). Separating out these two equations and isolating the v_1 portions gives

$$\begin{bmatrix} L_2\dot{j}_5 \\ L_3\dot{j}_6 \end{bmatrix} = \begin{bmatrix} 1 \\ -1 \end{bmatrix} v_1 + \begin{bmatrix} -1 & 0 \\ 0 & 1 \end{bmatrix} \begin{bmatrix} v_2 \\ v_3 \end{bmatrix} \tag{11.4.26}$$

Here the choices are to express v_1 either in terms of $\dot{j}_5$ and $\dot{j}_6$ or v_2 and v_3. The latter approach fails because v_1, v_2, and v_3 do not form a closed loop. The former approach gives

$$v_1 = L_1\dot{j}_1 \tag{11.4.27}$$

and, from the first Eq. (11.4.5),

$$j_1 = [0 \quad -1 \quad 1] \begin{bmatrix} j_4 \\ j_5 \\ j_6 \end{bmatrix} = -j_5 + j_6 \tag{11.4.28}$$

Combining Eqs. (11.4.27) and (11.4.28) yields

$$v_1 = [-L_1 \quad L_1] \begin{bmatrix} \dot{j}_5 \\ \dot{j}_6 \end{bmatrix} \tag{11.4.29}$$

As in the previous discussion, the zero (in general a square matrix of zeros) in Eq. (11.4.28) is significant and is a general result. It implies that no inductive tree-branch current can depend on a capacitive-link current because each capacitive-link current contributes only to capacitive tree-branch currents.

Inserting Eq. (11.4.29) into Eq. (11.4.26), collecting terms, and solving for $\dot{j}_5$ and $\dot{j}_6$, we obtain

$$\begin{bmatrix} \dot{j}_5 \\ \dot{j}_6 \end{bmatrix} = \begin{bmatrix} -\dfrac{L_1 + L_3}{L_p} & \dfrac{L_1}{L_p} \\ -\dfrac{L_1}{L_p} & \dfrac{L_1 + L_2}{L_p} \end{bmatrix} \begin{bmatrix} v_2 \\ v_3 \end{bmatrix} \tag{11.4.30}$$

where

$$L_p = L_1L_2 + L_2L_3 + L_1L_3 \tag{11.4.31}$$

Assembling Eqs. (11.4.24) and (11.4.30) into a single equation yields

$$\begin{bmatrix} \dot{v}_2 \\ \dot{v}_3 \\ \dot{j}_5 \\ \dot{j}_6 \end{bmatrix} \begin{bmatrix} & \mathbf{0} & \frac{C_1 + C_3}{C_p} & -\frac{C_1}{C_p} \\ & & \frac{C_1}{C_p} & -\frac{C_1 + C_2}{C_p} \\ -\frac{L_1 + L_3}{L_p} & \frac{L_1}{L_p} & & \mathbf{0} \\ -\frac{L_1}{L_p} & \frac{L_1 + L_2}{L_p} & & \end{bmatrix} \begin{bmatrix} v_2 \\ v_3 \\ j_5 \\ j_6 \end{bmatrix} \tag{11.4.32}$$

There are no resistive or conductive equations or variables to eliminate, as there were in Eq. (11.4.8), and so we are through. The four equations in Eq. (11.4.32) are a set of state equations for the network in Fig. 11.6*a*.

Consider next the network of Fig. 11.6*b* to see whether the same problems are present that proved bothersome in connection with the network of Fig. 11.6*a*. It *is* possible to draw a proper tree for this network; the tree in Fig. 11.6*c* is such a tree. Using this tree, we have the same fundamental-tie-set and cut-set matrices as before and the same branch-current and branch-voltage equations.

Equations similar to Eqs. (11.4.7) can be written for this network, with no integral terms present, unlike Eqs. (11.4.16). They are

$$\begin{bmatrix} j_1 \\ j_2 \\ j_3 \end{bmatrix} = \begin{bmatrix} C_1\dot{v}_1 \\ C_2\dot{v}_2 \\ C_3\dot{v}_3 \end{bmatrix} \qquad \begin{bmatrix} v_4 \\ v_5 \\ v_6 \end{bmatrix} = \begin{bmatrix} L_1\dot{j}_4 \\ L_2\dot{j}_5 \\ L_3\dot{j}_6 \end{bmatrix} \tag{11.4.33}$$

Proceeding directly, as for the network in Fig. 11.5, an equation similar to Eq. (11.4.8) can be written

$$\begin{bmatrix} C_1\dot{v}_1 \\ C_2\dot{v}_2 \\ C_3\dot{v}_3 \\ L_1\dot{j}_4 \\ L_2\dot{j}_5 \\ L_3\dot{j}_6 \end{bmatrix} \begin{bmatrix} & & & 0 & -1 & 1 \\ & \mathbf{0} & & 1 & 1 & 0 \\ & & & -1 & 0 & -1 \\ 0 & -1 & 1 & & & \\ 1 & -1 & 0 & & \mathbf{0} & \\ -1 & 0 & 1 & & & \end{bmatrix} \begin{bmatrix} v_1 \\ v_2 \\ v_3 \\ j_4 \\ j_5 \\ j_6 \end{bmatrix} \tag{11.4.34}$$

There are no resistive or conductive equations or variables to eliminate, but since $n = 4$ for the network in Fig. 11.6*b*, we must dispose of two constraint equations and the variables associated therewith. Any one of the voltage and any one of the current equations can be removed. Choosing the first and the fourth equations requires that the variables v_1 and j_4 be eliminated from the remaining four equations.

Consider first the elimination of j_4 from the second and third equations in Eq. (11.4.34). We can write

$$\begin{bmatrix} C_2\dot{v}_2 \\ C_3\dot{v}_3 \end{bmatrix} = \begin{bmatrix} 1 \\ -1 \end{bmatrix} j_4 + \begin{bmatrix} 1 & 0 \\ 0 & -1 \end{bmatrix} \begin{bmatrix} j_5 \\ j_6 \end{bmatrix} \tag{11.4.35}$$

which happens to be the same as Eq. (11.4.18) because the same tree is used in each network. Again it is necessary to express j_4 either in terms of $\dot{v}_2$ and $\dot{v}_3$ or j_5 and j_6. Before, the former worked and the latter did not. Here the reverse is true. A voltage equation around the inductive tie-set loop yields

$$0 = L_1\dot{j}_4 + L_2\dot{j}_5 + L_3\dot{j}_6 \tag{11.4.36}$$

Integrating once and solving for j_4 gives

$$j_4 = \frac{L_2}{L_1}j_5 + \frac{L_3}{L_1}j_6 \tag{11.4.37}$$

Inserting Eq. (11.4.37) into Eq. (11.4.35) yields

$$\begin{bmatrix} \dot{v}_2 \\ \dot{v}_3 \end{bmatrix} = \begin{bmatrix} \dfrac{L_1 + L_2}{L_1C_2} & \dfrac{L_3}{L_1C_2} \\ -\dfrac{L_2}{L_1C_3} & -\dfrac{L_1 + L_3}{L_1C_3} \end{bmatrix} \begin{bmatrix} j_5 \\ j_6 \end{bmatrix} \tag{11.4.38}$$

Following a similar procedure with the fifth and sixth of Eqs. (11.4.33) and noting that v_1 can be expressed in terms of v_2 and v_3 by writing a current equation at the capacitive cut-set node and then integrating, we arrive at the result

$$\begin{bmatrix} \dot{j}_5 \\ \dot{j}_6 \end{bmatrix} \begin{bmatrix} -\dfrac{C_1 + C_2}{C_1L_2} & -\dfrac{C_3}{C_1L_2} \\ \dfrac{C_2}{C_1L_3} & \dfrac{C_1 + C_3}{C_1L_3} \end{bmatrix} \begin{bmatrix} v_2 \\ v_3 \end{bmatrix} \tag{11.4.39}$$

Assembling Eqs. (11.4.38) and (11.4.39) into one equation gives a set of four state equations for the network in Fig. 11.6*b*

$$\begin{bmatrix} \dot{v}_2 \\ \dot{v}_3 \\ \dot{j}_5 \\ \dot{j}_6 \end{bmatrix} = \begin{bmatrix} \mathbf{0} & & \dfrac{L_1 + L_2}{L_1 C_2} & \dfrac{L_3}{L_1 C_2} \\ & & -\dfrac{L_2}{L_1 C_3} & -\dfrac{L_1 + L_3}{L_1 C_3} \\ -\dfrac{C_1 + C_2}{C_1 L_2} & -\dfrac{C_3}{C_1 L_2} & & \\ \dfrac{C_2}{C_1 L_3} & \dfrac{C_1 + C_3}{C_1 L_3} & \mathbf{0} & \end{bmatrix} \begin{bmatrix} v_2 \\ v_3 \\ j_5 \\ j_6 \end{bmatrix} \tag{11.4.40}$$

The reader should not lose sight of the differences in the procedure for writing state equations when inductive cut sets and capacitive tie sets are present (Fig. 11.6*a*) as opposed to when inductive tie sets and capacitive cut sets are present (Fig. 11.6*b*). In the former case a proper tree cannot be drawn; in the latter case one can. In the former case the redundant equations and variables are eliminated before equations of the form of Eqs. (11.2.2) are written; in the latter case they are eliminated after. In the former case, elimination of the unwanted current (voltage) variables is accomplished by expressing them in terms of the derivatives of the voltage (current) state variables. In the latter case, the current (voltage) variables to be deleted are expressed in terms of the current (voltage) state variables themselves, not their derivatives.

11.5 SOLUTION OF STATE EQUATIONS: *s* DOMAIN

The remainder of the chapter is devoted to obtaining solutions of state equations. In this section, the Laplace transform, partial-fraction-expansion solution technique presented in Chap. 10 will be extended to handle the matrix state equations. In later sections, a time-domain solution technique involving the use of convolution and *e* with a matrix exponent will be developed. Each of these techniques will present the total response of a linear system as the sum of the free (zero-input) and forced (zero-state) responses.

We begin by Laplace transforming the state equations (11.2.2)

$$sY(s) - Y_+ = AY(s) + BX(s) \tag{11.5.1}$$

Equation (11.5.1) is a matrix equation in which $Y(s)$ is an $n \times 1$ column matrix of transformed responses, Y_+ is an $n \times 1$ column matrix of initial conditions, and $X(s)$ is an $n \times 1$ column matrix of transformed excitations.

Collecting the $Y(s)$ terms in Eq. (11.5.1) yields

$$(sU - A)Y(s) = Y_+ + BX(s) \tag{11.5.2}$$

where U is an $n \times n$ identity matrix.

The characteristic equation of the system is derived by setting the right side of Eq. (11.5.2) equal to zero. Then, for a nonzero homogeneous solution to exist, it is necessary and sufficient that

$$|sU - A| = 0 \tag{11.5.3}$$

Equation (11.5.3) is the characteristic equation, as previously derived by assuming a homogeneous solution of the form $Y = Ce^{st}$ in Sec. 8.4. The n roots of Eq. (11.5.3) are the natural frequencies of the system.

To obtain the total solution for $Y(s)$ in Eq. (11.5.2), we write

$$Y(s) = (sU - A)^{-1}Y_+ + (sU - A)^{-1}BX(s) \tag{11.5.4}$$

The denominator of the Y_+ term in Eq. (11.5.4) is $|sU - A|$, the nth-degree characteristic polynomial of the system. The denominator of the $X(s)$ term consists of this same polynomial multiplied by the denominators of the excitation transforms [after each term in $X(s)$ is placed over a common denominator]. With these points in mind, a partial-fraction expansion of Eq. (11.5.4) takes the form

$$Y(s) = \frac{K_1}{s - s_1} + \cdots + \frac{K_n}{s - s_n} + \frac{K_a}{s - s_a} + \cdots + \frac{K_m}{s - s_m} \tag{11.5.5}$$

where s_1 through s_n are the n natural frequencies of the system and s_a through s_m are the poles of the transforms of the m excitations (it being assumed for ease of solution that all poles are simple and that each excitation has one pole). The K's, each an $n \times 1$ column matrix, are found in exactly the same way as in Sec. 10.3. K_1 through K_n are matrices of constants, whereas K_a through K_m contain constants and the initial-condition terms in Y_+.

The time-domain response is

$$Y(t) = K_1e^{s_1t} + \cdots + K_ne^{s_nt} + K_ae^{s_at} + \cdots + K_me^{s_mt} \tag{11.5.6}$$

The procedure will be illustrated through an example.

Example 11.6 For the given circuit (Fig. 11.7), initial conditions, and excitations, write state equations in the variables v and i; solve by the Laplace transform method.

Writing a KCL equation at a capacitive node and a KVL equation around an inductive loop, we have

$$i_s = \dot{v} + 2v + i$$

$$e_s = \tfrac{1}{2}\dot{i} + \tfrac{5}{2}i - v$$

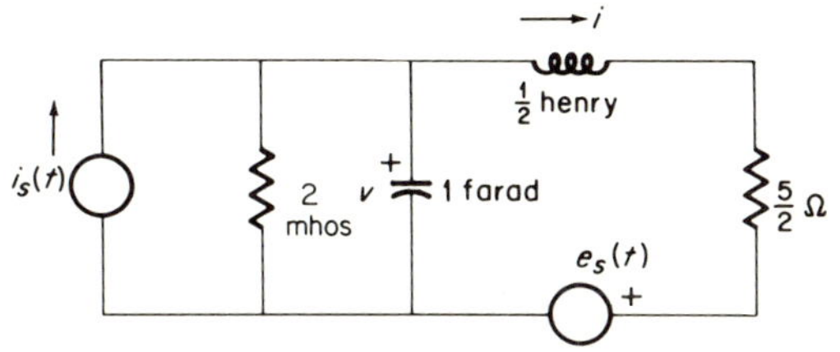

Fig. 11.7 $i_s(t) = u(t)$ amp, $e_s(t) = e^{-t}$ volts, $v_+ = 3$ volts, and $i_+ = 2$ amp.

Solving for $\dot{v}$ and $\dot{i}$ yields the state equations

$$\begin{bmatrix} \dot{v} \\ \dot{i} \end{bmatrix} = \begin{bmatrix} -2 & -1 \\ 2 & -5 \end{bmatrix} \begin{bmatrix} v \\ i \end{bmatrix} + \begin{bmatrix} 1 & 0 \\ 0 & -2 \end{bmatrix} \begin{bmatrix} i_s \\ e_s \end{bmatrix}$$

The Laplace transformation of these equations is

$$s \begin{bmatrix} V(s) \\ I(s) \end{bmatrix} - \begin{bmatrix} v_+ \\ i_+ \end{bmatrix} = \begin{bmatrix} -2 & -1 \\ 2 & -5 \end{bmatrix} \begin{bmatrix} V(s) \\ I(s) \end{bmatrix} + \begin{bmatrix} 1 & 0 \\ 0 & -2 \end{bmatrix} \begin{bmatrix} \dfrac{1}{s} \\ \dfrac{1}{s+1} \end{bmatrix}$$

Combining terms in $V(s)$ and $I(s)$ gives

$$\begin{bmatrix} s+2 & 1 \\ -2 & s+5 \end{bmatrix} \begin{bmatrix} V(s) \\ I(s) \end{bmatrix} = \begin{bmatrix} v_+ \\ i_+ \end{bmatrix} + \begin{bmatrix} 1 & 0 \\ 0 & -2 \end{bmatrix} \begin{bmatrix} \dfrac{1}{s} \\ \dfrac{1}{s+1} \end{bmatrix}$$

The characteristic equation of the system is

$$\begin{vmatrix} s+2 & 1 \\ -2 & s+5 \end{vmatrix} = s^2 + 7s + 12 = 0$$

yielding the natural frequencies $s = -3, -4$.

The next step is to solve for $V(s)$ and $I(s)$ by taking the inverse of the A matrix. The result is

$$\begin{bmatrix} V(s) \\ I(s) \end{bmatrix} = \frac{\begin{bmatrix} s+5 & -1 \\ 2 & s+2 \end{bmatrix} \begin{bmatrix} v_+ \\ i_+ \end{bmatrix}}{(s+3)(s+4)} + \frac{\begin{bmatrix} s+5 & -1 \\ 2 & s+2 \end{bmatrix} \begin{bmatrix} 1 & 0 \\ 0 & -2 \end{bmatrix} \begin{bmatrix} \dfrac{1}{s} \\ \dfrac{1}{s+1} \end{bmatrix}}{(s+3)(s+4)}$$

The first term on the right side is in proper form for partial-fraction expansion; the second term is not. To put the second term in proper form we rewrite each term in $X(s)$ over a common denominator and express that term as

$$\frac{\begin{bmatrix} s+5 & -1 \\ 2 & s+2 \end{bmatrix} \begin{bmatrix} 1 & 0 \\ 0 & -2 \end{bmatrix} \begin{bmatrix} s+1 \\ s \end{bmatrix}}{s(s+1)(s+3)(s+4)} = \frac{\begin{bmatrix} s+5 & -1 \\ 2 & s+2 \end{bmatrix} \begin{bmatrix} s+1 \\ -2s \end{bmatrix}}{s(s+1)(s+3)(s+4)}$$

Combining this term with the initial-condition term yields the final s-domain form

$$\begin{bmatrix} V(s) \\ I(s) \end{bmatrix} = \frac{\begin{bmatrix} s+5 & -1 \\ 2 & s+2 \end{bmatrix}\begin{bmatrix} (s+1)(sv_+ + 1) \\ s(s+1)i_+ - 2s \end{bmatrix}}{s(s+1)(s+3)(s+4)}$$

The partial-fraction expansion is

$$\begin{bmatrix} V(s) \\ I(s) \end{bmatrix} = \frac{K_1}{s+3} + \frac{K_2}{s+4} + \frac{K_a}{s} + \frac{K_b}{s+1}$$

Letting the numerator matrix product be $N(s)$, the coefficient matrices are

$$K_1 = \left.\frac{N(s)}{s(s+1)(s+4)}\right|_{-3} = \begin{bmatrix} 2v_+ - i_+ - \frac{5}{3} \\ 2v_+ - i_+ - \frac{5}{3} \end{bmatrix}$$

$$K_2 = \left.\frac{N(s)}{s(s+1)(s+3)}\right|_{-4} = \begin{bmatrix} -v_+ + i_+ + \frac{11}{12} \\ -2v_+ + 2i_+ + \frac{11}{6} \end{bmatrix}$$

$$K_a = \left.\frac{N(s)}{(s+1)(s+3)(s+4)}\right|_{0} = \begin{bmatrix} \frac{5}{12} \\ \frac{1}{6} \end{bmatrix}$$

$$K_b = \left.\frac{N(s)}{s(s+3)(s+4)}\right|_{-1} = \begin{bmatrix} \frac{1}{3} \\ -\frac{1}{3} \end{bmatrix}$$

Thus the total response is

$$\begin{bmatrix} v(t) \\ i(t) \end{bmatrix} = \begin{bmatrix} 2v_+ - i_+ - \frac{5}{3} \\ 2v_+ - i_+ - \frac{5}{3} \end{bmatrix} e^{-3t}u(t) + \begin{bmatrix} -v_+ + i_+ + \frac{11}{12} \\ -2v_+ + 2i_+ + \frac{11}{6} \end{bmatrix} e^{-4t}u(t)$$
$$+ \begin{bmatrix} \frac{5}{12} \\ \frac{1}{6} \end{bmatrix} u(t) + \begin{bmatrix} \frac{1}{3} \\ -\frac{1}{3} \end{bmatrix} e^{-t}u(t)$$

Inserting the values of the initial conditions gives

$$\begin{bmatrix} v(t) \\ i(t) \end{bmatrix} = \begin{bmatrix} \frac{7}{3} \\ \frac{7}{3} \end{bmatrix} e^{-3t}u(t) + \begin{bmatrix} -\frac{1}{12} \\ -\frac{1}{6} \end{bmatrix} e^{-4t}u(t) + \begin{bmatrix} \frac{5}{12} \\ \frac{1}{6} \end{bmatrix} u(t) + \begin{bmatrix} \frac{1}{3} \\ -\frac{1}{3} \end{bmatrix} e^{-t}u(t)$$

The reader may well wonder why the initial-condition values were not inserted much earlier to simplify the algebra. The answer is that by keeping the initial conditions identified until the total response is written, we can write down the zero-input and zero-state responses directly, without redoing the partial-fraction expansion. Thus, from the next to the last matrix equation, with v_+ and i_+ zero, the zero-state response is

$$\begin{bmatrix} v(t) \\ i(t) \end{bmatrix} = \begin{bmatrix} -\frac{5}{3} \\ -\frac{5}{3} \end{bmatrix} e^{-3t}u(t) + \begin{bmatrix} \frac{11}{12} \\ \frac{11}{6} \end{bmatrix} e^{-4t}u(t) + \begin{bmatrix} \frac{5}{12} \\ \frac{1}{6} \end{bmatrix} u(t) + \begin{bmatrix} \frac{1}{3} \\ -\frac{1}{3} \end{bmatrix} e^{-t}u(t)$$

and subtracting this from the total response yields the zero-input response as

$$\begin{bmatrix} v(t) \\ i(t) \end{bmatrix} = \begin{bmatrix} 4 \\ 4 \end{bmatrix} e^{-3t}u(t) + \begin{bmatrix} 1 \\ 2 \end{bmatrix} e^{-4t}u(t)$$

It should be noted carefully that the zero-state response contains terms produced by the poles of the excitation transforms and the natural frequencies of the network, not just the former. The zero-input response contains terms produced only by the natural frequencies.

We have yet to consider the multiple-root case. An example will suffice to show the few subtleties present in the Laplace solution procedure when multiple roots are present.

Example 11.7 Let the parameters in the circuit of Fig. 11.7 become $L = 1$ henry, $C = 1$ farad, $R = 2$ ohms, $G = 4$ mhos. With all excitations and initial conditions the same as before, repeat Example 11.6.

The state equations for the new parameters are

$$\begin{bmatrix} \dot{v} \\ \dot{i} \end{bmatrix} = \begin{bmatrix} -4 & -1 \\ 1 & -2 \end{bmatrix} \begin{bmatrix} v \\ i \end{bmatrix} + \begin{bmatrix} 1 & 0 \\ 0 & -1 \end{bmatrix} \begin{bmatrix} i_s \\ e_s \end{bmatrix}$$

Laplace transforming these equations and combining terms in $V(s)$ and $I(s)$ yields

$$\begin{bmatrix} s+4 & 1 \\ -1 & s+2 \end{bmatrix} \begin{bmatrix} V(s) \\ I(s) \end{bmatrix} = \begin{bmatrix} v_+ \\ i_+ \end{bmatrix} + \begin{bmatrix} 1 & 0 \\ 0 & -1 \end{bmatrix} \begin{bmatrix} \dfrac{1}{s} \\ \dfrac{1}{s+1} \end{bmatrix}$$

The characteristic equation is

$$\begin{vmatrix} s+4 & 1 \\ -1 & s+2 \end{vmatrix} = (s+3)^2 = 0$$

yielding the repeated natural frequency $s = -3$.

Solving for $V(s)$ and $I(s)$ gives

$$\begin{bmatrix} V(s) \\ I(s) \end{bmatrix} = \frac{\begin{bmatrix} s+2 & -1 \\ 1 & s+4 \end{bmatrix} \begin{bmatrix} v_+ \\ i_+ \end{bmatrix}}{(s+3)^2} + \frac{\begin{bmatrix} s+2 & -1 \\ 1 & s+4 \end{bmatrix} \begin{bmatrix} 1 & 0 \\ 0 & -1 \end{bmatrix} \begin{bmatrix} \dfrac{1}{s} \\ \dfrac{1}{s+1} \end{bmatrix}}{(s+3)^2}$$

Rewriting each term in $X(s)$ over a common denominator and combining this term with the initial-condition term, we have

$$\begin{bmatrix} V(s) \\ I(s) \end{bmatrix} = \frac{\begin{bmatrix} s+2 & -1 \\ 1 & s+4 \end{bmatrix} \begin{bmatrix} (s+1)(sv_+ + 1) \\ s(s+1)i_+ - s \end{bmatrix}}{s(s+1)(s+3)^2}$$

The partial-fraction expansion consists of

$$\begin{bmatrix} V(s) \\ I(s) \end{bmatrix} = \frac{K_1}{(s+3)^2} + \frac{K_2}{s+3} + \frac{K_a}{s} + \frac{K_b}{s+1}$$

Again letting the numerator matrix product be $N(s)$, the coefficient matrices are

$$K_1 = \left.\frac{N(s)}{s(s+1)}\right|_{-3} = \begin{bmatrix} -v_+ - i_+ - \frac{1}{6} \\ v_+ + i_+ + \frac{1}{6} \end{bmatrix}$$

$$K_a = \left.\frac{N(s)}{(s+3)^2(s+1)}\right|_{0} = \begin{bmatrix} \frac{2}{9} \\ \frac{1}{9} \end{bmatrix}$$

$$K_b = \left.\frac{N(s)}{(s+3)^2 s}\right|_{-1} = \begin{bmatrix} \frac{1}{4} \\ -\frac{3}{4} \end{bmatrix}$$

The coefficient K_2 can be evaluated by the methods of Sec. 10.3, either by differentiation according to

$$K_2 = \left.\frac{d}{ds}\frac{N(s)}{s(s+1)}\right|_{-3}$$

or by using the residue-sum theorem. The latter is much the easier method. We note that the highest degree of each row in $N(s)$ is 3 and that the denominator has a highest degree of 4, that is,

$$\begin{bmatrix} V(s) \\ I(s) \end{bmatrix} = \frac{\begin{bmatrix} v_+ s^3 + \cdots \\ i_+ s^3 + \cdots \end{bmatrix}}{s^4 + \cdots}$$

From the second of Eqs. (10.3.17), the sum of the residues is v_+ in the first row and i_+ in the second row. Thus

$$K_2 = \begin{bmatrix} v_+ \\ i_+ \end{bmatrix} - \begin{bmatrix} \frac{2}{9} \\ \frac{1}{9} \end{bmatrix} - \begin{bmatrix} \frac{1}{4} \\ -\frac{3}{4} \end{bmatrix} = \begin{bmatrix} v_+ - \frac{17}{36} \\ i_+ + \frac{23}{36} \end{bmatrix}$$

With the constants evaluated, the total response is

$$\begin{bmatrix} v(t) \\ i(t) \end{bmatrix} = \begin{bmatrix} -v_+ - i_+ - \frac{1}{6} \\ v_+ + i_+ + \frac{1}{6} \end{bmatrix} te^{-3t}u(t) + \begin{bmatrix} v_+ - \frac{17}{36} \\ i_+ + \frac{23}{36} \end{bmatrix} e^{-3t}u(t) + \begin{bmatrix} \frac{2}{9} \\ \frac{1}{9} \end{bmatrix} u(t) + \begin{bmatrix} \frac{1}{4} \\ -\frac{3}{4} \end{bmatrix} e^{-t}u(t)$$

Inserting the given initial conditions yields

$$\begin{bmatrix} v(t) \\ i(t) \end{bmatrix} = \begin{bmatrix} -\frac{31}{6} \\ \frac{31}{6} \end{bmatrix} te^{-3t}u(t) + \begin{bmatrix} \frac{91}{36} \\ \frac{95}{36} \end{bmatrix} e^{-t}u(t) + \begin{bmatrix} \frac{2}{9} \\ \frac{1}{9} \end{bmatrix} u(t) + \begin{bmatrix} \frac{1}{4} \\ -\frac{3}{4} \end{bmatrix} e^{-t}u(t)$$

The zero-state response is obtained by equating the initial conditions to zero in the next to the last equation

$$\begin{bmatrix} v(t) \\ i(t) \end{bmatrix} = \begin{bmatrix} -\frac{1}{6} \\ \frac{1}{6} \end{bmatrix} te^{-3t}u(t) + \begin{bmatrix} -\frac{17}{36} \\ \frac{23}{36} \end{bmatrix} e^{-3t}u(t) + \begin{bmatrix} \frac{2}{9} \\ \frac{1}{9} \end{bmatrix} u(t) + \begin{bmatrix} \frac{1}{4} \\ -\frac{3}{4} \end{bmatrix} e^{-t}u(t)$$

Subtracting this result from the total response yields the zero-input response as

$$\begin{bmatrix} v(t) \\ i(t) \end{bmatrix} = \begin{bmatrix} -5 \\ 5 \end{bmatrix} te^{-3t}u(t) + \begin{bmatrix} 3 \\ 2 \end{bmatrix} e^{-3t}u(t)$$

11.6 TIME-DOMAIN SOLUTIONS: PRELIMINARY

In Sec. 9.8, the real-convolution theorem was presented as an alternative to the Laplace transform technique for solving an nth-order differential equation. With convolution, the solution is carried out entirely in the time domain. In the next three sections, the use of convolution is extended to the solution of n first-order (state) differential equations.

To show the general form that a solution in the time domain of a set of state equations will take, let us review the use of an integrating factor in solving an ordinary first-order differential equation of the form

$$\frac{dy(t)}{dt} + py(t) = q(t) \tag{11.6.1}$$

where p is a constant for a non-time-varying system, $q(t)$ is the excitation, and $y(t)$ is the response. Let each side of the equation be multiplied by the integrating factor e^{pt}

$$e^{pt}\frac{dy}{dt} + e^{pt}py(t) = e^{pt}q(t) \tag{11.6.2}$$

The left side of Eq. (11.6.2) is the time derivative of $y(t)e^{pt}$, and thus we can write

$$\frac{d}{dt}\,e^{pt}y(t) = e^{pt}q(t) \tag{11.6.3}$$

Integrating from 0 to t yields

$$e^{pt}y(t) - y(0) = \int_0^t e^{pt}q(t)\,dt \tag{11.6.4}$$

giving

$$y(t) = y(0)e^{-pt} + e^{-pt}\int_0^t e^{pt}q(t)\,dt \tag{11.6.5}$$

The e^{-pt} term outside the integral in Eq. (11.6.5) can be placed inside by changing the variable of integration from t to τ. This change results in the final form

$$y(t) = y(0)e^{-pt} + \int_0^t e^{-p(t-\tau)}q(\tau)\,d\tau \tag{11.6.6}$$

We note that the response is composed of two parts. One contains the initial condition $y(0)$ and is the zero-input response. The integral term contains the excitation and is the zero-state response. We see from the integrand that the integral term is the convolution of e^{-pt} and $q(t)$.

We ask next whether a similar result can be obtained when we are dealing with n response variables rather than the one contained in Eq. (11.6.6).

Specifically, can we perform the steps from Eqs. (11.6.1) to (11.6.6) in solving matrix state equations?

To begin, Eq. (11.2.2) is rearranged to match the form of Eq. (11.6.1)

$$\dot{Y}(t) - AY(t) = BX(t) \tag{11.6.7}$$

By direct analogy with the first-order case, the integrating factor should be e^{-At}. This immediately introduces the question of what we mean by e raised to a matrix power. Setting that question aside for the moment, we premultiply both sides of Eq. (11.6.7) by e^{-At} and obtain

$$e^{-At}\dot{Y}(t) - e^{-At}AY(t) = e^{-At}BX(t) \tag{11.6.8}$$

We are tempted immediately to say that the left side of Eq. (11.6.8) is the time derivative of $e^{-At}Y(t)$, but this is true only if the matrix product $e^{-At}A$ is commutative and if the customary procedure for obtaining the derivative of a product is valid for matrices. Assuming these to be the case and deferring the proofs until later, we write

$$e^{-At}AY(t) = Ae^{-At}Y(t) \tag{11.6.9}$$

and obtain from Eq. (11.6.8)

$$\frac{d}{dt}[e^{-At}Y(t)] = e^{-At}BX(t) \tag{11.6.10}$$

Integrating from 0 to t yields

$$e^{-At}Y(t) - Y_+ = \int_0^t e^{-At}BX(t)\,dt \tag{11.6.11}$$

where Y_+ conforms to the previous notation in Sec. 11.5. Now premultiplying by e^{At} results in

$$Y(t) = e^{At}Y_+ + e^{At}\int_0^t e^{-At}BX(t)\,dt \tag{11.6.12}$$

Finally, we replace t by τ in the integrand to obtain

$$Y(t) = e^{At}Y_+ + \int_0^t e^{A(t-\tau)}BX(\tau)\,d\tau \tag{11.6.13}$$

The form of Eq. (11.6.13) indeed "matches" that of Eq. (11.6.6), but Eq. (11.6.13) is of little use to us until we learn what is meant by e^{At} and how to evaluate this quantity, given the matrix A. Some insight into the form of e^{At} can be gained through the following development. Consider a set of homogeneous state equations

$$\dot{Y} = AY \tag{11.6.14}$$

As in Sec. 8.4, assume a solution

$$Y = Ce^{st} = C_1e^{s_1t} + C_2e^{s_2t} + \cdots + C_ne^{s_nt} \tag{11.6.15}$$

where each s is a root of the characteristic equation of A and each C is an $n \times 1$ column matrix of constants. There are n^2 constants to evaluate in terms of the n initial conditions; thus n^2 equations in these constants are needed.

The first set of n equations is obtained by evaluating Eq. (11.6.15) at $t = 0_+$

$$Y_+ = C = C_1 + C_2 + \cdots + C_n \tag{11.6.16}$$

The second set of n equations results from evaluating the first time derivative of Eq. (11.6.15) at $t = 0_+$ and equating the result to Eq. (11.6.14) evaluated at $t = 0_+$

$$\dot{Y}_+ = sC = s_1C_1 + s_2C_2 + \cdots + s_nc_n = AY_+ \tag{11.6.17}$$

To obtain the third set of n equations, the second time derivative of Eq. (11.6.15) is evaluated at $t = 0_+$. This result is equated to the first time derivative of Eq. (11.6.14) evaluated at $t = 0_+$

$$\ddot{Y}_+ = s^2C = s_1{}^2C_1 + s_2{}^2C_2 + \cdots + s_n{}^2C_n = A\dot{Y}_+ = A^2Y_+ \tag{11.6.18}$$

Proceeding to the nth set of n equations, we have

$$s_1^{n-1}C_1 + s_2^{n-1}C_2 + \cdots + s_n^{n-1}C_n = A^{n-1}Y_+ \tag{11.6.19}$$

Comparing Eqs. (11.6.16) to (11.6.19) with the first term in Eq. (11.6.13), we begin to see the form that e^{At} must have, namely, a series expansion in A of the form

$$e^{At} = k_0U + k_1A + k_2A^2 + \cdots + k_{n-1}A^{n-1} \tag{11.6.20}$$

where the k's are functions of time. The first term in Eq. (11.6.20) arises from Eq. (11.6.16) (the identity matrix U is included to convert from $n \times 1$ to $n \times n$). The second term in Eq. (11.6.20) comes from Eq. (11.6.17), the third from Eq. (11.6.18), and so on.

Example 11.8 Consider the network of Example 11.3 with the excitations e_s and i_s zero and with $C_1 = 1$ farad, $G = 1$ mho, $R = 2$ ohms, $C_2 = \frac{1}{2}$ farad, $L = \frac{27}{2}$ henrys. With these values, the state equations in Example 11.3 become

$$\begin{bmatrix} \dot{v}_1 \\ \dot{v}_2 \\ \dot{i} \end{bmatrix} = \begin{bmatrix} -1 & 0 & -1 \\ 0 & -1 & 2 \\ \frac{2}{27} & -\frac{2}{27} & 0 \end{bmatrix} \begin{bmatrix} v_1 \\ v_2 \\ i \end{bmatrix}$$

As in Sec. 8.4, we assume that the solution is

$$\begin{bmatrix} v_1 \\ v_2 \\ i \end{bmatrix} = \begin{bmatrix} c_1 \\ c_2 \\ c_3 \end{bmatrix} e^{st}$$

Differentiating yields

$$\begin{bmatrix} \dot{v}_1 \\ \dot{v}_2 \\ \dot{i} \end{bmatrix} = s \begin{bmatrix} c_1 \\ c_2 \\ c_3 \end{bmatrix} e^{st} = s \begin{bmatrix} v_1 \\ v_2 \\ i \end{bmatrix}$$

The natural frequencies are obtained from $|A - sU| = 0$ [Eq. (8.4.9)].

$$\begin{vmatrix} -1 - s & 0 & -1 \\ 0 & -1 - s & 2 \\ 2/27 & -2/27 & -s \end{vmatrix} = 0 \qquad s = -1/3, \quad -2/3, \quad -1$$

Thus the solution takes the form

$$\begin{bmatrix} v_1 \\ v_2 \\ i \end{bmatrix} = \begin{bmatrix} c_{11} \\ c_{21} \\ c_{31} \end{bmatrix} e^{-1/3 t} + \begin{bmatrix} c_{12} \\ c_{22} \\ c_{32} \end{bmatrix} e^{-2/3 t} + \begin{bmatrix} c_{13} \\ c_{23} \\ c_{33} \end{bmatrix} e^{-t}$$

where it will be understood that each term is multiplied by $u(t)$.

There are nine coefficients to evaluate in terms of the initial conditions. This requires nine independent equations. Three of them come from Eq. (11.6.16)

$$\begin{bmatrix} v_1 \\ v_2 \\ i \end{bmatrix}_+ = \begin{bmatrix} c_{11} \\ c_{21} \\ c_{31} \end{bmatrix} + \begin{bmatrix} c_{12} \\ c_{22} \\ c_{32} \end{bmatrix} + \begin{bmatrix} c_{13} \\ c_{23} \\ c_{33} \end{bmatrix}$$

The second three equations arise from Eq. (11.6.17)

$$\begin{bmatrix} \dot{v}_1 \\ \dot{v}_2 \\ \dot{i} \end{bmatrix}_+ = -\frac{1}{3} \begin{bmatrix} c_{11} \\ c_{21} \\ c_{31} \end{bmatrix} - \frac{2}{3} \begin{bmatrix} c_{12} \\ c_{22} \\ c_{32} \end{bmatrix} - 1 \begin{bmatrix} c_{13} \\ c_{23} \\ c_{33} \end{bmatrix} = \begin{bmatrix} -1 & 0 & -1 \\ 0 & -1 & 2 \\ 2/27 & -2/27 & 0 \end{bmatrix} \begin{bmatrix} v_1 \\ v_2 \\ i \end{bmatrix}_+$$

The final three equations are provided by Eq. (11.6.18)

$$\begin{bmatrix} \ddot{v}_1 \\ \ddot{v}_2 \\ \ddot{i} \end{bmatrix}_+ = \frac{1}{9} \begin{bmatrix} c_{11} \\ c_{21} \\ c_{31} \end{bmatrix} + \frac{4}{9} \begin{bmatrix} c_{12} \\ c_{22} \\ c_{32} \end{bmatrix} + 1 \begin{bmatrix} c_{13} \\ c_{23} \\ c_{33} \end{bmatrix} = \begin{bmatrix} 25/27 & 2/27 & 1 \\ 4/27 & 23/27 & -2 \\ -2/27 & 2/27 & -2/9 \end{bmatrix} \begin{bmatrix} v_1 \\ v_2 \\ i \end{bmatrix}_+$$

The nine equations can be solved for the nine coefficients. Actually this is not as difficult as it might seem because the nine equations can be considered in sets of three equations of three coefficients each. For example, we can write

$$c_{11} + c_{12} + c_{13} = v_{1+}$$

$$-\tfrac{1}{3}c_{11} - \tfrac{2}{3}c_{12} - c_{13} = -v_{1+} - i_+$$

$$\tfrac{1}{9}c_{11} + \tfrac{4}{9}c_{12} + c_{13} = \tfrac{25}{27}v_{1+} + \tfrac{2}{27}v_{2+} + i_+$$

which yields

$$c_{11} = -\tfrac{1}{3}v_{1+} + \tfrac{1}{3}v_{2+} - 3i_+$$

$$c_{12} = \tfrac{2}{3}v_{1+} - \tfrac{2}{3}v_{2+} + 3i_+$$

$$c_{13} = \tfrac{2}{3}v_{1+} + \tfrac{1}{3}v_{2+}$$

The other two sets of three equations are similarly treated, after which the solution to the homogeneous state equations can be placed in either of the following two forms:

$$\begin{bmatrix} v_1 \\ v_2 \\ i \end{bmatrix} = \begin{bmatrix} -\tfrac{1}{3}v_{1+} + \tfrac{1}{3}v_{2+} - 3i_+ & \tfrac{2}{3}v_{1+} - \tfrac{2}{3}v_{2+} + 3i_+ & \tfrac{2}{3}v_{1+} + \tfrac{1}{3}v_{2+} \\ \tfrac{2}{3}v_{1+} - \tfrac{2}{3}v_{2+} + 6i_+ & -\tfrac{4}{3}v_{1+} + \tfrac{4}{3}v_{2+} - 6i_+ & \tfrac{2}{3}v_{1+} + \tfrac{1}{3}v_{2+} \\ \tfrac{2}{9}v_{1+} - \tfrac{2}{9}v_{2+} + 2i_+ & -\tfrac{2}{9}v_{1+} + \tfrac{2}{9}v_{2+} - i_+ & 0 \end{bmatrix} \begin{bmatrix} e^{-\frac{1}{3}t} \\ e^{-\frac{2}{3}t} \\ e^{-t} \end{bmatrix}$$

$$\begin{bmatrix} v_1 \\ v_2 \\ i \end{bmatrix} = \begin{bmatrix} -\tfrac{1}{3}e^{-\frac{1}{3}t} + \tfrac{2}{3}e^{-\frac{2}{3}t} + \tfrac{2}{3}e^{-t} & \tfrac{1}{3}e^{-\frac{1}{3}t} - \tfrac{2}{3}e^{-\frac{2}{3}t} + \tfrac{1}{3}e^{-t} & -3e^{-\frac{1}{3}t} + 3e^{-\frac{2}{3}t} \\ \tfrac{2}{3}e^{-\frac{1}{3}t} - \tfrac{4}{3}e^{-\frac{2}{3}t} + \tfrac{2}{3}e^{-t} & -\tfrac{2}{3}e^{-\frac{1}{3}t} + \tfrac{4}{3}e^{-\frac{2}{3}t} + \tfrac{1}{3}e^{-t} & 6e^{-\frac{1}{3}t} - 6e^{-\frac{2}{3}t} \\ \tfrac{2}{9}e^{-\frac{1}{3}t} - \tfrac{2}{9}e^{-\frac{2}{3}t} & -\tfrac{2}{9}e^{-\frac{1}{3}t} + \tfrac{2}{9}e^{-\frac{2}{3}t} & 2e^{-\frac{1}{3}t} - e^{-\frac{2}{3}t} \end{bmatrix} \begin{bmatrix} v_{1+} \\ v_{2+} \\ i_+ \end{bmatrix}$$

Note that the first solution form is $Y = Ce^{st}$ and the second is $Y = e^{At}Y_+$.

11.7 FUNCTIONS OF MATRICES

This section is a mathematical digression to discuss functions of matrices, in order to:

1. Derive the expansion of e^{At} in Eq. (11.6.20), which was arrived at rather intuitively in the last section
2. Learn how to compute the k's in Eq. (11.6.20) so that we can insert the expansion of e^{At} into Eq. (11.6.13) to obtain the total response of a system
3. Justify the assumptions that the matrix product $e^{-At}A$ commutes and that the customary differentiation of a product applies to matrices, requirements for the validity of Eq. (11.6.13)

The foundation of any study of functions of matrices is the *Cayley-Hamilton theorem*, which states simply that every square matrix satisfies its own characteristic equation. Let A be a square matrix of rank n. The characteristic equation of A, call it $W(s)$, is given by Eqs. (8.4.9) and (8.4.10) as

$$W(s) = |A - sU| = s^n + \alpha_{n-1}s^{n-1} + \cdots + \alpha_1 s + \alpha_0 = 0 \qquad (11.7.1)$$

Then the Cayley-Hamilton theorem states that

$$W(A) = A^n + \alpha_{n-1}A^{n-1} + \cdots + \alpha_1 A + \alpha_0 U = 0 \tag{11.7.2}$$

where U and 0 are $n \times n$ identity and null matrices, respectively.

We shall prove the Cayley-Hamilton theorem for the case of nonrepeated roots of the characteristic equation of A.† The proof proceeds from a definition of the *modal matrix* of A, denoted by M. This matrix, also $n \times n$, is constructed from the eigenvectors of A such that each eigenvector forms a column of M. The eigenvectors of A are the Y_i's of Eq. (8.4.11). They are solutions of the homogeneous set of Eqs. (8.4.8). For each eigenvalue, we have

$$(A - s_i U) Y_i = 0 \tag{11.7.3}$$

and for the entire set, we can write

$$AM - MD = 0 \tag{11.7.4}$$

where D is an $n \times n$ diagonal matrix of the form

$$D = \begin{bmatrix} s_1 & 0 & \cdots & 0 \\ 0 & s_2 & \cdots & 0 \\ \cdot & \cdot & \cdot & \cdot \\ 0 & 0 & \cdots & s_n \end{bmatrix} \tag{11.7.5}$$

Assume M to be nonsingular. This will always be true if the characteristic equation does not contain multiple roots, which is the case being considered in this proof of the Cayley-Hamilton theorem. Then Eq. (11.7.4) can be premultiplied by M^{-1} to obtain

$$M^{-1}AM = D \tag{11.7.6}$$

which shows, incidentally, that the modal matrix can be employed to diagonalize A.

Returning now to the Cayley-Hamilton theorem proof, we factor Eq. (11.7.2) into

$$W(A) = (A - s_1 U)(A - s_2 U) \cdots (A - s_n U) \tag{11.7.7}$$

Next Eq. (11.7.7) is premultiplied by M^{-1}, postmultiplied by M, and written in the form

$$M^{-1}W(A)M = M^{-1}(A - s_1 U)MM^{-1}(A - s_2 U)M \cdots M^{-1}(A - s_n U)M \tag{11.7.8}$$

† The theorem also holds in the general case. See M. Bocher, "Introduction to Higher Algebra," The Macmillan Company, New York, 1931.

The following manipulation of each of the products $M^{-1}(A - s_iU)M$ can be carried out:

$$M^{-1}(A - s_iU)M = M^{-1}AM - M^{-1}s_iUM = D - s_iU \qquad (11.7.9)$$

Thus each of these products is the diagonal matrix of the eigenvalues less the diagonal matrix whose main-diagonal entries are the ith eigenvalue. It follows that the ith row in $(D - s_iU)$ has all its elements zero.

Inserting Eq. (11.7.9) into Eq. (11.7.8) yields

$$M^{-1}W(A)M = (D - s_1U)(D - s_2U)\cdots(D - s_nU) \qquad (11.7.10)$$

Now each of the matrices on the right side of Eq. (11.7.10) has one row with all its elements zero, and thus the right side of Eq. (11.7.10) is identically zero. Since M is not zero, it follows that $W(A)$ is zero and the Cayley-Hamilton theorem is proved.

Example 11.9 Given the state equations

$$\begin{bmatrix} \dot{y}_1 \\ \dot{y}_2 \\ \dot{y}_3 \end{bmatrix} = \begin{bmatrix} 0 & \frac{1}{3} & -\frac{1}{3} \\ -3 & -3 & 0 \\ 3 & 0 & -3 \end{bmatrix} \begin{bmatrix} y_1 \\ y_2 \\ y_3 \end{bmatrix}$$

find the modal matrix of A and show that it satisfies Eq. (11.7.4).

The characteristic equation of A is

$$\begin{vmatrix} -s & \frac{1}{3} & -\frac{1}{3} \\ -3 & -3 - s & 0 \\ 3 & 0 & -3 - s \end{vmatrix} = 0$$

which yields

$$s^3 + 6s^2 + 11s + 6 = (s + 1)(s + 2)(s + 3) = 0$$

Thus the eigenvalues are $s = -1, -2, -3$, and the solution takes the form

$$\begin{bmatrix} y_1 \\ y_2 \\ y_3 \end{bmatrix} = \begin{bmatrix} c_{11} \\ c_{21} \\ c_{31} \end{bmatrix} e^{-t} + \begin{bmatrix} c_{12} \\ c_{22} \\ c_{32} \end{bmatrix} e^{-2t} + \begin{bmatrix} c_{13} \\ c_{23} \\ c_{33} \end{bmatrix} e^{-3t}$$

The matrix equation $AY = sY$, which results from assuming a solution $Y = Ce^{st}$, must be satisfied for each of the three eigenvalues. For $s_1 = -1$,

$$\begin{bmatrix} 0 & \frac{1}{3} & -\frac{1}{3} \\ -3 & -3 & 0 \\ 3 & 0 & -3 \end{bmatrix} \begin{bmatrix} y_1 \\ y_2 \\ y_3 \end{bmatrix} = -1 \begin{bmatrix} y_1 \\ y_2 \\ y_3 \end{bmatrix} \quad \text{or} \quad \begin{bmatrix} 1 & \frac{1}{3} & -\frac{1}{3} \\ -3 & -2 & 0 \\ 3 & 0 & -2 \end{bmatrix} \begin{bmatrix} y_1 \\ y_2 \\ y_3 \end{bmatrix} = 0$$

The determinant of this 3×3 set of homogeneous equations vanishes; hence two of the variables can be expressed in terms of the third, and no unique eigenvector is possible: the ratios of the elements of the eigenvector are prescribed but not their magnitudes.

With $y_1 = c_{11}e^{-t}$, the other two variables are

$$y_2 = -\tfrac{3}{2}c_{11}e^{-t} \qquad \text{and} \qquad y_3 = \tfrac{3}{2}c_{11}e^{-t}$$

and the eigenvector is

$$c_{11}\begin{bmatrix} 1 \\ -\tfrac{3}{2} \\ \tfrac{3}{2} \end{bmatrix} e^{-t}$$

Similar treatment for the other two eigenvalues yields the eigenvectors

$$c_{12}\begin{bmatrix} 1 \\ -3 \\ 3 \end{bmatrix} e^{-2t} \qquad \text{and} \qquad c_{23}\begin{bmatrix} 0 \\ 1 \\ 1 \end{bmatrix} e^{-3t}$$

where, in the e^{-3t} case, the y_1 term must be zero so we have used c_{23}, rather than c_{13}, as the constant.

Defining $c_1 = c_{11}$, $c_2 = c_{12}$, $c_3 = c_{23}$, the modal matrix is

$$M = \begin{bmatrix} c_1e^{-t} & c_2e^{-2t} & 0 \\ -\tfrac{3}{2}c_1e^{-t} & -3c_2e^{-2t} & c_3e^{-3t} \\ \tfrac{3}{2}c_1e^{-t} & 3c_2e^{-2t} & c_3e^{-3t} \end{bmatrix}$$

To verify that $AM = MD$, it is not necessary to carry along the c's and e^{st}'s, since these simply cancel out on each side of the equation. Accordingly we revise M to be

$$M = \begin{bmatrix} 1 & 1 & 0 \\ -\tfrac{3}{2} & -3 & 1 \\ \tfrac{3}{2} & 3 & 1 \end{bmatrix}$$

and write

$$\begin{bmatrix} 0 & \tfrac{1}{3} & -\tfrac{1}{3} \\ -3 & -3 & 0 \\ 3 & 0 & -3 \end{bmatrix}\begin{bmatrix} 1 & 1 & 0 \\ -\tfrac{3}{2} & -3 & 1 \\ \tfrac{3}{2} & 3 & 1 \end{bmatrix} = \begin{bmatrix} 1 & 1 & 0 \\ -\tfrac{3}{2} & -3 & 1 \\ \tfrac{3}{2} & 3 & 1 \end{bmatrix}\begin{bmatrix} -1 & 0 & 0 \\ 0 & -2 & 0 \\ 0 & 0 & -3 \end{bmatrix}$$

$$= \begin{bmatrix} -1 & -2 & 0 \\ \tfrac{3}{2} & 6 & -3 \\ -\tfrac{3}{2} & -6 & -3 \end{bmatrix}$$

The modal matrix in Example 11.9 has only three arbitrary constants, whereas the original formulation of the solution had nine such constants. Actually six of the nine constants ($n^2 - n$ in the general case) are not arbitrary; they are prescribed by the ratios of the elements of the eigenvectors, which in turn are determined by the relationships of the state variables as fixed by the system. In Example 11.8, instead of formulating and solving nine equations to obtain the nine coefficients, we first could have found each eigenvector within a multiplying factor and then used one set of three initial-condition equations, instead of three sets, to find the three multiplying factors.

Consider now functions of matrices. Drawing on our knowledge of functions of a single variable, it is reasonable to assume that a function of a matrix should be expandable into a series of powers of that matrix. Such a series would have the form

$$F(A) = C_0U + C_1A + C_2A^2 + \cdots + C_nA^n + C_{n+1}A^{n+1} + \cdots \tag{11.7.11}$$

assuming again that A is an $n \times n$ matrix. Using the Cayley-Hamilton theorem, Eq. (11.7.2) can be rewritten in the form

$$A^n = -\alpha_{n-1}A^{n-1} - \cdots - \alpha_1A - \alpha_0U \tag{11.7.12}$$

Thus A^n in Eq. (11.7.11) can be expressed in terms of A through A^{n-1}. Now premultiply Eq. (11.7.12) by A to obtain an expression for A^{n+1}.

$$\begin{aligned} A^{n+1} &= -\alpha_{n-1}A^n - \cdots - \alpha_1A^2 - \alpha_0A \\ &= -\alpha_{n-1}(-\alpha_{n-1}A^{n-1} - \cdots - \alpha_1A - \alpha_0U) - \cdots - \alpha_1A^2 - \alpha_0A \end{aligned} \tag{11.7.13}$$

which shows that A^{n+1} can be expressed in terms of A through A^{n-1}. The same is true of all other powers of A greater than $n - 1$. Thus Eq. (11.7.11) can be written in the revised form

$$F(A) = k_0U + k_1A + k_2A^2 + \cdots + k_{n-1}A^{n-1} = \sum_0^{n-1} k_iA^i \tag{11.7.14}$$

which is identical to the result obtained intuitively in Eq. (11.6.20).

In Eq. (11.7.11), replace the matrix A by the scalar s to obtain

$$F(s) = C_0 + C_1s + C_2s^2 + \cdots + C_ns^n + C_{n+1}s^{n+1} + \cdots \tag{11.7.15}$$

Now s^n can be expressed in terms of s through s^{n-1} by solving Eq. (11.7.1) for s^n. Following the same pattern, s^{n+1} and all higher powers of s can also be expressed in terms of s through s^{n-1}. Substituting in Eq. (11.7.15) yields

$$F(s) = k_0 + k_1s + k_2s^2 + \cdots + k_{n-1}s^{n-1} = \sum_0^{n-1} k_is^i \tag{11.7.16}$$

which is identical to Eq. (11.7.14) with A replaced by s. Equation (11.7.1) is satisfied for each eigenvalue of A. Thus Eq. (11.7.16) must be satisfied for each eigenvalue of A. This gives a means of solving for the k's in Eq. (11.7.14), which will be demonstrated later in Example 11.11.

Consider next the function $AF(A)$. We want to show that this matrix product commutes, which will validate Eq. (11.6.9). The commutativity can be proved by writing

$$AF(A) = A\left(\sum_0^{n-1} k_i A^i\right) = \sum_0^{n-1} k_i A^{i+1} = \left(\sum_0^{n-1} k_i A^i\right) A = F(A)A \qquad (11.7.17)$$

An additional relationship that should be verified is

$$\frac{d}{dt} F(At) = A \frac{dF(At)}{d(At)} \qquad (11.7.18)$$

This will justify Eq. (11.6.10) by establishing that $d/dt\, e^{At} = Ae^{At}$. The proof proceeds from Eq. (11.7.14) by replacing A by At

$$F(At) = \sum_0^{n-1} k_i A^i t^i \qquad (11.7.19)$$

and

$$\frac{d}{dt} F(At) = \sum_0^{n-1} ik_i A^i t^{i-1} = A \sum_0^{n-1} ik_i A^{i-1} t^{i-1} = A \frac{dF(At)}{d(At)} \qquad (11.7.20)$$

Example 11.10 Given $A = \begin{bmatrix} -2 & -1 \\ 2 & -5 \end{bmatrix}$, evaluate $F(A) = A^3 - A^2$.

$F(A)$ could be found directly by simply raising A to the appropriate powers and adding. However, we know from Eqs. (11.7.12) and (11.7.13) that powers of A greater than 1 can be expressed in terms of A to the first power.

The characteristic equation of A is

$$s^2 + 7s + 12 = (s + 3)(s + 4) = 0$$

A must satisfy this equation. Thus

$$A^2 + 7A + 12U = 0$$

giving

$$-A^2 = 7\begin{bmatrix} -2 & -1 \\ 2 & -5 \end{bmatrix} + 12\begin{bmatrix} 1 & 0 \\ 0 & 1 \end{bmatrix} = \begin{bmatrix} -2 & -7 \\ 14 & -23 \end{bmatrix}$$

Premultiplying each term in the characteristic equation by A yields

$$A^3 + 7A^2 + 12A = 0$$

giving

$$A^3 = 7\begin{bmatrix} -2 & -7 \\ 14 & -23 \end{bmatrix} - 12\begin{bmatrix} -2 & -1 \\ 2 & -5 \end{bmatrix} = \begin{bmatrix} 10 & -37 \\ 74 & -101 \end{bmatrix}$$

Finally

$$F(A) = \begin{bmatrix} 10 & -37 \\ 74 & -101 \end{bmatrix} + \begin{bmatrix} -2 & -7 \\ 14 & -23 \end{bmatrix} = \begin{bmatrix} 8 & -44 \\ 88 & -124 \end{bmatrix}$$

Example 11.11 Find the expansion of e^{At} when A is the 3×3 matrix of Example 11.8.

From Eq. (11.7.14), the expansion will have the form

$$e^{At} = k_0 U + k_1 A + k_2 A^2$$

where the k's are functions of time because $F(A)$ has a time dependence.

The k's can be obtained from Eq. (11.7.16) by realizing that, for each root of the characteristic equation of A, Eq. (11.7.16) must be satisfied. Thus we can write

$$e^{s_i t} = k_0 + k_1 s_i + k_2 s_i^2 \qquad i = 1, 2, 3$$

and obtain

$$e^{-1/3 t} = k_0 - 1/3 k_1 + 1/9 k_2$$

$$e^{-2/3 t} = k_0 - 2/3 k_1 + 4/9 k_2$$

$$e^{-t} = k_0 - k_1 + k_2$$

Solution of these three equations yields

$$k_0 = 3e^{-1/3 t} - 3e^{-2/3 t} + e^{-t}$$

$$k_1 = 15/2 e^{-1/3 t} - 12e^{-2/3 t} + 9/2 e^{-t}$$

$$k_2 = 9/2 e^{-1/3 t} - 9e^{-2/3 t} + 9/2 e^{-t}$$

The k's are then introduced into the e^{At} expansion to obtain

$$e^{At} = (3e^{-1/3 t} - 3e^{-2/3 t} + e^{-t})\begin{bmatrix} 1 & 0 & 0 \\ 0 & 1 & 0 \\ 0 & 0 & 1 \end{bmatrix}$$

$$+ (15/2 e^{-1/3 t} - 12e^{-2/3 t} + 9/2 e^{-t})\begin{bmatrix} -1 & 0 & -1 \\ 0 & -1 & 2 \\ 2/27 & -2/27 & 0 \end{bmatrix}$$

$$+ (9/2 e^{-1/3 t} - 9e^{-2/3 t} + 9/2 e^{-t})\begin{bmatrix} 25/27 & 2/27 & 1 \\ 4/27 & 23/27 & -2 \\ -2/27 & 2/27 & -2/9 \end{bmatrix}$$

Collecting terms gives a result identical to that in the second solution form at the close of Example 11.8.

11.8 SOLUTION OF STATE EQUATIONS: t DOMAIN

Equation (11.6.13) having been thoroughly justified, we wish to use it in the time-domain determination of the response of a linear system. Two examples will serve to demonstrate the procedure.

Example 11.12 It is desired to solve the state equations of Example 11.6 in the time domain through Eq. (11.6.13).

The expansion of e^{At} must be obtained first. The A matrix and its eigenvalues are

$$A = \begin{bmatrix} -2 & -1 \\ 2 & -5 \end{bmatrix} \qquad s_1 = -3 \qquad s_2 = -4$$

Then from Eqs. (11.7.14) and (11.7.16)

$$e^{At} = k_0U + k_1A$$

and

$$e^{-3t} = k_0 - 3k_1 \qquad e^{-4t} = k_0 - 4k_1$$

Solving for the k's yields

$$k_0 = 4e^{-3t} - 3e^{-4t} \qquad k_1 = e^{-3t} - e^{-4t}$$

and inserting in the e^{At} expansion gives

$$e^{At} = \begin{bmatrix} 2e^{-3t} - e^{-4t} & -e^{-3t} + e^{-4t} \\ 2e^{-3t} - 2e^{-4t} & -e^{-3t} + 2e^{-4t} \end{bmatrix}$$

The zero-input solution can now be written down directly as $e^{At}Y_+$. The result agrees with that obtained in Example 11.6, once the Ce^{st} form of the latter result is rearranged.

The zero-state response requires the convolution of e^{At} with the excitation $X(t)$. We write

$$\int_0^t e^{A(t-\tau)}BX(\tau)\,d\tau$$
$$= \int_0^t \begin{bmatrix} (2e^{-3(t-\tau)} - e^{-4(t-\tau)})u(t-\tau) & (-e^{-3(t-\tau)} + e^{-4(t-\tau)})u(t-\tau) \\ (2e^{-3(t-\tau)} - 2e^{-4(t-\tau)})u(t-\tau) & (-e^{-3(t-\tau)} + 2e^{-4(t-\tau)})u(t-\tau) \end{bmatrix}$$
$$\times \begin{bmatrix} 1 & 0 \\ 0 & -2 \end{bmatrix} \begin{bmatrix} u(\tau) \\ e^{-\tau}u(\tau) \end{bmatrix}$$

In the expansion of this matrix product each term will contain the factor $u(\tau)u(t-\tau)$. This factor has a value of 1 between 0 and t and zero elsewhere and thus is redundant in view of the 0 to t limits of integration. Eliminating the unit steps and expanding gives

$$\int_0^t \begin{bmatrix} 2e^{-3t}e^{3\tau} - e^{-4t}e^{4\tau} + 2e^{-3t}e^{2\tau} - 2e^{-4t}e^{3\tau} \\ 2e^{-3t}e^{3\tau} - 2e^{-4t}e^{4\tau} + 2e^{-3t}e^{2\tau} - 4e^{-4t}e^{3\tau} \end{bmatrix} d\tau$$

which when integrated and evaluated gives

$$\begin{bmatrix} \tfrac{2}{3}e^{-3t}e^{3\tau} - \tfrac{1}{4}e^{-4t}e^{4\tau} + e^{-3t}e^{2\tau} - \tfrac{2}{3}e^{-4t}e^{3\tau} \\ \tfrac{2}{3}e^{-3t}e^{3\tau} - \tfrac{1}{2}e^{-4t}e^{4\tau} + e^{-3t}e^{2\tau} - \tfrac{4}{3}e^{-4t}e^{3\tau} \end{bmatrix}_0^t$$
$$= \begin{bmatrix} (-\tfrac{5}{3}e^{-3t} + \tfrac{11}{12}e^{-4t} + \tfrac{1}{3}e^{-t} + \tfrac{5}{12})u(t) \\ (-\tfrac{5}{3}e^{-3t} + \tfrac{11}{6}e^{-4t} - \tfrac{1}{3}e^{-t} + \tfrac{1}{6})u(t) \end{bmatrix}$$

where we have reinstated the unit step to emphasize that the time response is zero prior to $t = 0$. Again the result agrees with that in Example 11.6.

As was the case with Laplace transforms, we have yet to consider the repeated-root situation. This will also be done through an example.

Example 11.13 The state equations in Example 11.7 will be solved in the time domain using Eq. (11.6.13).

The A matrix and its repeated eigenvalue are

$$A = \begin{bmatrix} -4 & -1 \\ 1 & -2 \end{bmatrix} \qquad s_1 = s_2 = -3$$

Again e^{At} is given by

$$e^{At} = k_0 U + k_1 A$$

and we must find k_0 and k_1. Since the root is repeated, only one equation in the k's is obtained from Eq. (11.7.16). A second equation is obtained by differentiating Eq. (11.7.16) with respect to s to obtain

$$F'(s) = k_1 + 2k_2 s + \cdots + (n-1)k_{n-1}s^{n-2} = \sum_0^{n-1} ik_i s^{i-1} \tag{11.8.1}$$

Then we have

$$e^{-3t} = k_0 - 3k_1 \qquad te^{-3t} = k_1$$

yielding

$$k_0 = e^{-3t} + 3te^{-3t}$$

and

$$e^{At} = \begin{bmatrix} e^{-3t} - te^{-3t} & -te^{-3t} \\ te^{-3t} & e^{-3t} + te^{-3t} \end{bmatrix}$$

The zero-input response is $e^{At}Y_+$ and follows directly. The result checks that in Example 11.7.

The zero-state response is formulated as follows:

$$\int_0^t \begin{bmatrix} [e^{-3(t-\tau)} - (t-\tau)e^{-3(t-\tau)}]u(t-\tau) & -(t-\tau)e^{-3(t-\tau)}u(t-\tau) \\ (t-\tau)e^{-3(t-\tau)}u(t-\tau) & [e^{-3(t-\tau)} + (t-\tau)e^{-3(t-\tau)}]u(t-\tau \end{bmatrix}$$

$$\times \begin{bmatrix} u(\tau) \\ -e^{-\tau}u(\tau) \end{bmatrix} d\tau = \int_0^t \begin{bmatrix} e^{-3t}e^{3\tau} - te^{-3t}e^{3\tau} + e^{-3t}\tau e^{3\tau} + te^{-3t}e^{2\tau} - e^{-3t}\tau e^{2\tau} \\ te^{-3t}e^{3\tau} - e^{-3t}\tau e^{3\tau} - e^{-3t}e^{2\tau} - te^{-3t}e^{2\tau} + e^{-3t}\tau e^{2\tau} \end{bmatrix} d\tau$$

$$= \begin{bmatrix} (2/9 - 17/36\, e^{-3t} - 1/6\, te^{-3t} + 1/4\, e^{-t})u(t) \\ (1/9 + 23/36\, e^{-3t} + 1/6\, te^{-3t} - 3/4\, e^{-t})u(t) \end{bmatrix}$$

which is also in agreement with Example 11.7.

REFERENCES

11.1. Bashkow, T. R.: The *A* Matrix: New Network Description, *IEEE Profess. Group Circuit Theory*, Sept., 1957, pp. 117–119. This original work shows how the *A* matrix can be formulated topologically when a proper tree can be drawn.

11.2. DeRusso, P. M., R. J. Roy, and C. M. Close: "State Variables for Engineers," John Wiley & Sons, Inc., New York, 1966. A clear treatment with reasonable detail of state variables, matrices, and transforms. Simulation diagrams and solutions of state equations in the *s* and *t* domains are discussed.

11.3. Gupta, S. C.: "Transform and State Variable Methods in Linear Systems," John Wiley & Sons, Inc., New York, 1966. A concise treatment of eigenvalues, eigenvectors, and functions of matrices. Choosing state variables is discussed.

11.4. Kuh, E. S., and R. A. Rohrer: The State-variable Approach to Network Analysis, *Proc. IEEE*, July, 1965, pp. 672–686. An excellent overview of state-variable concepts, including zero-state and zero-input responses, solution of state equations, and topological formulations.

11.5. Lipshutz, S.: "Theory and Problems of Linear Algebra," McGraw-Hill Book Company, New York, 1968. Chapter 9 considers functions of matrices. The Cayley-Hamilton theorem is proved, and numerous solved problems are included.

11.6. Pipes, L. A.: "Matrix Methods for Engineering," Prentice-Hall, Inc., Englewood Cliffs, N.J., 1963. An excellent presentation of the calculus and functions of matrices, eigenvalues, eigenvectors, and modal matrices.

11.7. Seeley, S.: "Dynamic Systems Analysis," Reinhold Publishing Corporation, New York, 1964. The last chapter gives a brief survey of the state-space approach. State equations are written for both electric and mechanical networks.

11.8. Zadeh, L. A., and C. A. Desoer: "Linear System Theory: The State Space Approach," McGraw-Hill Book Company, New York, 1963. An in-depth presentation of general system theory from the state point of view.

PROBLEMS

Drill

11.1. For the following homogeneous fourth-order differential equation, write n first-order differential equations and find solutions for y, $\dot{y}$, $\ddot{y}$, and $\dddot{y}$ within four arbitrary constants.

$$\ddddot{y} + 7\dddot{y} + 22\ddot{y} + 32\dot{y} + 16y = 0$$

11.2. For the network shown, write state equations using the capacitor voltages and inductor currents as state variables.

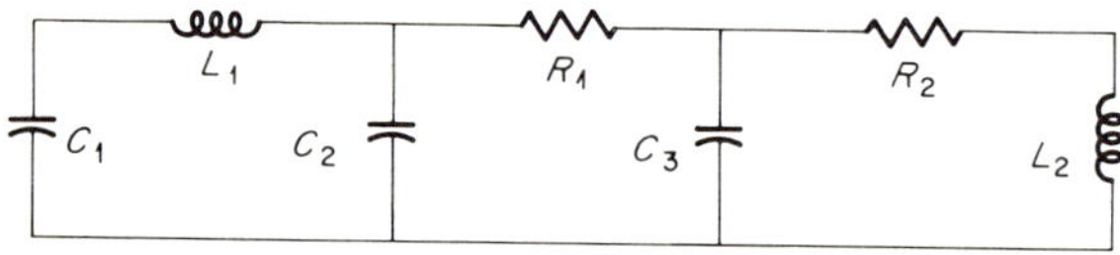

Fig. P11.2

11.3. For the network shown:

(*a*) Write state equations in the variables v_1 and v_2.
(*b*) Find whether or not i_1 and i_2 are an acceptable set of state variables.
(*c*) Obtain an expression for the output voltage v_L in terms of v_1, v_2, and e.

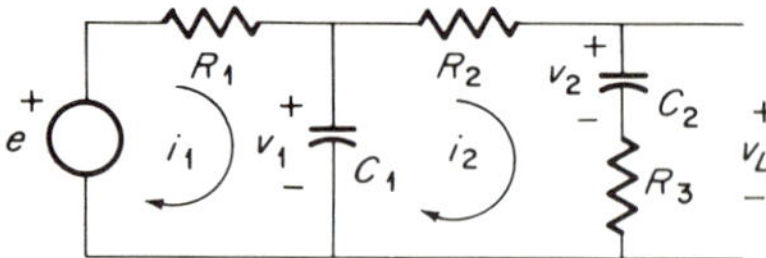

Fig. P11.3

11.4. For the network shown, write a set of state equations using the capacitor voltages and inductor currents as the state variables.

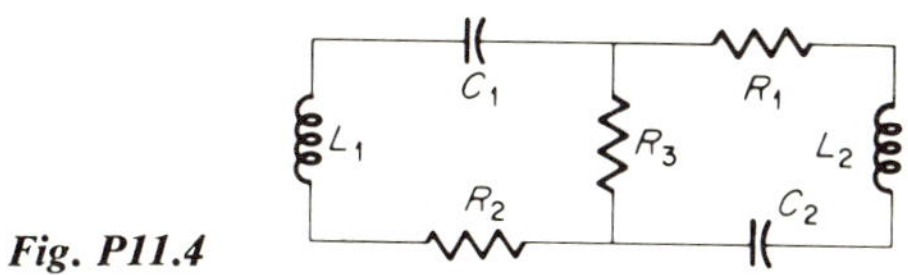

Fig. P11.4

11.5. For the network shown:

(*a*) Write state equations in v_C and i_L.

(*b*) Use the method of Sec. 11.2 to find state equations in v_1 and v_2 from the equations of part (*a*).

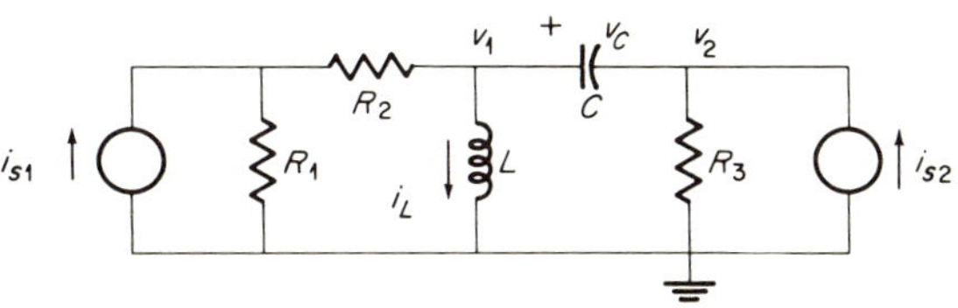

Fig. P11.5

11.6. For the given network:

(*a*) Write state equations in the capacitor voltages and inductor currents.

(*b*) With L_2 shorted and with all parameters unity, find the natural frequencies of the network.

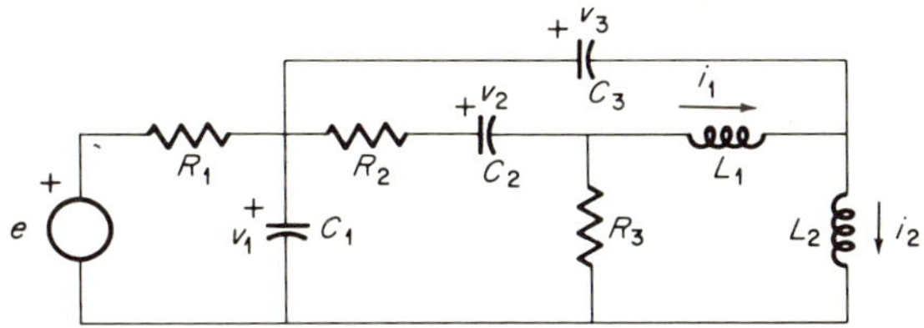

Fig. P11.6

11.7. The given network contains tie-set constraints. Write state equations using appropriate capacitor voltages and inductor currents as state variables.

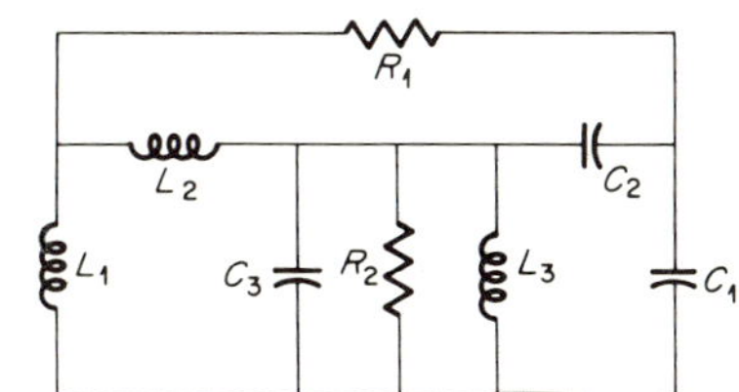

Fig. P11.7

11.8. Find the modal matrix of the A matrix in Example 11.8. Verify that Eq. (11.7.4) is satisfied in this case.

11.9. Find e^{At} using the matrix of Example 11.9.

11.10. Let the values of the parameters and excitations for the network in Fig. 11.2 be: $e_s = e^{-4t}u(t)$, $i_s = u(t)$, $R = 1$ ohm, $G = 1$ mho, $L = 3$ henrys, $C_1 = \frac{1}{3}$ farad, $C_2 = \frac{1}{3}$ farad.

(*a*) Write state equations in the variables v_1, v_2, and i.

(*b*) Find the zero-input, zero-state, and total responses by Laplace transforms.

(*c*) Repeat part (*b*) using Eq. (11.6.13).

11.11. Repeat Example 11.6 if e_s is $e^{-4t}u(t)$ volts, all other quantities remaining unchanged.

11.12. Repeat Prob. 11.10 for $R = \frac{1}{2}$ ohm, $G = 2$ mhos, $C_1 = C_2 = 1$ farad, and $L = 2$ henrys, the excitations remaining the same.

Theory and proofs

11.13. Show that Bashkow's procedure in Sec. 11.4 can be developed from $\beta_F V_B = 0$ and $\alpha_F J_B = 0$, instead of from $J_B = \beta_F{}^T I_L$ and $V_B = \alpha_F{}^T E_N$.

11.14. (*a*) Let A and B be $n \times n$ square matrices. Assuming that A and B commute, prove that $e^{A+B} = e^A e^B$.

(*b*) Use the result of part (*a*) to prove that $[e^A]^{-1} = e^{-A}$.

11.15. Given that $X = e^{At}X_+$ is the solution of $\dot{X} = AX$, assume that X can be represented by the infinite series

$$X = C_0 + C_1 t + C_2 t^2 + \cdots + C_n t^n + \cdots$$

and show that

$$e^{At} = \sum_{n=0}^{\infty} \frac{1}{n!} A^n t^n$$

11.16. A linear RLC network can be represented as shown, with all inductors and capacitors brought out to terminal pairs, leaving only resistors in the remaining network. Show that it is always possible to write state equations of the form

$$\dot{Y} = AY \qquad \text{or} \qquad \dot{y}_i = \sum_{j=1}^{n} a_{ij} y_j$$

for such a network, where the Y matrix contains all the capacitor voltages and inductor currents, it being assumed that there are no tie-set or cut-set constraint relations.

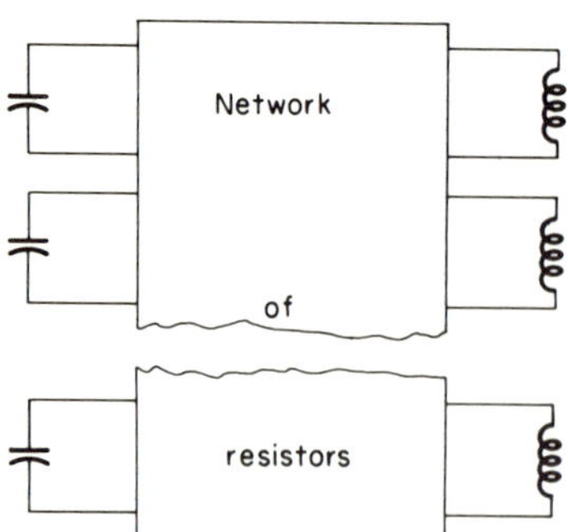

Fig. P11.16

The proof is one of finding the a_{ij}'s in terms of the L's and C's. This can be done by applying a unit step of voltage in series with the jth capacitor and finding expressions for the current into the ith capacitor and the voltage across the ith inductor. Also it is necessary to apply a unit step of current in parallel with the jth inductor and monitor the voltage across the ith inductor and current into the ith capacitor.

11.17. Prove that the inverse of a square $n \times n$ matrix A can be expressed in terms of the coefficients of the characteristic equation and powers of A from 0 to $n - 1$. Use the result to find the inverse of the A matrix in Example 11.9.

Application

11.18. Consider a homogeneous third-order set of state equations of the form $\dot{Y} = AY$. The zero-input solution will take the form

$$\begin{bmatrix} y_1 \\ y_2 \\ y_3 \end{bmatrix} = \begin{bmatrix} c_{11} \\ c_{21} \\ c_{31} \end{bmatrix} e^{s_1 t} + \begin{bmatrix} c_{12} \\ c_{22} \\ c_{32} \end{bmatrix} e^{s_2 t} + \begin{bmatrix} c_{13} \\ c_{23} \\ c_{33} \end{bmatrix} e^{s_3 t}$$

Show that by proper choice of the initial conditions y_{1+}, y_{2+}, y_{3+} it is always possible to make one of the column matrices of coefficients zero, thereby suppressing one natural frequency. For the set of equations in Example 11.8, choose v_{1+}, v_{2+}, and i_+ so as to make the $s = -\frac{1}{3}$ natural frequency disappear from the zero-input solution.

11.19. For the Anderson bridge network shown, write state equations in the variables v_C and i_L. Find the natural frequencies of the network from the state equations when the bridge is balanced, i.e., when

$$R_1R_4 = R_2R_3 \qquad \text{and} \qquad \frac{L}{C} = \frac{R_2}{R_1}(R_1R_3 + R_3R_5 + R_5R_1)$$

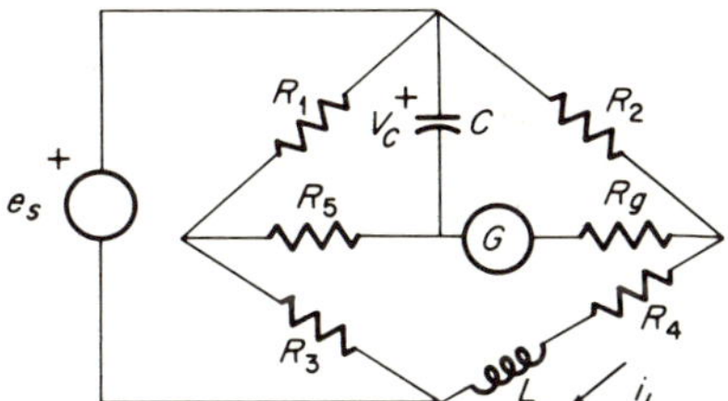

Fig. P11.19

11.20. For the bridged-T network with resistive load shown, write state equations using the capacitor voltages as state variables. From these equations and with $G_1 = G_2 = 1$ mho and $C_1 = C_2 = 1$ farad, find the natural frequencies of the bridged-T network as a function of G_L. Plot the natural frequencies vs. G_L for values of G_L from 0 to 1000 mhos.

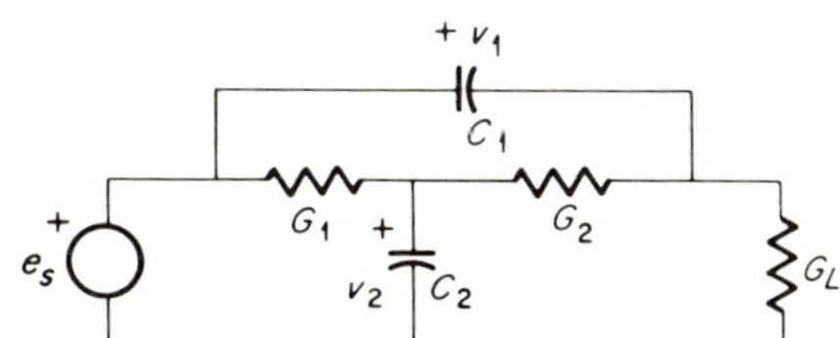

Fig. P11.20

12 COMPLEX-VARIABLE THEORY

12.1 INTRODUCTION

The theory of functions of a complex-variable and electric-network theory are inextricably tied together. For the student who wishes to have an understanding in depth of electric-network theory, a working knowledge of complex variables is essential.

No single chapter can hope to cover all facets of complex-variable theory. On the other hand, the portions of complex-variable theory pertaining directly to networks can be presented in sufficient depth in a single chapter, and this is our purpose here.

We begin by reviewing the properties of complex numbers and functions, restricting the discussion in the main to single-valued functions, the ones most often found in linear networks. Meanings of definitions and manipulative skills are stressed. The concepts of limit and continuity are discussed, and from them the derivative of a complex function is formulated and the Cauchy-Riemann conditions are derived. Regularity and analyticity are defined and compared, and singularities are classified. This concludes the first portion of the chapter.

In the latter part of the chapter, complex integration is featured. A sequence of theorems is derived and used that simplifies complex integration and leads directly to Cauchy's residue theorem for integrating around a contour enclosing a number of singularities. Thus, in turn, Green's theorem, Cauchy's integral theorem, Cauchy's integral formula, and the Taylor series expansion of a complex function are presented. Finally, the evaluation of the inverse Laplace transform integral through Cauchy's residue theorem is developed.

The chapter is not to be considered as a treatise on complex variables, but it does provide the student with the depth and manipulative skill he needs to proceed into other areas of network theory, particularly network synthesis.

12.2 COMPLEX NUMBERS AND VARIABLES

Since it is assumed that the reader has considerable familiarity with the properties and manipulations of complex numbers, the only purposes of this section are to establish the notation to be used in the remainder of the chapter and to summarize the basic properties and manipulative skills that will be assumed and drawn upon.

We shall use, for the most part, two complex variables s and z and shall express them in a variety of forms, both cartesian and polar, depending on the use to which they are to be put. These forms are

$$s = \sigma + j\omega = |s| \underline{/\alpha} = \rho e^{j\alpha} = \rho \cos \alpha + j\rho \sin \alpha \tag{12.2.1}$$

and

$$z = x + jy = |z| \underline{/\theta} = re^{j\theta} = r \cos \theta + jr \sin \theta \tag{12.2.2}$$

where $\rho = \sqrt{\sigma^2 + \omega^2}$ is the modulus (absolute value) of s, $\alpha = \tan^{-1} (\omega/\sigma)$ its argument (positively traced counterclockwise from the plus σ axis), σ its real part, and ω (not $j\omega$) its imaginary part, and similarly for the parts of z. The variable s represents a point in the complex plane (or a vector from the origin to that point) and will generally be the independent variable, whereas z will be the dependent variable and a function of s, represented by

$$z = f(s) \tag{12.2.3}$$

A number of basic properties of complex numbers and variables are listed in Table 12.1. Verification of each is easily obtained by direct expansion in cartesian or polar form and use of the properties of real numbers. The first five show that complex numbers obey the commutative, associative, and distributive laws of algebra. The next eight concern complex conjugates and

TABLE 12.1 Properties of complex variables

1. $s_1 + s_2 = s_2 + s_1$
2. $s_1 s_2 = s_2 s_1$
3. $s_1 + (s_2 + s_3) = (s_1 + s_2) + s_3$
4. $s_1(s_2 s_3) = (s_1 s_2) s_3$
5. $s_1(s_2 + s_3) = s_1 s_2 + s_1 s_3$
6. $ss^* = |s|^2 = \sigma^2 + \omega^2$
7. $(s_1 + s_2)^* = s_1^* + s_2^*$
8. $(s_1 s_2)^* = s_1^* s_2^*$
9. $s + s^* = 2\sigma$
10. $s - s^* = j2\omega$
11. $|s_1 s_2| = |s_1|\,|s_2|$
12. $|s_1 + s_2| \leq |s_1| + |s_2|$
13. $|s_1 - s_2| \geq \big||s_1| - |s_2|\big|$
14. $\operatorname{Re} s_1 s_2 \neq \operatorname{Re} s_1 \operatorname{Re} s_2$
15. $\operatorname{Im} (s_1 s_2) \neq \operatorname{Im} s_1 \operatorname{Im} s_2$
16. $s^n = \rho^n e^{jn\alpha}$
17. $s^{1/n} = \rho^{1/n} e^{j(\alpha + 2\pi k)/n} \qquad k = 0, 1, 2, \ldots, n-1$
18. $\dfrac{ds}{dt} = \dfrac{d\alpha}{dt} + j\dfrac{d\omega}{dt}$
19. $\operatorname{Re} \dfrac{ds}{dt} = \dfrac{d}{dt} \operatorname{Re} s$
20. $\operatorname{Im} \dfrac{ds}{dt} = \dfrac{d}{dt} \operatorname{Im} s$

magnitudes of complex quantities. The inequalities in 12 and 13 are important and can be verified most easily by graphs. Properties 14 and 15 are included in a precautionary sense. Properties 16 and 17 are two forms of De Moivre's theorem. The $2\pi k$ is included in the argument of 17 because $s^{1/n}$ has n values for each finite nonzero value of s. We shall discuss single- and multiple-valued functions in the next section. Finally, properties 18 to 20 are derivative relationships. The three operations listed are also valid for indefinite integrals of s.

12.3 COMPLEX FUNCTIONS

We have already noted that z is to be a dependent variable, functionally related to s as indicated by Eq. (12.2.3). By this functional relationship we mean that for each value of s there is a definite value or set of values of z.

A complex function of s may not be *defined* (may not *exist*) at one or more points in the s plane for one of several reasons. Consider the three functions

$$z_1 = \frac{1}{s} \qquad z_2 = \frac{s^2 + s}{s} \qquad z_3 = \operatorname{Arg} s \tag{12.3.1}$$

None is defined at $s = 0$: the first because z_1 is infinite there, the second because z_2 is indeterminate there, and the third because z_3 can have any value there.

A complex function may be either single- or multiple-valued. It is single-valued if there is only one value of z in the complex z plane corresponding to each value of s in the complex s plane. We then say that z has a *unique* value for each value of s. Figure 12.1 illustrates the situation. In Fig. 12.1*a*, we show two points in the s plane and the vectors they define. Now consider two functions of s

$$z_1 = s^2 \qquad \text{and} \qquad z_2 = s^{1/2} \tag{12.3.2}$$

The first is a single-valued function; for each s value we square its modulus and double its argument to obtain the corresponding single value of z_1, as illustrated in Fig. 12.1*b*.

Turning to the second function, we note that it is multiple-valued; there are two values of z_2 for each value of s. To obtain these values, we use property 17 in Table 12.1 with $k = 0$ and 1 for each value of s. For example, for $s = 2\underline{/120^\circ}$ in Fig. 12.1*a*, we obtain $z_2 = \sqrt{2}\,\underline{/60^\circ}$ and $\sqrt{2}\,\underline{/240^\circ}$ in Fig. 12.1*c*.

Throughout the remainder of the text we shall be dealing almost exclusively with single-valued functions. The reader interested in studying multiple-valued functions and the associated topics of branches, branch cuts, branch points, and Riemann surfaces is urged to read Refs. 12.1 and 12.3.

There are many types of complex functions, but basically they fall into two general categories, *algebraic* and *transcendental*. Any function obtained by performing the operations of addition, subtraction, multiplication, or division on s or raising s to a positive or negative power is an algebraic function. All others are transcendental. Algebraic functions can be *rational* or *irrational*, *integral* or *fractional*. A rational algebraic function contains only integer powers of s; an integral one only positive powers of s. A *polynomial* is a rational integral algebraic function and is a power series in s

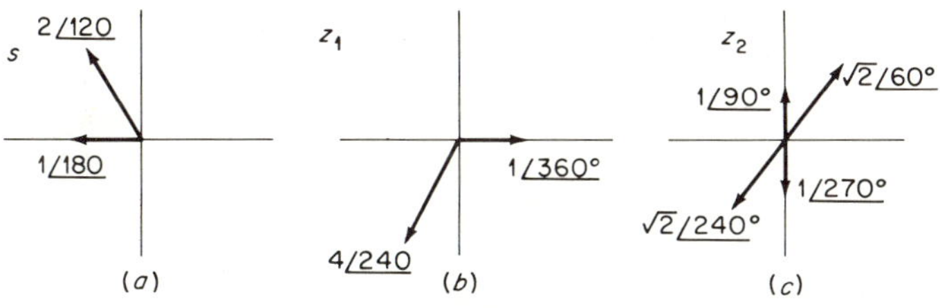

Fig. 12.1 Single- and multiple-valued functions.

of the form $z = \sum_{0}^{n} a_k s^k$. It is with ratios of polynomials that we are primarily concerned in network theory.

Transcendental functions include the exponential, trigonometric, hyperbolic, and logarithmic functions and their inverses. They also include non-algebraic functions of the type $z = s + s^*$ and $z = ss^*$. Defining relations for the former group follow:

$$e^s = e^{\sigma + j\omega} = e^\sigma(\cos \omega + j \sin \omega) \tag{12.3.3}$$

$$\cos s = \frac{e^{js} + e^{-js}}{2} = \cos \sigma \cosh \omega - j \sin \sigma \sinh \omega \tag{12.3.4}$$

$$\sin s = \frac{e^{js} - e^{-js}}{2j} = \sin \sigma \cosh \omega + j \cos \sigma \sinh \omega \tag{12.3.5}$$

$$\cosh s = \frac{e^{s} + e^{-s}}{2} = \cosh \sigma \cos \omega + j \sinh \sigma \sin \omega \tag{12.3.6}$$

$$\sinh s = \frac{e^{s} - e^{-s}}{2} = \sinh \sigma \cos \omega + j \cosh \sigma \sin \omega \tag{12.3.7}$$

$$\ln s = \ln \rho e^{j(\alpha + 2k\pi)} = \ln \rho + j(\alpha + 2k\pi) \tag{12.3.8}$$

$$\pm\cos^{-1} s = -j \ln [s + (s^2 - 1)^{1/2}] \tag{12.3.9}$$

$$\sin^{-1} s = -j \ln [js + (1 - s^2)^{1/2}] \tag{12.3.10}$$

$$\pm\cosh^{-1} s = \ln [s + (s^2 - 1)^{1/2}] \tag{12.3.11}$$

$$\sinh^{-1} s = \ln [s + (s^2 + 1)^{1/2}] \tag{12.3.12}$$

The representation of e^s in Eq. (12.3.3) is truly by definition. When s is real, each side of the equation reduces to e^σ. When s is imaginary, the well-known Euler equation results. Thus the definition is consistent with previous knowledge.

The definition of the exponential function leads to definitions of the trigonometric and hyperbolic functions in Eqs. (12.3.4) to (12.3.7). The tangent and other trigonometric and hyperbolic functions are obtained from these four in the same way as for real variables. Also, all the trigonometric and hyperbolic identities for real variables apply when the variables are complex, as do the customary differentiation and integration relationships. It is useful to note, from the defining relations, that the trigonometric and hyperbolic functions are related by

$$\cos s = \cosh js \qquad j \sin s = \sinh js \tag{12.3.13}$$

$$\cosh s = \cos js \qquad j \sinh s = \sin js \tag{12.3.14}$$

The exponential, trigonometric, and hyperbolic functions are single-valued. This is not true of the logarithmic and inverse trigonometric and hyperbolic functions. The logarithmic function in Eq. (12.3.8) has infinitely many values for each value of s. The value of $\ln s$ for $k = 0$ is called its *principal value* and is denoted by $\text{Ln } s$. The function $\text{Ln } s$ is a single-valued function. The inverse trigonometric and hyperbolic equations are derived from the trigonometric and hyperbolic equations. We shall illustrate some of the techniques in handling transcendental functions through four examples.

Example 12.1 Derive Eq. (12.3.12).

We begin by defining

$$z = \sinh^{-1} s$$

and thus, from Eq. (12.3.7),

$$s = \sinh z = \frac{e^z - e^{-z}}{2}$$

Forming a quadratic in e^z, we obtain

$$e^{2z} - 2se^z - 1 = 0$$

whose roots are

$$e^z = s + (s^2 + 1)^{1/2}$$

We do not write $\pm$ to indicate two roots because $(s^2 + 1)^{1/2}$ is already a double-valued function of s. Now taking the logarithm of e^z, we obtain

$$z = \sinh^{-1} s = \ln [s + (s^2 + 1)^{1/2}]$$

Example 12.2 Find the poles of $z = 1/(1 - 2 \cosh s)$.

The poles of z are the zeros of its denominator. Thus we wish to find the roots of

$$\cosh s = 1/2$$

which, from Eq. (12.3.6), requires that

$$\cosh \sigma \cos \omega + j \sinh \sigma \sin \omega = 1/2$$

yielding the equations

$$\cosh \sigma \cos \omega = 1/2 \qquad \sinh \sigma \sin \omega = 0$$

If we try to make the second equation zero by requiring that $\sin \omega = 0$, we are left with $\cosh \sigma = 1/2$, which cannot be satisfied for any real σ. Thus we require instead that $\sinh \sigma = 0$, yielding $\sigma = 0$ and $\cosh \sigma = 1$. This makes $\cos \omega = 1/2$, giving

$$\omega = \pm\left(\frac{\pi}{3} + 2k\pi\right) \qquad k = 0, 1, 2, \ldots$$

Finally the poles of the given function occur at

$$s = \pm j\left(\frac{\pi}{3} + 2k\pi\right) \qquad k = 0, 1, 2, \ldots$$

Example 12.3 Show that

$$|\cosh s|^2 = \sinh^2 \sigma + \cos^2 \omega$$

From Eq. (12.3.6)

$$\cosh s = \cosh \sigma \cos \omega + j \sinh \sigma \sin \omega$$

The magnitude squared is then given by

$$|\cosh s|^2 = \cosh^2 \sigma \cos^2 \omega + \sinh^2 \sigma \sin^2 \omega$$

Recalling the trigonometric identity

$$\sin^2 \omega = 1 - \cos^2 \omega$$

we have

$$|\cosh s|^2 = \cos^2 \omega (\cosh^2 \sigma - \sinh^2 \sigma) + \sinh^2 \sigma$$

The quantity in parentheses equals 1 from the familiar hyperbolic identity, giving the desired result.

Example 12.4 Find the poles of $z = 1/(1 - s^6)$.

We need the roots of $s^6 = 1$. From property 17 in Table 12.1,

$$1^{1/6} = 1e^{j(0+2\pi k)/6} \qquad k = 0, 1, \ldots, 5$$

which, from Eq. (12.3.3), gives

$$1^{1/6} = \cos \frac{\pi k}{3} + j \sin \frac{\pi k}{3} \qquad k = 0, 1, \ldots, 5$$

The roots lie on a unit circle locus about the origin and separated from each other by 60°.

12.4 REGIONS

In the work to follow, we shall be concerned with various types of regions in the complex plane. An appropriate description of the region involved could be given each time we discuss a new theorem or application, but it is simpler to define and name the important ones now.

We define an *open region* as any region whose boundary is not included in the region. Examples are

$$|s| < k_1 \qquad |s| > k_2 \qquad k_3 < |s| < k_4 \tag{12.4.1}$$

where the k's are positive real constants. The important property to note here is the absence of equal signs in the inequalities. A *closed region* is one whose boundary is included in the region. A region with two boundaries may be open with respect to one and closed with respect to the other. The third inequality in (12.4.1) would be this kind of region if an equals sign were added to one of the inequality signs.

A region is *bounded* if all its points lie within some circle centered at the origin of the complex plane. The first and third regions in (12.4.1) are bounded; the second is not.

A region is *connected* if any two points in the region can be joined by a continuous path wholly within the region. A region is *simply connected* if every closed path within the region does not encircle points outside the region.

Consider the regions (shaded areas) in Fig. 12.2. In Fig. 12.2*a*, the region is described by

$$|s| < 1 \qquad \text{and} \qquad \text{Re}\, s > 2 \tag{12.4.2}$$

whereas in Fig. 12.2*b*,

$$|s| > 1 \qquad \text{and} \qquad |s| < 3 \tag{12.4.3}$$

describes the region. The region in Fig. 12.2*a* is disconnected: a point in the left section of the region cannot be joined to a point in the right section of the region by a line wholly within the region. The region in Fig. 12.2*b* is doubly connected. Any path in the region that encircles the origin encircles points not in the region. It is doubly (rather than higher-order) connected because only *one* cut, as illustrated in Fig. 12.2*c*, is required to convert it to a simply connected region.

A *neighborhood* is a small open circular region defined by

$$|s - s_0| < \delta \tag{12.4.4}$$

where δ is a small positive real number. Note that s_0 is included in the neighborhood but points on the circle are not. If the point s_0 is excluded from the neighborhood, we have a *deleted neighborhood*.

12.5 LIMITS

Now that the complex plane and the algebra of complex numbers have been reviewed it is time to develop the calculus of complex variables. We begin with the concepts of limit and continuity.

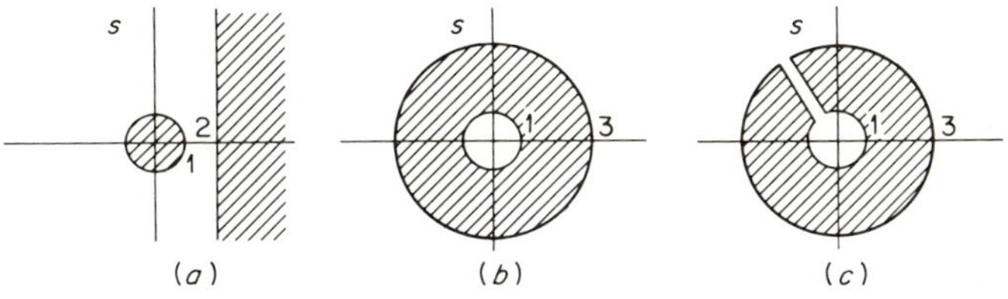

Fig. 12.2 Connectivity.

Let $z = f(s)$ be a single-valued function of s defined at all points in some deleted neighborhood of a point $s = s_0$. Then

$$\lim_{s \to s_0} z = L_0 \tag{12.5.1}$$

if, for every positive real number ϵ, there exists a positive real number δ such that

$$|z - L_0| < \epsilon \qquad \text{when } 0 < |s - s_0| < \delta \tag{12.5.2}$$

over any path within the deleted neighborhood. The limit can exist when $f(s_0)$ does not exist and also when $f(s_0)$ exists but is not equal to L_0. The zero in the second inequality in (12.5.2) permits these two situations. However, the function must be defined at all other points in the neighborhood of s_0 for the limit to exist.

We have indicated that the limit L_0 in Eq. (12.5.1) must be unique. If it were not, if, for example, two limits, L_{01} and L_{02}, existed, we would have to require

$$|z - L_{01}| - |z - L_{02}| \leq |z - L_{01}| + |z - L_{02}| < 2\epsilon \tag{12.5.3}$$

which would impose a constraint that we cannot satisfy for an arbitrarily small ϵ.

By recognizing that both z and L_0 in Eq. (12.5.1) are complex quantities, each composed of two real variables, we can rewrite Eq. (12.5.1) as

$$\lim_{\substack{\sigma \to \sigma_0 \\ \omega \to \omega_0}} (x + jy) = M_0 + jN_0 \tag{12.5.4}$$

which yields

$$\lim_{\substack{\sigma \to \sigma_0 \\ \omega \to \omega_0}} x = M_0 \qquad \lim_{\substack{\sigma \to \sigma_0 \\ \omega \to \omega_0}} y = N_0 \tag{12.5.5}$$

These results let us establish directly certain theorems on limits of two complex functions from the corresponding theorems for two real functions. Let z_1 and z_2 be two complex functions of s whose limits as $s \to s_0$ are L_{01} and L_{02}, respectively. Then

$$\lim_{s \to s_0} (z_1 + z_2) = L_{01} + L_{02} \tag{12.5.6}$$

$$\lim_{s \to s_0} z_1 z_2 = L_{01} L_{02} \tag{12.5.7}$$

$$\lim_{s \to s_0} \frac{z_1}{z_2} = \frac{L_{01}}{L_{02}} \qquad L_{02} \neq 0 \tag{12.5.8}$$

We shall illustrate the actual process of finding a limit using Eqs. (12.5.1) and (12.5.2) through an example.

Example 12.5 Given that

$$z = \frac{s^2 + 3s + 2}{s^2 + 4s + 3}$$

show that

$$\lim_{s \to -1} z = \tfrac{1}{2}$$

Note that z evaluated at $s = -1$ is indeterminate ($0/0$). What we wish to show, from Eq. (12.5.2), is that

$$\left| \frac{s+2}{s+3} - \frac{1}{2} \right| = \left| \frac{s+1}{2(s+3)} \right| < \epsilon \qquad \text{when } 0 < |s+1| < \delta$$

Try 4ϵ as the small positive real δ. Then, from Fig. 12.3,

$$0 < |s+1| < 4\epsilon \qquad \text{and} \qquad 2 - 4\epsilon < |s+3| < 2 + 4\epsilon$$

The small circle in Fig. 12.3 has a radius 4ϵ. The magnitude of $s + 1$ must be less than 4ϵ; thus the point s must lie within the circle. Then it is easily seen that the magnitude of $s + 3$ must be bounded between $2 - 4\epsilon$ and $2 + 4\epsilon$.

To show that $z - \frac{1}{2}$ is always less than ϵ, we let the numerator of $z - \frac{1}{2}$ take on its upper bound of 4ϵ and the denominator its lower bound of $2(2 - 4\epsilon)$, giving

$$\left| \frac{s+1}{2(s+3)} \right| < \frac{4\epsilon}{2(2 - 4\epsilon)}$$

As ϵ becomes very small, this quantity approaches ϵ in value. Thus for a positive real number ϵ, there is a positive real number $\delta = 4\epsilon$ such that

$$|z - \tfrac{1}{2}| < \epsilon \qquad \text{when } 0 < |s+1| < \delta$$

and therefore

$$\lim_{s \to -1} z = \tfrac{1}{2}$$

Example 12.6 Find the limit as $s \to j1$ and $s \to 0$ of

$$f(s) = \lim_{n \to \infty} \frac{1}{1 - s^{2n}}$$

This function is infinite-valued at $s = \pm j1$ but single-valued at $s = 0$. As $s \to 0$, the limit is 1. As $s \to \pm j1$, the limit does not exist because it is not unique. Approaching $s = j1$ from $s = 0$ yields a limiting value of 1. Approaching $s = j1$ from $s = j2$ yields a limiting value of 0.

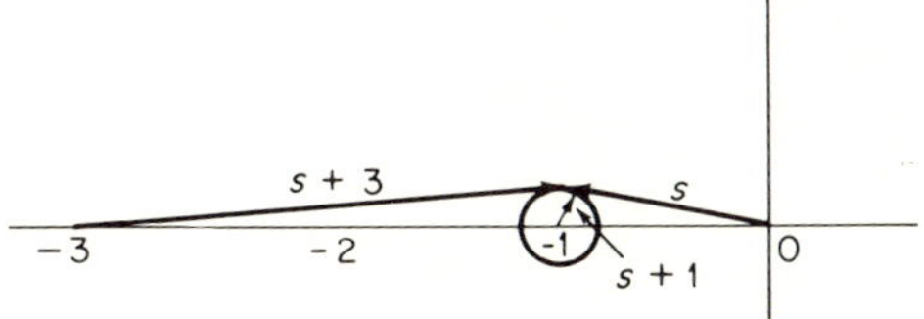

Fig. 12.3 Graphical interpretation of inequalities.

12.6 CONTINUITY AND THE DERIVATIVE

In the previous section it was pointed out that $\lim_{s \to s_0} f(s)$ could exist even though $f(s_0)$ did not exist or existed but did not equal the value of the limit. Now require that $f(s_0)$ does exist, i.e., that z be defined at s_0 so that the neighborhood is no longer a deleted one. Then z is continuous at s_0 if

$$\lim_{s \to s_0} f(s) = f(s_0) \equiv z_0 \tag{12.6.1}$$

Note carefully that Eq. (12.6.1) actually implies that three conditions are met, namely, that $\lim f(s)$ exists, that $f(s_0)$ exists, and that $\lim_{s \to s_0} f(s) = f(s_0)$.

The function in Example 12.5 is not continuous at $s = -1$ because the function is indeterminate there. On the other hand, the function in Example 12.6 is not continuous at $s = \pm j1$ because the function is infinite-valued there. This latter function is continuous at $s = 0$.

We can express Eq. (12.6.1) in terms of an equation similar to Eq. (12.5.2). We say that z is continuous at $s = s_0$ if, for every positive real number ϵ, there exists a positive real number δ such that

$$|z - z_0| < \epsilon \qquad \text{when } |s - s_0| < \delta \tag{12.6.2}$$

With the concepts of limit and continuity now in mind, we are prepared to discuss derivatives of complex functions. Let $z = f(s)$ be continuous at $s = s_0$. Then

$$f'(s_0) = \lim_{s \to s_0} \frac{f(s) - f(s_0)}{s - s_0} \tag{12.6.3}$$

is the derivative of $f(s)$ evaluated at $s = s_0$ if the limit on the right side exists. For it to exist, we know from Eq. (12.5.3) that it must be unique. Thus continuity of $f(s)$ at $s = s_0$ (which makes the quotient in Eq. (12.6.3) indeterminate rather than infinite) is a necessary condition for $f'(s_0)$ to exist but not a sufficient one. We shall illustrate this point through an example.

Example 12.7 Find the derivatives, if they exist, of $f(s) = s^2$ and $f(s) = ss^*$ at $s = s_0$.

(*a*) Forming Eq. (12.6.3) for $f(s) = s^2$, we have

$$\lim_{s \to s_0} \frac{s^2 - s_0^2}{s - s_0} = \lim_{s \to s_0} \frac{2s}{1} = 2s_0$$

or

$$\lim_{s \to s_0} \frac{s^2 - s_0^2}{s - s_0} = \lim_{\Delta s \to 0} \frac{(s_0 + \Delta s)^2 - s_0^2}{\Delta s} = \lim_{\Delta s \to 0} \frac{2s_0\,\Delta s + (\Delta s)^2}{\Delta s} = 2s_0$$

In the first evaluation, L'Hospital's rule is used to evaluate the limit; in the second evaluation $s - s_0$ is replaced by the small complex quantity Δs to obtain the limit.

In either case the limit is unique, and thus

$$f'(s_0) = 2s_0$$

(*b*) Use of Eq. (12.6.3) when $f(s) = ss^*$ yields

$$\lim_{s\to s_0} \frac{ss^* - s_0s_0^*}{s - s_0} = \lim_{s\to s_0} \left(s_0^* + s\,\frac{s^* - s_0^*}{s - s_0}\right)$$

Resorting to the polar form in Eq. (12.2.2), let $s - s_0 = \rho e^{j\alpha}$ and the limit expression becomes

$$\lim_{s\to s_0} (s_0^* + se^{-j2\alpha})$$

Except for $s_0 = 0$, the value of the quantity in parentheses as $s \to s_0$ depends on the path of approach of s to s_0. Thus the limit is not unique, and the derivative does not exist except at $s_0 = 0$, where the derivative is zero.

The fundamental differentiation formulas applicable to functions of real variables also apply to functions of complex variables. As in the case of the theorems on limits, we justify this conclusion by saying that a complex variable is composed of two real variables. Derivations could be carried out by following the procedures used for real variables. We give only the results here.

If z_1 and z_2 are two functions of s whose derivatives exist, then

$$\frac{d}{ds}(z_1 + z_2) = \frac{dz_1}{ds} + \frac{dz_2}{ds} \tag{12.6.4}$$

$$\frac{d}{ds} z_1z_2 = z_1\frac{dz_2}{ds} + z_2\frac{dz_1}{ds} \tag{12.6.5}$$

$$\frac{d}{ds}\frac{z_1}{z_2} = \frac{z_2\, dz_1/ds - z_1\, dz_2/ds}{z_2^{\,2}} \qquad z_2 \neq 0 \tag{12.6.6}$$

$$\frac{d}{ds}\frac{1}{z_2^{\,n}} = \frac{-n}{z_2^{n+1}}\frac{dz_2}{ds} \qquad \begin{array}{l} n = \text{positive integer} \\ z_2 \neq 0 \end{array} \tag{12.6.7}$$

$$\frac{d}{ds} z_1[z_2(s)] = \frac{dz_1}{dz_2}\frac{dz_2}{ds} \tag{12.6.8}$$

$$\frac{d}{ds} z_2^{\,n} = nz_2^{n-1}\frac{dz_2}{ds} \qquad n = \text{positive integer} \tag{12.6.9}$$

The notation $z_2 \neq 0$ in Eq. (12.6) is necessary because if z_2 were zero at $s = s_0$, the function z_1/z_2 would not be continuous there, and hence the derivative would not be defined.

12.7 CAUCHY-RIEMANN CONDITIONS

If we are given a function $z = f(s)$ and asked to find whether its derivative exists at a point $s = s_0$, we have no recourse but to evaluate the limit in Eq. (12.6.3). In this section an alternate, and often easier, procedure for ascertaining the existence or nonexistence of the derivative is developed.

Assume that the given function $z = f(s)$ does have a derivative at $s = s_0$. Then that derivative is unique and does not depend on the path over which s approaches s_0 in Eq. (12.6.3). We can use this fact to develop a useful set of conditions that $z(s)$ must satisfy.

Insert $z(s)$ and s in their cartesian forms into Eq. (12.6.3) to obtain

$$f'(s_0) = \lim_{\substack{\sigma \to \sigma_0 \\ \omega \to \omega_0}} \frac{x(s) + jy(s) - x(s_0) - jy(s_0)}{\sigma + j\omega - \sigma_0 - j\omega_0}$$

$$= \lim_{\substack{\Delta\sigma \to 0 \\ \Delta\omega \to 0}} \frac{\Delta x(s) + j\,\Delta y(s)}{\Delta\sigma + j\,\Delta\omega} \tag{12.7.1}$$

If s is required to approach s_0 along a horizontal line in the s plane, $\Delta\omega$ will be zero and Eq. (12.7.1) becomes

$$f'(s) = \lim_{\Delta\sigma \to 0} \frac{\Delta x(s) + j\,\Delta y(s)}{\Delta\sigma} = \frac{\partial x}{\partial\sigma} + j\frac{\partial y}{\partial\sigma} \tag{12.7.2}$$

where s_0 has been replaced by the more general label s. On the other hand, if s approaches s_0 along a vertical line in the s plane, $\Delta\sigma$ is zero

$$f'(s) = \lim_{\Delta\omega \to 0} \frac{\Delta x(s) + j\,\Delta y(s)}{j\,\Delta\omega} = \frac{\partial y}{\partial\omega} - j\frac{\partial x}{\partial\omega} \tag{12.7.3}$$

Now Eqs. (12.7.2) and (12.7.3) must yield the same values for $f'(s)$. Equating reals to reals and imaginaries to imaginaries in these two equations yields a pair of relations known as the *Cauchy-Riemann conditions*. They are

$$\frac{\partial x(s)}{\partial\sigma} = \frac{\partial y(s)}{\partial\omega} \qquad \text{and} \qquad \frac{\partial x(s)}{\partial\omega} = -\frac{\partial y(s)}{\partial\sigma} \tag{12.7.4}$$

If the derivative of $z = f(s)$ exists at $s = s_0$, the Cauchy-Riemann conditions must hold there. Thus satisfaction of the Cauchy-Riemann conditions is *necessary* for the existence of the derivative. If they are not satisfied, the derivative does not exist. On the other hand, if they are satisfied, we cannot as yet conclude that the derivative does exist. We must show whether or not satisfaction of the Cauchy-Riemann conditions is *sufficient* for the existence of the derivative. This is done through the total differential of z.

Assume that the four partial derivatives in Eq. (12.7.4) are continuous functions. Then the total differentials of x and y are

$$dx = \frac{\partial x}{\partial \sigma}\,d\sigma + \frac{\partial x}{\partial \omega}\,d\omega \qquad dy = \frac{\partial y}{\partial \sigma}\,d\sigma + \frac{\partial y}{\partial \omega}\,d\omega \tag{12.7.5}$$

and, from these, the total differential of z is

$$dz = dx + j\,dy = \left(\frac{\partial x}{\partial \sigma} + j\frac{\partial y}{\partial \sigma}\right) d\sigma + \left(\frac{\partial x}{\partial \omega} + j\frac{\partial y}{\partial \omega}\right) d\omega \tag{12.7.6}$$

Now the Cauchy-Riemann conditions permit Eq. (12.7.6) to be expressed in either of the following equivalent forms:

$$dz = \left(\frac{\partial x}{\partial \sigma} + j\frac{\partial y}{\partial \sigma}\right)(d\sigma + j\,d\omega) = \left(\frac{\partial y}{\partial \omega} - j\frac{\partial x}{\partial \omega}\right)(d\sigma + j\,d\omega) \tag{12.7.7}$$

From Eq. (12.7.7) we form the derivative $f'(s) = dz/ds$ and obtain

$$f'(s) = \frac{\partial x}{\partial \sigma} + j\frac{\partial y}{\partial \sigma} = \frac{\partial y}{\partial \omega} - j\frac{\partial x}{\partial \omega} \tag{12.7.8}$$

Thus if the Cauchy-Riemann conditions are satisfied by x and y, and if the first partial derivatives of x and y with respect to σ and ω are continuous, $z = x + jy$ has a derivative.

Example 12.8 Using the Cauchy-Riemann conditions, show that the first function in Example 12.7 possesses a derivative whereas the second one does not except at $s = 0$.

In cartesian form, the first function is

$$z = s^2 = (\sigma + j\omega)^2 = \sigma^2 - \omega^2 + j2\sigma\omega = x + jy$$

The Cauchy-Riemann relations yield

$$\frac{\partial x}{\partial \sigma} = 2\sigma = \frac{\partial y}{\partial \omega} \qquad \frac{\partial x}{\partial \omega} = -2\omega = -\frac{\partial y}{\partial \sigma}$$

and are seen to hold for all values of s. The second function in cartesian form is

$$z = ss^* = (\sigma + j\omega)(\sigma - j\omega) = \sigma^2 + \omega^2 + j0 = x + j0$$

giving

$$\frac{\partial x}{\partial \sigma} = 2\sigma \neq \frac{\partial y}{\partial \omega} = 0 \qquad \frac{\partial x}{\partial \omega} = 2\omega \neq -\frac{\partial y}{\partial \sigma} = 0$$

The Cauchy-Riemann conditions are seen to hold only at $s = 0$.

Example 12.9 The impedance of a lossless parallel resonant circuit is given by $Z = s/(s^2 + 1)$. Find the partial derivatives of the real part of Z with respect to σ for $\sigma = 0$ and of the imaginary part of Z with respect to ω for $\omega = 0$.

There are at least two ways to work this problem. If we let $s = \sigma + j\omega$ in Z and conjugate the resulting expression to display $\operatorname{Re} Z = R$ and $\operatorname{Im} Z = X$, we can then take the partial derivatives directly. However, the conjugated Z expression is

$$Z = \frac{\sigma(\sigma^2 + \omega^2 + 1) + j\omega(1 - \sigma^2 - \omega^2)}{(\sigma^2 - \omega^2 + 1)^2 + 4\sigma^2\omega^2}$$

and obtaining the partial derivatives directly from this expression is relatively tedious because we must first differentiate and then insert the $\sigma = 0$ and $\omega = 0$ constraints.

Alternatively, we note that $Z(s)$ has a derivative everywhere except at $s = \pm j1$. Hence the Cauchy-Riemann conditions hold except at those points. We wish to find

$$\left.\frac{\partial R}{\partial \sigma}\right|_{\sigma=0} \quad \text{and} \quad \left.\frac{\partial x}{\partial \omega}\right|_{\omega=0}$$

From the Cauchy-Riemann conditions

$$\left.\frac{\partial R}{\partial \sigma}\right|_{\sigma=0} = \left.\frac{\partial x}{\partial \omega}\right|_{\sigma=0} \quad \text{and} \quad \left.\frac{\partial X}{\partial \omega}\right|_{\omega=0} = -\left.\frac{\partial R}{\partial \sigma}\right|_{\omega=0}$$

It is relatively easy to compute the *right* sides of these equations because we can let σ in the first equation and ω in the second equation be zero *before* differentiating. Thus

$$\left.\frac{\partial X}{\partial \omega}\right|_{\sigma=0} = \frac{\partial[\omega/(1 - \omega^2)]}{\partial \omega} = \frac{1 + 2\omega - \omega^2}{(1 - \omega^2)^2}$$

and

$$\left.\frac{\partial R}{\partial \sigma}\right|_{\omega=0} = \frac{\partial[\sigma/(\sigma^2 + 1)]}{\partial \sigma} = \frac{(1 - \sigma)^2}{(1 + \sigma^2)^2}$$

and therefore

$$\left.\frac{\partial R}{\partial \sigma}\right|_{\sigma=0} = \frac{1 + 2\omega - \omega^2}{(1 - \omega^2)^2} \quad \text{and} \quad \left.\frac{\partial X}{\partial \omega}\right|_{\omega=0} = -\frac{(1 - \sigma)^2}{(1 + \sigma^2)^2}$$

12.8 REGULARITY, ANALYTICITY, AND SINGULARITIES

Regularity is a property of a complex function at a point in the s plane. If $z = f(s)$ has a derivative at every point in some neighborhood (not deleted) about $s = s_0$, then z is said to be *regular* at s_0 and s_0 is a *regular point.* If z does not possess a derivative at each point in the neighborhood of s_0, s_0 is said to be a *singular point*, or *singularity*, of z.

Analyticity is an overall property of a complex function. If $z = f(s)$ is regular at one point at least in the s plane, z is called an *analytic* function of s. Thus an analytic function can have points (singularities) at which it is not regular.

The function $z = s/(s^2 + 1)$ is an analytic function regular at all points except $s = \pm j1$, which are singularities. The function $z = ss^*$ is not an analytic function. It does possess a derivative at $s = 0$ but not in the neighborhood of $s = 0$. Thus the function has singularities everywhere, including $s = 0$.

If $f(s)$ is an analytic function, and if $g(s)$ is a second analytic function, then $g[f(s)]$ is an analytic function. In other words, an analytic function of an analytic function is analytic. For example, if

$$f(s) = s + \frac{2}{s} \qquad \text{and} \qquad g(s) = \frac{s}{s^2 + 1}$$

then

$$g[f(s)] = \frac{f(s)}{f^2(s) + 1} = \frac{s + 2/s}{s^2 + 4 + 4/s^2 + 1} = \frac{s(s^2 + 2)}{s^4 + 5s^2 + 4}$$

is analytic since $f(s)$ and $g(s)$ are analytic.

Note that we have two ways of checking whether or not a given complex function is analytic. The first way is by direct use of the definition of the derivative in Eq. (12.6.3). By showing the existence of the derivative in some region of the s plane we show that the function is regular there and hence analytic. The second way is through the Cauchy-Riemann conditions. If they hold in a region, the derivative exists in that region, and thus the function is analytic.

Three types of singularities may be encountered, namely, poles, essential singularities, and branch points. To define a *pole*, assume that $z = f(s)$ is single-valued but not regular at $s = s_0$. Next assume that $(s - s_0)^n z$ is regular at s_0 for some least positive integer n. Then z is said to have a pole of degree n at $s = s_0$. For example, let

$$z = \frac{s + 3}{(s + 1)(s + 2)^2}$$

a function that is not regular at $s = -1$ and $s = -2$. Now $(s + 2)^2 z$ is regular at $s = -2$, and $(s + 1)z$ is regular at $s = -1$. Thus z has a second-degree pole at $s = -2$ and a first-degree pole at $s = -1$.

To define an *essential singularity* we begin as in defining the pole, but this time there is no positive integer n that makes $(s - s_0)^n z$ regular at $s = s_0$. An example of such a function is $z = \cosh(1/s)$, which can be expanded into

$$z = 1 + \frac{1}{2!\,s^2} + \frac{1}{4!\,s^4} + \frac{1}{6!\,s^6} + \cdots$$

There is no value of n that will make $s^n z$ regular at $s = 0$. Therefore $s = 0$ is an essential singularity of z.

The third type of singularity is a *branch point*. It arises when we are dealing with multiple-valued functions; as mentioned in Sec. 12.3, such functions will not be considered further in this book.

As stated earlier, the predominant concern in network theory is with ratios of polynomials. Hence the singularity that will be of primary interest in the remainder of this chapter is the pole.

12.9 COMPLEX INTEGRATION

Assume a curved line in the complex plane joining two points s_1 and s_2. We define the line integral of a complex function $f(s)$ between these two points by

$$\int_{s_1}^{s_2} f(s)\,ds = \lim_{\substack{n\to\infty \\ \Delta_m\to 0}} \sum_{i=1}^{n} f(s_i)(s_i - s_{i-1}) \tag{12.9.1}$$

where Δ_m is the largest value of $s_i - s_{i-1}$ in the interval. The process described by Eq. (12.9.1) is diagrammed in Fig. 12.4. The vectors s_i and s_{i-1} are shown, as is their small difference vector $s_i - s_{i-1}$. We are instructed to multiply the value of $f(s)$ at s_i times this small difference vector and add all such products between s_1 and s_2. As we let the number of intervals become very large and the length of the largest interval become very small, the sum approaches and becomes, in the limit, an integral.

It is possible to transform Eq. (12.9.1) into the sum of real line integrals. To this end we use Eqs. (12.2.1) and (12.2.2) and write

$$\begin{aligned}\int_{s_1}^{s_2} f(s)\,ds &= \int_{s_1}^{s_2} (x + jy)(d\sigma + j\,d\omega) \\ &= \int_{s_1}^{s_2} (x\,d\sigma - y\,d\omega) + j\int_{s_1}^{s_2} (y\,d\sigma + x\,d\omega)\end{aligned} \tag{12.9.2}$$

where it is understood that x and y are, in general, functions of both σ and ω.

Example 12.10 Find the value of $\int f(s)\,ds$ clockwise around the closed circular contour c shown in Fig. 12.5 when $z = 1/(s + 1)$ and when $z = 1/(s + 3)$.

Integrations around circular contours are often carried out most easily by using the polar form of s. For $z = 1/(s + 1)$, if we let

$$s + 1 = \rho e^{j\alpha}$$

then only the angle of $s + 1$ will change as we integrate around the contour. It follows that

$$ds = j\rho e^{j\alpha}\,d\alpha$$

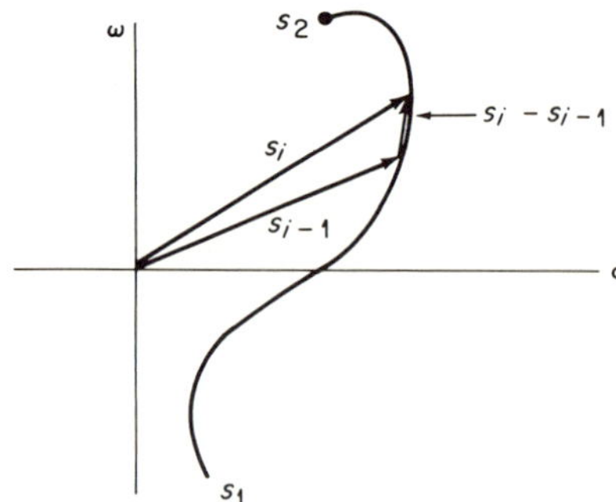

Fig. 12.4 Line integration.

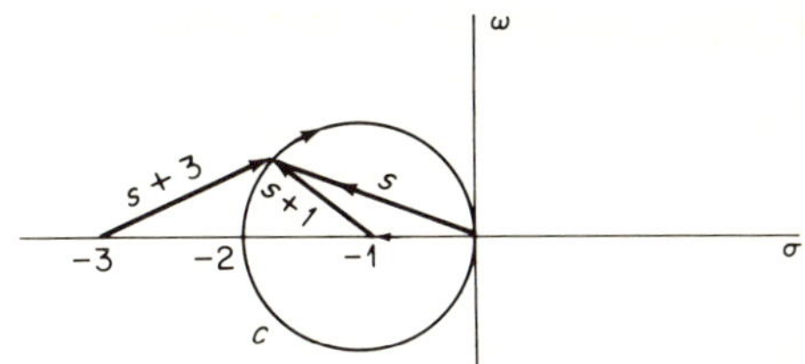

Fig. 12.5

and the line integral can be written

$$\oint_C \frac{ds}{s+1} = \int_{\pi}^{-\pi} \frac{j\rho e^{j\alpha}}{e^{j\alpha}}\, d\alpha = j\alpha \Big|_{\pi}^{-\pi} = -j2\pi$$

The limits on α require some explanation. We must begin and end the integration at the same point on the contour, and the choice of this point is arbitrary. We have chosen $\alpha = \pi$ as the starting point. As the integration proceeds, α changes from π to 0 and finally to $-\pi$. If we had started at $\alpha = \pi$ and proceeded counterclockwise, the final value of α would be 3π and

$$\oint_C \frac{ds}{s+1} = +j2\pi$$

For $z = 1/(s+3)$, if we let $s + 3 = \rho e^{j\alpha}$ and try to follow the previous procedure we quickly see that ρ is no longer constant during the integration and thus $ds \neq j\rho e^{j\alpha}\, d\alpha$. The integration becomes much more difficult. In Sec. 12.11, the Cauchy-integral theorem will be derived, from which the value of $\oint_C \frac{ds}{s+3}$ will immediately be seen to be zero, the reason being, as will be shown, that the singularity is outside the contour of integration; that is, z is regular within the contour.

Example 12.11 Find the value of $\int_0^{1+j} s^*\, ds$ (*a*) along the path $\sigma = \omega$ and (*b*) along the paths $\sigma = 0$ and $\omega = 1$.

(*a*) Putting the integral in cartesian form yields

$$\int_{0,0}^{1,1} (\sigma - j\omega)(d\sigma + j\, d\omega) = \int_{0,0}^{1,1} (\sigma\, d\sigma + \omega\, d\omega + j\sigma\, d\omega - j\omega\, d\sigma)$$
$$= 2\int_0^1 \sigma\, d\sigma = 1$$

where the fact that $\sigma = \omega$ and $d\sigma = d\omega$ along the path has been used to eliminate the ω variable.

(*b*) Here a two-part integration is required

$$\int_{0,0}^{1,1} (\sigma - j\omega)(d\sigma + j\, d\omega) = \int_0 \omega\, d\omega + \int_0^1 (\sigma - j1)\, d\sigma$$
$$= \frac{\omega^2}{2}\Big|_0^1 + \frac{\sigma^2}{2}\Big|_0^1 - j\sigma\Big|_0^1 = 1 - j$$

The value of the integral is seen to depend on the path chosen, the reason being that $z = s^*$ is not regular in the region bounded by the paths in (*a*) and (*b*).

12.10 GREEN'S THEOREM

The pattern for most of the remaining sections of this chapter is a series of derivations of integration properties of complex variables, each derivation requiring the previous one for its successful development. Ultimately Cauchy's residue theorem will be derived, which will permit an evaluation of the inverse Laplace integral, Eq. (9.5.9). The starting point is the derivation of Green's theorem for real line integrals.

Let $P(x,y)$ and $Q(x,y)$ be two functions of the real variables x and y. Assume that these functions and their first partial derivatives are continuous in a closed region R of the xy plane bounded by a closed curve c. Then, Green's theorem states that

$$\oint_c (P\,dx + Q\,dy) = \iint_R \left(\frac{\partial Q}{\partial x} - \frac{\partial P}{\partial y}\right) dx\,dy \tag{12.10.1}$$

To prove Eq. (12.10.1) let a function **F** be defined by

$$\mathbf{F} = P\,\mathbf{a}_x + Q\mathbf{a}_y \tag{12.10.2}$$

where $\mathbf{a}_x$ and $\mathbf{a}_y$ are unit vectors in the x and y directions. Now perform $\oint \mathbf{F}\cdot d\mathbf{l}$ around the small square path shown in Fig. 12.6*a*, from the starting point in the lower left-hand corner labeled P, Q. We note first that

$$\mathbf{l} = x\mathbf{a}_x + y\mathbf{a}_y \qquad d\mathbf{l} = dx\,\mathbf{a}_x + dy\,\mathbf{a}_y \tag{12.10.3}$$

Now the desired integral is

$$\oint \mathbf{F}\cdot d\mathbf{l} = P\,dx + \left(Q + \frac{\partial Q}{\partial x}\,dx\right) dy - \left(P + \frac{\partial P}{\partial y}\,dy\right) dx - Q\,dy \tag{12.10.4}$$

The first term is the integral along the bottom path in Fig. 12.6*a*. It results from forming the dot product of F and dx (dy is zero) and from assuming that the x component of F does not change from its initial value P over the infinitesimal path length. The second term in Eq. (12.10.3) is the integral along the rightmost path in Fig. 12.6*a*. It is the dot product of F and dy (dx now zero). The value of the y component of F along this infinitesimal path is assumed to remain constant at the value $Q + (\partial Q/\partial x)\,dx$. The third and fourth terms are obtained similarly. Collecting terms in Eq. (12.10.3) yields

$$\oint \mathbf{F}\cdot d\mathbf{l} = \left(\frac{\partial Q}{\partial x} - \frac{\partial P}{\partial y}\right) dx\,dy \tag{12.10.5}$$

Now consider many such tiny square regions adjacent to each other (four

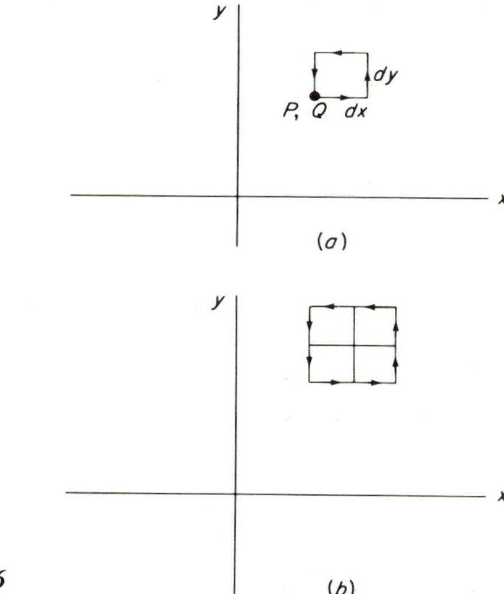

Fig. 12.6

are shown in Fig. 12.6*b*). If we perform the $\oint \mathbf{F} \cdot d\mathbf{l}$ around each of these squares and add the results, the integrations along all inner edges cancel. The left side of Eq. (12.10.5) becomes $\oint \mathbf{F} \cdot d\mathbf{l}$ around the boundary c of the nest of squares; the right side becomes the surface integral of the function in parentheses over the region R bounded by c. Recalling that the squares are infinitesimal in size, if we take a sufficient number of them and arrange them properly, we can produce a region of any shape we wish and arrive at Green's theorem in Eq. (12.10.1).

12.11 CAUCHY INTEGRAL THEOREM

Let $f(s) = z = x + jy$ be an analytic function that is regular in a closed region R of the s plane bounded by a closed curve c. Since the region is closed, the boundary is a part of the region. Performing $\oint f(s)\, ds$ around the boundary, using Eq. (12.9.2), yields

$$\oint_c f(s)\, ds = \oint_c (x\, d\sigma - y\, d\omega) + j \oint_c (y\, d\sigma + x\, d\omega) \tag{12.11.1}$$

The two integrals on the right side of Eq. (12.11.1) are real line integrals.

Using Green's theorem in Eq. (12.10.1), we can write them as

$$\oint_c (x\, d\sigma - y\, d\omega) = \iint_R \left(-\frac{\partial y}{\partial \sigma} - \frac{\partial x}{\partial \omega} \right) d\sigma\, d\omega$$

$$\oint_c (y\, d\sigma + x\, d\omega) = \iint_R \left(\frac{\partial x}{\partial \sigma} - \frac{\partial y}{\partial \omega} \right) d\sigma\, d\omega \qquad (12.11.2)$$

Now, through the second Cauchy-Riemann condition in Eq. (12.7.4), we see that the quantity in parentheses on the right side of the first of Eqs. (12.11.2) vanishes. Also the first Cauchy-Riemann condition causes the quantity in parentheses on the right of the second of Eqs. (12.11.2) to vanish. Thus Eq. (12.11.1) becomes

$$\oint_c f(s)\, ds = 0 + j0 = 0 \qquad (12.11.3)$$

The result in Eq. (12.11.3) is known as *Cauchy's integral theorem.* It states that the line integral of any analytic function around a closed curve within a region of regularity is zero. It also states that the line integral of the function between any two points in the region of regularity is independent of the integration path.

It is worth noting that application of the Cauchy-Riemann conditions in the derivation just completed implies that the two integrands on the left sides of Eqs. (12.11.2) are exact differentials. To verify this fact, assume that $x\, d\sigma - y\, d\omega$ is the exact differential of a function $f_1(s)$ such that

$$df_1(s) = \frac{\partial f_1}{\partial \sigma} d\sigma + \frac{\partial f_1}{\partial \omega} d\omega = x\, d\sigma - y\, d\omega \qquad (12.11.4)$$

Now we can write

$$\frac{\partial^2 f_1}{\partial \sigma\, \partial \omega} = \frac{\partial x}{\partial \omega} \qquad \frac{\partial^2 f_1}{\partial \omega\, \partial \sigma} = -\frac{\partial y}{\partial \sigma} \qquad (12.11.5)$$

These second partials must be equal, thus yielding the second Cauchy-Riemann condition.

The quantity $y\, d\sigma + x\, d\omega$ is to be the exact differential of a second function $f_2(s)$, given by

$$df_2(s) = \frac{\partial f_2}{\partial \sigma} d\sigma + \frac{\partial f_2}{\partial \omega} d\omega = y\, d\sigma + x\, d\omega \qquad (12.11.6)$$

The second partials are

$$\frac{\partial^2 f_2}{\partial \sigma\, \partial \omega} = \frac{\partial y}{\partial \omega} \qquad \frac{\partial^2 f_2}{\partial \omega\, \partial \sigma} = \frac{\partial x}{\partial \sigma} \qquad (12.11.7)$$

which, when equated, yield the first Cauchy-Riemann condition.

Use of Eqs. (12.11.4) and (12.11.6) permits Eq. (12.11.1) to be rewritten as

$$\oint_c f(s)\,ds = \oint_c df_1(s) + j\oint_c df_2(s) \tag{12.11.8}$$

which reduces to Eq. (12.11.3) directly because the integration of an exact differential around a closed contour (between two coincident end points) must yield a zero result.

There is one further subtlety in the derivation of Eq. (12.11.3) that should be mentioned. It would appear that in addition to requiring $f(s)$ to be regular in the region R we should also require that $f'(s)$ be continuous there. For only then, it would seem, could we be sure that x and y and their first partial derivatives would be continuous there, a requirement for the Green's theorem to hold. Actually this additional requirement is already embodied in the requirement that $f(s)$ be analytic, as will be shown later in Sec. 12.12.

Consider next an analytic function $f(s)$ that is regular everywhere in a simply connected region R except at $s = s_0$. It is desired to find $\int f(s)\,ds$ around the closed contour bounding the region. The situation is shown in Fig. 12.7, where c_2 is the boundary of the region under consideration. As it stands now, we cannot use the Cauchy integral theorem to evaluate the integral around c_2 because of the singularity enclosed.

Place a small circle c_1 around $s = s_0$. Now $f(s)$ is regular everywhere within the new region between the two closed curves, but the new region is doubly connected and hence there does not appear to be a continuous closed path around which to apply the Cauchy integral theorem. This difficulty is removed by the cut shown between the two curves, which converts the region back into a simply connected one. Now use of Eq. (12.11.3) yields

$$\oint_{c_2} f(s)\,ds + \int_a^b f(s)\,ds \oint_{c_1} f(s)\,ds + \int_c^d f(s)\,ds = 0 \tag{12.11.9}$$

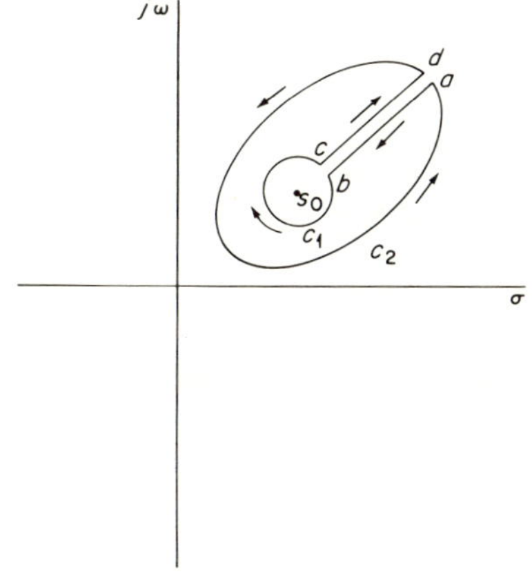

Fig. 12.7

We evaluate Eq. (12.11.9) by a limiting process, in order to ultimately enclose all the region by the curve c_2. First let the line *ab* approach the line *cd*. In the limit, as these two lines come so close together that they may be considered superimposed, the second and fourth integrals in Eq. (12.11.9) cancel, leaving us with

$$\oint_{c_2} f(s)\,ds = \oint_{c_1} f(s)\,ds \tag{12.11.10}$$

Thus the problem is reduced from one of integrating around the irregularly shaped closed contour c_2 to an integration around the small circle c_1.

Example 12.12 It is required to find $\oint_{c_2} f(s)\,ds$ in Fig. 12.7 when $f(s) = (s - s_0)^n$, with n taking on all plus and minus integer values.

(*a*) For $n \geqslant 0$, $f(s)$ is regular throughout the region bounded by c_2, and the required integral is zero by Eq. (12.11.3).

(*b*) For $n < -1$, $f(s)$ has a second- or higher-degree pole at $s = s_0$, and Eq. (12.11.10) must be used to evaluate the required integral. Thus

$$\oint_{c_2} (s - s_0)^n\,ds = \oint_{c_1} (s - s_0)^n\,ds$$

To evaluate the c_1 integral, let

$$s - s_0 = \rho e^{j\alpha}$$

Since c_1 is a circle, ρ is constant and

$$ds = j\rho e^{j\alpha}\,d\alpha$$

Inserting these quantities in the c_1 integral yields

$$\oint_{c_1} (s - s_0)^n\,ds = \int_0^{2\pi} \rho^n e^{jn\alpha} j\rho e^{j\alpha}\,d\alpha = j\rho^{n+1}\int_0^{2\pi} e^{j(n+1)\alpha}\,d\alpha$$

$$= \frac{\rho^{n+1}}{n+1}\left(e^{j2\pi(n+1)} - 1\right) = 0$$

The integral is zero for all $n < -1$, irrespective of the size of ρ; that is, it was not necessary to take the limit as $\rho \to 0$ to have the integral vanish.

(*c*) For $n = -1$, $f(s)$ has a simple pole at $s = s_0$. Using the polar forms of part (*b*),

$$\oint_{c_1} (s - s_0)^n\,ds = \int_0^{2\pi} \frac{1}{\rho e^{j\alpha}} j\rho e^{j\alpha}\,d\alpha = 2\pi j$$

The overall conclusion from parts (*a*) to (*c*) is

$$\oint_{c_2} (s - s_0)^n\,ds = \begin{cases} 2\pi j & n = -1 \\ 0 & n \neq -1 \end{cases} \tag{12.11.11}$$

where n is an integer and c_2 is a closed curve enclosing $s = s_0$.

The final result of Example 12.12 in Eq. (12.11.11) does not violate the Cauchy integral theorem; it merely goes beyond it for this particular function. The Cauchy integral theorem states that the integral of $f(s)$ is zero if $f(s)$ is regular in the region; it does not state that the integral is not zero if $f(s)$ is not regular in the region.

To illustrate the importance and usefulness of Eq. (12.11.11), consider the following example.

Example 12.13 Find $\oint f(s)\,ds$ around the two semicircular closed paths shown in Fig. 12.8 when

$$f(s) = \frac{1}{(s+1)^2(s+3)}$$

(*a*) Consider first the closed path c_1 from $-j2$ to $+j2$ along the $j\omega$ axis and from $+j2$ to $-j2$ along the smaller semicircle. The given function is regular everywhere within this contour except at $s = -1$, where it has a second-degree pole. As in Fig. 12.7, we place a small circle c_3 about the singularity and write

$$\oint_{c_1} f(s)\,ds = \oint_{c_3} f(s)\,ds = \oint_{c_3} \frac{ds}{(s+1)^2(s+3)}$$

The substitution $s + 1 = \rho e^{j\alpha}$ does not simplify the integration in this case. Try instead a partial-fraction expansion of the function $f(s)$ into three subfunctions. We have

$$f(s) = \frac{1/4}{s+3} + \frac{1/2}{(s+1)^2} - \frac{1/4}{s+1}$$

Thus

$$\oint_{c_3} f(s)\,ds = \oint_{c_3} \frac{1/4\,ds}{s+3} + \oint_{c_3} \frac{1/2\,ds}{(s+1)^2} + \oint_{c_3} \frac{-1/4\,ds}{s+1}$$

The first integral on the right side is zero by Eq. (12.11.3), since the subfunction is regular within the c_3 circle. The second integral is zero by Eq. (12.11.11); the third

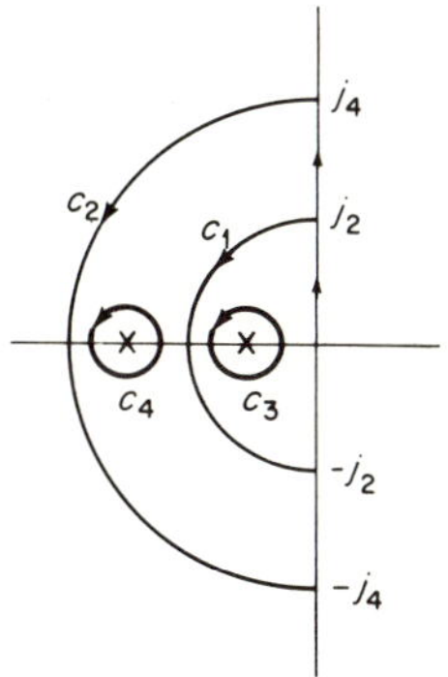

Fig. 12.8

integral is $-\frac{1}{2}\pi j$ by this same equation. Thus

$$\oint_{c_1} f(s)\, ds = -\tfrac{1}{2}\pi j$$

(*b*) Now consider the closed path c_2 formed by the $j\omega$ axis from $-j4$ to $+j4$ and the larger semicircle. The given function has singularities within the contour at $s = -1$ and $s = -3$. Placing small circles around each pole creates a triply connected region. By direct extension of Eq. (12.11.10) we can write

$$\oint_{c_2} f(s)\, ds = \oint_{c_3} f(s)\, ds + \oint_{c_4} f(s)\, ds$$

Using the partial-fraction expansion of $f(s)$, the first integral on the right side can be written exactly as it was in part (*a*).

$$\oint_{c_3} f(s)\, ds = \oint_{c_3} \frac{\frac{1}{4} ds}{s+3} + \oint_{c_3} \frac{\frac{1}{2} ds}{(s+1)^2} - \oint_{c_3} \frac{\frac{1}{4} ds}{s+1}$$

The arguments of part (*a*) apply, and

$$\oint_{c_3} f(s)\, ds = -\tfrac{1}{2}\pi j$$

The second integral on the right side can also be written as the sum of three integrals. We have

$$\oint_{c_4} f(s)\, ds = \oint_{c_4} \frac{\frac{1}{4} ds}{s+3} + \oint_{c_4} \frac{\frac{1}{2} ds}{(s+1)^2} - \oint_{c_4} \frac{\frac{1}{4} ds}{s+1}$$

The last two of these integrals vanish by Eq. (12.11.3) because the two subfunctions are regular within the c_4 circle. The first integral is $+\frac{1}{2}\pi j$ by Eq. (12.11.11). Thus

$$\oint_{c_4} f(s)\, ds = +\tfrac{1}{2}\pi j$$

and

$$\oint_{c_2} f(s)\, ds = -\tfrac{1}{2}\pi j + \tfrac{1}{2}\pi j = 0$$

Example 12.14 Find the value of $\displaystyle\int_{-j4}^{j4} \frac{ds}{s}$ in Fig. 12.9.

This example introduces two new subtleties of complex integration, namely, finding the integral over an open path from the known value of the integral over a closed path and integrating through a singularity.

The desired integral passes through the pole at $s = 0$. If we form a closed path by scribing semicircles c_δ of radius δ (called an indented contour) and c_ρ of radius 4 about $s = 0$, then by Eq. (12.11.3), the integral around this closed path is zero, permitting us to write

$$\int_{-j4}^{-j\delta} \frac{ds}{s} + \int_{c_\delta} \frac{ds}{s} + \int_{j\delta}^{j4} \frac{ds}{s} + \int_{c_\rho} \frac{ds}{s} = 0$$

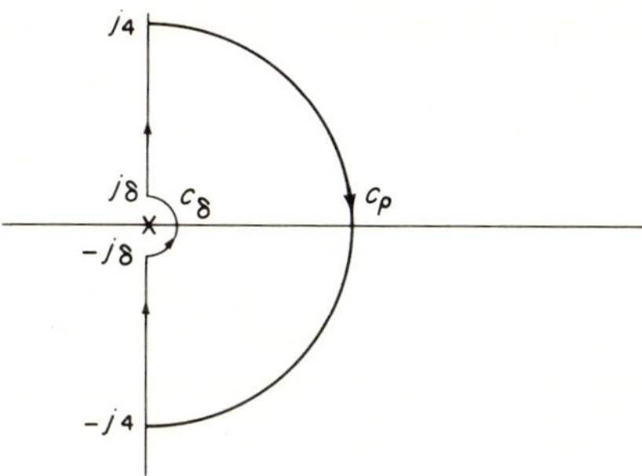

Fig. 12.9

The second and fourth of these integrals can be evaluated directly.

$$\oint_{c_\delta} \frac{ds}{s} = \int_{-\pi/2}^{\pi/2} \frac{j\delta e^{j\beta}\, d\beta}{\delta e^{j\beta}} = j\pi$$

$$\oint_{c_\rho} \frac{ds}{s} = \int_{\pi/2}^{-\pi/2} \frac{j\rho e^{j\alpha}\, d\alpha}{\rho e^{j\alpha}} = -j\pi$$

[Note that Eq. (12.11.11) can be used to obtain these evaluations by inspection.] Thus we are left with

$$\int_{-j4}^{-j\delta} \frac{ds}{s} + \int_{j\delta}^{j4} \frac{ds}{s} = 0$$

Is the sum of these two integrals the same as $\int_{-j4}^{j4} \frac{ds}{s}$? The answer lies in considering what is meant by the Cauchy *principal value* of an integral. Examine

$$\lim_{\substack{\delta_1 \to 0 \\ \delta_2 \to 0}} \left(\int_{-j4}^{-j\delta_2} \frac{ds}{s} + \int_{j\delta_1}^{j4} \frac{ds}{s} \right)$$

where $\delta_1 \neq \delta_2$. Performing the integration, we have

$$\lim_{\substack{\delta_1 \to 0 \\ \delta_2 \to 0}} \left(\ln \frac{\delta_2}{4} + \ln \frac{4}{\delta_1} \right) = \lim_{\substack{\delta_1 \to 0 \\ \delta_2 \to 0}} \ln \frac{\delta_2}{\delta_1}$$

The result depends on how δ_1 and δ_2 approach zero. Letting $\delta_1 = \delta_2 = \delta$, the limit is defined and equals zero. We then say that

$$P \int_{-j4}^{j4} \frac{ds}{s} = 0$$

where the symbol P denotes the Cauchy principal value of the integral.

12.12 CAUCHY INTEGRAL FORMULA

It has been shown through Example 12.12 and Eq. (12.11.11) that

$$\oint_{c_2} \frac{ds}{s - s_0} = 2\pi j$$

when the contour c_2 encloses the point $s = s_0$. We now wish to extend this result to show that if a function $f(s)$ is regular in a closed region bounded by a closed curve c_2, then

$$\oint_{c_2} \frac{f(s)\,ds}{s - s_0} = 2\pi j f(s_0) \tag{12.12.1}$$

where again the contour encloses the point $s = s_0$.

The situation is depicted by Fig. 12.7, the only difference between the present discussion and that in Sec. 12.11 being that now $f(s)$ is regular throughout the entire region, including s_0, whereas in Eq. (12.11.10) it was not.

From Eq. (12.11.10) and Fig. 12.7

$$\oint_{c_2} \frac{f(s)}{s - s_0}\,ds = \oint_{c_1} \frac{f(s)}{s - s_0}\,ds \tag{12.12.2}$$

Letting $s = s_0 + \rho e^{j\alpha}$ and $ds = j\rho e^{j\alpha}\,d\alpha$ in the c_1 integral, we have

$$\oint_{c_1} \frac{f(s)}{s - s_0}\,ds = \int_0^{2\pi} \frac{f(s_0 + \rho e^{j\alpha})}{\rho e^{j\alpha}}\,j\rho e^{j\alpha}\,d\alpha \tag{12.12.3}$$

Now Eq. (12.12.2) must hold for all small circles c_1 of radius ρ, and, in particular, in the limit as $\rho \to 0$. As this limit is taken, $f(s_0 + \rho e^{j\alpha})$ approaches $f(s_0)$, a constant that can be brought outside the integral. Thus the c_1 integral becomes in the limit

$$f(s_0)\int_0^{2\pi} j\,d\alpha = 2\pi j f(s_0) \tag{12.12.4}$$

and Eq. (12.12.1) is proved.

The result in Eq. (12.12.1) tells us that the value of a complex function at any interior point of a region of regularity is determined by the values the function has on the boundary of the region. Once the latter are prescribed, the former are fixed.

Example 12.15 For the function $z = (s^2 + as + b)/(s^2 + 2s + 4)$, find a and b such that

$$\oint_c \frac{z\,ds}{s} = \pi j \quad \text{and} \quad \oint_c \frac{z\,ds}{s - 1} = 2\pi j$$

where c is a circle of radius $3/2$ centered at $s = 0$.

Figure 12.10 shows the circle of radius $3/2$ around which the integrations are to be performed. The poles of z, given by

$$s_1, s_2 = -1 \pm j\sqrt{3}$$

are seen to lie outside this circle. Thus z is regular within the circle and the Cauchy integral formula can be applied.

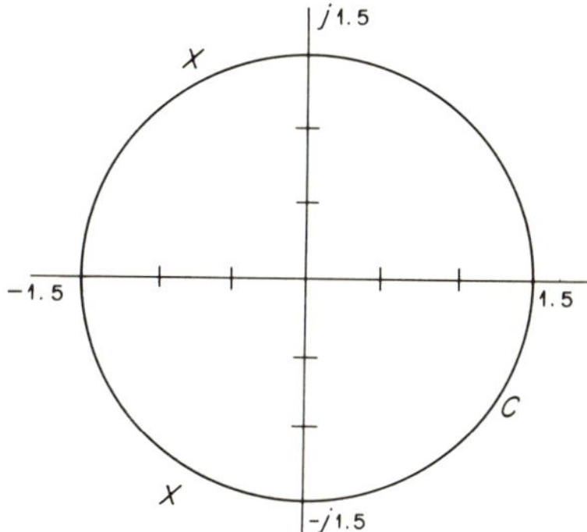

Fig. 12.10

For the first integral, we write

$$\oint_c \frac{z\,ds}{s} = 2\pi j z(0) = \pi j$$

Thus $z(0) = \frac{1}{2}$, and $b = 2$.

Proceeding to the second integral, we have

$$\oint_c \frac{z\,ds}{s-1} = 2\pi j z(1) = 2\pi j$$

Thus $z(1) = 1$, yielding

$$\frac{3+a}{7} = 1 \qquad a = 4$$

The desired z expression is

$$z = \frac{s^2 + 4s + 2}{s^2 + 2s + 4}$$

Cauchy's integral formula in Eq. (12.12.1) is useful when we wish to integrate around paths enclosing first-degree poles, as in Example 12.15. For multiple-degree poles an extension of Eq. (12.12.1) is needed. This will now be derived, and in the process it will be shown that (1) the derivatives of all orders of an analytic function are analytic and (2) the partial derivatives of all orders of the real and imaginary parts of an analytic function are continuous.

We begin with Cauchy's integral formula in Eq. (12.12.1)

$$f(s_0) = \frac{1}{2\pi j}\oint_{c_2} \frac{f(s)\,ds}{s - s_0} \tag{12.12.5}$$

where $f(s)$ is regular within and on c_2 and s_0 lies within c_2. To handle the multiple-pole case, we must somehow replace $s - s_0$ in the denominator of Eq. (12.12.5) by $(s - s_0)^n$, n a positive integer. It would appear that repeated differentiation of Eq. (12.12.5) with respect to s_0 would accomplish this objective. Accordingly we write

$$
\begin{aligned}
f'(s_0) &= \lim_{\Delta s_0 \to 0} \frac{f(s_0 + \Delta s_0) - f(s_0)}{\Delta s_0} \\
&= \lim_{\Delta s_0 \to 0} \left[\frac{1}{2\pi j} \oint_{c_2} \frac{f(s)\, ds}{(s - s_0 - \Delta s_0)\, \Delta s_0} \right. \\
&\qquad \left. - \frac{1}{2\pi j} \oint_{c_2} \frac{f(s)\, ds}{(s - s_0)\, \Delta s_0} \right] \\
&= \lim_{\Delta s_0 \to 0} \frac{1}{2\pi j} \oint_{c_2} \frac{f(s)\, ds}{(s - s_0 - \Delta s_0)(s - s_0)}
\end{aligned}
\tag{12.12.6}
$$

Now, in the limit, we have

$$
f'(s_0) = \frac{1}{2\pi j} \oint_{c_2} \frac{f(s)\, ds}{(s - s_0)^2} \tag{12.12.7}
$$

The result in Eq. (12.12.7) states that the value of the derivative of a complex function at any interior point of a region of regularity is determined by the values the function has on the boundary of the region. Observing the result in Eq. (12.12.7), we see that the formal procedure in Eq. (12.12.6) can be circumvented by simply differentiating Eq. (12.12.5) with respect to s_0 inside the integral sign. With this in mind the result can be extended by writing

$$
f''(s_0) = \frac{2!}{2\pi j} \oint_{c_2} \frac{f(s)\, ds}{(s - s_0)^3} \tag{12.12.8}
$$

and finally

$$
f^n(s_0) = \frac{n!}{2\pi j} \oint_{c_2} \frac{f(s)\, ds}{(s - s_0)^{n+1}} \tag{12.12.9}
$$

Equation (12.12.9) shows that derivatives of all orders of $f(s)$ exist at all points inside c_2.

Recall the definition of regularity in Sec. 12.8. If the derivative of a function exists at every point in the neighborhood of a given point, the function is regular at that given point. Now, $f^n(s_0)$ in Eq. (12.12.9) exists at every point within c_2 and thus at every point in the neighborhood of the point s_0. It follows that $f^{n-1}(s_0)$ is regular at s_0. We conclude that the derivatives of all orders of an analytic function are analytic.

Next note that if the derivative of a function exists, that function is necessarily continuous and so are its real and imaginary parts. Letting $f(s) = x(s) + jy(s)$, the existence of $f'(s)$ means that $f(s)$ and thus $x(s)$ and $y(s)$ are continuous. The existence of $f''(s)$ means that $f'(s)$ is continuous and, noting from Eqs. (12.7.2) and (12.7.3) that

$$f'(s) = \frac{\partial x}{\partial \sigma} + j\frac{\partial y}{\partial \sigma} = \frac{\partial y}{\partial \omega} - j\frac{\partial x}{\partial \omega} \tag{12.12.10}$$

it also means that the first partial derivatives of $x(s)$ and $y(s)$ with respect to σ and ω are continuous. Extending these results, we conclude that the partial derivatives of all orders of the real and imaginary parts of an analytic function are continuous.

Example 12.16 Let it be required to perform the contour integrations of Example 12.13 using Eqs. (12.12.1) and (12.12.7).

Let the given function $1/[(s + 1)^2(s + 3)]$ be renamed $g(s)$. It is regular inside contour c_1 in Fig. 12.8 except at $s = -1$, and it is regular inside c_2 except at $s = -3$ and $s = -1$. Thus, as before, we can write

$$\oint_{c_1} g(s)\,ds = \oint_{c_3} g(s)\,ds$$

and

$$\oint_{c_2} g(s)\,ds = \oint_{3} g(s)\,ds + \oint_{c_4} g(s)\,ds$$

In Example 12.13 it was necessary to perform a partial-fraction expansion of $g(s)$ in order to carry out the integrations around c_3 and c_4. Now, with the aid of Eq. (12.12.7), we can avoid the partial-fraction expansion and write for the c_3 integral

$$\oint_{c_3} \frac{1/(s + 3)}{(s + 1)^2}\,ds = 2\pi j f'(-1)$$

where $f(s) = 1/(s + 3)$ and thus $f'(s) = -1/(s + 3)^2$. Noting that $f'(-1) = -\frac{1}{4}$, we have

$$\oint_{c_1} g(s)\,ds = 2\pi j(-\tfrac{1}{4}) = -\frac{\pi}{2}j$$

in agreement with the result in Example 12.13.

For the c_2 contour, integrations around c_3 and c_4 are required. The c_3 integration is that just completed. The c_4 integration can be written, with the aid of Eq. (12.12.1), as

$$\oint_{c_4} \frac{1/(s + 1)^2}{s + 3}\,ds = 2\pi j f(-3)$$

where, in this case, $f(s) = 1/(s + 1)^2$ and therefore $f(-3) = \frac{1}{4}$. Then we have

$$\oint_{c_4} g(s)\,ds = 2\pi j(\tfrac{1}{4}) = \tfrac{1}{2}\pi j$$

and

$$\oint_{c_2} g(s)\,ds = -\tfrac{1}{2}\pi j + \tfrac{1}{2}\pi j = 0$$

12.13 TAYLOR SERIES EXPANSION

Let a function $f(s)$ be regular in an open region R bounded by a closed curve c_1. About a point s_0 within R, draw the largest radius circle c_2 wholly within R. Now $f(s)$ can be expanded into a Taylor series about s_0 that converges for all values of s *within* the circle c_2. The series has the form

$$f(s) = f(s_0) + f'(s_0)(s - s_0) + \frac{f''(s_0)}{2!}(s - s_0)^2 + \cdots \tag{12.13.1}$$

The situation under discussion is pictured in Fig. 12.11. What is desired is to express the function $f(s)$ at any point s inside c_2 in terms of its value $f(s_0)$ at the center of c_2.

We begin the derivation of Eq. (12.13.1) with Cauchy's integral formula in Eq. (12.12.1), rewritten with variables on the circle c_2 denoted by s' and those within the circle denoted by s

$$f(s) = \frac{1}{2\pi j}\oint_{c_2} \frac{f(s')\,ds'}{s' - s} \tag{12.13.2}$$

Let us define the complex variable

$$h = s - s_0 \tag{12.13.3}$$

and rewrite Eq. (12.13.2) in the form

$$f(s) = \frac{1}{2\pi j}\oint_{c_2} \frac{f(s')\,ds'}{(s' - s_0)[1 - h/(s' - s_0)]} \tag{12.13.4}$$

The quantity containing h in Eq. (12.13.4) can be expanded to yield

$$\left(1 - \frac{h}{s' - s_0}\right)^{-1} = 1 + \frac{h}{s' - s_0} + \frac{h^2}{(s' - s_0)^2} + \cdots \tag{12.13.5}$$

Inserting Eq. (12.13.5) into Eq. (12.13.4) gives

$$f(s) = \frac{1}{2\pi j}\oint_{c_2}\left[\frac{f(s')\,ds'}{s' - s_0} + \frac{hf(s')\,ds'}{(s' - s_0)^2} + \frac{h^2 f(s')\,ds'}{(s' - s_0)^3} + \cdots\right] \tag{12.13.6}$$

Use of Eqs. (12.12.9) and (12.13.3) then yields Eq. (12.13.1).

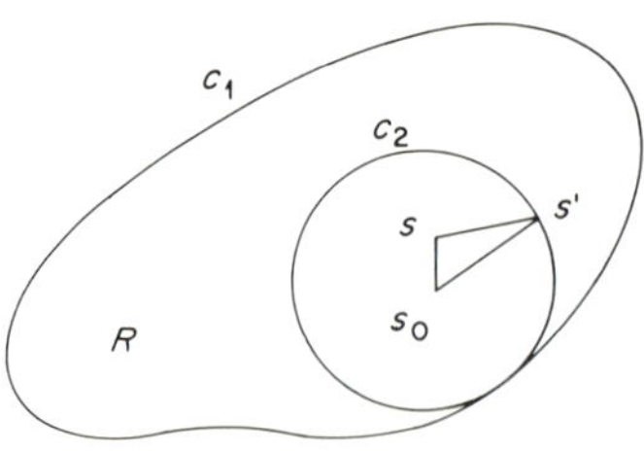

Fig. 12.11

An alternative and useful form of Eq. (12.13.1) is

$$f(s) = A_0 + A_1(s - s_0) + A_2(s - s_0)^2 + \cdots \tag{12.13.7}$$

where

$$A_n = \frac{1}{2\pi j}\oint_{c_2} \frac{f(s')\,ds'}{(s' - s_0)^{n+1}} = \frac{1}{n!}\frac{d^n f(s)}{ds^n}\bigg|_{s=s_0} \tag{12.13.8}$$

It remains to show that the Taylor series in Eq. (12.13.1) converges for all values of s within the circle c_2. This will be done by showing that it converges absolutely within c_2, that is, that the series consisting of the absolute values of each of its terms converges. If a series converges absolutely, it converges.

To prove absolute convergence, a ratio test can be performed. Let a series have the form

$$S = \sum_{i=0}^{\infty} a_i \tag{12.13.9}$$

This series converges absolutely if

$$\lim_{i\to\infty}\left|\frac{a_{i+1}}{a_i}\right| < 1 \tag{12.13.10}$$

Now for the series in Eq. (12.13.5),

$$a_i = \left(\frac{h}{s' - s_0}\right)^i = \left(\frac{s - s_0}{s' - s_0}\right)^i \tag{12.13.11}$$

Performing the ratio test of Eq. (12.13.10) yields

$$\lim_{i\to\infty}\left|\frac{s - s_0}{s' - s_0}\right| < 1 \tag{12.13.12}$$

which is satisfied for any s such that

$$|s - s_0| < |s' - s_0| \tag{12.13.13}$$

i.e., for any value of s *within*, but not *on*, c_2.

It is because the series in Eq. (12.13.5) converges within c_2 that we can replace the integrand in Eq. (12.13.2) by the series in Eq. (12.13.6), with the assurance that the latter equation represents $f(s)$ within c_2 just as surely as the former one did.

There is only one Taylor series expansion of a given function about a given point, as can be proved by trying to represent a given Taylor series about a point by a second Taylor series about the same point and showing that the coefficients of the two series must be identical. Thus a given Taylor series is unique, and we can obtain it in any convenient and easy way, not necessarily through Eq. (12.13.8). Several methods of obtaining a Taylor series are illustrated in the following examples.

Example 12.17 Find the Taylor series expansion of $f(s) = 1/(s + 1)$ about $s = 0$ (powers of s).

The situation is depicted in Fig. 12.12. The region of regularity is the entire complex plane with the exception of the point $s = -1$. The largest-radius circle c_2 that can be drawn about $s_0 = 0$ without including the pole at $s = -1$ has a radius of 1. Thus, from inequality (12.13.13), the series about $s = 0$ will converge for all $|s| < 1$.

To write the series, we use Eq. (12.13.1) directly. The required coefficients are

$$f(0) = 1 \qquad f'(0) = \left.\frac{-1}{(s+1)^2}\right|_0 = -1$$

$$f''(0) = \left.\frac{2}{(s+1)^3}\right|_0 = 2 \qquad f'''(0) = \left.\frac{-6}{(s+1)^4}\right|_0 = -6$$

and the series is

$$f(s) = 1 - s + s^2 - s^3 + \cdots$$

$$= \sum_{n=0}^{\infty} (-1)^n s^n \qquad |s| < 1$$

We note from Example 12.17 that the size of c_2 is determined by the distance from the point about which the series is to be written to the nearest singularity. With c_2 established, s in the series expansion must lie within c_2.

Example 12.18 Find the Taylor series expansion of $g(s) = 1/s(s + 1)$ about $s = 0$.

In this example, we are confronted with a new problem, namely, expanding a function about a singularity. There are two ways to proceed:

Technique (a)

Removal of the singularity Let $f(s) = sg(s)$. Now $f(s)$ is regular within c_2 in Fig. 12.12 and has the expansion in Example 12.17. It follows that

$$g(s) = \frac{1}{s} - 1 + s - s^2 + \cdots \qquad |s| < 1$$

Technique (b)

Partial-fraction expansion

$$g(s) = \frac{1}{s(s+1)} = \frac{1}{s} - \frac{1}{s+1}$$

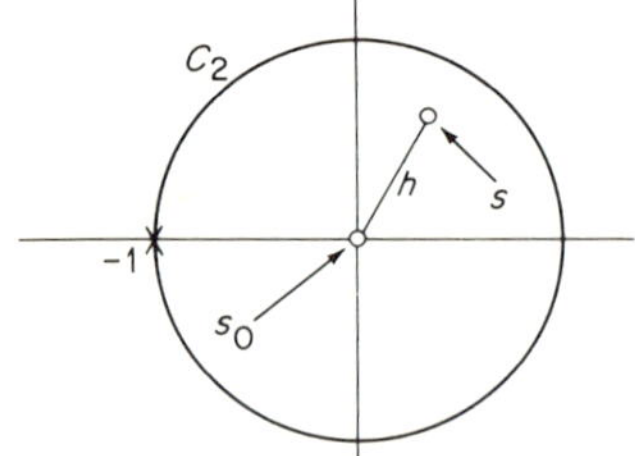

Fig. 12.12

The first term is already a power of s. The second term has the expansion of Example 12.17. Thus

$$g(s) = \frac{1}{s} - (1 - s + s^2 - \cdots)$$

$$= \frac{1}{s} - 1 + s - s^2 + \cdots \qquad |s| < 1$$

Techniques (*a*) and (*b*) are general techniques when the singularities are poles. To treat the general case, let

$$g(s) = \frac{P(s)}{(s - s_i)^n Q(s)} \tag{12.13.14}$$

where $P(s)$ and $Q(s)$ are polynomials in s neither of which has a root $s = s_i$. A Taylor series expansion about $s = s_i$ (powers of $s - s_i$) is desired. By technique (*a*), we multiply by $(s - s_i)^n$ to obtain

$$f(s) = (s - s_i)^n g(s) \tag{12.13.15}$$

expand $f(s)$ into a Taylor series, and divide the result by $(s - s_i)^n$. The series converges within a circle centered at s_i whose radius is the distance from s_i to the nearest zero of $Q(s)$.

By technique (*b*), we perform a partial-fraction expansion of $g(s)$ and obtain

$$g(s) = \frac{k_n}{(s - s_i)^n} + \frac{k_{n-1}}{(s - s_i)^{n-1}} + \cdots + \frac{k_1}{(s - s_i)} + g_1(s) \tag{12.13.16}$$

Now $g_1(s)$ can be expanded into a Taylor series about $s = s_i$ with the same circle of convergence as for technique (*a*).

Example 12.19 Find the Taylor series expansion of $g(s) = 1/s(s + 1)$ about $s = \infty$ (powers of $1/s$).

The expansion of this function about $s = \infty$ will be identical to the expansion of $g(p)$ about $p = 0$, where $p = 1/s$. Substituting in the $g(s)$ expression, we obtain

$$g(p) = p^2 \frac{1}{p + 1}$$

From Example 12.17, the expansion of the fraction is

$$\frac{1}{p + 1} = 1 - p + p^2 - p^3 + \cdots \qquad |p| < 1$$

Thus

$$g(p) = p^2 - p^3 + p^4 - p^5 + \cdots \qquad |p| < 1$$

and

$$g(s) = \frac{1}{s^2} - \frac{1}{s^3} + \frac{1}{s^4} - \frac{1}{s^5} + \cdots \qquad |s| > 1$$

As an alternative procedure, we can write from Example 12.18

$$\frac{1}{s(s+1)} = \frac{1}{s} - 1 + s - s^2 + \cdots \qquad |s| < 1$$

Letting $p = 1/s$ in both sides gives

$$\frac{1}{(1/p)(1/p+1)} = p - 1 + \frac{1}{p} - \frac{1}{p^2} + \cdots \qquad |p| > 1$$

which is an expansion about ∞. The left side is $p^3/p(p+1)$. Dividing both sides by p^3 yields

$$\frac{1}{p(p+1)} = \frac{1}{p^2} - \frac{1}{p^3} + \frac{1}{p^4} - \frac{1}{p^5} + \cdots \qquad |p| > 1$$

which is the desired series if p is replaced by s.

Example 12.20 Find the Taylor series expansion of $f(s) = 1/(\sinh s)$ about $s = 0$.

In this example we are asked to find the Taylor series expanison of a nonalgebraic function. The first task is to obtain the poles of $f(s)$. Using the technique of Example 12.2, we find that the poles are simple and are located at $s = \pm jn\pi$, $n = 0, 1, 2, \ldots$. Thus the function $g(s) = sf(s)$ is regular at $s = 0$ and can be expanded into a Taylor series about that point. The radius of the c_2 circle is π, which means that the series will converge for $|s| < \pi$.

The coefficients necessary to write the first few terms of the series are, from Eq. (12.13.1),

$$g(0) = \left.\frac{s}{\sinh s}\right|_0 = 1$$

$$g'(0) = \left.\frac{\sinh s - s\cosh s}{\sinh^2 s}\right|_0 = 0$$

$$g''(0) = \left.\frac{s\sinh^2 s + 2s - 2\sinh s\cosh s}{\sinh^3 s}\right|_0 = -\frac{1}{3}$$

Each of these coefficients requires the use of L'Hospital's rule in its evaluation. The first two terms of the series are

$$f(s) = \frac{1}{s} - \frac{1}{3}s$$

12.14 CAUCHY RESIDUE THEOREM

Let $f(s)$ be an analytic function regular everywhere in a region R of the s plane except at certain points where it has poles, as shown in Fig. 12.13. Draw a large boundary c enclosing all the poles and a small circle, $c_1, c_2, \ldots, c_n$, around each pole. For the case shown of three poles, we obtain, from Cauchy's integral theorem and the discussion in Sec. 12.11,

$$\oint_c f(s)\,ds = \oint_{c_1} f(s)\,ds + \oint_{c_2} f(s)\,ds + \oint_{c_3} f(s)\,ds \tag{12.14.1}$$

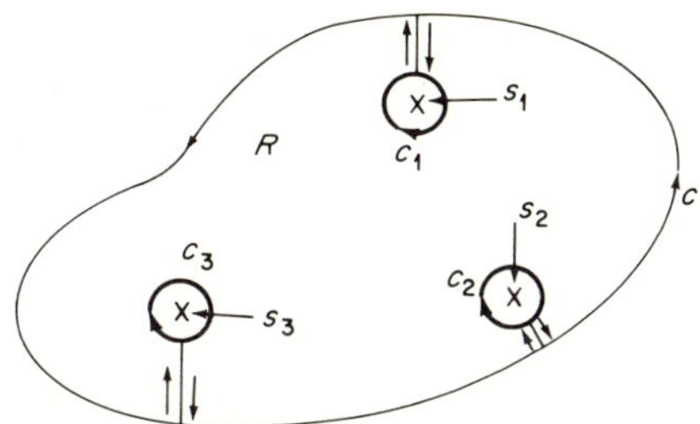

Fig. 12.13

Assume that the pole at $s = s_1$ is of order m. Then

$$g(s) = (s - s_1)^m f(s) \tag{12.14.2}$$

is regular at $s = s_1$ and can be expanded into a Taylor series about that point. The result is, from Eq. (12.13.8),

$$g(s) = A_0 + A_1(s - s_1) + A_2(s - s_1)^2 + \cdots \tag{12.14.3}$$

where, from Eq. (12.14.9),

$$A_n = \frac{1}{n!} \frac{d^n g(s)}{ds^n}\bigg|_{s=s_1} \tag{12.14.4}$$

Thus the expansion of $f(s)$ about $s = s_1$ is

$$f(s) = \frac{A_0}{(s - s_1)^m} + \frac{A_1}{(s - s_1)^{m-1}} + \cdots + \frac{A_{m-1}}{s - s_1} + A_m + A_{m+1}(s - s_1) + \cdots + A_n(s - s_1)^{n-m} \tag{12.14.5}$$

Now reference to Example 12.12, and particularly to Eq. (12.11.11), shows that if $f(s)$ in Eq. (12.14.5) is integrated term by term around the small circle c_1, all integrations yield a zero result except that of the first-degree pole term. Accordingly

$$\oint_{c_1} f(s)\, ds = 2\pi j A_{m-1} \tag{12.14.6}$$

The process is repeated for each of the other poles, in each case yielding the integral of $f(s)$ around the small circle surrounding the pole as $2\pi j A_{m-1}$, where m is the order of pole and A_{m-1} is obtained from Eq. (12.14.4).

Each of the A_{m-1} coefficients is termed a *residue*. Thus a residue is the coefficient of the first-degree pole term in a Taylor series expansion of $f(s)$ about a particular pole. It follows from Eqs. (12.14.1) and (12.14.6) that

$$\oint_{c} f(s)\, ds = 2\pi j \sum_{k=1}^{n} \text{Res } s_k \tag{12.14.7}$$

where n is the number of poles and Res s_k indicates the residue for the pole at $s = s_k$. This result is known as the *Cauchy residue theorem*.

In finding residues, we note from Eq. (12.14.4) that for a first-degree pole, $m = 1$ and

$$A_{m-1} \equiv A_0 = \frac{1}{0!}\frac{d^0 g(s)}{ds^0}\bigg|_{s=s_1} = g(s)\bigg|_{s=s_1} = (s - s_1)f(s)\bigg|_{s=s_1} \tag{12.14.8}$$

For a second degree pole, $m = 2$ and

$$A_{m-1} \equiv A_1 = \frac{1}{1!}\frac{d^1 g(s)}{ds^1}\bigg|_{s=s_1} = \frac{dg(s)}{ds}\bigg|_{s=s_1} \frac{d[(s - s_1)^2 f(s)]}{ds}\bigg|_{s=s_1} \tag{12.14.9}$$

The reader should note that Eqs. (12.14.8) and (12.14.9) are identical in form to Eqs. (10.3.5) and (10.3.15) (with $r = 2$), which are used to obtain the coefficients in a partial-fraction expansion. Hence these latter coefficients are properly termed residues also, and we should be able to use a partial-fraction expansion in obtaining the residues in Eq. (12.14.7). How this can be done will now be illustrated.

Using the same $f(s)$ as before (the function whose poles are depicted in Fig. 12.13), perform a partial-fraction expansion and obtain

$$f(s) = \frac{a_0}{(s - s_1)^m} + \frac{a_1}{(s - s_1)^{m-1}} + \cdots + \frac{a_{m-1}}{s - s_1} + \frac{b_0}{s - s_2} + \frac{c_0}{s - s_3} \tag{12.14.10}$$

where, for the purpose of illustration, we have assumed the poles at $s = s_2$ and $s = s_3$ to be first degree. Integrating $f(s)$ in Eq. (12.14.10) around the circle c_1 yields

$$\oint_{c_1} f(s)\, ds = 2\pi j a_{m-1} \tag{12.14.11}$$

because (1) all integrals around c_1 of terms involving s_1 are zero, except that of the a_{m-1} term from Example 12.12 [the result for the a_{m-1} term is the right side of Eq. (12.14.11)] and (2) because all integrals around c_1 of terms involving s_2 and s_3 are zero from the Cauchy integral theorem in Eq. (12.11.3) since these terms are regular within and on c_1. We conclude that

$$a_{m-1} = A_{m-1} \tag{12.14.12}$$

that is, the coefficient of the s_1 first-degree pole term in the partial-fraction expansion of $f(s)$ is identical to the coefficient of the s_1 first-degree pole term in the Taylor series expansion of $f(s)$ about $s = s_1$.

Thus, it is not necessary, in the case of ratios of polynomials, to use the Taylor series expansion at all in finding residues; we can just as readily use the partial-fraction-expansion technique. However, the Taylor series expansion is necessary when dealing with transcendental functions, such as $f(s)$ in Example 12.20.

Example 12.21 Find the value of

$$\oint_c \frac{s(s^2+2)\,ds}{(s-1)^2(s^2+9)}$$

where c is the circle $|s| = 4$ in Fig. 12.14.

The poles of the given function lie at $s = 1, j3$, and $-j3$, as shown in Fig. 12.14. Small circles are drawn about each of these pole locations. We must find the residue at each pole and sum according to Eq. (12.14.7).

The poles inside c_1 and c_2 are first degree. Use of Eq. (12.14.8) gives

$$\text{Res } j3 = (s-j3)f(s)\big|_{s=j3} = \frac{s(s^2+2)}{(s-1)^2(s+j3)}\bigg|_{s=j3} = \frac{28-j21}{100}$$

and

$$\text{Res } -j3 = \frac{28+j21}{100}$$

The pole inside c_3 is second degree, requiring Eq. (12.14.9)

$$\text{Res } 1 = \frac{d}{ds}\left[\frac{s(s^2+2)}{s^2+9}\right]_{s=1} = \frac{44}{100}$$

Thus

$$\oint_c \frac{s(s^2+2)\,ds}{(s-1)^2(s^2+9)} = 2\pi j\,\frac{56+44}{100} = 2\pi j$$

Equations (12.14.8) and (12.14.9) save us the trouble of having to find the A_{m-1} coefficients from an actual expansion, Taylor series or partial fraction, each time. However, it is interesting to compare the two expansions. To obtain a Taylor series expansion of $f(s)$ about $s = 1$, we write

$$g(s) = (s-1)^2 f(s) = \frac{s(s^2+2)}{s^2+9}$$

Expanding $g(s)$ yields

$$g(s) = \tfrac{3}{10} + \tfrac{44}{100}(s-1) + \tfrac{91}{500}(s-1)^2 + \cdots$$

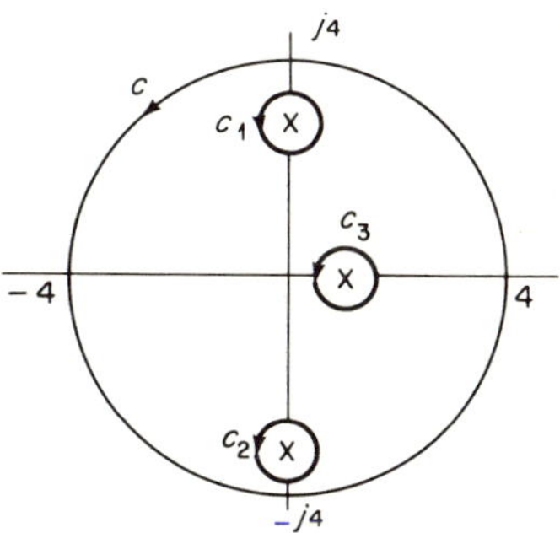

Fig. 12.14

and then

$$f(s) = \frac{3}{10(s-1)^2} + \frac{44}{100(s-1)} + \frac{91}{500} + \cdots$$

The residue is the coefficient of the first-degree pole term.

On the other hand, a partial-fraction expansion takes the form

$$\frac{s(s^2+2)}{(s-1)^2(s^2+9)} = \frac{3}{10(s-1)^2} + \frac{44}{100(s-1)} + \frac{28+j21}{100(s+j3)} + \frac{28-j21}{100(s-j3)}$$

which displays all three residues ($\frac{3}{10}$ is not a residue) in the one expansion.

The coefficients of the $s-1$ and $(s-1)^2$ terms in the two expansions are identical, but there the similarity ends. The Taylor series expansion contains a constant followed by terms in $s-1$ to positive integer powers; the partial-fraction expansion contains terms in the other poles. At $s = 1$, all terms following the constant term in the Taylor expansion of $f(s)$ vanish, and in the partial-fraction expansion

$$\left.\frac{28+j21}{100(s+j3)}\right|_{s=1} + \left.\frac{28-j21}{100(s-j3)}\right|_{s=1} = \left.\frac{28s+63}{50(s^2+9)}\right|_{s=1} = \frac{91}{500}$$

A general conclusion stemming from this last calculation is

$$A_{mj} = \sum_i \left.\frac{k_i}{s-s_i}\right|_{s=s_j} \tag{12.14.13}$$

which says that A_m in the Taylor series expansion about $s = s_j$ is equal to the summation, evaluated at $s = s_j$, of all terms in the partial expansion not containing s_j.

When we are given a ratio of polynomials in unfactored form, it is tedious to factor the denominator to obtain the poles of the function. Assume a ratio of polynomials given by

$$f(s) = \frac{P(s)}{Q(s)} \tag{12.14.14}$$

where $Q(s)$ has a root at $s = s_k$ and $P(s_k) \neq 0$. Write a Taylor series expansion for $P(s)$ and one for $Q(s)$, about $s = s_k$

$$f(s) = \frac{f(s_k) + P'(s_k)(s-s_k) + (1/2!)P''(s_k)(s-s_k)^2 + \cdots}{Q(s_k) + Q'(s_k)(s-s_k) + (1/2!)Q''(s_k)(s-s_k)^2 + \cdots} \tag{12.14.15}$$

The residue at $s = s_k$ is given by $(s-s_k)f(s)\,|_{s=s_k}$ if the root at $s = s_k$ is of first order. For this case, $Q(s_k) = 0$, but $Q'(s_k) \neq 0$. It follows that

$$(s-s_k)f(s)\,|_{s=s_k} = \frac{P(s_k)}{Q'(s_k)} \tag{12.14.16}$$

Example 12.22 For the $f(s)$ of Example 12.21, find the residue at $s = j3$ from Eq. (12.14.16).

$$f(s) = \frac{s(s^2 + 2)}{(s - 1)^2(s^2 + 9)} = \frac{s^3 + 2s}{s^4 - 2s^3 + 10s^2 - 18s + 9}$$

$$\text{Res}\, j3 = \frac{(s^3 + 2s)|_{s=j3}}{(4s^3 - 6s^2 + 20s - 18)|_{s=j3}}$$

$$= \frac{-j21}{36 - j48} = \frac{28 - j21}{100}$$

Example 12.23 Equation (12.14.16) is quite useful in finding the residues at simple poles of transcendental functions. To illustrate, find the residue of $f(s) = 1/(\sinh s)$ at the pole at $s = 0$.

From Example 12.20, where the first two terms of the Taylor series expansion of $f(s)$ about $s = 0$ were written, it is seen that the pole at $s = 0$ is simple with a residue of 1.

Alternatively, Eq. (12.14.16) gives

$$\text{Res}\,(0) = \left.\frac{1}{\cosh s}\right|_{s=0} = 1$$

The nonvanishing of $\cosh s$ at $s = 0$ confirms that the pole of $f(s)$ there is simple.

Example 12.24 Let $f(s)$ be given by

$$f(s) = \frac{a_n s^n + a_{n-1}s^{n-1} + \cdots + a_1 s + a_0}{b_n s^{n+r} + b_{n-1}s^{n+r-1} + \cdots + b_1 s + b_0}$$

Prove that

$$\sum \text{Res}\, f(s) = \begin{cases} 0 & \text{for } r > 1 \\ \dfrac{a_n}{b_n} & \text{for } r = 1 \end{cases}$$

This is the residue-sum theorem, presented without proof in Sec. 10.3.

The proof begins with Eq. (12.14.7),

$$\sum \text{Res}\, f(s) = \frac{1}{2\pi j} \oint_c f(s)\, ds$$

where, in this case, we shall choose c as a large circle of radius ρ about the origin enclosing all the poles of $f(s)$. Thus let

$$s = \rho e^{j\alpha} \qquad ds = j\rho e^{j\alpha}\, d\alpha$$

Then

$$\lim_{s\to\infty} f(s) = \lim_{s\to\infty} \frac{a_n}{b_n s^r} = \lim_{\rho\to\infty} \frac{a_n}{b_n \rho^r e^{jr\alpha}}$$

Performing the contour integration in the limit as $\rho \to \infty$ yields

$$\sum \operatorname{Res} f(s) = \frac{1}{2\pi j} \lim_{\rho \to \infty} \int_0^{2\pi} \frac{a_n j \rho e^{j\alpha}\, d\alpha}{b_n \rho^r e^{jr\alpha}}$$

$$= \frac{1}{2\pi} \lim_{\rho \to \infty} \int_0^{2\pi} \frac{a_n}{b_n} \rho^{1-r} e^{j(1-r)\alpha}\, d\alpha$$

$$= \frac{a_n}{2\pi b_n} \lim_{\rho \to \infty} \rho^{1-r} \int_0^{2\pi} e^{j(1-r)\alpha}\, d\alpha$$

For $r > 1$, ρ^{1-r} causes the right side to vanish as $\rho \to \infty$. For $r = 1$, the right side is independent of ρ and becomes a_n/b_n.

12.15 ABSOLUTE VALUES OF INTEGRALS AND ANALYTIC CONTINUATION

This section is a slight digression to consider a pair of topics that are relevant to the discussion of the inverse Laplace transform in Sec. 12.16, as well as being of general importance.

Consider a complex function $f(s)$ which is to be integrated around a closed contour c in the s plane. The inequality

$$\left| \oint_c f(s)\, ds \right| \leq \oint_c |f(s)|\, |ds| \tag{12.15.1}$$

must hold because the left member is the magnitude of a summation, which is less than or at most equal to the right member, which is the summation of magnitudes.

Let the maximum value of $|f(s)|$ on c be M. Then

$$\oint_c |f(s)|\, |ds| \leq M \oint_c |ds| = ML \tag{12.15.2}$$

where L is the length of c. (It should be noted that $\oint_c ds = 0$ but that $\oint_c |ds| = L$, the first being a summation of vectors, the second of scalars.) Now by combining (12.15.1) and (12.15.2), we have

$$\left| \oint_c f(s)\, ds \right| \leq ML \tag{12.15.3}$$

Let a function $f(s)$ be regular in an open region R bounded by a closed curve c_1. Pick a point s_0 within R and draw the largest-radius circle c_2 wholly within R about s_0. This situation is shown in Fig. 12.11. Cauchy's integral formula can be used to evaluate $f(s_0)$ in terms of values of $f(s)$ on c_2.

Denoting points on c_2 by s, rather than by s' as was done in Sec. 12.13, we have

$$f(s_0) = \frac{1}{2\pi j} \oint_c \frac{f(s)\,ds}{s - s_0} \tag{12.15.4}$$

The concern here is with the absolute value of $f(s_0)$. Using the relations (12.15.1) to (12.15.4), we can write

$$\begin{aligned} |f(s_0)| &= \left| \frac{1}{2\pi j} \oint_{c_2} \frac{f(s)\,ds}{s - s_0} \right| \le \frac{1}{2\pi} \oint_{c_2} |f(s)| \left| \frac{ds}{s - s_0} \right| \\ &\le \frac{1}{2\pi} M \oint_{c_2} \left| \frac{ds}{s - s_0} \right| = \frac{1}{2\pi} M 2\pi = M \end{aligned} \tag{12.15.5}$$

The first inequality follows from (12.15.1) and the second inequality from (12.15.2). The value of $\oint_{c_2} \left| \frac{ds}{s - s_0} \right|$ can be justified as follows.

Let

$$s - s_0 = \rho e^{j\alpha} \tag{12.15.6a}$$

giving

$$ds = j\rho e^{j\alpha}\,d\alpha \quad (12.15.6b) \quad \text{and} \quad \left| \frac{ds}{s - s_0} \right| = |j\,d\alpha| = |d\alpha| \quad (12.15.6c)$$

Then

$$\oint_{c_2} |d\alpha| = 2\pi \tag{12.15.6d}$$

We conclude from (12.15.5) that $f(s_0)$ must be less than or at most equal to the maximum value of $f(s)$ on c_2. The equals sign holds only if $f(s)$ is a constant. To extend this result, pick a new s_0 at the point on c_2 where $f(s)$ has its maximum value M, draw a new maximum-radius circle about this point, and reapply (12.15.5). Repeating the process, we eventually conclude that the maximum value of $f(s)$ must occur on the boundary c_1 of the region of regularity R. This result is known as the *maximum-modulus theorem.*

The process just described of continuing a function outside of a closed contour in the s plane is of general importance in complex-variable and Laplace transform theory and merits further discussion.

Consider the Taylor series in Eq. (12.13.7), which can be expressed in the closed form

$$f_0(s) = \sum_0^{\infty} A_k (s - s_0)^k \tag{12.15.7}$$

where the zero subscript indicates expansion about s_0. Assume that this series converges for all values of s within a circle c about s_0, as shown in Fig. 12.15. Call the region so enclosed R.

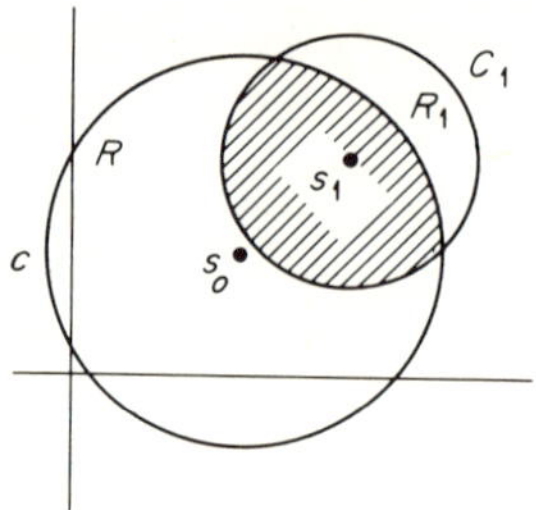

Fig. 12.15 Analytic continuation.

Choose a point s_1 within c and draw a second circle c_1 about s_1, defining the region within c_1 as R_1. As with c, make this second circle the largest possible such that no singularities are enclosed. Now $f(s)$ can be expanded into a second Taylor series about s_1 that converges within c_1. Let that series be represented by

$$f_1(s) = \sum_0^\infty B_k(s - s_1)^k \tag{12.15.8}$$

The two series have a common region of convergence, as indicated by the shaded area in Fig. 12.15. Because of the uniqueness of the Taylor series, it follows that

$$f_1(s) = f_0(s) \tag{12.15.9}$$

within this shaded area.

The process can be continued by choosing a point s_2 within R_1 but outside R, drawing a third maximum-radius circle c_2, and creating a Taylor series called $f_2(s)$. Eventually the entire s plane, except for singularities, is covered by the overlapping circles, and the function $f(s)$ can be defined by

$$f(s) = \begin{cases} f_0(s) & s \text{ within } R \\ f_1(s) & s \text{ within } R_1 \\ \cdots\cdots & \cdots\cdots \\ f_n(s) & s \text{ within } R_n \end{cases} \tag{12.15.10}$$

Each of the functions $f_0(s)$ through $f_n(s)$ is an *element* of $f(s)$, and $f(s)$ is the *analytic continuation* of any of the elements into the entire complex plane. For example, the function $[1 - h/(s' - s_0)]^{-1}$ in Eq. (12.13.5) is the analytic continuation of the Taylor series

$$1 + \frac{h}{s' - s_0} + \frac{h^2}{(s' - s_0)^2} + \cdots$$

in that same equation, the series being defined within the circle $|h| < |s' - s_0|$ whereas the function is defined throughout the complex plane except at the singularity $s = s'$.

The matter of analytic continuation has relevance in Laplace transform theory. Recall from the discussion in Sec. 9.7 that the Laplace transform of the function $e^{s_0 t}$ existed only to the right of the vertical line $\sigma = \sigma_0$, known as the *abscissa of convergence*, in the s plane. This was the least value of σ that ensured the convergence of the Laplace transform integral. The function $F(s) = 1/(s - s_0)$ is the analytic continuation of the Laplace transform of $e^{s_0 t}$ and exists everywhere in the s plane except at the singularity $s = s_0$. In general, $\mathscr{L}[f(t)]$, defined for $\sigma > \sigma_0$, can be considered as an element of the analytic function $F(s)$, defined throughout the s plane except at its singularities. Conversely, $F(s)$ is the analytic continuation throughout the s plane of $\mathscr{L}[f(t)]$.

12.16 INVERSE LAPLACE TRANSFORM

In Sec. 9.5 the Laplace transform pairs were derived. The forward transform Eq. (9.5.8) has been used extensively since, but we have postponed using the inverse transform Eq. (9.5.9) pending our development of complex-integration techniques in this chapter. We are now in a position to find inverse transforms through Eq. (9.5.9), which is repeated here in slightly revised form.

$$f(t) = \frac{1}{2\pi j} \lim_{\rho \to \infty} \int_{\sigma - j\rho}^{\sigma + j\rho} F(s) e^{st}\, ds \qquad (12.16.1)$$

One observation should be made about Eq. (12.16.1) before proceeding. The integration path from $\sigma - j\infty$ to $\sigma + j\infty$ is not prescribed until the proper value of σ is determined, and this, in turn, is determined by the abscissa of convergence, to the right of which the Laplace transform exists and to the left of which it does not. In performing the integration in Eq. (12.16.1), we must be careful that the path from $\sigma - j\infty$ to $\sigma + j\infty$ lies within the region of existence of $F(s)$. This will be ensured if this path lies to the right of all singularities of $F(s)$.

In Fig. 12.16 two contours to be used in evaluating Eq. (12.16.1) are shown. Each contains the path c from $\sigma - j\infty$ to $\sigma + j\infty$ (the end points become $\sigma - j\infty$ and $\sigma + j\infty$ when ρ becomes infinite). The first contour also includes a semicircular arc c_1 and the second a semicircular arc c_2. Typical pole locations of $F(s)$ are indicated by crosses.

Now using Cauchy's residue theorem in Eq. (12.14.7), we can write

$$\lim_{\rho \to \infty} \left[\int_{\sigma - j\rho}^{\sigma + j\rho} F(s) e^{st}\, ds + \int_{c_1 \text{ or } c_2} F(s) e^{st}\, ds \right] = \begin{cases} 2\pi j \sum \operatorname{Res}(s_k) & \text{for } c_1 \\ 0 & \text{for } c_2 \end{cases} \qquad (12.16.2)$$

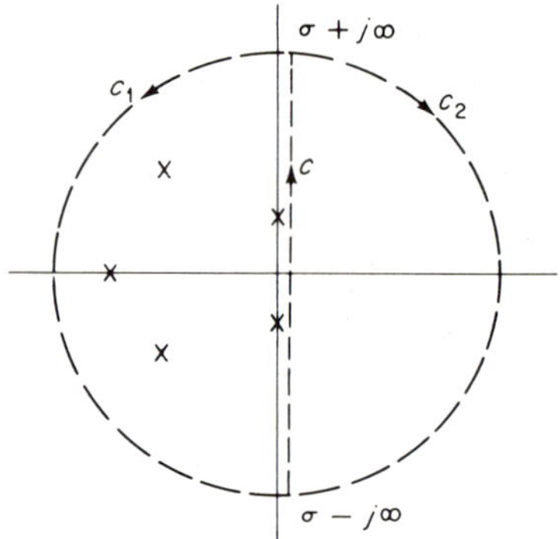

Fig. 12.16 Inverse Laplace transform integration contours.

where, on the right side, we are evaluating the residues of $F(s)e^{st}$ [not just $F(s)$] at the poles of $F(s)$ (e^{st} has no poles) that are enclosed by the contours. The right side is zero for the c_2 path because the path c must lie to the right of all the poles of $F(s)$. Correspondingly, all the residues of $F(s)e^{st}$ are included when the c_1 path is used.

The integrals along c_1 and c_2 need to be evaluated. There are two questions to answer: Which path, c_1 or c_2, do we use in evaluating Eq. (12.16.2), and what are the values of the integrals around c_1 and c_2 in the limit as ρ becomes infinite? Inequality (12.15.1) is used to answer these questions. From it, we obtain

$$\left| \int_{c_1 \text{ or } c_2} F(s)e^{st}\, ds \right| \leq \int_{c_1 \text{ or } c_2} |F(s)|\, |e^{st}|\, |ds| \tag{12.16.3}$$

On the paths c_1 and c_2,†

$$s = \rho e^{j\alpha} \qquad ds = j\rho e^{j\alpha}\, d\alpha$$

Thus

$$|ds| = \rho\, d\alpha \qquad \text{and} \qquad |e^{st}| = e^{t\rho\cos\alpha} \tag{12.16.4}$$

Assume $F(s)$ to be a ratio of polynomials given by

$$F(s) = \frac{a_n s^n + a_{n-1}s^{n-1} + \cdots + a_1 s + a_0}{b_m s^m + b_{m-1}s^{m-1} + \cdots + b_1 s + b_0} \tag{12.16.5}$$

where m and n are related by

$$m = n + r \tag{12.16.6}$$

† Actually $s = \sigma + \rho e^{j\alpha}$ and $|e^{st}| = e^{t(\sigma+\rho\cos\alpha)}$ on these paths, where σ is the displacement of the path c from the $j\omega$ axis. However, in the limit as $\rho \to \infty$, $s \to \rho e^{j\alpha}$ and $|e^{st}| \to e^{t\rho\cos\alpha}$.

with r a positive integer or zero. Then on the paths c_1 and c_2

$$|F(s)| \to \frac{a_n}{b_m \rho^r} \qquad \text{as } \rho \to \infty \tag{12.16.7}$$

Thus, as ρ becomes infinite, the integrand on the right side of (12.16.3) becomes

$$\lim_{\rho \to \infty} |F(s)|\, |e^{st}|\, |ds| = \lim_{\rho \to \infty} \frac{a_n}{b_m \rho^r} e^{t\rho \cos \alpha} \rho\, d\alpha \tag{12.16.8}$$

Whether the limit in Eq. (12.16.8) becomes infinite or vanishes as ρ approaches infinity depends on (1) whether t is positive or negative, (2) in which quandrant α lies, and (3) the value of r.

If $r \geqslant 1$, any values of t and α that make $t\rho \cos \alpha$ negative will cause the limit to be zero. For t negative, the limit will vanish if α is in the first or fourth quadrants (to the right of the path c). It will become infinite for values of α in the second and third quadrants.

If t is positive, the limit will vanish for second- and third-quadrant values of α (to the left of the path c) and become infinite when α lies in the first or fourth quadrant.

Thus in Eq. (12.16.2), we use the path c_1 for positive values of t and the path c_2 for negative values of t and obtain

$$\lim_{\rho \to \infty} \left[\int_{\sigma - j\infty}^{\sigma + j\infty} F(s) e^{st}\, ds \right] = \begin{cases} 2\pi j \sum \text{Res } s_k & \text{for } +t \\ 0 & \text{for } -t \end{cases} \tag{12.16.9}$$

and from Eq. (12.16.1)

$$f(t) = \begin{cases} \sum \text{Res } s_k & \text{for } +t \\ 0 & \text{for } -t \end{cases} \tag{12.16.10}$$

Note in Eq. (12.16.10) that $f(t)$ is zero for t less than zero, as we constrained it to be originally in deriving Eqs. (9.5.8) and (9.5.9).

The case when $r = 0$ in Eq. (12.16.8) has not as yet been considered. This is the case when the highest powers of s in the numerator and denominator of $F(s)$ in Eq. (12.16.5) are the same. We can deduce the form of $f(t)$ for this $F(s)$ from our previous experience with finding the inverse Laplace transform through a partial-fraction expansion. Recall that, before the expansion, a constant is extracted such that $F(s)$ takes the form

$$F(s) = \frac{a_n}{b_n} + F_1(s) \tag{12.16.11}$$

where $F_1(s)$ has the same denominator as $F(s)$ but a new numerator, whose highest power of s is 1 less than that of the denominator. The inverse transform of $F_1(s)$ can be found from Eq. (12.16.10). The inverse transform of a_n/b_n is simply $a_n/b_n\, \delta(t)$, an impulse at $t = 0$.

Where does the development leading to Eq. (12.16.10) fail when $r = 0$? To answer this question we need to reexamine Eq. (12.16.8). With $r = 0$, the ρ in the numerator of the right side of that equation is no longer eliminated by ρ^r in the denominator. The exponential term still causes the limit to be zero if $t\rho \cos \alpha$ is negative, which it is for α in the second or third quadrant and t positive. But for $t = 0$, the exponential term is unity, and as ρ approaches infinity, so does the limit. But this is exactly the definition of an impulse function, i.e., a function which is 0 for $t \neq 0$ and ∞ for $t = 0$. Thus, for $r = 0$, the integral along c_1 yields an impulse (actually a negative impulse), which, from Eq. (12.16.2), guarantees that the inverse Laplace transform will contain a positive impulse, as it should.

Example 12.25 Assume that the capacitor in Fig. 12.16*a* is initially uncharged. The switch S_1 closes at $t = 0$ and opens at $t = 1$ sec. The switch S_2 closes at $t = 1$ sec. Find $i(t)$ by complex integration.

We shall work the problem in two ways: (*a*) by finding $i(t)$ for $0 < t < 1$ and $t > 1$ separately and (*b*) by finding them together.

(*a*) The KVL equation in the time domain for $0 < t < 1$ is

$$E = iR + \frac{1}{C}\int_0^t i\,dt + v_{C0}$$

With the initial capacitor voltage zero, the transform equation in the s domain is

$$\frac{E}{s} = I(s)\left(R + \frac{1}{sC}\right)$$

giving

$$I(s) = \frac{E/R}{s + 1/RC}$$

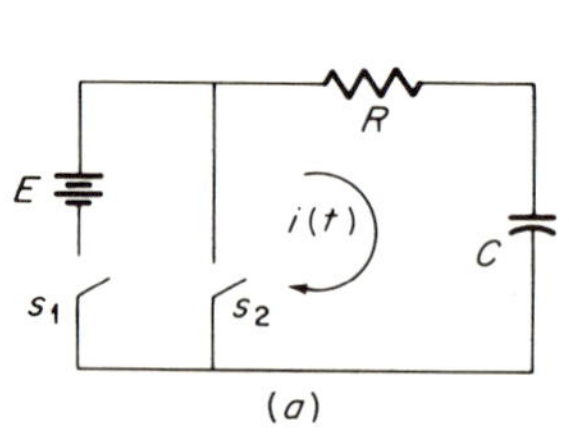

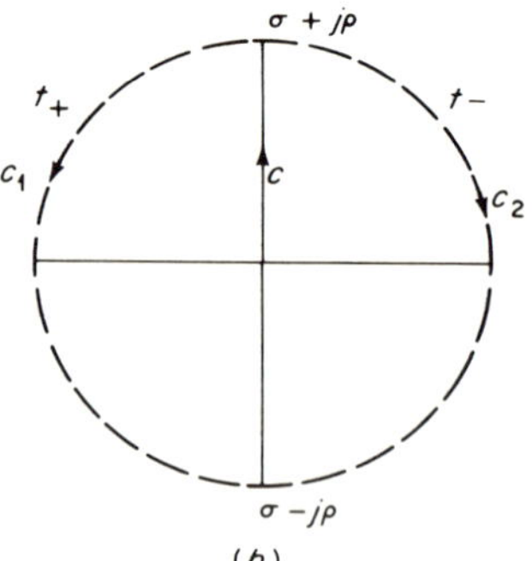

Fig. 12.17

$I(s)$ has a single pole at $s = -1/RC$. This is shown in Fig. 12.17*b*, along with the integration paths. Since $r = 1$, Eqs. (12.16.9) and (12.16.10) are valid. Thus $i(t)$ is zero for $t < 0$ because no poles are enclosed by the c and c_2 contour and the integral along c_2 in Eq. (12.16.2) vanishes. For $t > 0$, the integral along c_1 in Eq. (12.16.2) vanishes, and, there being only one pole,

$$i(t) = \text{Res}\left(-\frac{1}{RC}\right) = \frac{E}{R}e^{st}\bigg|_{-1/RC} = \frac{E}{R}e^{-t/RC} \qquad 0 < t < 1$$

where Eq. (12.14.8) has been used in evaluating the residue of $I(s)e^{st}$.

For $t > 1$, we have

$$0 = iR + \frac{1}{C}\int_0^{t'} i\,dt' + v_{C1}$$

where $t' = t - 1$ is used so that we may again have zero as the starting time and v_{C1} is the capacitor voltage at $t' = 0$ $(t = 1)$. Transforming gives

$$0 = I(s)\left(R + \frac{1}{sC}\right) + \frac{E(1 - e^{-1/RC}}{s}$$

yielding

$$I(s) = \frac{-(E/R)(1 - e^{-1/RC})}{s + 1/RC}$$

Then

$$\begin{aligned} i(t') &= \text{Res}\left(-\frac{1}{RC}\right) \\ &= \frac{E}{R}(1 - e^{-1/RC})e^{st'}\bigg|_{-1/RC} \\ i(t) &= -\frac{E}{R}(1 - e^{-1/RC})e^{-(t-1)/RC} \\ &= \frac{E}{R}e^{-t/RC} - \frac{E}{R}e^{-(t-1)/RC} \qquad t > 1 \end{aligned}$$

(*b*) In this procedure, we write

$$e(t) = iR + \frac{1}{C}\int_0^t i\,dt + v_{C0}$$

where v_{C0} is again zero, and

$$e(t) = Eu(t) - Eu(t - 1)$$

Transforming and solving for $I(s)$ gives

$$I(s) = \frac{(E/R)(1 - e^{-s})}{s + 1/RC}$$

We then have from Eq. (12.16.1)

$$i(t) = \frac{1}{2\pi j}\lim_{\rho\to\infty}\int_{\sigma-j\rho}^{\sigma+j\rho}\left[\frac{(E/R)e^{st}\,ds}{s + 1/RC} - \frac{(E/R)e^{s(t-1)}\,ds}{s + 1/RC}\right] = i_1(t) - i_2(t)$$

The first integral is the integral evaluated in the first part of (*a*). The result is

$$i_1(t) = \frac{E}{R} e^{-t/RC} \qquad t > 0$$

The second integral requires further comment. In the development leading to Eq. (12.16.10) we must replace each t by $t - 1$. Thus we are dealing with $e^{(t-1)\rho\cos\alpha}$ in Eq. (12.16.8). The integral along c_2 in Eq. (12.16.2) vanishes for $t < 1$, and the integral along c_1 in Eq. (12.16.2) vanishes for $t > 1$. The residue evaluation yields

$$i_2(t) = \text{Res}\left(-\frac{1}{RC}\right) = \frac{E}{R} e^{s(t-1)}\bigg|_{-1/RC} = (E/R)e^{-(t-1)/RC} \qquad t > 1$$

Combining the two responses gives the overall response as

$$i(t) = \frac{E}{R} e^{-t/RC} u(t) - \frac{E}{R} e^{-(t-1)/RC} u(t-1) \qquad \text{all } t$$

Example 12.26 The transfer function of a system is given by

$$T(s) = \frac{s^4 + 9s^3 + 30s^2 + 41s + 20}{(s+1)^2(s+2)^2}$$

Find the impulse response of the system.

The impulse response is the system weighting function, the inverse Laplace transform of the system transfer function. We shall find the weighting function through integration in the complex plane.

Because the numerator and denominator of $T(s)$ are of the same order, the limit in Eq. (12.16.8) does not vanish. Thus, before integrating, Eq. (12.16.11) is used to obtain

$$T(s) = 1 + \frac{3s^3 + 17s^2 + 29s + 16}{(s+1)^2(s+2)^2} = 1 + T_1(s)$$

Now Eq. (12.16.10) can be used to find the inverse transform of $T_1(s)$. Since the poles are second degree, Eq. (12.14.9) is used to find the residues

$$\begin{aligned}
\text{Res}\,(-1) &= \frac{d}{ds}\left(\frac{3s^3 + 17s^2 + 29s + 16}{s^2 + 4s + 4} e^{st}\right)\bigg|_{-1} \\
&= e^{st}\frac{d}{ds}\frac{3s^3 + 17s^2 + 29s + 16}{s^2 + 4s + 4}\bigg|_{-1} \\
&\quad + te^{st}\left(\frac{3s^3 + 17s^2 + 29s + 16}{s^2 + 4s + 4}\right)\bigg|_{-1} \\
&= (2e^{-t} + te^{-t})u(t)
\end{aligned}$$

$$\begin{aligned}
\text{Res}\,(-2) &= \frac{d}{ds}\frac{3s^3 + 17s^2 + 29s + 16}{s^2 + 2s + 1} e^{st}\bigg|_{-2} \\
&= (e^{-2t} + 2te^{-2t})u(t)
\end{aligned}$$

Thus the system's impulse response is

$$T(t) = (2e^{-t} + te^{-t} + e^{-2t} + 2te^{-2t})u(t) + \delta(t)$$

In connection with this example it is instructive to point out how the impulse response might be obtained through complex integration of the terms in the partial-fraction expansion of $T_1(s)$. Performing this expansion gives

$$T_1(s) = \frac{A_0}{(s+1)^2} + \frac{A_1}{s+1} + \frac{B_0}{(s+2)^2} + \frac{B_1}{s+2}$$

$$= \frac{1}{(s+1)^2} + \frac{2}{s+1} + \frac{2}{(s+2)^2} + \frac{1}{s+2}$$

In the expansion, A_1 and B_1 are residues of $T_1(s)$; A_0 and B_0 are not. Now using Eqs. (12.16.2) and (12.16.10) to obtain the individual time-domain responses for each of the terms in the expansion of $T_1(s)$, we have

A_0 term: $\frac{d}{ds} e^{st}|_{-1} = te^{-t}u(t)$

A_1 term: $2e^{st}|_{-1} = 2e^{-t}u(t)$

B_0 term: $\frac{d}{ds} 2e^{st}|_{-2} = 2te^{-2t}u(t)$

B_1 term: $e^{st}|_{-2} = e^{-2t}u(t)$

the sum of which checks the previous result.

REFERENCES

12.1. Churchill, R. V.: "Complex Variables and Applications," 2d ed., McGraw-Hill Book Company, New York, 1960. Presents the full spectrum of complex-variable theory in readable form, including many topics not included in this chapter.

12.2. Hildebrand, F. B.: "Advanced Calculus for Engineers," Prentice-Hall, Inc., Englewood Cliffs, N.J , 1948. Widely used by engineers. Complex variables are discussed in Chapter 9, with an excellent treatment of limiting and indented contours.

12.3. Le Page, W. R.: "Complex Variables and the Laplace Transform for Engineers," McGraw-Hill Book Company, New York, 1961. An excellent treatment in depth of complex-variable theory. The material of this chapter is contained primarily in Chapters 2, 4, and 5.

12.4. Peskin, E.: "Transient and Steady-state Analysis of Electric Networks," D. Van Nostrand Company, Inc., Princeton, N.J., 1961. Chapter 3 is devoted to complex variables with particular emphasis on evaluation of the Laplace inversion integral.

12.5. Stewart, J. L.: "Fundamentals of Signal Theory," McGraw-Hill Book Company, New York, 1960. Chapters 8 and 9 develop complex-contour integration leading to an evaluation of the Laplace inversion integral and other applications.

PROBLEMS

Drill

12.1. Define the branches of the function $Z = s^{3/4}$ and draw an s-plane diagram showing the branch cuts and branch point. Repeat for $Z = s^{4/3}$.

12.2. Find the poles of the following functions:

(*a*) $Z = \dfrac{1}{1 - k^2 s^6}$

(*b*) $Z = \dfrac{1}{1 - 4 \sinh s}$

(*c*) $Z = \dfrac{1}{1 - 4 \tanh s}$

(*d*) $Z = \dfrac{1}{3 + e^s}$

(*e*) $Z = \dfrac{1}{2 - \sin s}$

(*f*) $Z = \dfrac{1}{\ln s + 2 - j(\pi/3)}$

12.3. Describe each of the following regions and sketch the boundary of each on the s plane:

(*a*) $|s - 2| + |s + 2| \leq 5$

(*b*) $|s| < |s - 5|$

12.4. Given that $Z = (s^2 + 5s + 6)/(s^3 + 7s^2 + 14s + 8)$, show that $\lim_{s \to -2} Z = -\frac{1}{2}$.

12.5. Find $\oint \dfrac{s}{s^2 + 1}$ around each of two closed contours by first passing to the left of the two $j\omega$ axis poles and then to the right. The integral is to be found in the limit as $\rho \to \infty$.

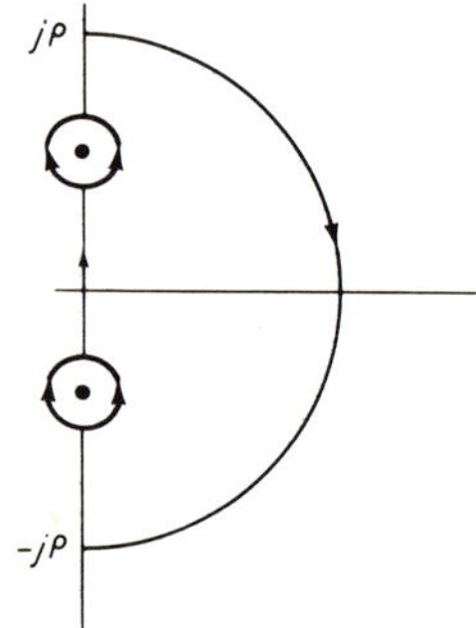

Fig. P12.5

12.6. Let $Z(s) = (s^2 + 6s + 8)/(s^2 + 4s + 3)$. Find $\oint_c \dfrac{Z(s)}{s}\, ds$, where c is a circle of radius 2 centered at the origin.

12.7. Let c be the circle $|s| = 2$. If

$$Z(s_0) = \oint_c \frac{2s^2 + s + 1}{s - s_0}\, ds$$

find $Z(1)$ in two ways and find $Z(3)$.

12.8. Given

$$Z(s) = \frac{(s+2)(s+4)(s+6)(s+8)(s+10)}{(s^2+6s+5)(s^2+6s+9)(s^2+2s+2)}$$

Find $\oint Z(s)\,ds$ around the two contours shown.

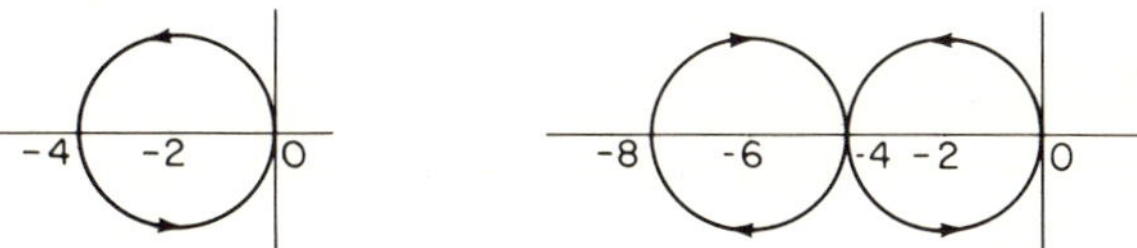

Fig. P12.8

12.9. Find $\oint_c \frac{Z(s)}{s-1}\,ds$ when c is a circle of radius 3/2 centered at the origin and when:

(*a*) $Z(s) = \frac{s^2+2s+1}{s^2+5s+6}$

(*b*) $Z(s) = \frac{s^2+5s+6}{s^2+2s+1}$

12.10. Find the expansion of $Z(s) = 1/s^2$ as a series of positive powers of (*a*) $s + 1$ and (*b*) $s - 2$. Indicate the region of convergence in each case.

12.11. Find the expansion of $Z(s) = s/(s^2 + 1)$ (*a*) as a series of positive powers of s and (*b*) as a series of negative powers of s. Indicate the region of convergence in each case.

12.12. Represent the function $Z(s) = s/[(s - 1)(s - 3)]$ by a series of positive and negative powers of $s - 1$ which converges to $Z(s)$ when $0 < |s - 1| < 2$.

12.13. Find Taylor series expansions of $Z(s) = 1/[s(s + 1)(s + 3)]$ in terms of (*a*) powers of $s + 1$, (*b*) powers of s, (*c*) k_1/s and powers of $s + 1$, (*d*) $k_2/(s + 1)$ and powers of s. Indicate the region of convergence in each case.

12.14. Find the first three terms of the Taylor series expansion about $s = -1$ of $T_1(s)$ in Example 12.26. Repeat for $s = -2$. Use these results directly to find $\oint_c T_1(s)\,ds$, where c is the circle $|s| = 4$. Also find $\oint_c T_1(s)\,ds$ using the partial-fraction expansion of $T_1(s)$ in Example 12.26.

12.15. Without evaluating the integral, show that $\left|\int_{-j}^{j} (\sigma^2 + j\omega^2)\,ds\right| \le 2$ when the integration path is along the ω axis; $\le \pi$ when the integration path is a right-half-plane semicircle of radius 1.

12.16. Use the method of Sec. 12.16 to find the inverse Laplace transforms of:

(*a*) $F(s) = \frac{1}{(s+a)(s+b)(s+c)}$

(*b*) $F(s) = \dfrac{s}{s^2 - a^2}$

(*c*) $F(s) = \dfrac{\omega}{(s + a)^2 + \omega^2}$

(*d*) $F(s) = \dfrac{(s + a)^2 - \omega^2}{[(s + a)^2 + \omega^2]^2}$

Theory and proofs

12.17. A function $H(s)$ is analytic in the right half of the s plane and on the $j\omega$ axis and $\lim_{s\to\infty} H(s) = 0$. Show that $H(j\omega_0)$ [the value of $H(s)$ at any point $j\omega_0$ on the $j\omega$ axis] is given by

$$H(j\omega_0) = \frac{jP}{\pi}\int_{-\infty}^{\infty} \frac{H(j\omega)\,d\omega}{\omega - \omega_0}$$

where P denotes principal value.

12.18. A function $H(s)$ is analytic on and everywhere to the right of the line $s = -\sigma_0$, σ_0 a positive real number, and $\lim_{s\to\infty} H(s) = 0$. Let $R(s)$ and $X(s)$ be the real and imaginary parts, respectively, of $H(s)$. Prove that

$$R(-\sigma_0 + j\omega_0) = -\frac{1}{\pi}\int_{-\infty}^{\infty} \frac{\sigma_0 R(\omega) + (\omega - \omega_0)X(\omega)}{\sigma_0{}^2 + (\omega - \omega_0)^2}\,d\omega$$

$$X(-\sigma_0 + j\omega_0) = \frac{1}{\pi}\int_{-\infty}^{\infty} \frac{-\sigma_0 X(\omega) + (\omega - \omega_0)R(\omega)}{\sigma_0{}^2 + (\omega - \omega_0)^2}\,d\omega$$

Repeat the derivation for $\sigma_0 = 0$ and show that the resulting expressions for $R(j\omega_0)$ and $X(j\omega_0)$ are the same as would be obtained by letting $\sigma_0 = 0$ in the preceding expressions. These latter results are the Hilbert transforms relating the real and imaginary parts of a complex function of ω.

12.19. Define a simply connected region R in the s plane bounded by a closed curve c. Show that the area of R is given by

$$A = \frac{1}{2}\oint (\sigma\,d\omega - \omega\,d\sigma)$$

12.20. (*a*) By actually performing the integration, show that

$$\frac{1}{2\pi j}\lim_{\rho\to\infty}\int_{\sigma-j\rho}^{\sigma+j\rho} Ke^{st}\,ds = K\delta(t)$$

This verifies that the inverse Laplace transform of a constant is an impulse function.

(*b*) By actually performing the integration, show that

$$\frac{1}{2\pi j}\lim_{\rho\to\infty}\oint_{c_1} Ke^{st}\,ds = -K\delta(t)$$

where c_1 is the contour indicated in Fig. 12.16. This also verifies that the inverse Laplace transform of a constant is an impulse function, since Eq. (12.16.2) must be satisfied and Σ Res $s_k = 0$.

12.21. The driving-point impedance of a lossless 1-port can be expressed as

$$Z(s) = \frac{1}{|I|^2}\left[sT(s) + \frac{1}{s}V(s)\right]$$

where I is the current into the 1-port and T and V are energy functions giving the average stored energy in the inductances and capacitances, respectively, of the network ($\frac{1}{2}L\,|I|^2$ and $\frac{1}{2}C\,|V|^2$). Use the Cauchy-Riemann conditions and the fact that the average stored energy in a network can never be negative to prove that the slope of $Z(s)$ is positive for all $s = j\omega$.

12.22. A semicircular arc c is shown. Assume that a function $f(s)$ approaches zero uniformly (independently of angular position) as σ approaches infinity. Prove that

$$\lim_{\sigma\to\infty} \int_c f(s)e^{jks}\,ds = 0 \qquad k > 0$$

This result is known as *Jordan's lemma.*

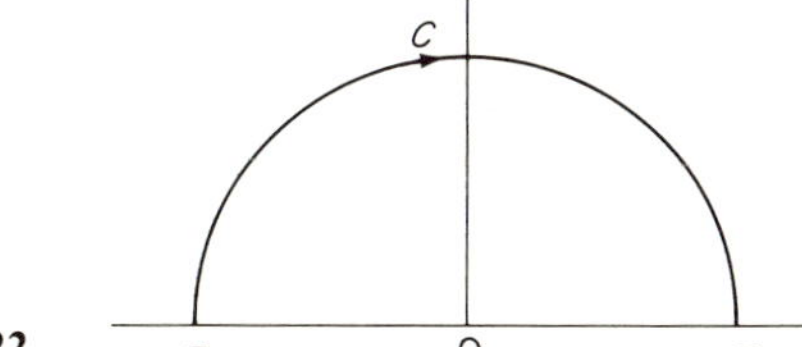

Fig. P12.22

12.23. An analytic function $Z(s)$ is regular in the entire right half of the s plane and on the $j\omega$ axis. Use the maximum-modulus theorem and the transformation $W(s) = e^{-Z(s)}$ to show that $\operatorname{Re} Z(s)$ has its minimum value on the $j\omega$ axis.

12.24. Let a function $Z(s)$ have a pole of degree n at a point $s = s_0$. Let it be required that $\operatorname{Re} Z(s)$ be positive everywhere in the right half of the s plane and nonnegative on the $j\omega$ axis. Show that $n = 0$ if s_0 is in the right half of the s plane and $n = 1$ if s_0 is on the $j\omega$ axis. This result shows that a driving-point impedance can have no right-half-plane poles and that $j\omega$ axis poles must be simple.

12.25. A function $Z(s)$ is positive real if and only if:

(*a*) $Z(s)$ is real for s real.
(*b*) $\operatorname{Re} Z(s) \geqslant 0$ when $\operatorname{Re} s = 0$.
(*c*) $\operatorname{Re} Z(s) > 0$ when $\operatorname{Re} s > 0$.

Use the results of Probs. 12.23 and 12.24 to show that an equivalent set of conditions for $Z(s)$ to be positive real is

(*a*) $Z(s)$ is real for s real.
(*b*) $j\omega$ axis poles of $Z(s)$ are simple with positive real residues.
(*c*) $Z(s)$ is analytic in the right half of the s plane.
(*d*) $\operatorname{Re} Z(s) \geqslant 0$ when $\operatorname{Re} s = 0$.

Application

12.26. The function $Z(s) = \coth \pi s$ is the input impedance of an open-circuited lossless transmission line. Find the Taylor series expansion of $Z(s)$ about $s = 0$. Also find the partial-fraction expansion of $Z(s)$. Show that one of these expansions can be used to represent the transmission line at low frequency by a finite number of lumped LC elements.

12.27. Let $T(s)$ be the transfer function of a low-pass filter with cutoff at ω_c rad/sec. Show that $T[G(s)]$, where $G(s) = (s^2 + \omega_0^2)/s$, is the transfer function of a band-pass filter centered about ω_0 rad/sec. Express the upper and lower cutoff frequencies of the band-pass filter in terms of ω_c and ω_0. What transformation would convert the low-pass filter to a band-stop filter centered at ω_0 rad/sec?

12.28. Cauchy's residue theorem can be used to evaluate real definite integrals of the form $I = \int_{-\infty}^{\infty} f(\sigma)\, d\sigma$, where $f(\sigma)$ is a rational function of σ whose denominator is of degree 2 or more greater than its numerator. Referring to the contour shown, the procedure is to write

$$\int_{-\rho}^{\rho} f(\sigma)\, d\sigma + \int_{c} f(s)\, ds = 2\pi j \sum \operatorname{Res} f(s)$$

The residues of $f(s)$ are evaluated, as is the integral along c, in the limit as $\rho \to \infty$ in order to obtain the desired integral.

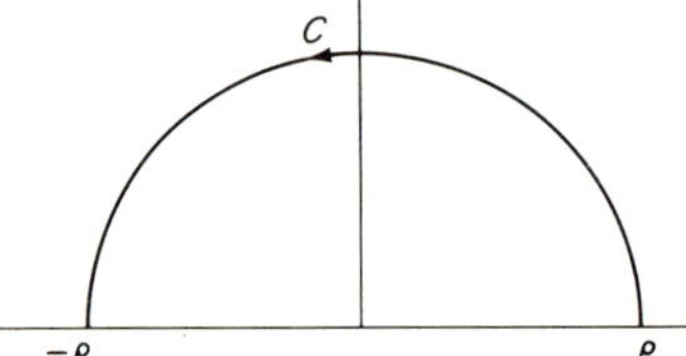

Fig. P12.28

Use this technique to find:

(a) $\displaystyle\int_{-\infty}^{\infty} \frac{d\sigma}{1 + \sigma^4}$

(b) $\displaystyle\int_{-\infty}^{\infty} \frac{\sigma^2\, d\sigma}{(\sigma^2 + 1)^2}$

12.29. Real definite integrals of the form

$$I_1 = \int_{-\infty}^{\infty} f(\sigma) \sin \sigma\, d\sigma \qquad \text{and} \qquad I_2 = \int_{-\infty}^{\infty} f(\sigma) \cos \sigma\, d\sigma$$

where $f(\sigma)$ is a rational function of σ, can also be evaluated by the technique of Prob. 12.28. In this case sin σ and cos σ are replaced by $e^{j\sigma}$. Then I_1 and I_2 are the imaginary and real parts, respectively, of $\int_{-\infty}^{\infty} f(\sigma)e^{j\sigma}\, d\sigma$. Use this technique to find I_1 and the principal value of I_2 (see Example 12.14) when $f(\sigma) = 1/\sigma$.

12.30. Cauchy's residue theorem can be used to evaluate real definite integrals of the form

$$I = \int_{0}^{2\pi} f(\sin \alpha, \cos \alpha)\, d\alpha$$

where f is a rational and finite function of α. The integration is carried out by letting $s = e^{j\alpha}$ and recognizing that s traverses a unit circle as α goes from 0 to 2π.

Evaluate the following definite integrals:

$$(a)\quad \int_0^{2\pi} \frac{\cos^2 3\alpha \, d\alpha}{5 - 4\cos 2\alpha}$$

$$(b)\quad \int_0^{\pi} \frac{\cos 2\alpha \, d\alpha}{1 + k^2 - 2k\cos\alpha}$$

12.31. Assume that $i(t) = 1$ amp at $t = 0_-$. Use the method of Sec. 12.16 to find $i(t)$ for $t > 0$ for the given input voltage.

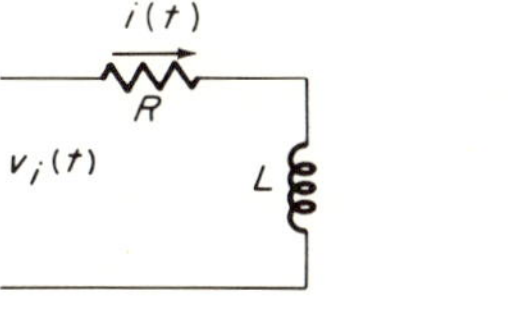

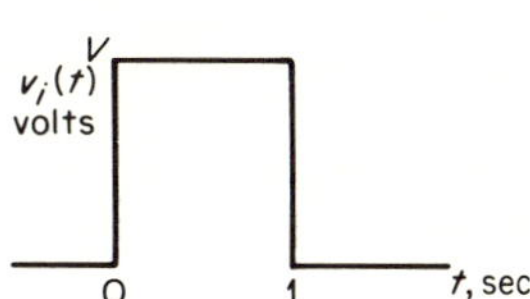

Fig. P12.31

12.32. For the network and the input voltage shown, find $v_0(t)$ using the method of Sec. 12.16. All initial conditions are zero.

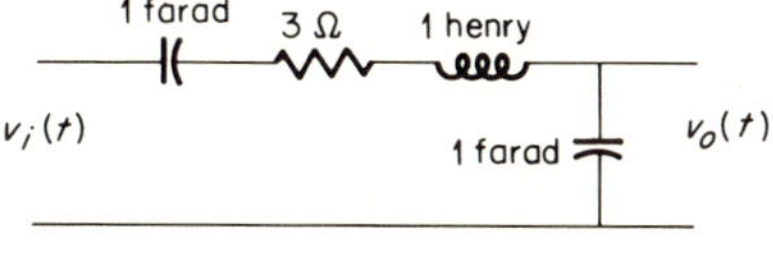

Fig. P12.32

12.33. For the network and $v_1(t)$ given, find $v_0(t)$ by the method of Sec. 12.16.

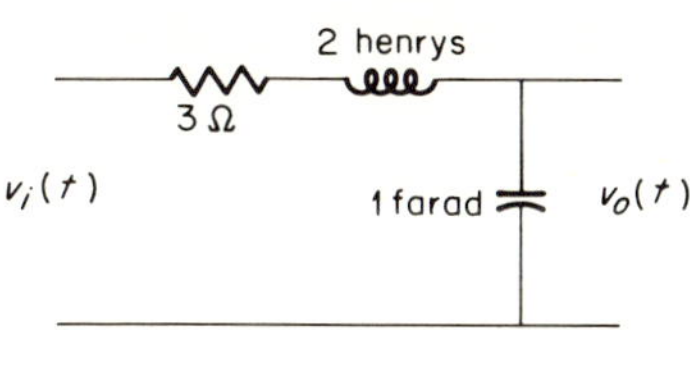

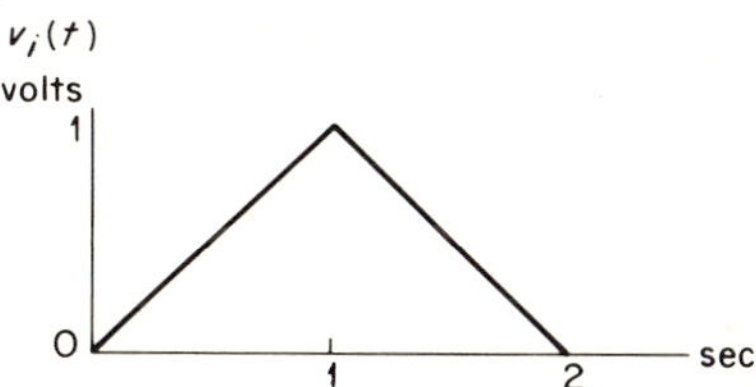

Fig. P12.33

12.34. The Fourier transform of the voltage function shown is given by $V(\omega) = (2 \sin \omega)/\omega$. Assume now that $V(\omega)$ is known and it is desired to find $v(t)$. In Example 9.3, the inverse Fourier transform was obtained directly in terms of sine integrals. As an alternative procedure, $V(s)$, the analytic continuation of $V(\omega)$ can be formed by replacing ω by s/j. Then the inverse Fourier transform can be found using the method of Sec. 12.16. Carry out this latter approach and obtain $v(t)$.

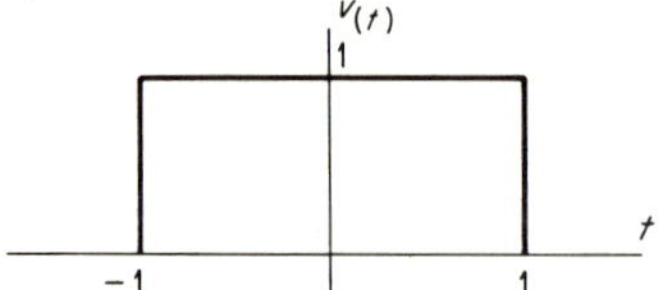

Fig. P12.34

12.35. The real part of an impedance as a function of ω is shown. Use the results of Prob. 12.18 to find the imaginary part of the impedance. Sketch the imaginary part vs. ω.

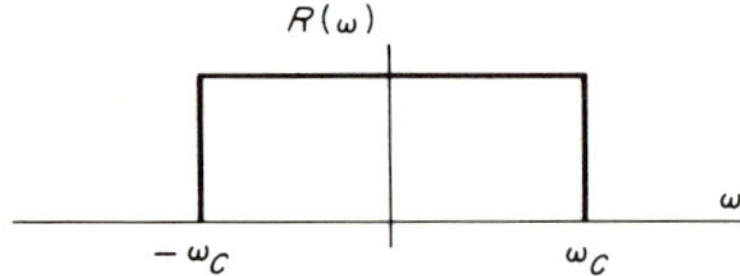

Fig. P12.35

INDEX